Muscle Strength

Edited by

Shrawan Kumar

University of Alberta
Edmonton AB, Canada

CRC Press
Taylor & Francis Group
Boca Raton London New York

CRC Press is an imprint of the
Taylor & Francis Group, an **informa** business

CRC Press
Taylor & Francis Group
6000 Broken Sound Parkway NW, Suite 300
Boca Raton, FL 33487-2742

First issued in paperback 2019

© 2004 by Taylor & Francis Group, LLC
CRC Press is an imprint of Taylor & Francis Group, an Informa business

No claim to original U.S. Government works

ISBN-13: 978-0-415-36953-4 (hbk)
ISBN-13: 978-0-367-39435-6 (pbk)

Library of Congress Cataloging-in-Publication Data

Muscle strength / edited by Shrawan Kumar.
      p. cm.
Includes bibliographical references and index.
  ISBN 0-415-36953-3 (alk paper)
  1. Muscle strength. I. Kumar, Shrawan.
QP321.M8946 2004
612.7'.4—dc21   2003013279

Library of Congress Card Number 2003013279

**Visit the Taylor & Francis Web site at
http://www.taylorandfrancis.com**

**and the CRC Press Web site at
http://www.crcpress.com**

Dedication

It is with profound respect and love that I dedicate this book to:

My parents — Mr. Tribeni Shankar and Mrs. Dhairyvati Shankar

for their unconditional love and instilling uncompromising human and moral values in me.

My wife — Mrs. Rita Kumar

for her love and companionship and support through thick and thin.

My children — Rajesh, Zoe, and Sheela Kumar

for their highest values and outstanding brilliance, unbounded love, and

most of all for being my children.

Troy Jones, BScPT, PhD Candidate
Editorial Assistant

Preface

The editors and authors feel duty bound to write prefaces to orient the reader to the light in which their book may be read. An unscientific poll revealed that prefaces are rarely read. Is this then a futile exercise? It all depends. Depends on what? Depends on reader's approach and need, you say. However, any information can be presented from more than one point of view. For a serious reader, knowing the purpose of the book is essential. It will allow the reader to make an independent judgment regarding the light in which it may be interpreted. For a casual user it may be of less importance, but the preface may add value to the fragments of information sought. In any case, faced with the foregoing reality, I am on the horns of a dilemma. The decision is bound to be wrong for some, but I will venture a brief write-up.

In the zoological domain (we are all animals of some sort), irritability is a property of life that is demonstrated in lower forms by reacting to stimuli. The reaction responses in most protozoans are manifested by movement of the entire organism. This may involve streaming of protoplasm or beating of cilia or undulation of flagella. The net result of all these mechanisms is the same, i.e., motion. The protozoans and planktons float in an aquatic medium, deriving their nutrition from and delivering their waste to it. With organizational complexity and growth in size of the higher-form organisms, a fixed position would quickly result in depletion of nutritional resources. Furthermore, accumulated waste could poison the creature. Hence, motion of the entire organism became a prerequisite for survival. Escaping predators added the need for speed. In order to provide structural support to organisms, skeletal structures were formed, initially through differentiation of notochord. Laying down of bones occurred segmentally to allow the development of joints that allowed motion of segments of the body with respect to the entire body. Development of contractile muscles segmentally (metamerism) between bones across joints gave the ability to move joints. Bilateral symmetry acted as an advanced and well-designed body plan to allow for what we call agonism and antagonism. In primitive animals like fish, it allowed lateral undulations, which acted as alternate compressive thrusts on the medium that the animals inhabited (water). Amphibians retained this property as they ventured on land. The feature of lateral flexion is seen in reptiles and almost all quadruped mammals even today. Combined with flexion extension, it has served as an excellent and rapid mode of locomotion. However, locomotion or any movement of an organism could not be possible without contractile muscles, which act as engine of locomotion.

The foregoing is the evolutionary perspective of one of the fundamental properties of life. In human beings, it has been complicated significantly due to diversification of purpose of movement. The body motion now no longer is used solely for acquiring food and finding a mate. The cerebral nature of the species, endowed with intelligence and rationality, changed its perspective on life. Survival alone was no longer sufficient. Order, security, mating territory, supplies in plenty, and an ever-increasing appetite for artificial aesthetics and power through technology became the norm. Money was invented, trade was developed, industries were found, and technology was developed. All these activities progressively geared the utilization of human motion capability in an increasingly narrow domain with varying magnitude. Because they evolved later, humans combine many advantageous features of the rest of the animal kingdom. This makes their bodies versatile and able to cope with a multitude of circumstances through the entire range of motion of their joints and strength of muscular contractions. Therefore, humans are capable of compensating for poor design and difficult situations. While this is an industrial asset, occasionally it overexerts the body. Overexertion is one of the most common factors associated with injury.

In current developed and civilized society, industrial productivity is desirable for our economy. However, the injuries mediated by these activities are not desirable for the economy, workers, or the society. Since muscle exertion is at the heart of these partially complementary and partially conflicting interests, it will serve the ergonomists to have a better understanding of the nature, magnitude, and limitations of muscle strength. Muscle contraction manifests itself in the production of force. Since the contraction produces force, looking into force magnitudes in various configurations and postures allows us to gage muscle capability.

Since each muscle in the body is laid down in an orientation that meets the primary motion need best, it is ill adapted to perform other activities. The combinations of circumstances are infinite, and the body does not have sets of muscles to match each situation. Therefore, a number of muscles or muscle groups may be recruited simultaneously to obtain the net resultant force required, but each of them will be acting at varying to considerable disadvantage, causing significantly larger loads on the joint in question. Thus the job may get done, but the action results in a significantly higher level of effort and stress for a proportionally much smaller task. It is obvious that any given posture or configuration does not transfer to another posture or configuration. Nevertheless, for ease of application, frequently a value is chosen in an optimal condition and is used for others.

Given the complex musculoskeletal system with components such as muscles, tendons, ligaments, cartilages, and bones, it would be simplistic to choose one standard for all circumstances. In order to control injuries, we need to know the mechanical properties of all tissues involved, with precision and individual variation in the context of occupational tasks. Currently this has not been achieved, and perhaps it will take some time before a clearer picture emerges. However, common stressor or stimulus is the muscle force generated. It has been demonstrated by numerous epidemiological studies that force variables are associated with frequency and severity of injuries. Taking these and other experimental observations, Kumar proposed four theories of occupational musculoskeletal injury causation: (a) multivariate interactive theory, (b) cumulative load theory, (c) differential fatigue theory, and (d) overexertion theory. A review of the four theories indicates that the role of force exertion is common to all, though other factors may modify the response. Thus it is important to understand the context of the occupational activity and the role of force generated in it. It is with these considerations that a distinguished panel of scientists was assembled to address specific aspects of muscle force generation. Among these are the pattern and magnitude of the force data in many different circumstances for readers looking for application usages. We have considered the fundamental phenomena, databases for the normal, and application in people with disabilities. Finally, a look to the future is attempted. The book begins with an introductory chapter by Kumar on terminology related to muscle strength. In this chapter, strength is defined, and its ergonomic relevance is discussed. The mechanism of contraction is briefly mentioned, and the evolution of strength in humans is described. The terminology used in physiological and applied fields is discussed. Finally, the role of strength in injury causation in occupational settings is addressed. In the next two chapters, Bell and Syrotuik deal with the physiology and biochemistry of strength generation, and Haykowsky writes about the cardiac effects of strength training. Bell and Syrotuik describe the structure and architecture of the muscle, which is followed by the mechanism of contraction. Subsequently, they cover fiber types, motor unit types, and their recruitment. The anatomical and physiological description is followed by muscle metabolism, energy system approach, and regulation of biochemical pathways. Bell and Syrotuik go on to describe the classical mechanical factors that limit the force of a contraction in muscles, e.g., muscle length and joint angle, force velocity, role of preloading, stretch shortening cycle, etc. Pursuing the physiological and biochemical theme, Haykowsky looks at hemodynamic response during strength exercise, which is related to most heavy work situations. He also tackles the effect of exercise on cardiac morphology, the chronic effect of strength training on cardiac volumes, and hemodynamics during lifting. Finally, Haykowsky discusses the cardiac effects of strength training on older adults and people with heart disease. The latter has special significance in an aging workforce and a workforce with cardiac impairments.

In Chapter 4, Herzog addresses the issues related to determinants of muscle strength. Herzog takes a fresh and progressive look at myofilaments, some aspects of which are subjects of controversy and topics of current active research. In this context, he deals with myosin, actin, other proteins and excitation–contraction coupling. The effects of muscle length and speed of contraction on strength are discussed. The role of age on strength generation is of particular importance to practitioners in an aging workforce. The chapter on endurance, by MacIntosh and Devrome, defines endurance and provides methods for quantification of endurance. An interesting discussion on the physiological view of endurance deals with differences between efficiency and economy, the energy cost of endurance, and homeostasis and its disturbances. The authors then describe the factors that influence endurance and methods of enhancing endurance. From the foregoing discussion, it is obvious that muscle has limitations, and under prolonged contraction it fatigues. The fatigue has been a phenomenon of considerable interest to ergonomists and has been extensively employed in a number of studies. Many variables, both objective and subjective, have been chosen to measure fatigue. However, their relative merit and accuracy have not been compared. Kumar et al., in Chapter 6, tackle this issue based on their experimental findings.

In both basic and applied research, different authors have chosen different methodologies to measure strength. Depending on the methodology used, different aspects of strength are measured. In Chapter 7, Dempsey addresses the theoretical aspects of psychophysics. On the other hand, Gallagher et al. describe the protocols and devices for isometric, isoinertial, and psychophysical methodology in Chapter 8. Dvir provides a detailed discussion of isokinetic dynamometry, with particular reference to protocol, devices, basic principles, reproducibility, and effort optimality in Chapter 9.

The next large section of the book, from Chapters 10 through 15, discusses the strengths of different parts of the body, giving significant data. In Chapter 10, Fagarasanu and Kumar detail aspects of hand strength, including grip and pinch forces, factors that affect hand strength. An important factor is how hand strength changes due to differences in shoulder, elbow, and wrist positions. The authors also describe finger strengths. The effect of aging on this strength is dealt with, and methods of measurement are described. Kim and Bishu deal with grasp at submaximal strength, with emphasis on gender, glove, hand orientation, grasping direction, and surface friction, in Chapter 11. In Chapter 12, Amell describes kinematics and kinetics of the shoulder, elbow, and forearm. In Chapter 13, Ferrari tackles cervical strength. Typically, cervical strength is not used to carry out a task. However, many current occupations result in cervical symptoms; hence, considering cervical strength is important. In addition, many occupations require driving vehicles as the central task. Of late, an explosion in whiplash cases also warrants a consideration of this topic. Ferrari describes measuring methods and cervical muscle strength in clinical populations. In Chapter 14, Kumar deals with trunk and lifting strengths in separate sections of the chapter. The author describes functional need and gives a rationale for measurement of trunk and lifting strengths. Trunk and lifting strengths are described separately, providing databases for each. In Chapter 15, Andres tackles the important aspects of pushing and pulling strengths. He relates these to manual material-handling and measurement techniques and reviews the research in the area. Kroemer provides some design applications of the strength data, discussing the principles involved, in the next chapter.

Frequently, the technique of electromyography (EMG) is used as an indirect method to estimate force applied or strength exerted. Sommerich and Marras discuss electromyography and force in isometric exertions, measurement requirements, and use of EMG in EMG-assisted biomechanical models in Chapter 17. EMG does not directly measure force, and a degradation of contraction leading to fatigue can be discerned by changes in EMG. Merletti et al. describe myoelectric manifestation of muscle fatigue in both isometric and dynamic contractions in Chapter 18. They discuss the repeatability of estimates of EMG and fatigue indices.

Frequently in the workplace, ergonomists deal with injured workers. Often the tool employed to measure the physical demands of a task, compared with those of the injured worker, is not given as much thought and consideration as it deserves. Jones and Kumar, in Chapter 19, summarize several physical-demands analysis techniques. These are then critiqued, and a comparative synthesis is provided. Bloswick and Hinckley deal with job accommodations for impaired workers and discuss regulatory frameworks in Chapter 20. In Chapter 21, Koontz et al. address a relatively less explored issue of disability and strength. Their work encompasses adjustment to disability, programmatic barriers, and ergonomic relevance. Finally, the authors deal with orthopedic, neurologic, aging, metabolic, mental, respiratory, and coronary impairments and children with disabilities. The authors provide general recommendations for strength training and dealing with environmental barriers.

In the last chapter (22), Kumar discusses gaps in knowledge in the strength field and reflects on the future of the field. This includes issues both for basic understanding and application. Several issues, which are central to injury control and good ergonomic practice, are raised, such as margin of safety, capability contours, force transmission with accurate estimates of load sharing, and *in vivo* load-deformation issues.

This work has been both interesting and revealing to me. I have enjoyed the development of the book through many logistic challenges and hope the reader finds the book reader friendly and of some value in the domains of basic understanding as well as application. If this hope is realized, this effort will have been worthwhile.

Contributors

Fabrisia Ambrosio   Human Engineering Research Laboratories, Veteran Affairs Pittsburgh HealthCare System, Pittsburgh, Pennsylvania

Tyler Amell   Department of Physical Therapy, Faculty of Rehabilitation Medicine, University of Alberta, Edmonton, Alberta, Canada

Robert O. Andres   Ergonomic Engineering, Inc., Pelham, Massachusetts

Julianna Arva   Permobil Inc., Lebanon, Tennessee

G.J. Bell   Faculty of Physical Education and Recreation, University of Alberta, Edmonton, Alberta, Canada

R.R. Bishu   Department of Industrial Engineering, University of Nebraska, Lincoln, Nebraska

Donald S. Bloswick   Department of Mechanical Engineering, Ergonomics and Safety Program, *and* Rocky Mountain Center for Occupational and Environmental Health, University of Utah, Salt Lake City, Utah

Mary Ellen Buning   Department of Rehabilitation Science and Technology, University of Pittsburgh, Pittsburgh, Pennsylvania

Rory A. Cooper   Human Engineering Research Laboratories, Veteran Affairs Pittsburgh HealthCare System, Pittsburgh, Pennsylvania

Patrick G. Dempsey   Liberty Mutual Research Center for Safety, Hopkinton, Massachusetts

Andrea N. Devrome   Faculty of Kinesiology and Faculty of Medicine, University of Calgary, Calgary, Alberta, Canada

Zeevi Dvir   Department of Physical Therapy, Sackler Faculty of Medicine, Tel Aviv University, Ramat Aviv, Israel

Mircea Fagarasanu   Department of Physical Therapy, Faculty of Rehabilitation Medicine, University of Alberta, Edmonton, Alberta, Canada

Dario Farina   Laboratorio di Ingegneria del Sistema Neuromuscolare, Dipartimento di Elettronica, Politecnico di Torino, Torino, Italy

Robert Ferrari   Department of Medicine, University of Alberta Hospital, Edmonton, Alberta, Canada

Sean Gallagher National Institute for Occupational Safety and Health, Pittsburgh Research Laboratory, Pittsburgh, Pennsylvania

Mark J. Haykowsky Faculty of Rehabilitation Medicine, University of Alberta, Edmonton, Alberta, Canada

Walter Herzog Faculty of Kinesiology, University of Calgary, Calgary, Alberta, Canada

Dee Hinckley HealthSouth, Inc., Salt Lake City, Utah

Troy Jones Faculty of Rehabilitation Medicine, University of Alberta, Edmonton, Alberta, Canada

B.J. Kim Industrial Engineering Program, Louisiana Tech University, Ruston, Louisiana

Alicia M. Koontz Human Engineering Research Laboratories, Veteran Affairs Pittsburgh HealthCare System, Pittsburgh, Pennsylvania

Karl H.E. Kroemer Industrial and Systems Engineering Department, Virginia Tech, Blacksburg, Virginia

Shrawan Kumar Ergonomics Research Laboratory, Department of Physical Therapy, University of Alberta, Edmonton, Alberta, Canada

Brian R. MacIntosh Faculty of Kinesiology and Faculty of Medicine, University of Calgary, Calgary, Alberta, Canada

William S. Marras Institute for Ergonomics, The Ohio State University, Columbus, Ohio

Roberto Merletti Laboratorio di Ingegneria del Sistema Neuromuscolare, Dipartimento di Elettronica, Politecnico di Torino, Torino, Italy

J. Steven Moore Department of Occupational and Environmental Medicine, School of Rural Public Health, Texas A&M University, College Station, Texas

Yogesh Narayan Department of Physical Therapy, Faculty of Rehabiltation Medicine, University of Alberta, Edmonton, Alberta, Canada

Narasimha Prasad Department of Mathematical Sciences, Faculty of Science, University of Alberta, Edmonton, Alberta, Canada

Alberto Rainoldi Laboratorio di Ingegneria del Sistema Neuromuscolare, Dipartimento di Elettronica, Politecnico di Torino, Torino, Italy

Carolyn M. Sommerich Institute for Ergonomics, The Ohio State University, Columbus, Ohio

Aaron L. Souza Human Engineering Research Laboratories, Veteran Affairs Pittsburgh HealthCare System, Pittsburgh, Pennsylvania

Terrence J. Stobbe   Embry-Riddle Aeronautical University, Prescott, Arizona

D.G. Syrotuik   Faculty of Physical Education and Recreation, University of Alberta, Edmonton, Alberta, Canada

Contents

1

Introduction and Terminology

Shrawan Kumar

CONTENTS

1.1 Ergonomic Relevance of Strength

Industrial activity around the world is performed by two-thirds of the world population over 10 years of age spending one third of their lives on work (WHO, 1995). This enormous workforce, through its contribution, generates a wealth of 21.3 trillion U.S. dollars, which sustains the socioeconomic fabric of the international society. Given the uneven distribution of population between developed and developing countries, a large majority of humanity is resorting to mechanical force application to accomplish their tasks. Additionally, in spite of the progression of technology, there are a great number of jobs in developed countries that require muscle force application (e.g., mining, forestry, agriculture, drilling and exploration, manufacturing and service sector). In many recreational activities and sports, as well as industry, force application is the primary requirement after skill. Thus, force application is essential to getting things done, more in some activities than others. The quantity, duration, and frequency of force application at a job provide an essential tool to gauge the job demands. A determination of such demands, when compared against the worker's ability to exert force (strength), provides ergonomists a meaningful measure of stress on workers. Even in developed countries a large number of tasks are not automated, due to the cost and complexity. Consequently, the responsibility of strength exertion falls on the workers. It has been estimated that, in order to produce one ton of product, a worker has to lift or manually handle anywhere between 80 and 320 tons of raw materials. Workers in an industrial economy are a very important resource, and for industrial health, they must be protected. The values for strength requirement and strength available are important data, which need to be used in job design for worker health and safety and industrial productivity.

As early as 1978 Chaffin et al. reported that worker safety is seriously jeopardized when the job strength requirement exceeds the isometric strength of the worker. The opinion for using strength data, especially in designing manual materials handling jobs, is particularly well supported (Ayoub and McDaniel, 1974; Chaffin and Park, 1973; Chaffin et al., 1978; Davis and Stubbs, 1980; Garg et al., 1980; Kamon et al., 1982; Keyserling et al., 1980; Kroemer, 1983; Kumar, 1991a and b, 1995; Kumar and Garand, 1992; Mital et al., 1993).

1.2   Strength

1.2.1   What Is It?

Strength in the context of this book is human ability to exert physical force and is measured as the maximum force one can exert in a single maximal voluntary contraction (MVC). However, it must be recognized that the overall manifestation of force will be dependent on the length of the lever arm on which the force is acting. The product of the force and moment arm is designated as torque, and the product of force and time is called impulse. If one measures the product of torque and time, it is called impulse over time or cumulative load.

$$\text{Force (F)} = \text{Newtons (N)}$$

$$\text{Torque (T)} = \text{N} \times \text{meters (Nm)}$$

1.2.2   How Is It Created?

A detailed description of the mechanism of force generation is provided in Chapters 2 and 4. However, a very brief version follows. The force is generated by contraction of skeletal muscles. In fact, force is generated by smooth (involuntary) muscles as well. However, since smooth muscles are postural muscles and under the control of autonomic nervous system, we do not exercise them for industrial use. Therefore, the entire discussion of muscle strength in this book relates to skeletal (voluntary) muscles. Contraction of these muscles can be produced at will in any configuration. The fundamental property of life "irritability" is expressed in the physical domain through these muscle contractions. Such muscle contractions allow us to conduct all our voluntary activities — work, leisure, and activities of daily living.

The mechanism of contraction of skeletal muscle has been a subject of vigorous debate in recent years. The initial mechanism (proposed by Huxley, 1957) suggested that the actin filaments in each sarcomere, at the time of contraction, slide to the middle by repeated bonding, pulling in, release, and rebonding of the actin and myosin filaments. This caused shortening of each of the sarcomeres and, in turn, the entire muscle fiber. The phenomenon of contraction involves electrical impulses coming via the nerves as commands, releasing the neurotransmitters at motor end plate in the myoneural cleft. This alters the ionic balance of the sarcolemma (cell membrane) of the muscle fiber, causing the sarcoplasmic reticulum to release calcium ions, which are essential to bind actin and myosin. This bondage triggers a ratchet action, pulling the actin past the myosin. The bondage is released by adenosine triphosphate (ATP) to repeat the process.

The above-mentioned shortening of all sarcomeres in any given muscle fiber can reduce the length of muscle to anywhere between two-thirds to half its original length. Since muscles have a stable point of origin from one bone and are inserted across, at least, one joint on another bone, they exert a force on the joint to move through their connective tissue attachment (tendons). This application of force across a joint moves that joint. By a complex series of coordinated movements, we are able to do anything we want.

The alternate theory and supporting evidence of muscle contraction are presented in Chapter 4. The evidence for the alternate theory is mounting but it is by no means fully established and universally accepted.

1.3 Evolution of Strength

Ability to exert strength is not one of the fundamental properties of life. During the course of evolution a series of adaptations occurred in organisms, which were primarily driven by the necessity of food. In some very primitive organisms (which lived in ocean or other aquatic media) the nutritional needs were met by medium itself, for example planktons. The protozoans could also exhaust the nutritional supply by just staying at one place. They achieved their locomotion through streaming of their protoplasm and rolling along to reach additional sources. In others, for example paramecia, there were multitudes of cilia, which beat in a coordinated and rhythmic manner to cause the organisms to move from one place to the other more rapidly than amoebae. Yet another higher level of organization among protozoans was demonstrated by development of colonies, as seen in volvox. However, these colonies did not have any functional differentiation, but by increasing their sizes they were able to achieve even faster movement.

It was not until the multicellular organisms developed that the phenomenon of functional differentiation began to manifest. Clearly, if each cell of a multicellular organism were to do everything, it would have been very inefficient. Thus, differentiation of neuron and nervous system; stomodeum, alimentary canal, and proctodeum; gills and lungs; gonads and locomotor systems took place. With reference to the locomotor system, it became more essential due to bilaterian body plan (Shankland and Seaver, 2002). Also, the growing sizes of the organisms could not be supported by external cuticle as exoskeleton. With the increase of the size of organisms there was a need for an internal skeleton, which many structures could be anchored to or suspended from. The notochord changed into the axial skeleton to provide support and protection to the dorsal nervous system (Kent, 1992; Walker and Liem, 1994). With a need to locomote to different surroundings in search of food, the fishes developed fins, and other classes of the vertebrate phylum differentiated further their pectoral and pelvic girdles, with specialized skeletal limbs to suit their habitat. However, skeleton alone would not have been much use without skeletal muscles, which are the engines of movement. Whereas an undulating and sequential concentric contraction and expansion can be seen in many invertebrates, e.g., annelids, the contractile structure differentiated into a sophisticated skeletal muscle. These muscles have the architecture, physiology, and mechanical properties described in Chapters 2 and 4. In addition, there was a need for the organisms to stay in one place as intact organisms when they were not locomoting. To meet this requirement, development of involuntary muscles took place, which in all probability preceded the development of the skeletal muscle. As such, we find in our muscles slow and fast (or Type I and Type II) fibers, where division of labor has been such that postural loads are borne by Type I, and motion and force application is largely achieved through the Type II muscles.

In ergonomics, we are mostly interested in the Type II muscle and the internal load it develops. However, postural muscle's contribution for the maintenance of posture over a long period of time is also of considerable consequence. In activities where we involve Type II muscles, in static contractions, over and above the level provide by Type I, muscles have proven to be of considerable consequence.

1.4   Terminology Used in Strength

1.4.1   Scientific Classification

The scientific classification of strength had been based entirely on physiological state of the muscle in contraction. The two categories of strength described are: (a) static or isometric (from Greek, *iso* = equal; *metrien* = measure or length), and (b) dynamic or isotonic (from Greek, *iso* = equal; *tonic* = muscle tone) (Vander et al., 1975). It was, therefore, initially thought that in isometric (static) efforts the length of the muscle remained constant. Externally there is no discernible motion of limb segments or body parts involved in this contraction. This was the primary reason why this contraction was named isometric. However, the question arises that, if the muscle is not shortening, how is it exerting force on the external object. Thus, it is clear and universally accepted that the muscle does contract and shorten because the muscle exerts its force (or torque) on the bone into which its tendon is inserted across a joint. Tendons elongate between 4 and 8% of their initial short length to transmit the force. Since this length is almost miniscule compared to muscle length, there does not appear to be significant change in its length while exerting force in isometric mode. In spite of the fact the term isometric represents the exertion inaccurately, it continues to be used due to its universal popularity, and also because people understand what is meant.

Similarly, the dynamic or isotonic strength or contraction was used on the logic of constant tension. Again, it is well established that, with change in the length of any muscle, its tension (and tone) changes. A dynamic contraction entails a change in joint angle and a change of body parts in space. These changes alter muscle length and hence tension. Therefore, the term isotonic is also a misnomer.

Even a cursory examination of any given situation will reveal the inconsistencies in terms used. As a manual materials handler moves an object from point A to B, the resistance offered by the object may continually change as a result of changing mechanical disadvantage of the load, even if the mass of the object does not change. Thus, the magnitude of the effort required to handle the load (resistance) does not remain constant.

As the object is supported by the worker through the range, the tension has to vary with the resistance, hence defying the term isotonic. On the muscle side, in addition to the change in load with respect to the relevant bony articulation, not only does the muscle length change with changing joint angle, but also the disadvantage for the muscle will continually change. The muscle will have to compensate for its varying length and mechanical advantage/disadvantage and also compensate for varying torque of the load. Clearly, such a dynamic *in situ* condition cannot be managed by a constant tension in the muscle. To complicate the system further, with the preponderance of third-class lever systems in body, the velocity of motion may significantly change with natural changes in muscles contraction. This further changes the inertial property of the object being handled (Kroemer, 1970; Mital and Kumar, 1998a,b). An activity of true constant tension will be an isometric effort where the force is not varied. Furthermore, in all probability, a true isometric condition may not exist at all, as any contraction will change the length of a given muscle.

Initially, any strength measurement involving motion was assigned as a dynamic motion. However, dynamic motion is of two kinds: (a) concentric — shortening contraction, and (b) eccentric — lengthening contraction. The concentric contraction is generally characterized by approximation of proximal and distal segments across a joint, where the length of the contracting muscle continues to shorten through the range of motion. This is exemplified by arm curl, where a load is picked up by the hand and is brought closer to the shoulder by contraction of elbow flexors. In contrast, the eccentric contraction is a lengthening contraction when a muscle or a group of muscles attempts to shorten by contraction but is lengthened by external force. This contraction is exemplified by lowering the weight from shoulder to knuckle height in the arm curl exercise. Thus, lifting is concentric, and lowering is eccentric. Because of differences in mechanical and physiological nature, these two types of isotonic contractions have different demands of the human operator. Concentric contraction requires approximately three times the physiological cost (Astrand and Rohdahl, 1977) and generates three times electromyographic (EMG) activity (Basmajian and De Luca, 1985) in comparison to eccentric contraction. Thus, in ergonomic sense, it is desirable to use eccentric contraction instead of concentric contractions wherever one can.

There is a contradiction between mechanical and physiological domains with respect to strength application. The work done mechanically is given by the product of force and displacement. However, if one engages in static force exertion, there is actually no mechanical work done, because there is no displacement. In physiological terms, indices of work done can be described in two ways. First, the percent MVC force exerted over time is a qualitative measure of work done. Second, the amount of oxygen consumed during this exertion, over and above resting level, will be a quantitative measure of physiological work. In dynamic strength exertion, one can obtain both the mechanical work done as well as physiological work. This is because in every dynamic work there is movement of body parts as well as displacement of objects, and also the muscle work due to force application, which consumes oxygen.

1.4.2 Applied Terminology

Applied terminology is a derivation of convenience from scientific nomenclature. As indicated above, even the scientific terms in the field of strength are misnomers. The derivations are no different. However, they serve an important purpose of being commonly understood descriptive terms suitable for field application. This classification is based on Mital and Kumar (1998a).

1.4.2.1 Static Strength

Generally, the static strength has been most commonly reported for any joint or set of joints concerned in an optimum and standardized posture (Chaffin et al., 1978). Though the devices and protocol (Chapters 8 and 9) will provide details of methodology, in this introductory chapter only the conceptual basis is presented.

1.4.2.1.1 Job-Simulated Static Strength

Whereas the conventionally measured values of static strength, as described, provide an idea of capability of people in optimal conditions, a direct relationship with the job remains obscure. Such information is essential for controlling job-mediated musculoskeletal injuries. Hence, the static strength measured with mechanical conditions of the job, simulated accurately in terms of posture, direction of strength, coupling with object, and any other unique variable affecting the measured strength is called Job-Simulated Static Strength (Mital and Kumar, 1998a). The measurement of Job-Simulated Static Strength is also done according to the strength measuring protocol of Caldwell et al. (1974) and Chaffin (1975) for a period of 5 seconds.

1.4.2.1.2 Repetitive Static Strength

The job parameters can vary widely, and a single exertion of 5 seconds may be misleading, if the job requires intermittent force application at a fixed or variable frequency. This strength category describes the capability of workers to provide maximal voluntary contractions at a specified frequency (Mital and Kumar, 1998a).

1.4.2.1.3 Continuous Static Strength

This category of static strength is more accurately described by the term endurance at the level of contraction. The endurance concept will be discussed in detail in Chapter 5. A classical endurance graph by Rohmert (1966) describes the pattern of behavior of the combination of strength and the duration for which it can be held. It will suffice to say here that maximal voluntary contraction can be held only for about five seconds. Rohmert (1966) also suggested that 15 to 20% of MVC could be held indefinitely. In between these values, as one increases the percent MVC contraction value, the duration for which it can be exerted declines exponentially. More recent studies of Hagberg (1984) and Sjogaard (1986) have provided evidence that if one were to hold at 8% MVC for 60 minutes, fatigue sets in. It is fair to say that while this category of strength provides a description of industrially relevant capability, it may not be physiologically trouble free.

1.4.2.2 Dynamic Strength

As described before, the maximal voluntary strength exerted through a range of postures is chosen to represent the dynamic strength. As soon as one considers choosing one value from a range of values, and in particular the peak value, a range of inaccuracies creeps in. To overcome these difficulties Kumar (1991a,b), Kumar and Garand (1992), Kumar (1997), and Kumar et al. (1998) chose to develop a strength contour around the human body at different mechanical disadvantages and varying velocities. Clearly the differences obtained were not only significant ($p < 0.01$) but also considerable. In any event, the dynamic strength is of three types as described below.

1.4.2.2.1   Isotonic Strength

In industry and the field of ergonomics, this is dynamic strength with no constraint of load or velocity. It simply has to be measured in a condition where a body part is moving a constant external mass over a distance or through a range of motion. As indicated in Section 1.5.1, this strength is not isotonic at all, because the tension in muscle changes continuously. Due to the foregoing reason, this category of strength is inapplicable in industry and has not received any significant attention.

1.4.2.2.2   Isokinetic Strength

Isokinetic strength, in principle at least, is meant to represent the maximum force that can be exerted when either the muscle involved is shortening at a constant velocity or the joint concerned is moving at a constant angular velocity. Of course, in practice, what is measured is entirely different. What really happens is that an external object or a handle is moved through a range of postures executing a linear path at constant linear velocity controlled by an external device. Here too, as in other categories, the term isokinetic is inaccurate. The strength terminology is in principle based on the muscle state. If we examine any isokinetic test, it will become evident immediately that the articulated human body at any joint does not move at a constant angular velocity while carrying the handle with a constant linear velocity. This results in constantly changing inertia of limb segments and body members. The rate of muscle contraction and the angular velocity of the joints are also constantly changing to accommodate the constant velocity linear displacement of the handle. Interestingly the isokinetic strength can also be considered, only in mechanical terms, a special case of static strength. In static strength, the rate of change of the velocity of the object equals zero. In isokinetic mode of testing, since we keep the linear velocity of the handle constant, the change or rate of change of velocity is also zero. However, we must remember that measurement of strength and its terminology is based on the muscle state, which is entirely different between static and isokinetic tests.

1.4.2.2.3   Isoinertial Strength

The term isoinertial was the latest to be introduced in strength literature by Kroemer (1983). The term is based on the premise that it determines the maximal load a person can displace through a preprescribed distance. It was argued that since the load remains constant through the distance, it is considered of constant inertia. However, the logic is flawed, because the inertia changes during the exertion due to the changing mass distribution of the system. Additionally, as the velocity of lift or displacement changes, the inertia changes again. Therefore, like other terms, this term is erroneous as well.

1.4.2.3   Psychophysical Strength

Psychophysical strength refers to a strength value that is obtained by one determining cognitively as to what is the load level he or she is willing to lift or manually handle. This method has been largely used in the context of lifting and was initially advocated most vociferously by Snook (1978) and Ayoub et al. (1978). It is worth noting that psychophysical strength of the same individual can significantly change if the instructions to subjects or the information regarding the task changes. In this sense it is the most variable strength among all.

1.5   Strength Parameters and Transmission

1.5.1   Peak vs. Average Strength

It has been customary in the ergonomic field to measure peak strengths of people in optimal posture with verbal encouragement to exert their maximal force. The peak strength thus recorded is considered as 100% of the person's capability, and other efforts are normalized against this value. Additionally, averaging all values of strength obtained over the entire contraction is given the label of average strength. This method of determination of strength is no different than the method used in the field of sports physiology.

Unfortunately, strength is not an objective parameter in the same way as body mass, body temperature, blood pressure, vision, hearing, and most other variables of human performance. First, the effort is readily modified by the motivation with which one performs. It is also affected by the previous history of exertions, hydration, and nutritional state of the person. The interaction of these variables on manifestation of strength is not known quantitatively. To further obscure the relationship, verbal encouragement strongly and significantly modifies a manifestation of strength. It is well known that people can perform extraordinary feats in special circumstances for which they may be extremely motivated. This makes the strength value considerably variable, especially for the parameter of peak.

In athletic and sports environments, where the performance time is brief and psychological pressure to perform is immense, the foregoing technique of strength measurement is entirely appropriate. But in industrial environments, where ergonomic factors are constantly at play and the performance time for most people is 8 hours per day every day of their entire working lives, the athletic and sports model is unrealistic and inappropriate. This is not only because of the physiological parameters at play in sports and work but also due to their respective tools. In the sports arena, there is an incessant effort to improve the tools of action to shave off a fraction of a second or increase the speed or distance by small amounts. In contrast, the workers work, in majority of cases, with uncustomized tools or environments for long periods of time every day. It would therefore appear inappropriate to import the athletic and sports model wholesale in the work environment. A different paradigm is needed. It has been therefore suggested by Kumar (a) to modify the protocol of strength measurement and its application in the field and (b) in such tests, which must almost always be activity specific, to give the test subjects the appropriate information on the test and ask them to provide their maximal voluntary contraction. They should be reminded of the need to provide their maximal effort but must never be issued verbal encouragement. Verbal encouragement not only modifies the response, but it can also modify it variably, depending on the tone and vigor of the command issued at the time of the test. This may not always be kept constant. These factors then render the test for standardization substandard. In light of the foregoing, it is considered that a test conducted with information on the purpose of test and standardized variables with no verbal encouragements is more appropriate for work situations. From such a test, the average value should be taken and used as 100% for the purposes of normalization. In summary, the average strength is more appropriate for ergonomic application than the peak strength.

1.5.2   Force Transmission

Forces generated by muscles during their contractions need to be transmitted to the bone on which the muscle is inserted across the joint. During their contractions, the muscles undergo shortening. The force with which the muscle contracts is transmitted through the tendon, which happens to be at the other end. While this is the explanation provided in most cases, this information needs to be elaborated for a better understanding. Skeletal muscles may contain thousands of individual muscle fibers, which are organized in fascicles, which in turn collectively make the muscle. The muscle structure is provided in Chapters 2 and 4. However, a brief description will help here. Each single muscle fiber is encased in a collagen sheath called endomysium. The collection of muscle fibers, which make a fasciculus, are in turn surrounded by another larger and thicker collagen sheath called perimysium. It is interesting to note that the collagen fibers of the endomysia also contribute to the formation of the perimysium. Additionally, the collagen fibers internal to the fascicle are also linked together, making a mechanically tight and continuous unit. Similarly, the fascicles of muscles are surrounded by an overall collagen harness, termed epimysium, which is developed by the contribution of perimysia. We can thus visualize that the entire contractile machine of muscles is surrounded by an extensive and intricate system of collagen harness which is continuous. Interestingly, to make a mechanical continuity the sarcolemma and collagen fibers are biochemically bonded with no difference in density. Thus, when a muscle contracts, it pulls on its sarcolemma bonded with collagen. Therefore the entire harness of the collagen is pulled toward the point of origin of the muscle, shortening the muscle length and transmitting the force through the tendon. Tendon, in fact, is nothing more than the aggregation of endomysia, perimysia, and the epimysium. Collagen is most suitable to transmit force, as it is a substance of high modulus of elasticity with its physiological length of elongation around 4 to 6% of its entire length. Since muscles can contract up to 50% of their initial length, and tendons (which are much shorter) are capable of elongating by 4 to 6%, even after compensation for elongation of tendon there is a large residual shortening in length, which is responsible for motion.

1.6 Occupational Injury Considerations

There is a vast epidemiological literature associating work with injuries (for reviews, see Bernard and Fine, 1997; Keyserling, 2000a,b; Punnet, 2000). Work is almost always achieved through application of force through range of motion required by the work concerned. Overwhelming evidence from around the world has demonstrated injuries to body parts that are most stressed in this force and motion environment. Based on the scientific evidence published in the literature and his own work, Kumar (2001) proposed four theories of occupational musculoskeletal injury causation. Three of these theories are associated with force or strength application. These are briefly described below.

Central to all theories is an assumption that all occupational musculoskeletal injuries are biomechanical in nature. Disruption of mechanical order of a biological system is dependent on the architecture of the structure and the individual components and their mechanical properties. Occupational activities are designed to meet occupational demands and rarely to optimize the biological compatibility. For these activities to be of economic and industrial value, they have to be repetitive. This sets the stage for injuries through repetitive force application through a range of motion over a long period of time.

1.6.1 Differential Fatigue Theory

Kumar (2001) proposed this theory on the basis of the fact that all industrial activities employ numerous muscles at various joints in industrially relevant human motions. These motions are frequently asymmetric in nature (Garg and Badger, 1986; Kumar, 1987, 1996; Kumar and Garand, 1992; Waters et al., 1993). In these activities, different joints may bear different loads. Depending on the motion executed and force applied, different muscles are differentially loaded as well. The differential loading will be largely determined by the task and not the capability of the muscles involved. Such differential prolonged or repeated loading may result in joint kinematics and loading patterns different from the optimum and natural. The differential demand is likely to cause differential fatigue of the muscles involved, resulting in a kinetic imbalance potentiating precipitation of an injury. Because the mechanical factor most at play in this scenario is the differential fatigue, as shown by Kumar and Narayan (1998), the theory was named "Differential Fatigue Theory."

1.6.2 Cumulative Load Theory

Biological tissues have, like all other physical materials, a finite life and are subject to wear and tear. They undergo mechanical degradation with repeated and prolonged usages. Also, all biological tissues are viscoelastic, and their prolonged loading may result in permanent deformation (Noyes, 1977; Viidik, 1973; Woo and Buckwalter, 1988). Repeated load application may also result in cumulative fatigue, reducing their stress-bearing capacity. Such changes may reduce threshold stress at which mechanical failure can occur. Furthermore, Kumar (1990) reported a strong association between cumulative load (biomechanical load and exposure time integral over entire work life) and low back pain/injury ($p < 0.01$). Based on foregoing scientific rationale and empirical evidence Kumar (2001) proposed the "Cumulative Load Theory" of injury causation at work.

1.6.3 Overexertion Theory

In simple terms, when exertion exceeds the tolerance limit, it is designated as overexertion. It can ensue by one supramaximal activity or can take place by repetitive action over long time. Since occupational activities require application of physical force from point A to point B, causing motion for a certain length of time (duration), overexertion in these circumstances will be a function of "force," "duration," "posture," and "motion." While all these variables are complex and described in detail in Kumar (1994) and Kumar (2001), it will suffice to say that the interaction between these four factors could bring a biological tissue to the brink of mechanical failure. This mode of injury precipitation was captured in "Overexertion Theory" by Kumar (2001).

In summary, then, force application in asymmetrical configurations may cause differential fatigue, load over time could cause cumulative load effect, and exertion of force at a frequency over a certain time requiring motion may result in overexertion. It is therefore clear that application of force is clearly a common denominator and a strong factor in the causation of occupational injuries.

1.7 Summary

Strengths (or forces/torques) are of considerable significance for ergonomics for both productivity and injury control. They are created by muscles and transmitted through tendons. The need for strength was felt by animal organisms as they grew in size and needed to move around in search of food and mates. The original reason for development of strength has been overshadowed by the needs of industrial economy. There are many advantages of work, including creation of wealth to sustain the society, but its requirements have been instrumental in precipitation of multitudes of occupational musculoskeletal injuries both in variety and body parts. Three theories of musculoskeletal injury causation have been briefly described. A rationale for a modified method to generate ergonomic standards has been argued.

References

Astrand, P.O., and Rohdahl, K. 1977. *Textbook of Work Physiology* (New York: McGraw-Hill).

Ayoub, M.M., and McDaniel, J.N. 1974. Effects of operator stance on pushing and pulling tasks, *AIIE Trans.*, Sept., 185–195.

Ayoub, M.M., Bethea, N.J., Deivanayangam, S., Asfour, S.S., Bakken, G.M., Liles, D., Mital, A., and Sherif, M. 1978. Determination and modeling of the lifting capacity, Final Report, DHEW (NIOSH), Grant 5 RO1 OH 00545-02. Cincinnati.

Basmajian, J., and De Luca, C. 1985. *Muscles Alive* (Baltimore, MD: Williams & Wilkins).

Bernard, B.P., and Fine, L.J. 1997. Musculoskeletal disorders and workplace factors, DHHS, (NIOSH) Publication No. 97B141, Cincinnati, OH.

Caldwell, L.S., Chaffin, D.B., Du Bobos, D., Kroemer, K.H.E., Laubach, L.L., Snook, S., and Wasserman, D.E. 1974. A proposed standard procedure of static muscle strength testing, *AIHA J.*, 35, 201–206.

Chaffin, D.B. 1975. Ergonomics guide for the assessment of human static strength, *AIHA J.*, 505–511.

Chaffin, D.B., and Park, K.S. 1973. A longitudinal study of low back pain as associated with occupational weight lifting factors, *AIHA J.*, 34:513–525.

Chaffin, D.B., Herrin, G.D., and Keyserling, W.M. 1978. Pre-employment strength testing: An updated position, *J. Occup. Med.*, 20, 403–408.

Davis, P.R., and Stubbs, D.A. 1980. *Force Limits in Manual Work* (Guildford, UK: IPC Science and Technology Press).

Garg, A., and Badger, D. 1986. Maximum acceptable weights and maximum voluntary isometric strengths for asymmetric lifting, *Ergonomics*, 29, 879–892.

Garg, A., Mital, A., and Asfour, S.S. 1980. A comparison of isometric strength and dynamic lifting capability, *Ergonomics*, 23, 13–27.

Hagberg, M. 1984. Occupational musculoskeletal stress and disorders of the neck and shoulder: A review of possible pathophysiology, *Int. Arch. Occup. Environ. Health*, 53, 269–278.

Huxley, A.F. 1957. Muscle structure and theories of contraction, *Prog. Biophys. Chem.*, 7, 255–318.

Kamon, E., Kiser, D., and Pytel, J. 1982. Dynamic and static lifting capacity and muscular strength of steelmill workers, *AIHA J.*, 43, 853–857.

Kent, G.C. 1992. *Comparative Anatomy of the Vertebrates* (St. Louis, MO: Mosby Year Book).

Keyserling, W.M. 2000a. Workplace risk factors and occupational musculoskeletal disorders, Part 1: A review of biomechanical and psychophysical research on risk factors associated with upper extremitiy disorders, *AIHA J.*, 61, 39–50.

Keyserling, W.M. 2000b. Workplace risk factors and occupational musculoskeletal disorders, Part 2: A review of biomechanical and psychophysical research on risk factors associated with upper extremity disorders, *AIHA J.*, 61, 231–243.

Keyserling, W.M., Herring, G.D., and Chaffin, D.B. 1980. Isometric strength testing as a means of controlling medical incidents on strenuous jobs, *J. Occup. Med.*, 22, 332–336.

Kroemer, K.H.E. 1970. Human strength: Terminology, measurements and interpretation of data, *Hum. Factors*, 12, 297–313.

Kroemer, K.H.E. 1983. An isionertial technique to assess individual lifting capability, *Hum. Factors*, 25, 493–506.

Kumar, S. 1987. Arm lift strength at different reach distances, in S. Asfour (Ed.), *Trends in Ergonomics/ Hum. Factors IV.* (Holland: Pub North).

Kumar, S. 1991a. *A Research Report on Functional Evaluation of Human Back.* (Edmonton, AB: University of Alberta).

Kumar, S. 1990. Cumulative load as a risk factor for low-back pain, *Spine*, 15, 1311-1316.

Kumar, S. 1991b. Arm lift strength in work space, *Appl. Ergon.*, 22, 317–328.

Kumar, S. 1994. A conceptual model of overexertion, safety, and risk of injury in occupational settings, *Hum. Factors*, 36, 197–209.

Kumar, S. 1995. Upper body push-pull strength of normal young adults in sagittal plane at three heights, *Int. J. Ind. Ergon.*, 15, 427–436.

Kumar, S. 1996. Isolated planar trunk strengths measurement in normals: Part III — results and database, *Int. J. Ind. Ergon.*, 17, 103–111.

Kumar, S. 1997. Axial rotation strength in seated neutral and prerotated postures of young adults, *Spine*, 22, 2213-2221.

Kumar, S. 2001. Theories of musculoskeletal injury causation, *Ergonomics*, 44, 17–47.

Kumar, S., and Garand, D. 1992. Static and dynamic strength at different reach distances on symmetrical and asymmetrical planes, *Ergonomics*, 35, 861–880.

Kumar, S., and Narayan, Y. 1998. Spectral parameters of trunk muscles during fatiguing isometric axial rotation in neutral posture, *J. Electromyogr. Kinesiol.*, 8, 257–267.

Kumar, S., Narayan, Y., and Zedka, M. 1998. Trunk strength in combined motions of rotation and flexion extension in normal young adults, *Ergonomics*, 41, 835–852.

Mital, A., and Kumar, S. 1998a. Human muscle strength definitions, measurement, and usage: Part I — guidelines for the practitioner, *Int. J. Ind. Ergon.*, 22, 101–121.

Mital, A., and Kumar, S. 1998b. Human muscle strength definitions, measurement, and usage: Part II — the scientific basis (knowledge basis) for the guide, *Int. J. Ind. Ergon.*, 22, 123–144.

Mital, A., Garg, A., Karwowski, W., Kumar, S., Smith, J.L., and Ayoub, M.M. 1993. Status in human strength research and application, *IIE Trans.*, 25, 57–69.

Noyes, F.R. 1977. Functional properties of knee ligaments and alterations induced by immobilization: A correlative biomechanical and histological study in primates, *Clin. Ortho. Rel. Res.*, 123, 210–242.

Punnett, L. 2000. Commentary on the scientific basis of the proposed Occupational Safety and Health Administration ergonomics program standard. *J. Occup. Environ. Med.*, 42, 970–981.

Rohmert, W. 1966. Maximal forces of men within the reach envelope of the arms and legs, Research Report No. 1616, State of Northrhine–Westfalia, Westdeutscher Verlag Koelm–Opladen.

Shankland, M., and Seaver, E.C. 2002. Evolution of bilaterian body plan: What have we learned from annelids? *Proc. Natl. Acad. Sci. U.S.A.*, 97, 4434–4437.

Sjogaard, G. 1986. Intra-muscular changes during long term contraction. In E.N. Corlett, J. Wilson and I. Manenica (Eds.), *The Ergonomics of Working Postures* (London: Taylor and Francis), pp. 136–143.

Snook, S.H. 1978. The design of manual handling tasks, *Ergonomics*, 21, 963–983.

Vander, A.J., Sherman, J.H., and Luciano, D.S. 1975. *Human Physiology: The Mechanisms of Body Function* (New York: McGraw-Hill).

Viidik, A. 1973. Functional properties of collagenous tissues, *Int. Rev. Connect. Tissue Res.*, 6, 127–215.

Walker, W.F., and Liem, K.F. 1994. *Functional Anatomy of Vertebrates: An Evolutionary Perspective* (Fort Worth, TX: Sanders College Publishing).

Waters, T., Putz-Anderson, V., Garg, A., and Fine, L.J. 1993. Revised NIOSH equation for the design and evaluation of manual lifting tasks, *Ergonomics*, 36, 749-776.

WHO, 1995. Global strategy for occupational health for all (Geneva: WHO).

Woo, S.L.Y., and Buckwalter, J.A. 1988. *Injury and Repair of the Musculoskeletal Soft Tissues* (Park Ridge, IL: American Academy of Orthopaedic Surgeons).

2

Physiology and Biochemistry of Strength Generation and Factors Limiting Strength Development in Skeletal Muscle

G.J. Bell and D.G. Syrotuik

CONTENTS

ABSTRACT   Muscular strength refers to the maximal force capabilities of a muscle or muscle group that can be exerted against a resistance in a particular movement pattern and at a defined velocity of movement. Muscular strength is one component of musculoskeletal fitness that is an important aspect of a variety of activities for daily living, certain occupations, general health, rehabilitation, and sport performance. There are several physiological and biochemical factors underlying the expression of muscular strength. Ergonomists should have a basic understanding of these concepts to further their awareness of how these associated factors can assist with the optimization of human well-being and work-related performance, as well as for designing prescription programs to improve strength as it may relate to systems, human factors, and productivity. This information will also be useful when making decisions or setting standards for work rate and load in various work applications.

2.1   Introduction

Skeletal muscle strength has been defined as "the maximal force a muscle or muscle group can generate at a specified velocity" (Knuttgen and Kramer, 1987). Harman (1993) has defined strength as "the ability to exert force under a given set of conditions defined by body position, the body movement by which force is applied, movement type, and movement speed." Regardless of the definition, the expression of muscular strength has the components of force, velocity, movement pattern, and the position the body is in. Muscular strength must also be distinguished from muscular power and endurance. Power can be defined as the product of force and velocity (Herzog, 1996), or it can also be expressed as work divided by time (Foss and Keteyian, 1998). Muscular endurance is the ability to maintain a muscle contraction or continue repetitive muscle contractions for a prolonged period of time to a defined endpoint.

The physiology or function of the skeletal muscle system and the biochemistry or chemical processes underlying the expression of muscular strength are important for ergonomics. Within the field of ergonomics, it is necessary to conceptualize the way muscle functions, as well as the interaction with the energy provision sources and pathways during various work-related activities, activities of daily living, rehabilitation activities, and recreation and leisure-time pursuits. The physiological and biochemical factors within skeletal muscle underlie our ability to maintain posture, perform a variety of movement patterns, resist fatigue, and be considered physically fit and healthy (Kell et al., 2001). A general understanding of the physiological and biochemical factors involved in muscle strength expression and the important factors that limit strength is the theme for this chapter.

2.2   Review of Skeletal Muscle Physiology and Biochemistry Underlying Strength Generation

Knowledge of the skeletal muscle system requires an understanding of structure and function (McComas, 1996; McMahon, 1984). It is also an advantage to understand the biochemistry of muscle contraction as it relates to energy provision for muscular work (Houston, 1995). This will underscore the physiological and biochemical aspects of submaximal and maximal force, power, or endurance applications in work situations and relate them to the use of various technologies in the workplace.

2.2.1   Connective Tissue Component of Skeletal Muscle

Skeletal muscle varies in size, shape, content, and function. It is highly adaptable and includes improvements in structure and function due to such things as physical training, or decrements in these aspects due to immobilization or spinal cord injury. Generally, a connective tissue sheath called the epimysium, composed primarily of collagen fibers, surrounds each whole muscle. The perimysium is also connective tissue, and it divides the muscle into fascicles (bundles) of fibers. Each fasciculus differs in shape and encompasses a varying number of muscle fibers. The perimysium is mostly composed of collagen and provides structural support to vessels and nerves that run along the length of the muscle. The endomysium component is also composed of collagen and surrounds each muscle fiber within the fascicle, and this sheath can interact with the perimysium. It also interacts with the basement membrane of the muscle fiber. The terminal ends of skeletal muscle generally taper due to a narrowing of the muscle fibers, especially in a fusiform muscle. The internetwork of connective tissue sheaths coalesces with the tendon that provides the attachment point of the muscle to the bone at each end. The connective tissue components of skeletal muscle provide structure and support, resist stretch, and allow for the distribution of tension development to avoid structural damage and assist with the transmission of contractile tension to the tendon. They are also involved in storing potential energy during certain types of muscle contractions.

2.2.2   Consideration of Muscle Architecture

Each skeletal muscle is made of individual muscle fibers or cells that are unique in shape, since they are elongated, as opposed to a spherical shape typical of many other cell types. Fusiform muscle fibers are usually arranged in a parallel fashion and run the entire length of the muscle (e.g., biceps brachii). Other types of muscle contain fibers that can run obliquely, and this is termed pennation (Figure 2.1). Unipennate muscles are arranged in such a way that the fibers run obliquely to a tendon that exists along one side (e.g., extensor digitorum longus). Bipennate muscles have a central tendon that runs up the center of the muscle with the fibers extending outward from the central origin (e.g., gastrocnemius). Multipennate muscles have two sets of obliquely running fibers attached to tendon material, which extend from one end of the muscle to another (e.g., deltoid). Muscles that are pennated can provide more tension than muscles with parallel fibers of the same volume. This is due to the architectural arrangement that allows a larger number of fibers per cross-sectional area to develop force in a pennated muscle.

2.2.3   Consideration of Muscle Fiber Structure and Content

Each muscle fiber is bound by a membrane known as the sarcolemma, which is a bilipid layer containing particular extrinsic and intrinsic proteins that function as receptors, transport systems, enzymes, and kinases, as well as having a structural role. Within each muscle fiber, a cytoskeletal framework exists, as well as various organelles, intracellular systems, proteins, metabolites, and multiple nuclei (unlike other uninucleated cell types). Each fiber also contains many myofibrils, which are long thin structures that are somewhat separated from one another by intracellular structures (e.g., sarcoplasm reticulum, T-tubule system, and mitochondria). The myofibril itself contains a series of sarcomeres that produce the striated appearance of skeletal muscle. These are composed primarily of myosin and actin proteins arranged in a highly organized fashion. The sarcomere forms

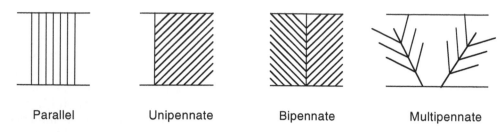

Parallel	Unipennate	Bipennate	Multipennate

FIGURE 2.1

Schematic illustration of skeletal muscle pennation patterns. [Adapted from McMahon, T.A., *Muscles, Reflexes, and Locomotion,* (Princeton, NJ: Princeton University Press), 1984.]

bands that are defined by the way in which they refract light. The dark A-band is anisotropic to polarized light and contains the thick protein filaments termed myosin. The light I-bands are composed of the actin protein filaments and are isotropic. Each I-band has a dark central region termed the Z-disc or Z-line. The area formed from the myosin that does not overlap with the actin in the central portion of the A-band is termed the H-zone. The M-line or region is a narrow dark band in the center of the A-band that contains protein filaments to support the myosin proteins and an enzyme, creatine kinase. Several other proteins associated with the sarcomere are involved in the contractile process and provide structure and support to the skeletal muscle cytoskeleton. The sarcomere is known as the basic structural unit of skeletal muscle.

Each myosin molecule within the sarcomere is composed of two strands (light meromyosin) arranged in a double alpha-helical structure. The structure terminates in two globular heads (heavy meromyosin) that are associated with four smaller protein strands called light chains. This combination of myosin heavy and light chains is the primary determinant of the amount and speed of force development within muscle. The head of the myosin molecule extends outward from the base strand and will form a cross-bridge attachment with an active site on actin. Actin exists in two forms; globulous G-actin polymerizes to form a double helical chain called fibrous F-actin. Associated with the actin are two other proteins, troponin and tropomyosin. Tropomyosin is arranged in an alpha helix and is contained in the groove formed between the actin chains. Troponin is composed of three molecules termed C, T, and I. Troponin is associated with the ends of the tropomyosin on the actin. A cross-sectional view of the actin and myosin arrangement reveals a hexagonal array such that six actin molecules surround one myosin molecule. Enhancing the number of cross-bridges that can be formed will directly influence the amount of force a muscle is capable of developing.

2.2.4 The Process of Muscle Contraction

The most widely accepted theory of muscle contraction is the sliding filament theory (Huxley and Hansen, 1954). This theory simply states that muscle shortening occurs though interaction of the myosin heads with active sites on the actin, revealed through a conformational change in the tropomyosin/troponin complex initiated by calcium interaction. The myosin cross-bridges that are formed swivel and pull the actin over the myosin in an energy-consuming process (Figure 2.2). The energy for muscle contraction is supplied by the hydrolysis of adenosine triphosphate (ATP).

The process of voluntary muscle contraction is initiated by an alpha motoneuron. A successful excitatory nerve impulse is transmitted to the muscle through the release of acetylcholine from terminal vesicles into the synaptic cleft formed between the nerve and muscle at the motor end plate (neuromuscular junction). The resultant depolarization of

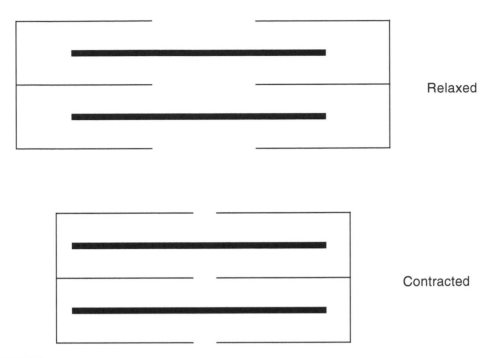

FIGURE 2.2
Schematic illustrating the sliding filament theory of skeletal muscle contraction within a sarcomere. Thick lines represent myosin, thin horizontal lines represent actin, and thin vertical lines represent Z-lines.

the sarcolemma (opening of the voltage-gated sodium channels) spreads along the muscle fibers in both directions (propagation). This depolarization is also transferred to the T-tubule system surrounding the muscle fibers that extend inward throughout the muscle. This stimulates the release of calcium from the sarcoplasmic reticulum through two types of channels [dihydropyridine (DHP) and ryanodine]. Calcium binds to a site on troponin, producing a conformational change in the tropomyosin/troponin complex and revealing an active site on the actin that can interact with the myosin cross-bridge during the active state of the muscle fiber.

Excitation–contraction coupling refers to the combined processes of electrical activation of the contractile apparatus, muscle contraction, and the involved metabolic energy dynamics. At rest, the cross-bridge formation between myosin and actin is maintained in a weak binding state. With the release of calcium and exposure of the active site on actin, a much stronger binding state is formed at the cross-bridge. Inorganic phosphate (P_i) is released, producing a conformational change in the orientation of the cross-bridge that allows the "power stroke" to proceed, initiating the process of muscle tension development. Adenosine diphosphate (ADP) is then released from myosin, completing the cross-bridge movement. Another molecule of ATP now binds to myosin, releasing the cross-bridge. Adenosine triphosphatase (ATPase) is the enzyme responsible for hydrolyzing ATP, liberating energy from the breaking of the terminal phosphate bond of the ATP molecule. This "energizes" the myosin cross-bridge, providing further energy to continue the muscle-shortening process. This process will continue until stimulation of the muscle ceases and calcium is removed through active pumping back into the sarcoplasmic reticulum. At this point, the tropomyosin/troponin protein complex returns to its original inhibitory conformation, essentially "turning off" the active site on actin. Muscle can shorten up to 60% of its resting length, but this varies depending on the type, function, architecture, and orientation of muscle in the skeletal system.

2.2.5   Muscle Motor Unit Types and Properties

Muscle contractile properties are partly dependent on the differing proportions of motor unit types that exist in the skeletal muscle system. A motor unit is a single alpha motoneuron and all the muscle fibers that it innervates. The motor unit is described as the basic functional unit of the muscle. The amount of force and power expressed by skeletal muscle is directly related to the type, number, and size of motor units available and recruited in the muscle. The number of muscle fibers that exist within a single motor unit varies widely, depending on the function of the muscle, from a eye muscle that has low fiber-to-nerve ratio for fine motor control, to a leg extensor muscle that has a high fiber-to-nerve ratio to elicit more gross movement patterns. There are thought to be three different primary motor unit types in the skeletal muscle system. The classification and terminology differs in the literature, but motor unit types are usually described based on their twitch characteristics, tension characteristics, and fatigability (Table 2.1).

Slow (S) twitch motor units have a small alpha motoneuron and the lowest threshold for activation (recruitment threshold), low force production, slow contraction/relaxation speed, and high fatigue resistance and contain muscle fibers that have a high oxidative or "aerobic" capacity (Type I muscle fibers). These motor units are recruited for light- to moderate-intensity activities.

Fast twitch motor units exist in two main types. Fast fatigue resistant (FR) motor units have a large alpha motoneuron that results in a later recruitment threshold compared to slow twitch motor units. They are capable of high force production and fast contraction/relaxation speed, are fatigue resistant, and contain muscle fibers that tend to have a high capacity for both glycolytic (anaerobic) and oxidative metabolism (Type II a fibers).

Fast fatigable (FF) motor units have a large motoneuron and are recruited later than FR and S motor units (highest recruitment threshold). They are capable of the greatest force production and fastest contraction/relaxation speed, fatigue easily, and contain muscle fibers that have a high glycolytic capacity but low oxidative capacity (Type II b fibers).

Activities that require rapid, high force and/or high power movements progressively require recruitment of FR and ultimately FF motor units (McArdle et al., 2001).

TABLE 2.1

Skeletal Muscle Motor Unit Classification and Physiological Properties

Classifications	Motor Unit Type		
	Slow Twitch or Slow Fatigable	**Fast Twitch or Fast Fatigue Resistant**	**Fast Twitch or Fast Fatigable**
Neuronal properties	Small soma and axon diameter	Large soma and axon diameter	Largest soma and axon diameter
	Early recruitment threshold	Intermediate recruitment threshold	Late recruitment threshold
Fatigue properties	Fatigue resistant	Intermediate fatigable	Fast fatiguing
Contractile properties	Slow time to peak tension	Fast time to peak tension	Fastest time to peak tension
Fiber properties	Smallest fiber diameter	Large fiber diameter	Largest fiber diameter
	Low ATPase activity	High ATPase activity	Highest ATPase activity
	High oxidative capacity	Intermediate oxidative capacity	Low oxidative capacity
	Low glycolytic capacity	Intermediate glycolytic capacity	High glycolytic capacity

2.2.6 Consideration of Muscle Fiber Types

Muscle fiber type refers to the specific properties of the individual fibers that exist within a motor unit. There have been several different classification schemes proposed for muscle fiber types, but the most common is based on the predominant myosin heavy chain (MHC) type that exists within the fiber. Type I fibers contain MHC I, Type II a fibers contain MHC II a, and Type II b fibers contain MHC II b (Figure 2.3). More recently, a new type of MHC has been identified and termed Type "d" or "x" by some researchers (Pette and Staron, 2001; Schiaffino et al., 1989). Type II d(x) fibers contain MHC II d(x) and are believed by some to be the actual type of fiber that exists in certain human muscles that were previously classified as Type II b fibers, and true Type II b fibers may in fact be scarce or nonexistent in human skeletal muscle. Further research will be required before this problem is settled.

The predominant fiber type that exists in a muscle is determined by the motor unit that it belongs to and by the function of that muscle. For example, postural or antigravity muscles such as the soleus, which require prolonged tonic muscle contractions, contain mostly slow twitch Type I fibers. Other muscles, such as the gastrocnemius muscle, contain a mixture of Type I and II muscle fibers. It is also important to note that muscle fiber types not only vary between different muscles in the same individual (e.g., soleus vs. gastrocnemius), but the same muscle of two different individuals may vary in the proportion of

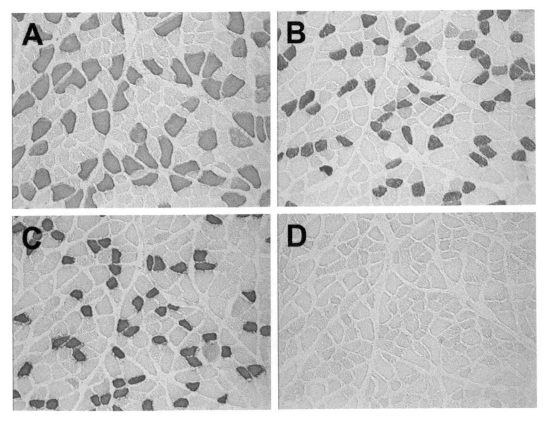

FIGURE 2.3
Mammalian skeletal muscle fiber types. Dark stained fibers in A are Type II b, in B are Type II a, and in C are Type I. D represents control. (Courtesy of Dr. Ted Putman and Greg Hansen, University of Alberta.)

fiber types in the same muscle (e.g., 80% Type I vs. 50% Type I fibers in the vastus lateralis muscle of two different individuals). This is primarily due to genetic variation and/or activity level (McArdle et al., 2001; McComas, 1996).

Fiber types are known to be highly adaptable, depending on the change in the stimulus provided to the muscle. For example, physical training has been shown to produce a shift in the metabolic profile and MHC expression away from the Type II b or d(x) fiber types toward an increase in Type II a fibers. Spinal cord injury (Burnham et al., 1997), space flight (Edgerton et al., 1995), or limb immobilization (Dudley et al., 1992) can produce a shift to a greater proportion of Type II fiber properties in human muscle. Thus size, contractile properties, and metabolic and fatigue properties of muscle fibers can be shifted with a training stimulus or lack of stimulus.

2.2.7   Muscle Motor Unit Recruitment

Henneman's size principle (Henneman et al., 1965) states that the general order of motor unit recruitment during voluntary skeletal muscle contractions is S → FR → FF. This principle is quite robust for both slowly developing and highly forceful or powerful voluntary muscle contractions. This is because the slow twitch motoneuron has an earlier recruitment threshold than fast twitch motor units. However, during high force and/or high power contractions the relative contribution of slow twitch motor units to the overall force production is less than the fast twitch motor units (Figure 2.4). It is possible that some reflex muscle contractions, which are not under central nervous system control but form reflex arcs through interneurons within the spinal cord, may be exceptions to Henneman's motor unit size principle. Use of functional electrical stimulation applied to the surface of the skin over a muscle may also reverse this normal order of recruitment. This is because the electrical stimulus is directed to the axon of the motoneuron, and fast twitch motor units have larger axon diameters that are known to conduct neural impulse faster than the smaller slow twitch motor unit (Sale, 1987).

2.2.8   Muscle Contraction Types

There are three main types of muscle contractions: isometric, isotonic, and isokinetic. In the case of isometric, muscle develops tension without shortening. Isotonic refers to muscle developing a constant tension, and isokinetic is a muscle contraction at a constant velocity. A muscle can develop tension while shortening (concentric) or lengthening (eccentric), and this can happen with isotonic and isokinetic contractions (Foss and Keteyian, 1998; McArdle et al., 2001).

A neural stimulus that depolarizes a single motor unit produces a twitch. If a second stimulus arrives before complete relaxation, the second twitch can summate with the first and produce greater force. If the stimulus becomes repeated at a maximal rate, the twitches will summate and fuse to a tetanus state. When all motor units are recruited and firing at optimal frequency, the maximal force output of the muscle is achieved. This will continue with prolonged stimulation until fatigue occurs, resulting in a decline of force (Figure 2.5).

2.2.9   Skeletal Muscle Metabolism

The sole source of energy supply that can be directly used for muscle contraction is adenosine triphosphate (ATP). The energy is stored in the bonds formed between the

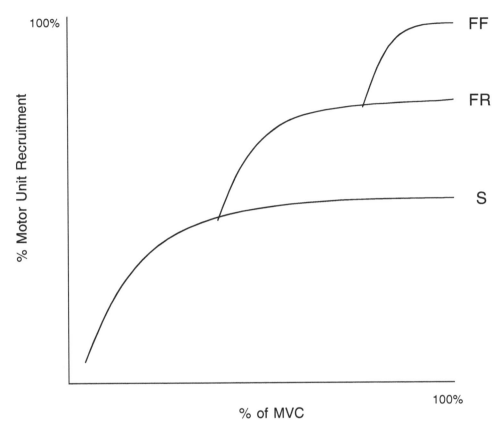

FIGURE 2.4
Motor unit recruitment pattern expressed as a percent of maximal voluntary contraction (MVC). (Adapted from Sale, 1987.) S = slow twitch motor unit, FR = fast twitch fatigue resistant motor unit, FF = fast twitch fatigable motor unit.

phosphate groups attached to the adenosine–ribose molecule. ATP is used for energy in a variety of cellular sites that require energy and contain an ATPase (e.g., sodium potassium ATPase, sarcoplasmic reticulum ATPase), but it is the specificity of myosin ATPase for ATP that supplies the energy for muscle cross-bridge cycling. ATP is in limited supply in muscle ($\approx$ 4 to 6 mmol/kg of muscle) and lasts for only a few seconds of muscle contractions. This is the primary energy source for a ballistic muscle contraction such as a single knee extension on an isokinetic dynamometer or a vertical jump movement. If this was the sole source of energy available to the human body, we would only be able to move for a very brief time — albeit a very fast movement! Fortunately, we have other metabolic sources to rely on. Thus, muscle contraction can continue beyond a few seconds, and this is accomplished with other energy sources that essentially transfer their energy to resynthesize the depleted ATP stores. The ATP reaction is summarized below.

$$\text{ATPase}$$

$$\text{ATP} + \text{H}_2\text{O} \longleftrightarrow \text{ADP} + \text{P}_i + \text{energy}$$

The most immediate source of this energy support comes from the bond energy within creatine phosphate (CP), also termed phosphocreatine (PCr). Creatine kinase (CKase) catalyzes the breakdown of CP to creatine (Cr) and P_i (see below for reaction). The

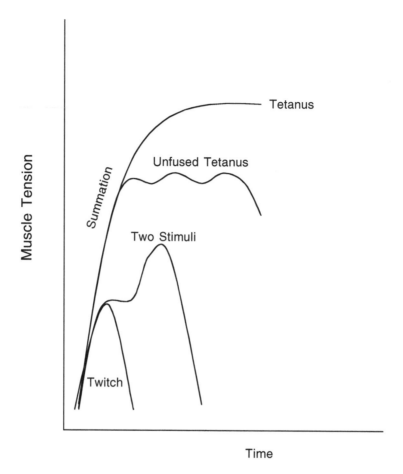

FIGURE 2.5
Muscle twitch characteristics. [Adapted from Foss, M.L. and Keteyian, S.J., *Fox's Physiological Basis for Exercise and Sport*, 6th ed. (Boston: WCB McGraw Hill), 1998.]

energy liberation allows the attachment of a phosphate group to ADP, forming ATP. The concentration of CP in skeletal muscle is about 15 to 17 mmol/kg and transfers enough energy to synthesize more ATP for an additional 8 to 12 seconds of energy. This can allow for further rapid muscular contractions to support such activities as a 100-meter sprint.

$$\text{CKase}$$

$$CP + ADP \longleftrightarrow Cr + ATP$$

During high rates of ATP turnover, such as during high-intensity exercise beyond a 100-meter sprint, ADP concentration increases as ATP and CP become rapidly depleted. This forms the basis for activation of adenylate kinase (AK) or myokinase in skeletal muscle to produce a small amount of additional ATP energy quite quickly. This reaction involves two molecules of ADP, catalyzed by AKase, to form one molecule of ATP and adenosine monophosphate (AMP; see below). AMP can then be deaminated (via adenylate deaminase) to form inosine monophosphate (IMP) and ammonia. This reaction can add a few more seconds to the short-duration, high-intensity exercise bout described previously.

AKase

$$ADP + ADP \longleftrightarrow ATP + AMP$$

To perform a moderately high rate of muscle work over a longer period of time, such as 30 seconds to 2 minutes, energy must be supplied using a more complex biochemical pathway that relies on a substrate or food source (e.g., glucose or glycogen). This is accomplished through a stepwise pathway termed glycolysis (*glyco* = glucose; *lysis* = breakdown) or glycogenolysis (lysis of glycogen) that can release bond energy for ATP regeneration at two different sites in this series of reactions. When this pathway is maximally engaged, such as during high-intensity exercise conditions lasting 30 seconds to 2 minutes, lactic acid is produced. This latter metabolic byproduct of "anaerobic" (meaning without oxygen) glycolysis is the major cause of fatigue for this type of muscular exercise. This is because lactic acid rapidly dissociates upon production in the aqueous medium of the muscle sarcoplasm, into lactate and hydrogen ion (H^+). The accumulation of H^+ can decrease muscle pH, resulting in a decrease in muscle contractile force. During the incomplete breakdown of glucose or glycogen, this pathway can release energy that can be used to resynthesize ATP. This pathway is termed anaerobic when the muscular work requirement is of high intensity and results in the production of lactic acid. As will be noted below, under low-intensity muscular work little lactic acid is produced, and pyruvate and nicotinamide dinucleotide (NADH) produced during glycolysis/glycogenolysis can enter the mitochondria through a protein channel (pyruvate) or can be transported using a shuttle system (NADH), resulting in further energy production through the involvement of other biochemical pathways presented later.

If the intensity of muscular work lasts from several minutes to hours, such as during aerobic endurance exercise or long-duration physical work environments, there is time for pyruvate produced during the final step in glycolysis/glycogenolysis to enter the mitochondria. Once in the mitochondria, the pyruvate is converted to acetyl coenzyme A (CoA) and continues through the Krebs cycle (also known as the tricarboxylic acid cycle), producing a small amount of additional energy for ATP resynthesis within the pathway and also some "reduced" (attached hydrogen and associated electrons) compounds [NADH and flavin adenide dinucleotide ($FADH_2$)] that can enter the electron transport chain, resynthesizing a much larger quantity of ATP.

Fats can also be taken up by skeletal muscle and begin to be metabolized in a pathway termed Beta oxidation. Note that "oxidation" refers to the removal of hydrogen and associated electrons. This pathway supplies NADH and $FADH_2$ bound for the electron transport chain and acetyl CoA that enters the Krebs cycle, reaping further energy benefits through more ATP, NADH, and $FADH_2$ production. As will be explained later, the NADH and $FADH_2$ will be used to produce more ATP. Thus, fats supply a relatively large amount of ATP energy. It is also important to note that a combination of fat and glucose/glycogen is utilized during the aerobic energy production of ATP, and the proportion of each varies in contribution, depending on the intensity and duration of the muscular work.

It should also be noted that under various circumstances of metabolic energy provision, certain amino acids can undergo oxidative "deamination" (removal of amine group) or "transamination" (transfer of amine group) to form some Krebs cycle intermediaries or, in some cases, pyruvate. Depending on the type of amino acids and where they enter the pathway, the amount of ATP energy provision varies. However, amino acids do not normally contribute significantly to energy production during muscular work but can provide up to 10% of the total energy requirements during long duration endurance exercise. Amino acids may become an important energy source during fasting or other types of nutrition deprivation.

2.2.10   Energy Systems Approach to Muscular Work

Each pathway contains key controlling points, such as rate-limiting enzyme reactions that determine how quickly and to what extent the pathway can proceed through its inherent steps. These biochemical pathways within skeletal muscle can be conveniently grouped into three systems (Foss and Keteyian, 1998). The anaerobic ATP–CP system includes energy from ATP and CP stores and the adenylate kinase reaction. The anaerobic glycolytic system includes energy from the incomplete breakdown of glucose and glycogen to lactic acid. Finally, the aerobic system includes the complete breakdown of glucose/glycogen, fat, or amino acids through "aerobic" glycolysis, Beta oxidation, Krebs cycle, and the electron transport chain. Each system has a maximum rate (power) and a maximum capacity (duration) of ATP energy provision for muscular work (Figure 2.6). These rates and capacities are summarized in Table 2.2.

The two most important determinants of which energy system is predominantly utilized are the intensity and duration of the muscular work bout. The available energy system is also strongly dependent on the motor unit and fiber type profile of the involved muscle. For example, if the muscle contains mostly Type II fibers, glycolytic breakdown of glucose and glycogen for energy will be greater than in a muscle composed of predominantly Type I fibers. Force and power development is also directly related to fiber type, with muscle containing more Type II fibers being able to deliver more force and power than a muscle with a higher proportion of Type I muscle fibers. However, if the involved muscle

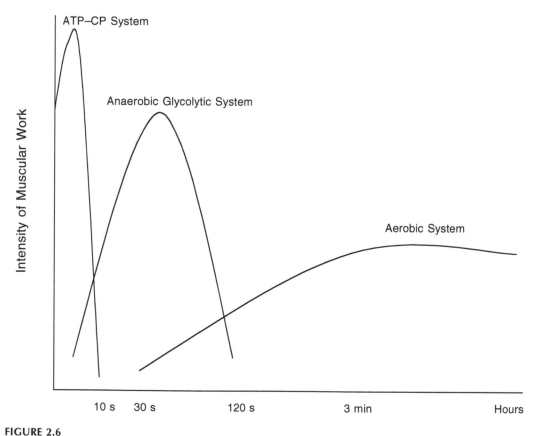

FIGURE 2.6
Schematic illustrating the intensity (power/rate) and duration (capacity) of the three metabolic energy systems operating in skeletal muscle. [Adapted from MacDougall, J.D., Wenger, H.A., and Green, H.J., *Physiological Testing of the High-Performance Athlete*, 2nd ed. (Champaign, IL: Human Kinetics), 1991.]

TABLE 2.2

Energy Systems Approach to the Power and Capacity for ATP Energy Provision during Muscular Work

Energy System	Power	Capacity	Intensity Factor for Power	Intensity Factor for Capacity
Anaerobic ATP CP system	0.1 – 3 s (< 5 s)	8 to 12 s (< 15 s)	Explosive	Maximal
Anaerobic glycolytic system	15 to 30 s	45 to 120 s (> 45 s)	Maximal	Maximal
Aerobic system	3 to 5 min	hour(s)	Near maximal	Submaximal

or muscle groups are recruited to prolonged work, the Type I fibers with their high fatigue resistance and capacity for aerobic metabolic energy supply would be advantageous (McArdle et al., 2001; McComas, 1996).

2.2.11   Regulation of the Biochemical Pathways for Skeletal Muscle Metabolism

There are several important regulatory factors that influence which biochemical pathway is predominantly utilized to maintain ATP levels during muscle work. The first consideration must be the demand for ATP (Figure 2.7). If this demand requires a high rate of ATP turnover, then the immediate stores of ATP and CP that exist directly in the muscle fiber are the only source that can meet this very rapid demand. This is the case for high-force, high-speed movements such as a kicking motion, throwing motion, or jumping. However, this rapid energy supply is short-lived, and muscle force and power output will begin to decline over the 8- to 12-second capacity of ATP and CP energy supply as these high-energy phosphagens become depleted. If ATP demand is moderately high for muscular work, the combination of ATP–CP and glycolytic energy contribution is utilized, with glycolysis becoming the most important source of ATP energy as the duration of muscular work extends to 2 minutes. Again, if the intensity is high, muscle force and power will decline over the 2-minute duration, primarily due to the production of fatigue-related glycolytic byproducts such those that cause a decrease in muscle pH. If ATP demand is submaximal in nature, the aerobic system becomes the predominant energy support pathway for muscular work. This system can continue for prolonged periods, but the depletion of the muscle and liver store of glycogen will eventually limit the level at which muscular work can be sustained using the aerobic system.

Substrate availability is also an important control factor for the biochemical pathways. This can include the available stores of ATP and CP, blood glucose (via liver or through digestion) and muscle glycogen stores for glycolysis, and fats (possibly amino acids) for aerobic metabolism. If stores of energy substrates are depleted, the power and capacity of the involved energy system will be compromised.

The activation of key enzyme complexes and rate-limiting enzymes at important sites in the various metabolic pathways can partly control the amount and rate of ATP energy provision. For example, phosphofructokinase (PFK) is thought to be the rate-limiting enzyme in glycolysis. Therefore, this biochemical pathway cannot proceed any faster than the ability of PFK to convert fructose 6-phosphate to fructose 1,6-diphosphate, a key regulatory step. PFK requires ATP energy input to proceed, so it is also an energy-consuming step. PFK is activated by increases in molecules such as ADP, AMP, and NH^{4+} and is inhibited by increases in ATP and citrate within the sarcoplasm. Other examples of important enzymes thought to be rate limiting are isocitrate dehydrogenase in the Krebs

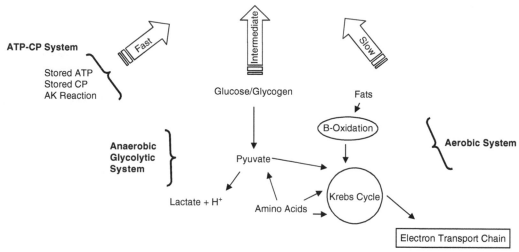

FIGURE 2.7
Schematic representation of the metabolic pathways with respect to the energy system approach of ATP energy supply for muscle force production.

cycle, cytochrome oxidase in the electron transport chain, and beta ketothiolase in the Beta oxidation pathway.

Certain hormones can contribute to the control of biochemical pathways in skeletal muscle. Epinephrine released from the adrenal medulla and glucagon released from the pancreas are known to indirectly activate phosphorylase through a second messenger system. The active form of phosphorylase is the enzyme responsible for the breakdown of glycogen prior to entry into the glycolytic energy pathway. These two hormones have also been shown to indirectly stimulate lipase activity to initiate the breakdown of trig-lycerides so that their products may enter the Beta oxidation pathway. Many other examples of hormonal control of biochemical pathways also exist.

The "oxidative state" is also an important regulatory factor in biochemical pathways. The availability of cellular oxygen to the electron transport chain and the ratio of NADH to NAD^+ and $FADH_2$ to FAD^+ partially control the rate of energy flux through aerobic metabolic pathways. In times when the intensity of exercise exceeds the rate at which the aerobic metabolic pathway can keep up, or when any other factor decreases the supply of oxygen (altitude, lower ambient O_2 concentration, industrial gases such as hydrogen sulfide), the proportion of ATP energy supply from aerobic sources will decline. This will require the activation of anaerobic energy sources if muscular work is continued and, ultimately, an earlier onset of fatigue.

2.3   Factors Limiting Strength Development in Skeletal Muscle

There are several important factors that can influence the force expression of muscle (Baechle and Earle, 2000). These are important for the ergonomist to consider, as they may play a role in developing the equipment, systems, tasks, or jobs that are effective and within the capabilities of the worker. Modifications to equipment, stations, or the way in which work is done may be necessary if movement patterns involving muscular strength are required.

2.3.1 Muscle Length and Joint Angle Relationship

The maximum force that a muscle or muscle group can generate depends on the length of that muscle. In an isolated muscle, maximum force will be generated when the muscle is at a length that maximizes cross-bridge formation. However, the optimal force–length relationship of muscle or muscle groups in the human body is difficult to determine because of the relationship between the various lever-arm moments formed by the different joints in the human body and the architecture of the muscle (Wickiewicz et al., 1984). As a result, the maximum force capabilities of muscle will also depend on the angle of the joint. Thus, each muscle movement pattern across a joint or series of joints will have maximum force output at a particular muscle length, depending on the joint(s) involved. For example, the maximum force capabilities of the knee extensors may occur at approximately 60 degrees lower than full knee extension (considered to be the anatomical neutral position or 0 degrees). Knee flexion maximum force occurs at approximately 30 degrees from full knee extension. Thus, body and limb position will partly limit and, therefore, determine muscular force and power expression during any type of muscle movement pattern.

2.3.2 Muscle Force Velocity, Power Velocity Relationship

The maximum muscle force that can be generated is greatest during an isometric contraction, provided it is performed at an optimal joint angle. Muscle force will decrease with increasing speed of muscle shortening. This is because, as the velocity of movement increases, the motor unit activation time decreases, thus limiting the amount of force generated at fast limb-movement velocities (Perrine and Edgerton, 1978). Conversely, during an eccentric muscle contraction, force increases with an increase in speed (Herzog, 1996). It has been suggested that during a maximal eccentric contraction there is greater neural activation and maximal engagement of the noncontractile properties of skeletal muscle (connective tissue).

Concentric power increases with an increase in the speed of contraction until approximately one third of maximum shortening velocity (Figure 2.8). In human movement, this may be influenced by the motor unit properties and architecture of the involved musculature and joint. Muscle power output is a product of force and velocity (distance/time), and since the decay in force over a particular distance is less than the decay in time, there is an increase in power output. This occurs until the time of contraction increases to a point where little force can be developed. When this occurs, muscular power will plateau and begin to decrease (Jones et al., 1986; MacDougall et al., 1991).

Thus, the speed at which a limb or body movement pattern is performed during an activity may limit the amount of muscular force or power that can be generated. This necessitates that movements requiring a high degree of force be performed at a slow speed, and those requiring a high level of power output will require a faster speed (Bell and Werger, 1992).

2.3.3 Motor Unit and Muscle Fiber Type

Regardless of velocity of muscle shortening, muscle force and muscle power output will be greater in muscle or muscle groups that contain a greater proportion of fast twitch

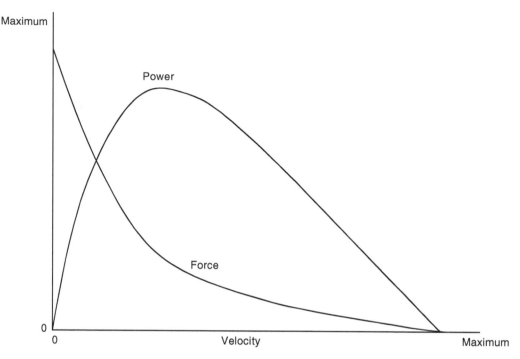

FIGURE 2.8

Schematic of skeletal muscle force-velocity, power-velocity relationship. [Adapted from Herzog, W., *Am. J. Sports Med.*, 24(6), S14–S19, 1996; and Vogel, S., *Prime Mover: A Natural History of Muscle*, (New York: W.W. Norton & Company), 2001.]

motor units. This is because fast twitch motor units contain Type II fibers that have a faster time to peak tension, greater sarcoplasmic reticulum network for calcium release and uptake, and faster metabolic energy supply pathways (highly anaerobic). Fast twitch motor units also contain muscle fibers with a larger cross-sectional area (CSA) that can produce greater force at a faster rate. Thus, the fiber type distribution of the muscle or muscle groups involved in a particular movement pattern may limit the amount of force and power generation.

2.3.4   Neural Tension-Limiting Mechanism

The skeletal muscle system contains neural tension-limiting mechanism(s) that act as safety devices within muscle during maximal force development to avoid damage. One such example of this is the Golgi tendon organ (GTO) that senses tension in the tendon as a result of muscle contraction. As muscle tension increases, the neural receptors of the GTO fire, and a neural feedback loop is created through an inhibitory interneuron in the spinal cord that causes an inhibition of the motor neuron stimulation of the muscle. There can also be an activation of the antagonist muscle group. These mechanisms can limit the degree of tension that the muscle can generate. This is thought to be an important factor that limits maximal force and power output of a muscle or muscle group.

2.3.5   Role of Preloading

Under resting conditions, skeletal muscle is generally preactivated or ready to produce force. Although the initiation of the muscle sarcomere movement is initiated with the release of calcium ions, this process does not occur instantaneously in all motor units or at all actin–myosin complexes. There is a slight time delay between the release of calcium and the activation of all the available actin–myosin filaments. This, in turn, slightly delays the force generation of the activated muscle. This delay in the development of high force production may be critical to some forms of strength expression and may hinder conditions where high tension development is necessary early in the range of motion. Kovaleski and colleagues (1995) suggest that "preloading" the muscle by performing an isometric contraction prior to initiating a concentric muscle action may enhance the ability to generate high tension at the start of movement or early in the range of motion. There is also support that using preloading during strength exercises can assist with force development early in a range of motion, especially at high velocities of contraction (Hunter and Culpepper, 1988, 1995). Thus, effective force production may be limited in the very early portion of the range of motion, especially in fast movements, but this may be somewhat overcome with a prior isometric contraction.

2.3.6 Stretch Shortening Cycle

Performing an eccentric muscle action immediately prior to recruiting the same muscle for a concentric contraction has been shown to augment force production during the subsequent contraction (Komi, 1992; Komi and Bosco, 1978). This phenomenon is referred to as the stretch-shortening cycle (SSC). The enhanced concentric contraction is a combined effect of the elastic energy stored in the tendon, connective tissue sheaths, and possibly the sarcomere and from the myotatic reflex of the muscle (Komi, 1992). The result is a potentiation of concentric muscle force output if followed by an imposed stretch such as that elicited with a prior eccentric contraction of the same muscle group. Without evoking the SSC, concentric force action may be somewhat attenuated.

2.3.7 Exercise-Induced Muscle Damage

Intense activity or exercise can induce skeletal muscle damage and induce muscle pain that is referred to as the delayed-onset of muscle soreness or DOMS. DOMS normally peaks 24 to 72 hours following exercise. Although the exact mechanism for DOMS is under investigation, eccentric muscle actions appear to induce the greatest discomfort. Documented structural abnormalities linked to DOMS include disruption of sacromere Z-lines, degradation of the titin filaments, and mitochondrial swelling (Figure 2.9) (McArdle et al., 2001; Waterman-Storer, 1991). Symptoms associated with DOMS include an inflammation-induced increase of fluid within the muscle and release of inflammatory factors (e.g., prostaglandins) that contribute to the muscle soreness. As a result of DOMS, there is a loss of voluntary force production by the muscle (McArdle et al., 2001) until recovery is attained.

2.3.8 Other Limiting Factors

There are many other contributing factors that can limit muscle force, power, and endurance production. These include aging that results in a decrease of muscle size (atrophy), and de-innervation resulting in a loss of muscle fiber numbers, and a general "deconditioning" leading to a loss of strength, power, and endurance. Gender differences also exist. Men have a greater absolute strength and power compared to women, probably due to the larger muscle mass and, possibly, fiber type differences. Environmental conditions can also lead

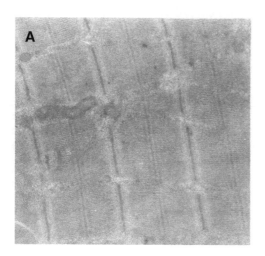 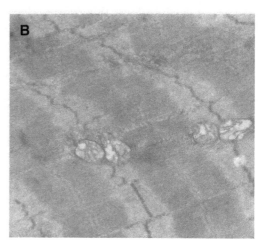

FIGURE 2.9
A normal electron micrograph of several muscle sarcomeres. (A) A muscle showing damage. (B) Note the Z-line streaming and swollen mitochondria. (Photographs courtesy of Ian Maclean, University of Alberta.)

to a decrement in force/power/endurance production. Heat, cold, humidity, wind, and air quality can all influence the expression of muscular force. Nutritional and hydration factors also underlie the metabolic support for muscular work. Proper nutrition and hydration practices are important in many work and domestic situations and become even more important when working in certain adverse environments (heat, cold, altitude). Both peripheral and central fatigue factors must also be considered when evaluating muscle performance factors. Finally, considerations such as the way in which an activity is performed, posture, use and set up of particular equipment, and psychological state of the individual can alter muscle performance. These topics will be discussed more fully in other chapters.

2.4   Summary and Conclusions

The physiology and biochemistry of the skeletal muscle system underlie the development of force, power, and endurance. It is clear that the way in which the body is positioned, moved, stabilized, or forced to react to various situations is partly determined by the structure and functioning of the skeletal muscle system. This is important in the interpretation, evaluation, application, and rehabilitation of the way in which the human body responds to its environment and the factors within that environment. Activities of daily living, occupation, recreation, and leisure pursuits all require proper function of the skeletal muscle system. It is also important to note that the skeletal muscle system is highly adaptable. It can alter its structure and functioning as a result of chronically imposed demand, resulting in better overall muscle performance and reduced fatigue. Conversely, there can be a loss of structure and function as a result of the removal of a particular demand, such as becoming too sedentary, experiencing spaceflight, or becoming immobilized as a result of an injury or as a result of aging. This knowledge helps to address the energy and muscular effort requirements of work, the underlying role of nutrition to support these efforts, and the interaction with the environment of the physical working conditions, and helps set optimal standards for acceptable work rates and loads. Applications of this information can also be useful for domestic situations in everyday life.

References

Baechle, T.R. and Earle, R.W. 2000. *Essentials of Strength Training and Conditioning,* 2nd ed. (Champaign, IL: Human Kinetics).

Bell, G., and Wenger, H.A. 1992. Physiological adaptations to velocity-controlled resistance training, *Sports Med.,* 13(4), 234–244.

Burnham, R., Martin, T., Stein, R., Bell, G., MacLean, I., and Steadward, R. 1997. Skeletal muscle fibre type transformation following spinal cord injury, *Spinal Cord,* 35, 86–91.

Dudley, G.A., Duvoisin, M.R., Adams, G.R., Meyer, R.A., Belew, A.H., and Buchanan, P. 1992. Adaptations to unilateral lower limb suspension in humans. *Aviat. Space Environ. Med.,* 63, 678–683.

Edgerton, V.R., Zhou, M., Ohira, Y., Klitgaard, H. Jiang, B., Bell, G., Harris, B., Saltin, B., Gollnick, P., Roy, R., Day, M., and Greenisen, M. 1995. Human fiber size and enzymatic properties after 5 and 11 days of spaceflight. *J. Appl. Physiol.,* 78(5), 1733–1739.

Foss, M.L. and Keteyian, S.J. 1998. *Fox's Physiological Basis for Exercise and Sport,* 6th ed. (Boston: WCB McGraw Hill).

Harman, E. 1993. Strength and power: a definition of terms, *Nat. Strength Cond. Assoc. J.,* 15(6), 18–20.

Henneman, E., Somjen, G., and Carpenter, D.O. 1965. Functional significance of cell size in spinal motoneurones, *J. Neurophysiol.,* 28, 560–580.

Herzog, W. 1996. Muscle function in movement and sports, *Am. J. Sports Med.,* 24(6), S14–S19.

Houston, M.E. 1995. *Biochemistry Primer for Exercise Science* (Champaign, IL: Human Kinetics).

Hunter, G.R. and Culpepper, M.I. 1988. Knee extension torque joint position relationships following isotonic fixed resistance and hydraulic resistance training, *J. Athl. Train.,* 23(1), 16–22.

Hunter, G.R. and Culpepper, M.I. 1995. Joint angle specificity of fixed mass versus hydraulic resistance knee flexion training, *J. Strength Cond. Res.,* 9(1), 13–16.

Huxley, H. and Hansen, J. 1954. Changes in the cross-striations of muscle during contraction and stretch and their structural interpretation, *Nature,* 173, 973–976.

Jones, N.L., McCartney, N., and McComas, A.J. 1986. *Human Muscle Power* (Champaign, IL: Human Kinetics).

Kell, R., Bell, G., and Quinney, H.A. 2001. Musculoskeletal fitness, health outcomes and quality of life, *Sports Med.,* 31(12), 863–873.

Knuttgen, H.G. and Kraemer, W.J. 1987. Terminology and measurement in exercise performance, *J. Appl. Sport Sci. Res.,* 1(1), 1–10.

Komi, P.V. 1992. The stretch-shortening cycle, in *Strength and Power in Sport,* P.V. Komi, Ed. (Boston: Blackwell Scientific), pp. 169–179.

Komi, P.V. and Bosco, C. 1978. Utilization of stored elastic energy in leg extensor muscle by men and women, *Med. Sci. Sports,* 10, 261–265.

Kovaleski, J.E., Heitman, R.H., Trundle, T.L., and Gilley, W.F. 1995. Isotonic preload versus isokinetic knee extension resistance training, *Med. Sci. Sports Exerc.,* 27(6), 895–899.

McArdle, W.D., Katch, F.I., and Katch, V. 2001. *Exercise Physiology Energy, Nutrition, and Human Performance,* 5th ed. (Baltimore, MD: Lippincott Williams & Wilkins).

McComas, A.J. 1996. *Skeletal Muscle Form and Function* (Champaign, IL: Human Kinetics).

MacDougall, J.D., Wenger, H.A., and Green, H.J. 1991. *Physiological Testing of the High-Performance Athlete,* 2nd ed. (Champaign, IL: Human Kinetics).

McMahon, T.A. 1984. *Muscles, Reflexes, and Locomotion* (Princeton, NJ: Princeton University Press).

Perrine, J. and Edgerton, V.R. 1978. Muscle force-velocity and power-velocity relationships under isokinetic loading, *Med. Sci. Sports,* 10(3), 159–166.

Pette, D. and Staron, R.S. 2001. Transitions of muscle fiber phenotypic profiles, *Histochem. Cell Biol.,* 115, 359–372.

Sale, D. 1987. Influence of exercise and training on motor unit activation, *Exerc. Sport Sci. Rev.,* 15, 95–151.

Schiaffino, S., Gorza, L., Sartore, S., Saggin, L., and Ausoni, S. 1989. Three myosin heavy chain isoforms in type II skeletal muscle fibers. *J. Muscle Res. Cell Motil.,* 10, 197–205.

Waterman-Storer, C.M., 1991. The cytoskeleton of skeletal muscle: is it affected by exercise? *Med. Sci. Sports Exerc.*, 23(11), 1240–1249.

Wickiewicz, T., Roy, R., Powell, P., Perrine, J., and Edgerton, V.R. 1984. Muscle architecture and force-velocity relationships in humans, *J. Appl. Physiol.*, 57(2), 435–443.

3

Cardiac Effects of Strength Training

Mark J. Haykowsky

CONTENTS

ABSTRACT Submaximal or maximal strength exercise is associated with an acute increase in heart rate, cardiac output, ejection fraction, and systolic blood pressure. The consequence of the heightened pressure load is that it may be an important stimulus to alter the size and shape of the heart. For example, a series of investigations have shown that strength-trained athletes have a larger left ventricular wall thickness and mass compared to age-matched sedentary individuals. Moreover, the strength-training-mediated alteration in left ventricular morphology may have physiologic and functional significance, because strength-trained athletes have been shown to have a greater stroke volume, ejection fraction, and cardiac output, with a concomitant reduction in heart rate and systolic blood pressure when performing submaximal isometric or dynamic resistance exercise. Despite these beneficial cardiovascular adaptations, a limitation of previous investigations that have examined the effects of strength training on left ventricular morphology and function was the primary focus on healthy younger individuals. The few investigations performed to date have found that strength training is an effective intervention to increase muscle strength and mass in healthy older individuals or cardiac patients without negatively altering the size or shape of their heart. This chapter will review the acute and chronic effects of strength training on left ventricular morphology and function in healthy individuals and strength athletes. In addition, the cardiac effects of strength training in female athletes, healthy older individuals, and older individuals with impaired left ventricular systolic function will also be discussed.

KEY WORDS: strength training, left ventricular hypertrophy, cardiac output.

3.1   Introduction

Strength training is an effective form of exercise to increase muscle strength and function. Moreover, strength training is an important mode of exercise to mitigate the senescent or disuse-mediated decline in muscle mass (Fiatarone et al., 1990). Despite the benefits of strength training on skeletal morphology, the effects that this type of exercise has on altering cardiac morphology and function remains uncertain. For example, a long-held belief in exercise science is that the strength-training-mediated increase in systolic blood pressure results in a concomitant elevation in left ventricular wall stress (the force that resists ventricular fiber shortening) which, in accordance with the load-induced hypertrophy hypothesis, is an important stimulus to increase left ventricular wall thickness and mass (Colan, 1992; Colan et al., 1987). However, this hypothesis may not be tenable, because a number of cross-sectional and short-term longitudinal investigations have shown that strength training does not alter the size or shape of the heart (Haykowsky et al., 1998, 2000a,b,c, 2002). In addition, a series of investigations have shown that left ventricular end–systolic wall stress is not elevated above resting values during submaximal isometric handgrip exercise or during submaximal or maximal or leg-press exercise (Galanti et al., 1992; Haykowsky et al., 2001). Thus, it is possible that the heightened left ventricular wall stress, if it occurs, may be too brief in duration or of insufficient magnitude to induce cardiac hypertrophy.

Currently, there is limited information regarding the acute hemodynamic and cardiac volume responses during resistance exercise in chronically strength-trained athletes. However, the few studies that have been done have found that strength-trained athletes have a greater stroke volume and ejection fraction and a lower heart rate, systolic blood pressure, and systemic vascular resistance compared to sedentary individuals during submaximal isometric or dynamic resistance exercise (Fisman et al., 1997; Fleck and Dean, 1987). This adaptive response may have functional and ergonomic significance, because a reduction in systolic blood pressure and systemic vascular resistance will allow recreational and occupational tasks that have a strength component to them to be performed with reduced stress on the heart.

The purpose of this chapter will be to examine the acute and chronic effects of strength training on left ventricular morphology and function in healthy individuals and strength athletes. In addition, a discussion regarding the cardiac effects of strength training in female athletes, healthy older individuals, and cardiac patients with impaired left ventricular systolic function will also be explored. Throughout this chapter, strength-trained athletes are defined as bodybuilders, weightlifters (athletes whose competitive events include the snatch and clean-and-jerk) and powerlifters (athletes whose competitive events include the squat, bench press, and deadlift). In addition, when reviewing the short-term effects of exercise training on left ventricular morphology and function, only those investigations incorporating pure strength exercise as the primary form of training or intervention will be reviewed.

3.2 Acute Hemodynamic Responses during Strength Exercise

Repetitive submaximal or maximal strength exercise is associated with a transient and marked increase in heart rate, systolic blood pressure, diastolic blood pressure, and mean arterial pressure (Table 3.1; Haykowsky et al., 2001; Lentini et al., 1993; MacDougall et al., 1985, 1992). Haykowsky et al. (2001) have shown that younger healthy males are capable of generating systolic and diastolic blood pressures > 300/230 mmHg when performing submaximal leg-press exercise to volitional exhaustion. Moreover, peak systolic pressures as high as 480 mmHg have been recorded during leg-press exercise (MacDougall et al., 1985). MacDougall et al. (1992) have demonstrated that the magnitude of the change in systolic and diastolic blood pressure during lifting was independent of the absolute weight lifted or size of the exercising musculature. Therefore, individuals with reduced muscle mass and strength (i.e., older individuals or deconditioned chronic stable heart failure patients) are able to generate peak systolic pressures as high as 189 to 247 mmHg when performing submaximal strength exercise (McCartney et al., 1993; McKelvie et al., 1995). The mechanisms responsible for the strength-training-mediated rise in heart rate and arterial pressure are related to: (a) a transient increase in total systemic vascular resistance associated with performing maximal muscular contractions, (b) an increase in intrathoracic pressure associated with performing a brief (2 to 3 second) Valsalva maneuver (a forced expiration against a closed glottis), and (c) neural-mediated mechanisms that stimulate the cardiovascular control center in the ventrolateral medulla (MacDougall, 1994; Mitchell, 1990).

3.3 Acute Effects of Strength Exercise on Systolic Pressure: Role of Increased Intramuscular Pressure

Sejersted et al. (1984) found that younger healthy individuals are capable of generating intramuscular pressures as great as 570 mmHg when performing maximal knee extensor

TABLE 3.1

Acute Hemodynamic and Left Ventricular Volume Responses during Dynamic Strength Exercise Performed with a Brief (2 to 3 sec) Valsalva Maneuver

Variable	Change from Rest
End-diastolic volume (ml)	↓
End-systolic volume (ml)	↓↓
Stroke volume (ml)	↓
Ejection fraction (%)	↑
Heart rate (beat/min)	↑↑
Cardiac output (L/min)	↑
Systolic blood pressure (mmHg)	↑↑↑
Intrathoracic pressure (mmHg)	↑↑↑
Left ventricular transmural pressure (mmHg)	↔
Left ventricular end-systolic wall stress (dynes/cm^2)	↔

Note: ↓, decrease from rest; ↑, increase from rest; ↔, no change from rest.

exercise. Moreover, Sylvest and Hvid (1959) reported a mean vastus medialis intramuscular pressure of 479 mmHg during maximal muscular contractions. Of possibly greater interest, intramuscular pressure increased to 1025 mmHg in one subject. The consequence of this extreme elevation in intramuscular (extravascular) pressure is that it occludes intravascular blood flow, resulting in an increase in total systemic vascular resistance and a concomitant rise in systolic, diastolic, and mean arterial blood pressure during lifting.

3.4   Acute Effects of Strength Exercise on Systolic Pressure: Role of Increased Intrathoracic Pressure

A second mechanism responsible for the elevated systolic and diastolic blood pressure during dynamic strength exercise is related to the increased intrathoracic pressure associated with performing a brief (2 to 3 second) Valsalva maneuver. Hamilton et al. (1936) demonstrated that the increased intrathoracic pressure associated with performing a Valsalva maneuver was transmitted directly to the arterial vasculature as an increase in systolic blood pressure. More importantly, they also found that the pressure to which the heart was exposed (i.e., left ventricular transmural pressure, derived as left ventricular pressure minus intrathoracic pressure) was not elevated above baseline values. Thus, a Valsalva paradox occurs during dynamic strength exercise, whereby the increased intrathoracic pressure associated with performing a brief (2 to 3 second) Valsalva maneuver results in a rapid and marked elevation in systolic blood pressure; however, the heightened intrathoracic pressure also prevents an increase in left ventricular transmural pressure. Although a Valsalva maneuver is often discouraged during lifting, MacDougall et al. (1992) found that this maneuver was a natural response when lifting a weight ≥80% of maximal strength or during the final few repetitions of lower-intensity exercise performed to volitional exhaustion.

In summary, acute heart-lung and skeletal muscle–peripheral vascular interactions occur during dynamic repetitive lifting that result in an increase in heart rate, systolic blood pressure, diastolic blood pressure, and mean arterial pressure.

3.5   Acute Effects of Strength Exercise on Cardiac Volumes and Output

At present, only two investigations have extensively examined the acute effects of strength exercise on cardiac volumes and systolic function in healthy younger individuals. Lentini et al. (1993) evaluated the left ventricular volume and cardiac output response during submaximal (95% of maximal strength) leg-press exercise performed with a brief Valsalva maneuver. More recently, Haykowsky et al. (2001) examined the acute hemodynamic, left ventricular volume, and end-systolic wall stress responses during submaximal (80 and 95% of maximal strength) and maximal leg-press exercise performed with a brief Valsalva maneuver. In both investigations, submaximal and maximal leg-press exercise performed with a brief Valsalva maneuver resulted in a reduction in end-diastolic volume, end-systolic volume, and stroke volume with a concomitant increase in heart rate, left ventricular contractility, ejection fraction, and cardiac output (Table 3.1). Haykowsky et al. (2001)

also reported that left ventricular end-systolic wall stress was 7 to 15% lower during lifting compared to resting values. The latter finding is consistent with the previous observation that left ventricular end-systolic wall stress was 25% lower during a Valsalva maneuver (without lifting) compared to rest in healthy individuals (Fuenmayor et al., 1992). These findings are important because a widely held, but unsubstantiated, belief in exercise science is that the heightened systolic blood pressure associated with performing a brief Valsalva maneuver (with or without lifting) results in a concomitant increase in left ventricular end-systolic wall stress. However, the findings of Lentini et al. (1993) and Haykowsky et al. (2001) suggest that the heightened systolic blood pressure during exertion is secondary to the increased intrathoracic pressure associated with performing a brief (2 to 3 second) Valsalva maneuver, since left ventricular transmural pressure was markedly lower than the recorded systolic blood pressure alone. Moreover, Galanti et al. (1992) found that favorable changes in left ventricular geometry (i.e., increased systolic ventricular wall thickening combined with a reduction in end-systolic cavity volume) occur during lifting that prevent an acute increase in left ventricular wall stress.

In summary, although few studies have been done, evidence indicates that repetitive submaximal or maximal strength exercise is associated with an acute decline in end-diastolic volume and stroke volume. This decline is offset by an increase in left ventricular contractility and ejection fraction that, combined with an increased heart rate, results in an elevated cardiac output during lifting. Moreover, isometric or dynamic strength exercise (with or without a brief Valsalva maneuver) does not appear to elevate left ventricular end-systolic wall stress.

3.6 Chronic Effects of Strength Training on Resting Cardiac Morphology: Role of the Type of Strength Training Performed

A number of investigations have examined the chronic effects of strength training on resting cardiac morphology in strength-trained athletes. The major finding from these studies is that strength training does not result in a homogeneous alteration in left ventricular morphology. For example, some investigators have found that strength training can increase left ventricular wall thickness and mass (Colan et al., 1987; Deligiannis et al., 1988; Fagard, 1996; Lattanzi et al., 1992; Pluim et al., 1999; Suman et al., 2000; Vinereanu et al., 2002). However, others have shown that strength training does not alter the size or shape of the heart (Haykowsky et al., 1998, 2000a,b,c, 2002; Wernstedt et al., 2002). The discrepancy between studies may be due to the type of strength athletes studied. For example, Murray et al. (1989), using impedance cardiography, compared the acute stroke volume and cardiac output responses during submaximal (50, 70, 80, and 90% of maximal strength) and maximal squatting exercise in 10 elite male bodybuilders and powerlifters. The main finding of this investigation was that the bodybuilders had a significantly greater stroke volume (bodybuilders overall mean: 78 ml/beat vs. powerlifters overall mean: 61 ml/beat) and cardiac output (bodybuilders: 12 l/min vs. powerlifters: 9 l/min) compared to the powerlifters when lifting. If both strength-trained groups had similar ejection fractions during lifting, then the bodybuilders' end-diastolic volume (and left ventricular end-diastolic wall stress) would be 28% greater than that of powerlifters. The consequence of the bodybuilding-mediated diastolic volume overload is that it may be a greater stimulus for cardiac hypertrophy compared to powerlifting training. This hypothesis is consistent with a previous finding that bodybuilders had a significantly greater resting left ventricular diastolic cavity dimension (equal to a 24 to 29% increase in end-diastolic volume) and left ventricular mass compared to powerlifters or weightlifters (Pelliccia et al., 1993). Furthermore, Haykowsky et al. (2002) recently reported that when strength training does alter the size and shape of the heart, the most common pattern that occurs is concentric hypertrophy (i.e., increased left ventricular wall thickness and mass with minimal alteration in left ventricular diastolic cavity dimension), which is commonly seen in weightlifters. The second most common cardiac pattern found was eccentric hypertrophy (i.e., increased left ventricular mass secondary to increases in left ventricular wall thickness and left ventricular cavity dimension) that was evident in bodybuilders. Finally, 38% of all athletes had a normal geometric pattern (i.e., no echocardiographic evidence of cardiac hypertrophy). Of interest, the majority of strength athletes with normal geometry were powerlifters. The above findings suggest that the pattern of cardiac hypertrophy that may occur with training appears to be related to the type of strength training performed. Moreover, strength training does not appear to negatively alter resting left ventricular systolic function (Haykowsky et al., 2002).

3.7 Chronic Effects of Strength Training on Cardiac Volumes and Hemodynamics during Lifting

Currently, there is a paucity of investigations that have examined the acute hemodynamic and cardiac volume responses during dynamic strength exercise in resistance-trained athletes. Fleck and Dean (1987) reported that bodybuilders had a lower peak heart rate, systolic and diastolic blood pressure compared to novice weight trainers or sedentary individuals when performing upper and lower extremity strength exercise. McCartney et al. (1993) and Sale et al. (1994) examined the acute hemodynamic responses (at the same absolute workload) before and after 12 to 19 weeks of strength training in younger (mean age: 22 years) and healthy older (mean age: 66 years) individuals. The main finding from these investigations was that peak systolic blood pressure, diastolic blood pressure, and heart rate were lower after training when lifting the same submaximal load. The mechanism responsible for this response is not well known, however, it may be related to a reduction in central command (i.e., a strength-training-mediated increase in muscle mass and strength, combined with enhanced motor unit recruitment, allows the same absolute load to be lifted with less effort, reducing sympathetic outflow via the cardiovascular control center). Regardless of the underlying mechanism, the reduced heart rate, blood pressure and rate pressure product during lifting will decrease myocardial oxygen demand. This adaptive response may have functional and ergonomic significance, since a reduction in systolic blood pressure associated with training will allow recreational and occupational tasks that have a strength component to them to be performed with reduced stress on the heart.

The long-term effects of strength training on cardiac volumes and function during dynamic strength exercise have not been examined. However, two investigations provide indirect evidence regarding how the strength-trained heart may function during exertion. Fisman et al. (1997), compared the acute cardiac volume and output responses during submaximal (50% maximal voluntary contraction) isometric deadlift exercise in younger (mean age: 23 years) weightlifters with > 6 years training experience and age-matched sedentary individuals. The unique finding from this investigation was that the weightlifters had a 22% greater left ventricular ejection fraction during exertion than the sedentary controls. Moreover, the athletes' heightened systolic function was due to an enhanced contractile reserve (and concomitant reduction in end-systolic volume) and to the reduced systemic vascular resistance during exercise. Suman et al. (2000) examined the acute hemodynamic, cardiac volume, and wall stress responses during dobutamine (a positive inotropic agent) administration in bodybuilders (mean age: 33 years, with a minimum of 5 years of training experience) and age-matched controls. The major finding of this novel investigation was that the bodybuilders had a significantly greater fractional shortening (an index of left ventricular systolic function) than the controls during dobutamine administration. Moreover, the bodybuilders' enhanced cardiac performance was secondary to an augmented inotropic response during dobutamine challenge. The findings of Fisman et al. (1997) and Suman et al. (2000) suggest that long-term strength training may result in favorable ventricular and vascular adaptations that result in enhanced cardiac performance during isometric exercise or pharmacologic stress. Whether these beneficial adaptive responses occur during repetitive dynamic upper or lower extremity strength exercise requires further study. A limitation of the majority of investigations that have examined the acute and chronic effects of strength training on cardiac morphology and performance was the primary focus on younger healthy males. Therefore, the cardiac effects of strength training in females, healthy older individuals, and cardiac patients are not well known. The remainder of this chapter will briefly explore these issues.

3.8   Cardiac Effects of Strength Training in Female Athletes

The role that strength training may play in altering the size and shape of the athletic female heart has received very little attention. George et al. (1998) found that younger (mean age: 25 years) elite female weightlifters had a greater posterior wall thickness, ventricular septal wall thickness, relative wall thickness (i.e., wall thickness to cavity dimension ratio), and left ventricular mass compared to age-matched sedentary women. Wernstedt et al. (2002), using echocardiographic and magnetic resonance imaging techniques, found no significant difference for left ventricular posterior wall thickness, ventricular septal wall thickness, relative wall thickness, diastolic cavity dimension, or left ventricular mass between elite female weightlifters (mean age: 30 years, with a minimum of 2 years of training experience) and sedentary controls. Consistent with these results, Haykowsky et al. (1998) found that elite female powerlifters (mean age: 31 years, with 6 years' training experience) had similar left ventricular cavity dimensions, wall thickness, and mass compared to age-matched controls. Thus, the limited evidence to date suggests, not unlike the observations found in males, that strength training does not result in a consistent alteration in the size and shape of the athletic female heart. Further investigations with a larger sample size are required to determine if there is a relationship between the type of strength training performed and subsequent alteration in cardiac morphology in women.

3.9   Cardiac Effects of Strength Training in Healthy Older Individuals

Strength training is an important intervention that can offset the age- and deconditioning-induced loss in muscle mass and strength. For example, researchers at Tufts University have found that two to three months of high-intensity (80% of maximal strength) lower extremity strength training resulted in a significant and marked increase in maximal muscular strength and muscle mass in individuals between 60 and 96 years of age (Fiatarone et al., 1990; Frontera et al., 1988). Despite these benefits, the role that strength training plays in altering left ventricular morphology and systolic function in healthy older adults has not been well studied. Sagiv et al. (1989) found that 3 months of low-intensity (30% of maximal muscular strength) strength training did not alter resting left ventricular posterior wall thickness, ventricular septal wall thickness, or ejection fraction in older (mean age: 68 years) males. Haykowsky et al. (2000) compared the effects of 4 months of moderate-to-high intensity (60 to 80% of one-repetition maximum) upper and lower extremity strength training or no training on resting left ventricular morphology, systolic function, and wall stress in healthy sedentary older (mean age: 68 years) men. The primary finding of this investigation was that 16 weeks of training resulted in a significant increase in upper and lower extremity maximal strength compared to the control subjects. Furthermore, the strength training program did not alter resting left ventricular systolic or diastolic posterior wall thickness, ventricular septal wall thickness, cavity dimension, ejection fraction, estimated left ventricular mass, or end-systolic wall stress. In a similar study, Hagerman et al. (2000) demonstrated that 4 months of high-intensity (85 to 90% of maximal strength) lower extremity strength training resulted in a significant increase in maximal muscular strength without altering left ventricular dimensions or mass in males between 60 and 75 years of age. Taken together, the few investigations performed to date suggest that moderate-to-high-intensity strength training is enough of a stimulus to increase skeletal muscle mass but is an insufficient stimulus to alter the size or shape of the senescent heart.

3.10 Cardiac Effects of Strength Training in Individuals with Heart Disease and Impaired Left Ventricular Systolic Function

Strength training has gained acceptance as an effective form of exercise to improve muscle strength and endurance in low-risk clinically stable cardiac patients (Pollock et al., 2000). Despite these proven benefits, the acute and chronic effects of this form of exercise on left ventricular systolic function have not been well studied in individuals with underlying cardiac disease. McKelvie et al. (1995), using invasive arterial blood pressure monitoring combined with two-dimensional transthoracic echocardiography, examined the acute hemodynamic and left ventricular volume responses during submaximal (70% of maximal strength for 2 sets of 10 repetitions) unilateral leg-press exercise in individuals with ischemic cardiomyopathy and reduced left ventricular systolic function (mean resting ejection fraction equal to 27%). Compared to baseline, submaximal leg-press exercise was associated with a significant increase in heart rate, systolic blood pressure, diastolic blood pressure, rate pressure product, systolic blood pressure to end-systolic volume ratio (an indirect measure of left ventricular contractility), and cardiac output with no significant alteration in end-diastolic volume, end-systolic volume, stroke volume, ejection fraction, or total peripheral resistance. In addition, no adverse symptoms, untoward arrhythmias, or evidence of myocardial ischemia occurred during lifting. In a similar investigation, Meyer et al. (1999) studied the acute hemodynamic responses during repetitive submaximal (60 and 80% of maximal voluntary contraction) bilateral leg-press exercise in nine males (mean age: 50 years) with clinically stable heart failure. The key finding of their investigation was that submaximal leg-press exercise was associated with an increase in heart rate, systolic blood pressure, mean arterial pressure, stroke volume, and cardiac output with a concomitant decline in total systemic vascular resistance. In addition, no adverse symptoms, arrhythmias, or pulmonary congestion was associated with performing the strength tests. The findings of McKelvie et al. (1995) and Meyer et al. (1999) suggest that submaximal unilateral or bilateral leg-press exercise does not negatively alter left ventricular systolic function in clinically stable individuals with heart failure.

Although dynamic strength exercise does not appear to acutely alter left ventricular systolic performance, the chronic effects of this form of training on global left ventricular performance remains uncertain. Koch et al. (1992) compared the effects of 3 months of strength training or no training on maximal muscular strength, exercise tolerance time, cardiac dimensions, and function in 23 clinically stable heart failure patients (mean resting ejection fraction: 26%). The major finding of this investigation was that strength training resulted in an increase in muscle strength, exercise tolerance time, and quality of life. More importantly, short-term strength training did not alter resting left ventricular diastolic cavity dimension, fractional shortening, or ejection fraction. Magnusson et al. (1996) found that 8 weeks of high-intensity (80% of one repetition maximum) isolated quadriceps strength training resulted in a significant increase in knee extensor work rate, dynamic, and isometric knee extensor strength without altering left ventricular ejection fraction in individuals with chronic heart failure. More recently, Pu et al. (2001) compared the effects of 10 weeks of high-intensity (80% of one repetition maximum) upper and lower extremity strength training on the distance walked in six minutes, peak aerobic power, muscle mass, maximal muscular strength and endurance, and left ventricular systolic function and diastolic filling in older women (mean age: 77 years) with stable chronic heart failure. The novel finding of this investigation was that short-term high-intensity strength training was a sufficient stimulus to increase upper and lower extremity maximal muscular strength, endurance, distance walked in six minutes without altering peak aerobic power, muscle mass, left ventricular ejection fraction, or diastolic filling. Therefore, the few studies that have been done with stable cardiac patients with impaired left ventricular systolic function suggest that supervised strength training appears to be an effective form of exercise that can improve muscle strength and function without altering the size or function of the heart.

3.11 Future Directions

A large number of investigations have examined the chronic effects of strength training on cardiac function and morphology in younger individuals and strength athletes. Additionally, there is greater awareness of the short-term effects of strength training on cardiac morphology and function in healthy older males. However, there is a lack of information regarding the acute and chronic effects of strength training on left ventricular morphology and function in healthy postmenopausal women or older women with underlying heart disease. Therefore, future investigations are required to examine the cardiac effects of strength training in healthy older women or older women with cardiac disease.

3.12 Summary

Strength training is an effective form of exercise to increase skeletal muscle strength and mass. The effect that this form of exercise has on altering cardiac morphology and function appears to be related to the type of strength training performed. More specifically, when cardiac changes occur with training they appear to be specific to weightlifters and body-builders. This adaptive response is most likely related to the heightened volume load associated with these types of training. In addition, it is also likely that some athletes may incorporate a brief (2 to 3 second) Valsalva maneuver during exertion that may attenuate the increase in left ventricular end-systolic wall stress, which, in accordance with the load-induced hypertrophy hypothesis, would reduce the stimulus for cardiac growth. The role that chronic strength training plays on altering cardiac performance during acute stress has not been well studied. However, favorable ventricular-vascular adaptations (i.e., increased left ventricular contractility and reduced systolic blood pressure and total systemic vascular resistance) that result in enhanced left ventricular systolic function have been found after long-term strength training. These beneficial adaptations may have functional and ergonomic significance, because they will allow recreational, vocational, or sport-related activities that have a strength component to them to be performed with lower stress being placed on the heart.

At the present time, strength training has become a popular and effective rehabilitative intervention to attenuate the muscle wasting and strength loss associated with aging. Moreover, strength training has gained popularity in cardiac rehabilitation settings as an effective form of exercise to improve muscular strength and endurance in stable cardiac patients with impaired left ventricular systolic function. To date, the paucity of studies suggest that this form of training does not negatively alter the size or shape of the senescent heart. Supervised strength training does not appear to alter left ventricular systolic function in clinically stable heart failure patients with impaired left ventricular systolic function. However, further investigations with a larger sample size are required to determine the long-term effects of this form of training on cardiac remodeling in these individuals. Furthermore, additional studies are required to examine the acute and chronic effects of strength training on cardiac morphology and function in healthy older women and older women with underlying cardiovascular disease.

References

Colan, S.D. 1992. Mechanics of left ventricular systolic and diastolic function in physiologic hypertrophy of the athlete heart, *Cardiol. Clin.*, 10, 227–240.

Colan, S.D., Sanders, S.P., and Borow, K.M. 1987. Physiologic hypertrophy: effects on left ventricular systolic mechanics in athletes, *J. Am. Coll. Cardiol.*, 9, 776–783.

Deligiannis, A., Zahopoulou, E., and Mandroukas, K. 1988. Echocardiographic study of cardiac dimensions and function in weight lifters and body builders, *Int. J. Sports Cardiol.*, 5, 24–32.

Fagard, R.H. 1996. Athletes heart: a meta-analysis of the echocardiographic experience. *Int. J. Sports Med.*, Suppl. 3, S140–S144.

Fiatarone, M.A., Marks, E.C., Ryan, N.D., Meredith, C.N., Lipsitz, L.A., and Evans, W.J. 1990. High-intensity strength training in nonagenarians: effects on skeletal muscle, *JAMA*, 263, 3029–3034.

Fisman, E.Z., Embon, P., Pines, A., Tenenbaum, A., Dror, Y., Shapira, I., and Motro, M. 1997. Comparison of left ventricular function using isometric exercise Doppler echocardiography in competitive runners and weightlifters vs. sedentary individuals, *Am. J. Cardiol.*, 79, 355–359.

Fleck, S.J. and Dean, L.S. 1987. Resistance-training experience and the pressor response during resistance exercise. *J. Appl. Physiol.*, 63, 116–120.

Frontera, W.R., Meredith, C.N., O'Reilly, K.P., Knuttgen, H.G., and Evans, W.J. 1988. Strength conditioning in older men: skeletal muscle hypertrophy and improved function, *J. Appl. P.*, 64, 1038–1044.

Fuenmayor, A.J., Fuenmayor, A.M., Winterdaal, D.M., and Londono, G. 1992. Cardiovascular responses to Valsalva maneuver in physically trained and untrained normal subjects, *J. Sports Med. Phys. Fitness*, 32, 293–298.

Galanti, G., Comeglio, M., Vinci, M., Cappelli, B., Toncelli, L., and Bamoshmoosh, M. 1992. Non-invasive left ventricular wall stress evaluation during isometric exercise in trained subjects, *Int. J. Sports Cardiol.*, 1, 89–93.

George, K.P., Batterham, A.M., and Jones, B. 1998. Echocardiographic evidence of concentric left ventricular enlargement in female weight lifters, *Eur. J. Appl. Physiol.*, 79, 88–92.

Hagerman, F.C., Walsh, S.J., Staron, R.S., Hikida, R.S., Gilders, R.M., Murray, T.F., Toma, K., and Ragg, K.E. 2000. Effects of high-intensity resistance training on untrained older men. I. Strength, cardiovascular, and metabolic responses, *J. Gerontol. A. Biol. Sci. Med. Sci.*, 7, B336–B346.

Hamilton, W.F., Woodbury, R.A., and Harper, H.T. 1936. Physiologic relationships between intrathoracic, intraspinal and arterial pressures, *JAMA*, 107, 853–856.

Haykowsky, M.J., Dressendorfer, R., Taylor, D., Mandic, S., and Humen, D. 2002. Resistance training and cardiac hypertrophy: unravelling the training effect, *Sports Med.*, 32, 837–849.

Haykowsky, M., Gillis, R., Quinney, A., Ignaszewski, A., and Thompson, C. 1998. Left ventricular morphology in elite female resistance-trained athletes, *Am. J. Cardiol.*, 82, 912–914.

Haykowsky, M., Humen, D., Teo, K., Quinney, A., Souster, M., Bell, G., and Taylor, D. 2000a. Effects of 16 weeks of resistance training on left ventricular morphology and systolic function in healthy men >60 years of age, *Am. J. Cardiol.*, 85, 1002–1006.

Haykowsky, M., Quinney, H.A., Gillis, R., and Thompson, C.R. 2000b. Left ventricular morphology in junior and master resistance trained athletes, *Med. Sci. Sports Exerc.*, 32, 349–352.

Haykowsky, M., Taylor, D., Teo, K., Quinney, A., and Humen, D. 2001. Left ventricular wall stress during leg-press exercise performed with a brief Valsalva maneuver, *Chest*, 119, 150–154.

Haykowsky, M., Teo, K., Quinney, A., Humen, D., and Taylor, D. 2000c. Effects of long term resistance training on left ventricular morphology, *Can. J. Cardiol.*, 16, 35–38.

Koch, M., Douard, H., and Broustet, J.P. 1992. The benefit of graded physical exercise in chronic heart failure, *Chest*, 101, 231S–235S.

Lattanzi, F., Di Bello, V., Picano, E., Caputo, M.T., Talarico, L., Di Muro, C., Landini, L., Santoro, G., Giusti, C., and Distante, A. 1992. Normal ultrasonic myocardial reflectivity in athletes with increased left ventricular mass. A tissue characterization study, *Circulation*, 85, 1828–1834.

Lentini, A.C., McKelvie, R.S., McCartney, N., Tomlinson, C.W., and MacDougall, J.D. 1993. Left ventricular response in healthy young men during heavy-intensity weight-lifting exercise, *J. Appl. Physiol.*, 75, 2703–2710.

MacDougall, J.D. 1994. Blood pressure responses to resistive, static and dynamic exercise, in *Cardiovascular Response to Exercise*, G.F. Fletcher, Ed. (New York: Futura Publishing), pp. 155–173.

MacDougall, J., McKelvie, R., Moroz, D., Sale, D., McCartney, N., and Buick, F. 1992. Factors affecting blood pressure during heavy weight lifting and static contractions, *J. Appl. Physiol.*, 73, 1590–1597.

MacDougall, J.D., Tuxen, D., Sale, D.G., Moroz, J.R., and Sutton, J.R. 1985. Arterial blood pressure response to heavy resistance exercise, *J. Appl. Physiol.*, 58, 785–790.

Magnusson, G., Gordon, A., Kaijser, L., Sylven, C., Isberg, B., Karpakka, J., and Saltin, B. 1996. High intensity knee extensor training, in patients with chronic heart failure: major skeletal muscle improvement, *Eur. Heart J.*, 17, 1048–1055.

McCartney, N., McKelvie, R.S., Martin, J., Sale, D.G., and MacDougall, J.D. 1993. Weight-training-induced attenuation of the circulatory response of older males to weight lifting, *J. Appl. Physiol.*, 74, 1056–1060.

McKelvie, R.S., McCartney, N., Tomlinson, C., Bauer, R., and MacDougall, J.D. 1995. Comparison of hemodynamic responses to cycling and resistance exercise in congestive heart failure secondary to ischemic cardiomyopathy, *Am. J. Cardiol.*, 76, 977–979.

Meyer, K., Hajric, R., Westbrook, S., Haag-Wildi, S., Holtkamp, R., Leyk, D., and Schnellbacher, K. 1999. Hemodynamic responses during leg press exercise in patients with chronic congestive heart failure, *Am. J. Cardiol.*, 83, 1537–1543.

Mitchell, J.H. 1990. Neural control of the circulation during exercise, *Med. Sci. Sports Exerc.*, 22, 141–154.

Murray, T.F., Falkel, J.E., and Fleck, S.J. 1989. Differences in cardiac output responses between elite powerlifters and bodybuilders during resistance exercise. *Med. Sci. Sports Exerc.*, 21, S114.

Pelliccia, A., Spataro, A., Caselli, G., and Maron, B.J. 1993. Absence of left ventricular wall thickening in athletes engaged in intense power training, *Am. J. Cardiol.*, 72, 1048–1054.

Pluim, B.M., Zwinderman, A.H., van Der Laarse, A., and van Der Wall, E.E. 1999. The athletes heart: a meta-analysis of cardiac structure and function, *Circulation*, 100, 336–344.

Pollock, M.L., Franklin, B.A., Balady, G.J., Chaitman, B.L., Fleg, J.L., Fletcher, B., Limacher, M., Pina, I.L., Stein, R.A., Williams, M., and Bazzarre, T. 2000. Resistance exercise in individuals with and without cardiovascular disease: benefits, rationale, safety, and prescription, *Circulation*, 101, 828–833.

Pu, C.T., Johnson, M.T., Forman, D.E., Hausdorff, J.M., Roubenoff, R., Foldvari, M., Fielding, R.A., and Singh, M.A. 2001. Randomized trial of progressive resistance training to counteract the myopathy of chronic heart failure, *J. Appl. Physiol.*, 90, 2341–2350.

Sagiv, M., Fisher, N., Yaniv, A., and Rudoy J. 1989. Effect of running vs. isometric training programs on healthy elderly at rest, *Gerontology*, 35, 72–77.

Sale, D.G., Moroz, D.E., McKelvie, R.S., MacDougall, J.D., and McCartney, N. 1994. Effect of training on the blood pressure response to weight lifting, *Can. J. Appl. Physiol.*, 19, 60–74.

Sejersted, O.M., Hargens, A.R., Kardel, K.R., Blom, P., Jensen, O., and Hermansen, L. 1984. Intramuscular fluid pressure during isometric contraction of human skeletal muscle, *J. Appl. Physiol.*, 56, 287–295.

Suman, O.E., Hasten, D., Turner, M.J., Rinder, M.R., Spina, R.J., and Ehsani, A. 2000. Enhanced inotropic response to dobutamine in strength-trained subjects with left ventricular hypertrophy, *J. Appl. Physiol.*, 88, 534–539.

Sylvest, O. and Hvid, N. 1959. Pressure measurements in human striated muscles during contraction, *Acta Rheumatol. Scand.*, 5, 216–222.

Vinereanu, D., Florescu, N., Sculthorpe, N., Tweddel, A.C., Stephens, M.R., and Fraser, A.G. 2002. Left ventricular long-axis diastolic function is augmented in the hearts of endurance-trained compared with strength-trained athletes, *Clin. Sci. (Lond.)*, 103, 249–257.

Wernstedt, P., Sjostedt, C., Ekman, I., Du, H., Thuomas, K.A., Areskog, N.H., and Nylander, E. 2002. Adaptation of cardiac morphology and function to endurance and strength training: a comparative study using MR imaging and echocardiography in males and females, *Scand. J. Med. Sci. Sports*, 12, 17–25.

4

Determinants of Muscle Strength

W. Herzog

CONTENTS

ABSTRACT This chapter deals with determinants of muscle strength. Muscle strength is defined here as the force a muscle can exert as a function of its contractile conditions, where contractile conditions refers to the length of the muscle, its instantaneous speed of shortening, and the history of its length change. In the ergonomic system, muscle force translates into a moment about a joint. For example, if the vastus lateralis (a one-joint knee extensor muscle) produces a force of 100 N, and its moment arm about the knee joint is 5 cm (0.05 m), its action will be reflected as a knee extensor moment of 5 N·m (100 N × 0.05 m).

In order to understand how muscle force is affected by contractile conditions, some basic principles of muscle structure and innervation will be reviewed. Then muscle force production as a function of muscle length, speed of shortening and stretch, and contractile history will be described. While establishing these foundations, the influence of fiber type distribution on force, power, and fatigability will be outlined. Similarly, architectural features influencing the contractile speed, force, power, and work potential of muscle will be discussed. Finally, the effect of age on muscle force will be briefly addressed.

4.1 Introduction

One of the basic concerns in ergonomic research is whether a person can perform a physical job requirement without undue fatigue and without risking a traumatic or a repetitive-motion injury. It has been argued that the ratio of the absolute muscle strength and the strength required for a physical task is a potent predictor for injury. In other words, the stronger a person, or the smaller the strength required to perform a work task, the better the chance that injury is avoided.

This chapter primarily deals with the basic determinants of muscle strength. These include structural considerations, the length of the muscle, and the speed of muscle shortening and stretch as the primary factors. From a structural point of view, it is obvious that a "bigger" muscle is also "stronger." But muscles of equal size (volume) may have completely different strength capabilities, depending on the arrangement of muscle fibers within the muscle. Also, the instantaneous length of a muscle is a crucial determinant of strength. Length is directly influenced by body position (joint angles). For example, strength in the knee extensor muscles is about 5–10 times greater with the knee flexed by about 60–70° compared to the knee near full extension (e.g., about 10° of flexion). Therefore, when performing a physical task, body position and limb orientation play a crucial role in the available muscle strength.

Similarly, the speed of muscle contraction is a powerful determinant of muscle strength, especially at small speeds of shortening or stretch. For example, shortening of a muscle at a mere 10% of its maximal speed of shortening causes approximately 50% loss of strength from the isometric force. The speed of muscle contraction is directly related to the speed of movement and the speed at which a physical work task is performed. Therefore, the speed of performing a physical task critically influences the available muscle strength.

4.2 Ergonomic Relevance

At the workplace, muscular strength may be important in tasks that require hard labor and are associated with musculoskeletal injury. For example, luggage carriers in the air travel industry are known to suffer a high rate of low back problems. The best predictor for incurring (or avoiding) injury appears to be absolute strength (i.e., stronger workers are less likely to incur injuries). This example may be construed as rendering the rest of this chapter irrelevant, as I will discuss the influence of muscle length (or joint position), speed of muscle contraction (or movement speed), as well as the history of contraction (or what a worker did prior to lifting a piece of luggage) on muscle strength; whereas the above story implies that absolute isometric strength is all that counts. Although, on a superficial level, this may be correct, I believe that the mechanics of movement execution, including body position and speed of movement, may influence the incidence of injury, although this may be hard to show scientifically in a workplace situation.

Also, many work-related injuries do not occur because of great loading of the muscles and the musculoskeletal system. Carpal tunnel syndrome in computer workers and people in the meatpacking industry (typical for Alberta) is thought to be associated with low-load, high-repetition movements. Here, the absolute loading of the muscle and tendon is minimal compared to the maximal strength of the involved muscle, but the great number of repetitions appears to cause inflammation of the muscle and tendon. Although it has been speculated that such "repetitive motion disorders" are associated with fatigue that accumulates because of the great number of movement cycles, it appears that fatigue may not be the only cause for the observed injuries. For example, if a stereotypical movement pattern, as it occurs in typing, was replaced with a movement pattern of similar loading and similar number of repetitions, but a nonstereotypical movement, many of the "repetitive motion disorders" might not occur. Thus, there appears to be a factor that one might call "stereotype" factor that indicates the similarity of repeat movements that may be strongly related to observed musculoskeletal problems and pain at the workplace.

Here, discussion of these issues is largely circumvented, because this is not the purpose of the current chapter. However, when reviewing some of the factors that affect muscle strength, the reader should keep in mind that muscle strength is merely one component of a biomechanical evaluation of the effectiveness, fatigability, and injury potential of an ergonomic task.

4.3 Muscle Structure, Excitation, and Strength

Skeletal muscles are organized in an intricate way, cross-sectionally and longitudinally. The entire muscle is surrounded by a layer of connective tissue called fascia and by a further connective tissue sheath known as the epimysium (Figure 4.1). The next smaller structure is the muscle bundle (fascicle), which consists of a number of muscle fibers surrounded by a connective tissue sheath called perimysium. Then comes the muscle fiber, an individual muscle cell surrounded by a thin sheath of connective tissue (endomysium), which connects the individual fibers within a fascicle. Muscle fibers are cells with a delicate membrane, called the sarcolemma. Muscle fibers are made up of myofibrils (discrete bundles of myofilaments) lying parallel to one another. The systematic arrangement of the myofibrils gives muscle its typical striated pattern, which is visible under the light microscope. The repeat unit in this pattern is a sarcomere (Figure 4.2). Sarcomeres are the basic contractile units of skeletal muscle; they are bordered by the Z-lines (*Zwischenscheibe*). Z-lines are thin strands of protein extending perpendicular to the long axis of the myofibrils. Sarcomeres contain thick (myosin) and thin (actin) filaments. The Z-lines intersect the thin myofilaments at regular intervals.

4.3.1 The Thick Filament

Thick filaments are typically located in the center of the sarcomere. They cause the dark band of the striation pattern in skeletal muscles, referred to as the A- (anisotropic) band (Figure 4.2). A thick filament is made up of approximately 180 myosin molecules. Each myosin molecule has a molecular weight of 5 kDa and contains a long tail portion consisting primarily of light meromyosin and a globular head portion consisting of heavy meromyosin. The heads extend outward from the thick filament in pairs (Figure 4.3). They contain a binding site for actin and an enzymatic site that catalyzes the hydrolysis of

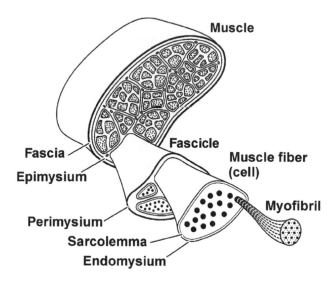

FIGURE 4.1
Schematic illustration of different structures and substructures of a muscle and a muscle fiber.

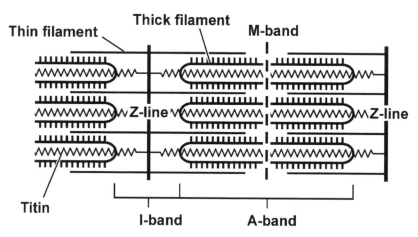

FIGURE 4.2
Schematic illustration of the basic contractile unit of the muscle, the sarcomere.

adenosine triphosphate (ATP), which releases the energy required for muscular contraction. Since the myosin heads have the ability to establish a link between the thick and thin filaments, they have been termed cross-bridges.

The myosin molecules in each half of the thick filament are arranged with their tail ends directed toward the center of the filament. Therefore, the head portions are oriented in opposite directions for the two halves of the filament, and upon contraction (i.e., when myosin heads — cross-bridges — attach to the thin filament) the myosin heads pull the actin filaments toward the center of the sarcomere.

The cross-bridges on the thick filament are about 14.3 nm apart longitudinally and are offset from each other by a 60° rotation (Figure 4.4). Since two cross-bridge pairs are offset by 180°, cross-bridge pairs with identical orientation are approximately 42.9 nm apart (3×14.3 nm).

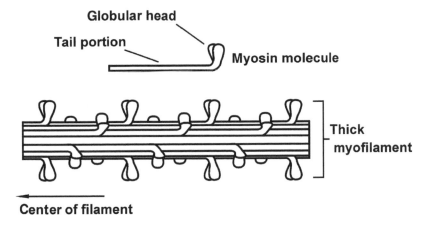

FIGURE 4.3
Schematic illustration of the thick myofilament.

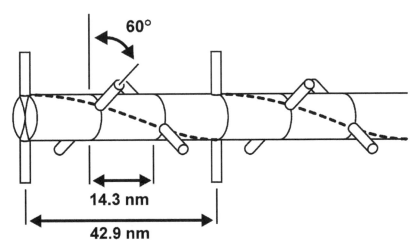

FIGURE 4.4
Schematic illustration of the arrangement of the cross-bridges on the thick filament.

4.3.2 The Thin Filament

The Z-lines bisect the thin filaments (Figure 4.2). Thin filaments appear light in the striation pattern, and the light band formed between the opposite ends of two thick filaments is called the I- (isotropic) band. The backbone of the thin filament consists of two helically interwoven chains of actin globules (Figure 4.5), whose diameter is about 5–6 nm. Thin filaments also contain the proteins tropomyosin and troponin. Tropomyosin is a long fibrous protein that lies in the grooves formed by the actin chains (Figure 4.5). Troponin is located at intervals of approximately 38.5 nm along the thin filament. Troponin is composed of three subunits: troponin C, which contains sites for Ca^{2+} binding; troponin T, which contacts tropomyosin; and troponin I, which is thought to physically block the cross-bridge attachment site in the resting (i.e., in the absence of Ca^{2+}) state.

When performing a cross-sectional cut through the zone of overlap between the thick and thin filaments in the sarcomere, it is revealed that each thick filament is surrounded by six thin filaments in a perfect hexagon (Figure 4.6). The cross-sections of the thick and

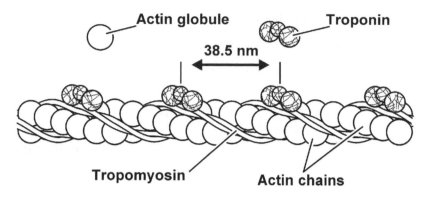

FIGURE 4.5
Schematic illustration of the thin myofilament, consisting of two helically interwoven chains of actin globules, tropomyosin, and troponin.

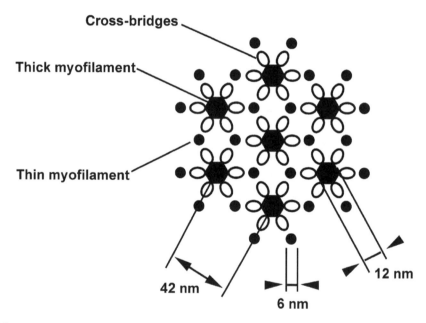

FIGURE 4.6
Schematic illustration of thick and thin myofilament arrangement in a cross-sectional view through the myofilament overlap zone.

thin filaments are approximately 12 and 6 nm in diameter, respectively. The distance between adjacent thick filaments is approximately 42 nm and varies with sarcomere length.

4.3.3 Other Protein Filaments in the Sarcomere

Aside from the contractile proteins actin and myosin, skeletal muscle sarcomeres contain a variety of other proteins that are associated with structural and passive functional properties of the sarcomere, rather than active force production. Arguably, the most important of these proteins, from a functional point of view, is titin.

Titin is a huge (mass 3MDa) protein that is found in abundance in myofibrils of vertebrate (and some invertebrate) striated muscle. Within the sarcomere, titin spans from the

Z-line to the M-band (i.e., the center of the thick filament) (Figure 4.2). Although the exact functional role of titin remains to be elucidated, it is generally accepted that titin acts as a molecular spring that develops tension when sarcomeres are stretched. Titins' location has prompted the idea that it might stabilize the thick filament within the center of the sarcomere (Figure 4.2). Such stabilization may be necessary to prevent the thick filament from being pulled to one side of the sarcomere when the forces acting on each half of the thick filament are not exactly equal.

Evidence for the role of titin in thick filament centering has been provided by Horowits and colleagues (Horowits and Podolsky, 1987, 1988; Horowits et al., 1989) who showed that, upon prolonged activation in chemically skinned rabbit psoas fibers, the thick myofilaments could easily be moved away from the center of the sarcomere at short (< 2.5 µm) but not at long (> 2.8 µm) sarcomere lengths when the titin "spring" presumably was tensioned and so helped center the thick myofilament.

4.3.4 Muscle Shapes

The organizational structure of fibers, sarcomeres, and myofilaments is extraordinary, and so is the organizational structure of the fibers within a muscle. Skeletal muscles contain fibers from a few millimeters to several centimeters in length, which are arranged either parallel to the longitudinal axis (i.e., parallel or fusiform muscles) or at a distinct angle to the longitudinal axis of the muscle (pennate muscles — Figure 4.7). Depending on the number (n) of distinct fiber directions, a muscle is called unipennate ($n = 1$), bipennate ($n = 2$), or multipennate ($n \geq 3$).

The variety of muscle shapes indicates the variety of functional tasks that need to be satisfied within an agonistic group of muscles. For example, the primary ankle extensor muscles in the cat are the two heads of the gastrocnemius, the soleus, and the plantaris. The soleus is essentially parallel-fibered. The medial head of the gastrocnemius and the plantaris are excellent examples of unipennate fiber arrangements, whereas the lateral head of the gastrocnemius is multipennate. The pennate muscles, by design, have shorter fibers than the parallel-fibered soleus (Sacks and Roy, 1982). Therefore, their absolute length range over which force can be generated (which is directly related to fiber lengths) is smaller compared to soleus. This arrangement makes perfect sense, considering that

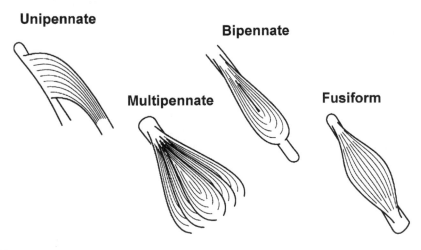

FIGURE 4.7
Classification of muscles into fusiform, unipennate, bipennate and multipennate.

the soleus is a one-joint muscle which should ideally accommodate the entire range of ankle motion. The gastrocnemius and the plantaris cross the ankle and the knee, and during locomotion, flexion and extension of these two joints occur in such a way that a shortening of these muscles at one joint is (at least partly) offset by an elongation at the other joint, therefore requiring less excursion than the one-joint soleus.

4.3.5 Motor Unit

Muscles receive the commands for force production from nerves. A single muscle nerve contains afferent and efferent axons. The afferent axons deliver information about the contractile status of the muscle to the central nervous system; the efferent axons deliver signals for contraction from the central nervous system to the muscle. The primary efferent pathways are called α motoneurons. Each α motoneuron innervates a number of muscle fibers. This functional unit (one α motoneuron and all the muscle fibers it innervates) is called a motor unit. A motor unit is the smallest control unit of a muscle, because all fibers belonging to the same motoneuron will always contract and relax in a synchronized manner. The force in a muscle can be increased in two ways: (a) by increasing the number of active motor units, or (b) by increasing the frequency of stimulation to a motor unit.

When a motor unit receives a single stimulation pulse from its motoneuron, the corresponding force response is a single twitch. The average twitch duration is approximately 200 ms in a purely slow-twitch fibered mammalian muscle, such as the cat soleus (Figure 4.8). Therefore, if the soleus receives less than about five equally spaced stimulation pulses per second, it will show a series of individual twitch responses. When the stimulation frequency exceeds about 5 Hz, a second pulse will stimulate the muscle before the force effects of the first pulse have completely subsided. In this case, the force begins to add, and with increasing frequencies, the force response becomes larger in magnitude and smoother (Figure 4.8). Therefore, aside from changing the number of active motor units, a muscle can adjust its force production by the frequency of stimulation of the motor units. For relatively low frequencies of motor unit stimulation (< about 20 Hz for slow motor units, and < about 50 Hz for fast motor units), there will be some force relaxation between stimulation pulses. The corresponding force time history of such contractions has force "ripples" [Figure 4.8 traces 6 (Hz), 10 (Hz), and 12.5 (Hz)]. Such contractions are called unfused tetanic contractions. With increasing stimulation frequencies, the corresponding force time traces become smoother until the force oscillations disappear; such contractions are referred to as fused tetanic contractions.

Sometimes contractions are referred to as submaximal, maximal, or supramaximal. During voluntary contractions, a maximal contraction corresponds to a maximal voluntary effort, and a submaximal contraction is any contraction that is less than maximal. During artificially elicited contractions of muscle, for example by stimulating a muscle nerve using a stimulation electrode, force production may exceed that which a subject may produce voluntarily. Such artificial contractions are called supramaximal. They may be achieved by employing a stimulation frequency of all motor units (e.g., 100–150 Hz) that cannot be achieved by voluntary effort.

Motor units are composed of muscle fibers with similar biochemical and twitch properties. According to these properties, motor units are classified as fast or slow (there are more fiber types than just fast and slow, and the difference in the properties varies not discretely but in a continuous manner; however, these detailed differences are of no importance in the present context). Fast motor units, as the name implies, have a high maximal speed of shortening and a brief twitch duration. Slow motor units have a slower maximal speed of shortening and an increased twitch duration compared to the fast motor

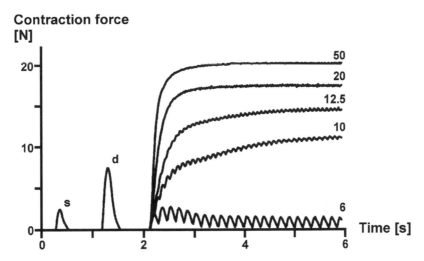

FIGURE 4.8
Single twitch (s), doublet twitch (d, two twitches separated by a 10 ms interval), and unfused and fused contractions of a cat soleus muscle at stimulation frequencies 6, 10, 12.5, 20, 50 Hz.

units. Furthermore, fast motor units are better equipped than slow motor units to produce muscle forces via anaerobic (not involving oxygen) pathways, whereas slow motor units are better suited than fast motor units to produce the energy for muscular contraction via aerobic pathways (involving oxygen). Slow motor units are typically much more fatigue resistant than fast motor units.

Morphologically, slow motor units are innervated by small-diameter motoneurons, and they contain fewer muscle fibers than the corresponding fast motor units. This structural arrangement by itself is not particularly fascinating, except when viewed in the context of muscle force control. Henneman et al. (1965) and Henneman and Olson (1965) revealed the significance of the structural differences in slow and fast motor units. They showed that, during a graded increase of force in a muscle, the small motoneurons innervating the small and slow motor units were recruited first. With increasing force demands, larger motoneurons, innervating progressively larger motor units with increasingly fast-type properties, were recruited. Therefore, a graded increase in force was accomplished by recruiting the smallest and slowest motor units first and the largest and fastest motor units last. This pattern of motor unit recruitment means that over a period of time there is a greater dependence on the slow motor units; the fast motor units are only recruited when particularly high forces are required. Since the small motor units typically have a large aerobic capacity and great endurance properties, and the fast motor units typically have little aerobic capacity and fatigue quickly, the order of recruitment of motor units makes perfect sense; it is referred to as the size principle of motor unit recruitment.

The size principle was formulated and tested for the recruitment order of motor units within a given muscle. It is interesting to observe that there exist vast differences in the fiber type distribution of skeletal muscles, even within the same functional group. Again, the cat ankle extensors are a perfect example. The cat soleus is composed of primarily slow motor units [95–100% (Ariano et al., 1973; Burke et al., 1974)], whereas the medial gastrocnemius contains predominantly fast motor units [70–80% (Ariano et al., 1973; Burke and Tsairis, 1973)]. During quiet standing, it has been observed that the soleus produces substantial forces, while the medial gastrocnemius may be silent (Hodgson, 1983). With increasing speeds of locomotion, the peak soleus forces remain about constant and the medial gastrocnemius forces increase several times (Walmsley et al., 1978). Finally, during a rapid paw shake action (a movement which occurs at a frequency of about 10 Hz), soleus forces are low (or even zero), whereas the medial gastrocnemius activity is high (Abraham and Loeb, 1985; Smith et al., 1980). This series of experiments illustrates the change in the functional role of the soleus and medial gastrocnemius for a variety of different movements. Likely, the changes in the functional roles are strongly associated with the distribution of fiber types in these two muscles.

4.3.6   Excitation–Contraction Coupling

The process of excitation–contraction coupling involves the transmission of signals along nerve fibers, across the neuromuscular junction (Figure 4.9, the place where the end of the nerve meets the muscle fiber), and along muscle fibers. At rest, nerve and muscle fibers maintain a negative charge inside the cell compared to the outside (i.e., the membrane is polarized). Nerve and muscle fibers are excitable, which means that they can change the local membrane potential in a characteristic manner when stimuli exceed a certain threshold. When a muscle membrane becomes depolarized beyond a certain threshold, there is a sudden change in membrane permeability, particularly to positively charged sodium ions, whose concentration outside the cell is much higher than that inside the cell. The resulting influx of sodium ions causes the charge inside the cell to become more positive. The membrane then decreases permeability to sodium and increases permeability to potassium ions, which are maintained at a much higher concentration inside than outside the cell. The resulting outflow of the positively charged potassium ions causes a restoration of the polarized state of the excitable membrane. This transient change in membrane potential is referred to as an action potential and lasts approximately 1 ms. In the muscle fiber, this action potential propagates along the fiber at a speed of about 5–10 m/s (Figure 4.10). In the α motoneuron, action potentials propagate at speeds in proportion to the diameter of the neuron, with the largest neurons (in mammals) conducting at 120 m/s.

The neuromuscular junction (Figure 4.9) is formed by an enlarged nerve terminal, known as the presynaptic terminal, that is embedded in small invaginations of the muscle cell membrane, the motor endplate, or postsynaptic terminal. The space between presynaptic and postsynaptic terminal is the synaptic cleft.

When an action potential of a motoneuron reaches the presynaptic terminal, a series of chemical reactions takes place that culminates in the release of acetylcholine (ACh) from synaptic vesicles located in the presynaptic terminal. Acetylcholine diffuses across the synaptic cleft, binds to receptor molecules of the membrane on the postsynaptic terminal, and causes an increase in permeability of the membrane to sodium (Na^+) ions. If the depolarization of the membrane due to sodium ion diffusion exceeds a critical threshold, then an action potential will be generated that travels along the stimulated muscle fiber. In order to prevent continuous stimulation of muscle fibers, acetylcholine is rapidly broken down into acetic acid and choline by acetylcholinesterase, which is liberally distributed in the postsynaptic membrane.

The action potential of the muscle fiber is not only propagated along and around the fiber, but also reaches the interior of the muscle fiber at invaginations of the cell membrane called T-tubules (Figure 4.11). Depolarization of the T-tubules causes the release of calcium (Ca^{2+}) ions from the terminal cisternae of the sarcoplasmic reticulum (membranous sac-like structure which stores calcium) into the sarcoplasm surrounding the myofibrils. Ca^{2+} ions bind to specialized sites on the troponin molecules of the thin myofilaments and remove an inhibitory mechanism that otherwise prevents cross-bridge formation in the relaxed state (Figure 4.12). Cross-bridges then attach to the active sites of the thin filaments and, through the breakdown of ATP into ADP plus a phosphate ion (P_i), the necessary

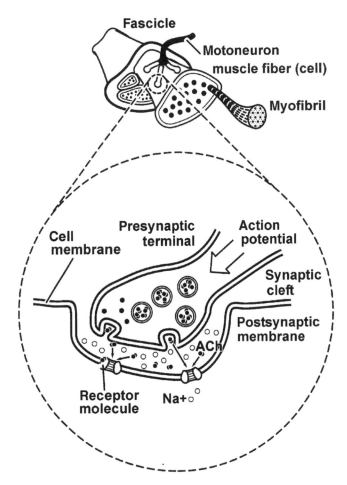

FIGURE 4.9
Schematic illustration of the neuromuscular junction with motoneuron and muscle cell membrane.

energy is provided to cause the cross-bridge head to move and so attempt to pull the thin filaments past the thick filaments (Figure 4.13). At the end of the cross-bridge movement, an ATP molecule is thought to attach to the myosin portion of the cross-bridge so that the cross-bridge can release from its attachment site, go back to its original configuration, and be ready for a new cycle of attachment. This cycle repeats itself as long as the muscle fiber is stimulated. When stimulation stops, Ca^{2+} ions are actively transported back into the sarcoplasmic reticulum, resulting in a decrease of Ca^{2+} ions in the sarcoplasm. As a consequence, Ca^{2+} ions diffuse away from the binding sites on the troponin molecule, and cross-bridge cycling stops.

4.4 Muscle Length and Strength

The force–length property of a muscle is defined by the maximal isometric force a muscle can exert as a function of its length. The fact that force production in skeletal muscle is length-dependent has been known for a long time (Blix, 1894). Force–length properties

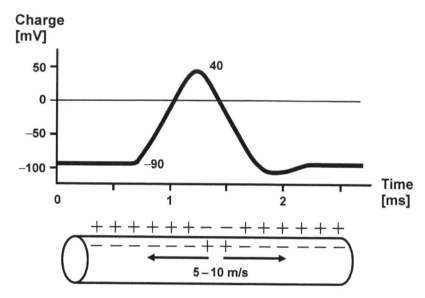

FIGURE 4.10
Schematic illustration of a single muscle fiber action potential (top) and the corresponding propagation of the action potential along the muscle fiber (bottom).

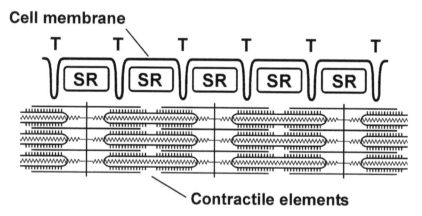

FIGURE 4.11
Schematic illustration of T-tubules (T) in a section of a muscle fiber and its association with the sarcoplasmic reticulum (SR) and the contractile myofilaments.

have been derived for sarcomeres (Gordon et al., 1966), for isolated fibers (ter Keurs et al., 1978), and for entire muscles (Goslow and Van DeGraaff, 1982; Herzog and ter Keurs, 1988; Herzog et al., 1992b).

Force–length properties may be described in the simplest case by a symmetric curve and the corresponding peak force (F_0) and working range (L) (Woittiez et al., 1984). An example of two force–length curves with different peak force values and working ranges is shown in Figure 4.14. Muscle 1 has a large cross-sectional area and corresponding peak force, F_{01}, and short fibers and a corresponding small working range, L_1; muscle 2 has a small peak force, F_{02}, and a large working range, L_2.

The maximal isometric force a muscle can exert is primarily governed by its physiological cross-sectional area (PCSA), where

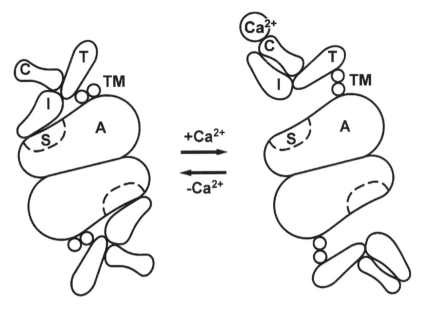

FIGURE 4.12
Schematic illustration of the inhibitory/excitatory regulation of cross-bridge attachment on the actin filament (A). Without calcium (left), the tropomyosin (TM) and troponin complex (troponin T, C, and I) are in a configuration that blocks the cross-bridge attachment site (S). Adding calcium (Ca^{2+}) to the calcium-binding site of the troponin (troponin C) changes the configuration of the tropomyosin-troponin complex in such a way that the cross-bridge attachment site is exposed and cross-bridge attachment is possible.

$$PCSA = \frac{\text{Muscle Volume}}{\text{Fiber Length}}$$

and Muscle Volume is the volume of the muscle, and Fiber Length is the mean length of muscle fibers at optimal sarcomere lengths. [Optimal sarcomere length is the length at which a sarcomere can produce maximal force — about 2.1 μm, 2.4 μm, and 2.7 μm for frog, cat, and human skeletal muscles, respectively (Herzog et al., 1992a; Walker and Schrodt, 1973)]. The physiological cross-sectional area, therefore, may be thought of as the sum of the cross-sectional area of all fibers arranged in parallel within a muscle. The maximal isometric force, F_0, is typically calculated from the known PCSA using a proportionality factor, i.e.,

$$F_0 = k \cdot PCSA$$

where k is typically taken as 20–40 N/cm², with an median value of about 25 N/cm² for mammalian muscles at body temperature.

The working range of the muscle, L, may be approximated by

$$L = \textit{Optimal Fiber Length}$$

indicating that the working range is approximately equal to the mean length of the muscle fibers at optimal sarcomere lengths. This relation between the working range and optimal fiber length makes sense when realizing that the working range of a sarcomere corresponds

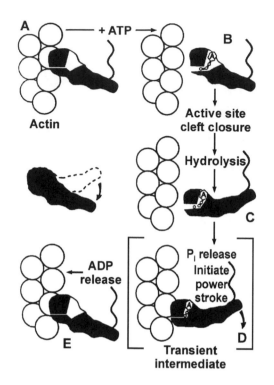

FIGURE 4.13
Schematics of the mechanics and biochemistry of a cross-bridge cycle as envisioned by Rayment et al. (1993). Note here, that the power stroke is assumed to occur about a point in the neck region of the cross-bridge. Note, further, how one cross-bridge cycle is associated with one cycle of ATP hydrolysis. (Reprinted with permission from Rayment, I., Holden, H.M., Whittaker, M., Yohn, C.B., Lorenz, M., Holmes, K.C., and Milligan, R.A, *Science*, 261, 58–65, 1993, copyright 2003 American Association for the Advancement of Science.)

approximately to the optimal length of a sarcomere (Gordon et al., 1966), and that fibers contain sarcomeres in series along their entire length.

The sarcomere force–length relation depends on the lengths of the thick and thin myofilaments. Once these lengths are known, the exact relation between sarcomere length and force can be calculated based on the cross-bridge theory (Gordon et al., 1966; Herzog et al., 1992a; Huxley, 1957). For frog skeletal muscle, the equations relating to sarcomere lengths (SL) in micrometers and normalized force (F) (the force divided by the maximal isometric force) are described by four straight lines (Figure 4.15).

$$F = -2.667 + 2.1\, SL; \quad 1.27\ \mu m \leq SL < 1.67\ \mu m$$

$$F = 0.04 + 0.48\, SL; \quad 1.67\ \mu m \leq SL < 2.00\ \mu m$$

$$F = 1.0; \quad 2.00\ \mu m \leq SL < 2.25\ \mu m$$

$$F = 1.592 - 0.71\, SL; \quad 2.25\ \mu m \leq SL < 3.65\ \mu m$$

For entire muscles, the relation between normalized force (F) and muscle fiber lengths (L) has been approximated by (Figure 4.16)

$$F = -6.25(L/L_0)^2 + 12.5(L/L_0) - 5.25 \tag{1}$$

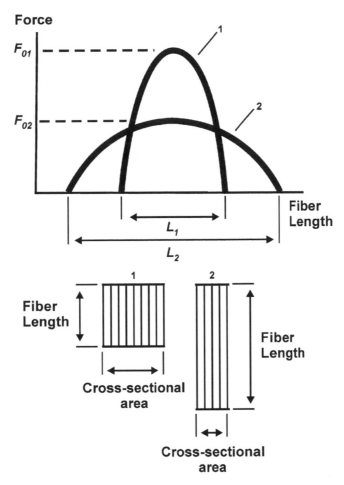

FIGURE 4.14
Schematic force–length relation of two muscles with different cross-sectional areas and fiber lengths.

where L_0 is the optimal muscle length (Woittiez et al., 1984). Note that Equation (1) gives a working range, R, of the muscle of $0.8\ L_0$.

In the sarcomere force–length relationship shown in Figure 4.15, the region of positive slope (i.e., for sarcomere lengths ranging from 1.27 µm to 2.00 µm) is called the ascending limb of the force–length relationship; the area of zero slope (2.0–2.25 µm) is the plateau region; and the region of negative slope (2.25 µm–3.65 µm) is the descending limb of the force–length relationship. These regions have also been defined (albeit less rigorously) for the force–length relationships of entire muscles (Figure 4.16).

When considering the human musculoskeletal system, or parts thereof, it is useful to know not only the generic force–length properties of skeletal muscle, but also the force–length property of a specific muscle within the anatomical constraints of the skeleton. For example, the human elbow joint can go through an angular displacement of approximately 150°. Questions that are relevant to the force–length properties of elbow flexor muscles include: Can the elbow flexors produce force over the entire range of elbow movements? At what joint angles do the elbow flexors produce maximal force? On what part of the force–length relation do the elbow flexors operate? Are the force–length properties of all elbow flexors about the same, or do they vary significantly? Are the force–length properties of the elbow flexors the same across individuals, or do they adapt

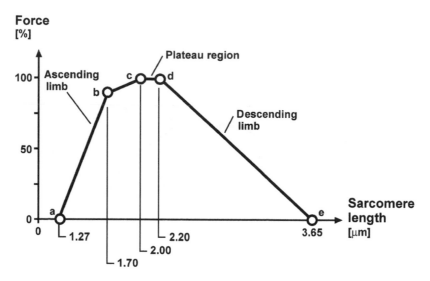

FIGURE 4.15
The sarcomere force–length relation of frog skeletal muscle. (Adapted from Gordon, A.M., Huxley, A.F., and Julian, F.J., *J. Physiol.*, 184, 170–192, 1966, with permission.)

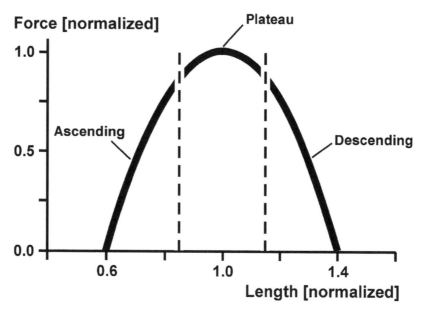

FIGURE 4.16
The normalized force–length relation of mammalian skeletal muscle obeying Equation 1.

to the specific needs of an individual? How are the force–length properties of the elbow flexors influenced by changes in the neighboring joint angles?

It is not our intent to elaborate on all of these questions here; however, some selected comments will be made regarding the operating range of muscles within the constraints of the skeleton, the force–length properties of agonist muscles (muscles producing moments in the same direction at a given joint), and the variations of force–length properties across subjects.

From the point of view of force production, the optimal operating range of a muscle is centered around the plateau region (i.e., a muscle is working on the plateau region in the middle of the working range, working on the descending limb when extremely elongated, and working on the ascending limb when extremely shortened) (Figure 4.16). Such a working range has been found for a variety of muscles, for example, the frog semimembranosus (Lutz and Rome, 1993) and the human rectus femoris (Herzog and ter Keurs, 1988). However, other muscles have been found to work predominantly on the ascending limb and plateau region, or the plateau region and descending limb of the force–length relation (Figure 4.16). For example, during frog jumping the knee and hip angles go from full flexion to full extension, and for this movement the semitendinosus was found to operate primarily on the plateau and descending limb region (Mai and Lieber, 1990). Muscles that have been found to work primarily on the ascending limb and the plateau regions include the triceps surae muscles of the striped skunk (Goslow and Van DeGraaff, 1982), the triceps surae and plantaris of the cat (Herzog et al., 1992b), and the human gastrocnemius (Herzog et al., 1991b). It is not clear why the operating range within the anatomical range of joint movements varies across muscles; however, some insight may be gained by studying the changes of the force–length properties through chronic alterations of the movement demands.

When considering joint function, the question arises as to whether or not the force–length properties of all agonistic muscles are similar, and therefore may be thought of as a scaled version of the force– (or moment) angle relation observed for the entire agonistic group. If the force–length properties of all agonistic muscles are similar, they reach their maximal force potential at a similar joint angle, and therefore the peak forces that can be achieved are high; however, the operating range is restricted. If the force–length properties are not similar, and the muscles reach their maximal force potential at different joint angles, the peak forces are not as high as for the previous case; however, the working range is increased compared to the previous possibility. Experimental work in this area is sparse; however, it appears that agonistic muscles tend to have similar force–length properties (Goslow and Van DeGraaff, 1982; Herzog et al., 1992b). Similar, in the present context, refers to the shape of the force–length curve and not to the absolute force values. This result suggests that muscles with similar functions have similar force–length properties, or it might even suggest that the function of a muscle dictates its force–length properties.

It has typically been assumed that the shape of the force–length property of specific muscles does not vary across subjects (Bobbert and van Soest, 1992; Pandy and Zajac, 1991; Pandy et al., 1990; van Leeuwen and Spoor, 1992; van Soest et al., 1993). For human or animal subjects that use muscles primarily for normal everyday tasks such as locomotion, force–length properties across subjects are probably similar. However, what about these properties if subjects place different demands on muscles? In a study on the force–length properties of the rectus femoris (RF) in elite cyclists and runners, it was found that RF of the cyclists was operating on the descending, and RF of the runners was operating on the ascending limb of the force–length relation (Figure 4.17). The adaptations were as one would expect; the cyclists who use RF in a chronically shortened position, because of the small hip angle associated with cycling, were relatively strong at short compared to long RF lengths; the runners who use RF at longer lengths than the cyclists, were relatively strong at long compared to short RF lengths. The possible mechanisms of adaptation of the force–length properties in this situation are not known; however, a shift of the force–length relationship in the way observed for these two groups of athletes could be achieved through a change in the number of sarcomeres arranged in series in the RF fibers (Herzog et al., 1991a). Such adaptations to the everyday functional demands, as observed here in athletes, may very well also occur in ergonomic situations with stereotypical movement demands.

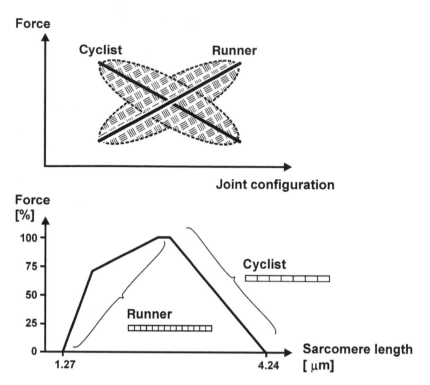

FIGURE 4.17
Schematic summary of the force–length results determined for the rectus femoris of elite runners and elite cyclists. The runners appeared to use the rectus femoris on the ascending limb, and the cyclists on the descending limb of the force–length relationship (top graph). This result could be explained if the number of sarcomeres in series in the rectus femoris fibers was greater in the runners compared to the cyclists (bottom graph). This would cause a shift of the force–length relationship of the runners to the left of that for the cyclists. Interestingly, force–length relationships of rectus femoris from nonathletes appear to be in between those shown in this figure.

4.5 Speed of Contraction and Strength

It has been known for almost a century that the speed of contraction affects the maximal force that can be produced by a muscle. Hill (1938) popularized what is commonly referred to as the force–velocity relationship. The force–velocity relationship describes the maximal, steady-state force of a muscle as a function of its speed of shortening and stretch (Figure 4.18). For shortening, Hill (1938) described the force–velocity relationship using part of a rectangular hyperbola:

$$(F + a)(v + b) = (F_0 + a)b \tag{2}$$

where F is the steady-state force for shortening at a velocity, v; F_0 is the maximal, isometric force at optimal contractile element length, and a and b are constants with units of force and velocity, respectively.

Solving for the force, F, Equation (2) becomes

$$F = \frac{F_0 b - av}{b + v} \tag{3}$$

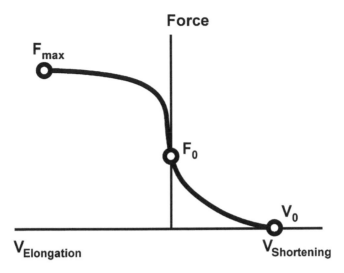

FIGURE 4.18
Schematic illustration of the force–velocity relationship of skeletal muscle.

F can be calculated if F_0, v, a, and b are known. For shortening at the maximal velocity, v_0, the force F becomes zero (Figure 4.18); therefore Equation (3) may be rewritten for this special case as

$$F_0 b = a v_0$$

or

$$a/F_0 = b/v_0 = 0.25 \tag{4}$$

a/F_0 and b/v_0 are dimensionless quantities of approximate value of 0.25 for many muscles across species and temperatures (Hill, 1970), including human fast twitch fibers at 37°C (Faulkner et al., 1980, 1986).

Force–velocity properties are available for a variety of animal muscles (Close and Hoh, 1967; Edman, 1979; Edman and Reggiani, 1983; Hill, 1938; Katz, 1939; Spector et al., 1980), and moment-angular velocity relationships have also been derived for agonistic groups of human skeletal muscles during maximal voluntary contractions (Perrine and Edgerton, 1978; Thorstensson et al., 1976). Force–velocity descriptions of individual human skeletal muscles are rare. Faulkner et al. (1980, 1986) determined force–velocity properties of isolated fiber bundle segments (10–25 mm long) from human fast and slow twitch muscles and found values for a/F_0 of 0.25 and 0.15 for fast and slow fibers, respectively. Therefore, knowing the physiological cross-sectional area (PCSA) of a muscle, the maximal isometric force, F_0, may be calculated using $F_0 = k \cdot PCSA$, with $k = 25$ N/cm², and the constant "a" may then be calculated as $a = 0.25 F_0$.

For a given cross-sectional area, slow and fast twitch fibers produce similar levels of isometric force, but the maximal velocity of shortening differs by a factor of 3 or 4. For human skeletal muscles, Faulkner et al. (1986) reported values for v_0 of 6 and 2 fiber length/s for fast and slow fiber bundle preparations, respectively. Compared to other reports of mammalian skeletal muscle at 37°C, these v_0-values appear low. Values obtained from fast and slow muscles in the mouse and rat range from 9.4 to 24.2 length/s and 6.5 to 12.7 length/s, respectively (Close, 1964, 1965, 1972; Luff, 1981; Ranatunga, 1981). One of the reasons why the human v_0-values reported by Faulkner et al. (1986) are relatively low may be associated with the fact that they did not measure v_0, but approximated v_0 by extrapolating Hill's equation to a value of zero force. Edman (1979) showed that Hill's equation gives good approximations of experimental force–velocity relations in a range of approximately 5–80% of the isometric force, but that it overestimates the actual values of F_0 and underestimates the actual values of v_0 considerably.

Measuring the maximal velocity of shortening in human skeletal muscles is difficult; however, a rough estimate may be obtained by determining the maximal power output of a group of muscles as a function of movement speed, and knowing that maximal power is achieved at 31% of v_0, if it is assumed that $a/F_0 = b/v_0 = 0.25$ (Herzog, 1994). For human knee extensor muscles, Suter et al. (1993) found that maximal power was produced at knee angular velocities of about 240°/s and 400°/s for predominantly slow-twitch fibered subjects (more than 55% slow twitch fibers in the vastus lateralis) and predominantly fast-twitch fibered subjects (more than 60% fast twitch fibers in the vastus lateralis). Therefore, the maximal speeds of contraction for these two subject groups are approximately 774°/s and 1290°/s, respectively. Assuming a moment arm of the knee extensor muscles about the knee axis of 0.05 m (Herzog and Read, 1993) at the instant of peak force production, the speeds of shortening for the two groups are about 0.68 m/s and 1.13 m/s. Knowing that the average fiber length in the knee extensor muscles is approximately 0.08 m (Wickiewicz et al., 1983), the maximal speeds of shortening expressed in terms of the contractile element lengths are about 8 length/s and 14 length/s for the slow-twitch and the fast-twitch fibered groups, respectively. These values for v_0 agree better with those obtained for slow and fast mammalian skeletal muscles (Close, 1964, 1965, 1972; Luff, 1981; Ranatunga, 1981), and they are considerably larger than those obtained by Faulkner et al. (1986) for human slow and fast fibers. The estimated v_0-values of 8 and 14 length/s must be considered with some reservation, because they were extrapolated from measurements of mixed muscles. It is expected that the v_0-value of a purely fast-twitch fibered muscle should be higher than 14 length/s, and the v_0-value of a purely slow-twitch fibered muscle should be lower than 8 length/s.

Once an estimate of F_0 and v_0 is obtained, the constants in Hill's (1938) equation, a and b, can be calculated using Equation (4). Knowing F_0, v_0, a, and b, the maximal, steady-state force, F, as a function of the shortening velocity, v, at optimal contractile element length is given by Equation (3).

So far, we have primarily discussed the force–velocity properties of shortening muscle. When a muscle is stretched at a given speed, its force exceeds the maximal isometric force, F_0, reaching an asymptotic value of about 2 F_0 at speeds of stretching much lower than the maximal velocity of shortening (Lombardi and Piazzesi, 1992). Also, using isotonic (Katz, 1939) or isokinetic stretches (Edman et al., 1978), researchers have found that there appears to be a discontinuity in the force–velocity relation across the isometric point: the rise in force associated with slow stretching is much larger than the fall in force associated with the corresponding velocities of shortening. In contrast to the force–velocity relation during shortening, the force–velocity relation during stretching is rarely described using a standard equation, such as the hyperbolic relation proposed by Hill (1938) for shortening. The primary reason for this discrepancy is the fact that force–velocity properties during stretching have been investigated much less, and that these properties are not as consistent as those obtained during shortening.

When modeling the force–velocity properties of skeletal muscles, the fiber type distribution must be accounted for (Faulkner et al., 1986; Hill, 1970; MacIntosh et al., 1993). Slow and fast twitch fibers of the same cross-sectional area have similar F_0-values, but they differ substantially in the maximal velocity of shortening (Figure 4.19a). Skeletal muscles are typically composed of a mixture of slow and fast twitch fibers. Hill (1970) made an attempt to model the force–velocity properties of a muscle containing 82 "fibers"

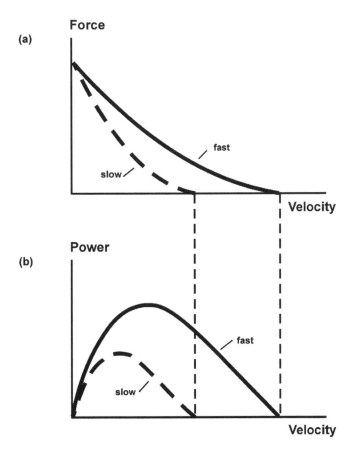

FIGURE 4.19
(a) Schematic illustration of the force–velocity properties, and (b) power-velocity properties of a fast and a slow skeletal muscle fiber.

with ten different intrinsic speeds of shortening. He found that the regular force–velocity equation fit the mixed-fibered muscle well, except at forces less than 5% of F_0.

The difference in fiber-type distribution of skeletal muscles is particularly important for power production. Power *(P)* is defined as

$$P = \mathbf{F} \cdot v$$

where $\mathbf{F}$ is the force (vector), $\mathbf{v}$ is the velocity (vector), and $\mathbf{F} \cdot \mathbf{v}$ represents the scalar (or dot) product. For muscles, power is typically approximated as the product of the muscle's force magnitude (scalar) and the speed of contraction (scalar). The power of a fast twitch fiber is higher at all speeds of shortening than that of the corresponding slow twitch fiber (Figure 4.19b). Also, the peak power is achieved at a higher absolute speed of shortening in the fast compared to the slow fiber.

In a study aimed at determining the power output of slow and fast human skeletal muscle fibers, Faulkner et al. (1986) found that slow fibers of equal isometric force as fast fibers produced peak powers that were 25% of those of the fast fibers (Figure 4.20). Also, a muscle composed of 50% slow and 50% fast twitch fibers was estimated to have a peak power value of only 55% of that of a corresponding 100% fast-twitch fibered muscle. Similar results were found by other investigators (MacIntosh et al., 1993).

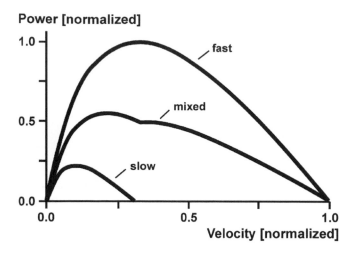

FIGURE 4.20
Power-velocity relation of human fast, slow, and mixed (50% fast, 50% slow) fibered skeletal muscle. [Adapted and reprinted with permission from Faulkner, J.A., Claflin, D.R., and McCully, K.K, in *Human Muscle Power*, N.L. Jones, N. McCartney, and A.J. McComas, Eds. (Champaign, IL: Human Kinetics), pp. 81–91, 1986.]

Up to now, the force–length and the force–velocity properties were treated as separate entities. Once both these properties have been determined, the question arises as to how they should be combined in a description of contractile element properties. It has been suggested that Hill's (1938) equation of the force–velocity relation can be used for all muscle lengths, provided that the F_0-value (the maximal isometric force at optimal length) is replaced by the maximal isometric force at the length of interest [F_0 (*l*), Equation (2)]. Using this suggestion, Equation (3) becomes

$$F = \frac{F_0(l)b - av}{b + v} \qquad (5)$$

Assuming that *v* is zero (i.e., we have an isometric contraction) $F = F_0$ (*l*). The maximal velocity of shortening, v_0, according to Equation (5) becomes

$$v_0 = \frac{F_0(l)b}{a} \qquad (6)$$

because *F* is zero in this situation. Since *a* and *b* are constants in Equation (6), v_0 depends directly on F_0 (*l*). This result suggests that the maximal speed of shortening is highest at optimal muscle length and becomes progressively smaller at lengths other than optimal. However, Edman (1979) showed convincingly that v_0 does not depend directly on the contractile element length. He performed experiments on single fibers of frogs aimed at measuring v_0 as a function of sarcomere length. The results of these experiments are summarized in Figure 4.21. In the range of sarcomere lengths from 1.65 μm to 2.70 μm, v_0 was nearly constant. Below sarcomere lengths of 1.65 μm, v_0 fell because of a presumed force resisting shortening at these sarcomere lengths (Gordon et al., 1966). Beyond sarcomere lengths of 2.7 μm, v_0 increased with increasing sarcomere lengths. This increase was associated with passive elastic forces, which produce rapid shortening

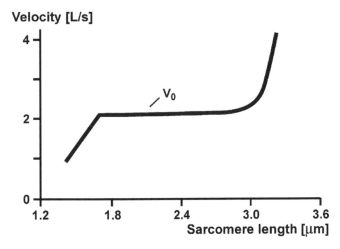

FIGURE 4.21

Maximal velocity (v_o) of shortening of frog skeletal muscle fibers as a function of sarcomere length. Note, that v_o remains nearly constant for sarcomere lengths ranging from 1.65 to 2.70 μm. (Adapted and reprinted with permission from Edman, K.A.P, *J. Physiol.*, 291, 143–159, 1979.)

of unstimulated muscle fibers at these lengths and which appear to enhance v_0 in the stimulated fiber.

Since v_0 appears to be constant for a large range of contractile element lengths, a better representation of the force–length-velocity relation than that shown in Equation (5) may be achieved by multiplying Hill's (1938) force–velocity relation by a normalized length factor, at least for contractile element lengths ranging from 1.65 to 2.70 μm.

$$F = \left(\frac{F_0 b - av}{b + v} \right) f(l) \qquad (7)$$

where $f(l)$ represents the normalized force according to the force–length properties of the muscle; $f(l)$ ranges from 0 to 1.0. For sarcomere lengths outside 1.65–2.70 μm, a different representation of the force–length-velocity relation is required.

Combining the force–velocity and force–length relationships also causes a conceptual problem. By definition, the force–length properties of skeletal muscles are determined under isometric conditions by fully activating the muscle at a series of discrete lengths. Therefore, the force–length relationship consists of a series of discrete force measurements obtained at discrete muscle lengths. In contrast to the force–length property, the force–velocity relationship is obtained for a series of dynamic contractions, typically performed under isotonic or isokinetic conditions. When combining the two relationships, the force–length property is treated like a continuous, dynamic function. This treatment of the force–length property causes (at least) two conceptual problems. First, force–length relationships are valid for isometric contractions; Equations (5) and (7), however, are aimed at describing dynamic processes. Second, the force–length property is treated like an instantaneous rather than a steady-state property which causes stability problems on the descending part of the force–length relationship, as discussed previously. The issue of instability has been thoroughly treated elsewhere (Epstein and Herzog, 1998).

4.6   Contractile History and Strength

It is well accepted that the steady-state isometric forces following active muscle shortening are smaller than the corresponding isometric forces, and that the steady-state isometric forces following active muscle stretching are greater than the corresponding isometric forces (e.g., Abbott and Aubert, 1952). These phenomena are typically referred to as force depression and force enhancement, respectively. These properties have received little attention in the biomechanics literature, presumably because they are not contained in typical cross-bridge explanations of muscle contraction. Furthermore, they have rarely, if ever, been considered in ergonomic applications, because these properties have only been observed for electrically stimulated preparations until recently (Lee and Herzog, 2002); thus, the relevance for normal movements has not been established. At the time of this writing this chapter, there has been no evidence of force enhancement or force depression for submaximal, voluntary contractions. However, pilot observations suggest that the isometric forces following submaximal, voluntary muscle shortening are decreased by 20–25% compared to the corresponding purely isometric contractions (E. Rousanoglou and W. Herzog, unpublished). Therefore, I believe that the contractile history in an ergonomic situation may affect the available muscle strength considerably. However, research in this area is lacking. Below, the observations and the thinking associated with skeletal muscle force depression and force enhancement are considered briefly.

4.6.1   Force Depression

Force depression is defined here as the absolute (or percentage) decrease in the steady-state isometric force following a shortening contraction compared to the purely isometric force at the corresponding length (Figure 4.22). For reasons of brevity, I will only consider single, constant-speed shortening contractions, and thereby leave out multiple-step shortening tests, stretch-shortening, or any other cyclic testing, and tests performed at variable speeds.

It is well accepted, and has been observed repeatedly in preparations ranging from single fibers to *in vivo* muscles, that the steady-state isometric force following muscle shortening is decreased compared to the corresponding purely isometric force. This force depression has been shown to increase (a) with increasing magnitudes of shortening (Herzog and Leonard, 1997; Maréchal and Plaghki, 1979), (b) with decreasing speeds of shortening (Abbott and Aubert, 1952; DeRuiter et al., 1998), and (c) with increasing force during the shortening phase (Herzog and Leonard, 1997). Combining observations (a) and (c) from the previous sentence, it is obvious that force depression should also increase in proportion with the amount of mechanical work done during the shortening phase, which it does (Figure 4.23; Herzog et al., 2000). Finally, Sugi and Tsuchiya (1988) provided convincing evidence, from work on single frog fibers, that force depression following shortening is associated with a proportional decrease in fiber stiffness, thereby suggesting that force depression is caused by a decrease in the number of attached cross-bridges, rather than a decrease in the average force per cross-bridge. The above observations are well accepted and have not been challenged seriously.

Based on the available experimental evidence, the most likely mechanism for force depression following muscle shortening is a stress-induced inhibition of cross-bridge attachment within the newly formed overlap zone during muscle shortening. This idea was first proposed by Maréchal and Plaghki (1979), who did not provide an explanation

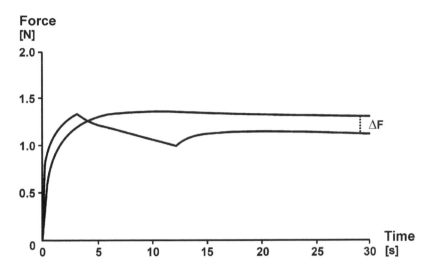

FIGURE 4.22
Illustration of force depression. The steady (top) line represents an isometric force at a given muscle length. The unsteady (bottom) line represents an isometric force (0–3 s), followed by a shortening contraction to the muscle length of the purely isometric contraction (3–12 s), followed by an isometric contraction at the muscle length of the purely isometric contraction. ΔF represents the isometric steady-state force depression following the shortening contraction.

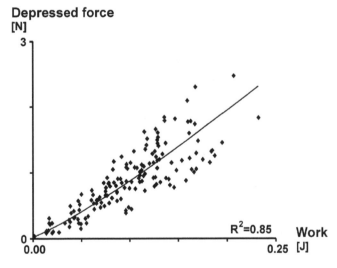

FIGURE 4.23
Force depression as a function of the work performed during the shortening phase of muscle contraction. Shown are the raw, nonnormalized values for six cat soleus muscles obtained using six different experimental protocols. Note that 85% of the variation in force depression is explained by the corresponding variation in the mechanical work.

for the stress-induced cross-bridge inhibition. However, with more recent evidence suggesting that the thick and thin myofilaments might possess considerable compliance within the range of physiological loads (Goldman and Huxley, 1994; Higuchi et al., 1995; Huxley et al., 1994; Kojima et al., 1994; Nishizaka et al., 1995; Wakabayashi et al., 1994), stress-induced longitudinal and rotational deformation of myofilaments could cause a stress-induced inhibition of cross-bridge attachment (Herzog, 1998).

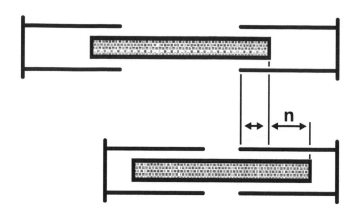

FIGURE 4.24
Schematic illustration of a sarcomere that has been shortening. After shortening, thick and thin myofilament overlap becomes greater than it was before shortening. The newly formed overlap zone is indicated by "n," and it is in this newly formed overlap zone that cross-bridge attachments are assumed to be inhibited because of the stress induced deformation of the myofilaments, particularly the thin filament.

Stress-induced inhibition of cross-bridge attachment may be explained as follows: A muscle is isometrically activated at some long length and then allowed to shorten. During activation, the compliant thick and thin myofilaments are stretched because of the increasing stress. For the thin filaments, which consist of two helically interwoven F-actin chains, such stretching would result in an angular distortion of the filament, specifically its binding sites (Daniel et al., 1998). The greater the stress, the greater the angular distortion, and this angular distortion might cause an inhibition of cross-bridge attachment. Such distortions would be greatest in areas of no support (i.e., no cross-bridge attachments), such as the I-band zone of the actin filament (Forcinito et al., 1997). If the muscle now shortens, and stress on the thin filament is maintained, cross-bridge attachment in the newly formed overlap zone containing these "deformed" actin filaments may be partly inhibited, and the stress may cause a decrease in the number of attached cross-bridges in the overlap zone formed during muscle shortening (Figure 4.24).

Based on this theory, the following testable hypotheses can be formulated, all of which have been supported in the literature:

- Force depression should increase with increasing shortening distance (shown by Abbott and Aubert, 1952; DeRuiter et al., 1998; Maréchal and Plaghki, 1979; Morgan et al., 2000).
- Force depression should increase with increasing force during the shortening phase (shown by Herzog and Leonard, 1997).
- Force depression should increase with the amount of mechanical work produced during the shortening phase [this is merely a combination of the first two hypotheses and was shown to be correct by Herzog et al., (2000)].
- Force depression should be long-lasting (Abbott and Aubert, 1952; Herzog et al., 1998).
- Force depression should be abolished immediately upon full stress release. This has been shown to be correct for stress release using deactivation protocols (Abbott and Aubert, 1952; Herzog and Leonard, 1997; Maréchal and Plaghki, 1979) but has not been confirmed for mechanical stress release, an experiment that should be performed in the future.

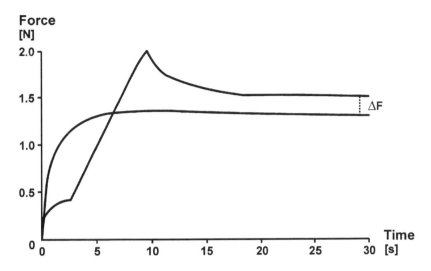

FIGURE 4.25
Illustration of force enhancement. The steady (bottom) line represents an isometric force at the final reference muscle length. The unsteady (top) line represents an isometric force (0–2.5 s), followed by a stretch (2.5–8 s), followed by an isometric contraction at the final reference length (8–30 s). ΔF represents the isometric steady-state force enhancement following the muscle stretch.

- Finally, force depression should be proportional to muscle (fiber) stiffness, which it appears to be, for single fiber preparations (Sugi and Tsuchiya, 1988) and for human adductor pollicis (Lee and Herzog, 2002).

4.6.2 Force Enhancement

Force enhancement is defined here as the absolute (or percentage) increase in the steady-state isometric force following a stretch contraction compared to the purely isometric force at the corresponding length (Figure 4.25). As for force depression, for reasons of brevity, only single-stretch contractions performed at a constant speed will be considered in the following.

It is well accepted and has been observed repeatedly in a variety of preparations that the steady-state isometric force following muscle stretch is increased compared to the corresponding purely isometric force. This force enhancement becomes greater for increasing amplitudes of stretch (Edman et al., 1978, 1982), and has been shown to depend on the speed of stretch in whole muscle (Abbott and Aubert, 1952), but was found to be independent of the speed of stretch in some fiber preparations (Edman et al., 1978). Furthermore, force enhancement is long lasting (Figure 4.25; Abbott and Aubert, 1952) and does not appear to be associated with a change in stiffness compared to isometric reference contractions in single fibers (Sugi and Tsuchiya, 1988), but stiffness is increased in the force-enhanced state compared to isometric reference contractions in cat soleus (Herzog and Leonard, 2002).

Force enhancement following muscle stretch has typically been explained on the basis of the sarcomere length nonuniformity theory (Morgan, 1990). This theory is based on the idea that, during muscle stretch, not all sarcomeres are stretched by the same amount. Rather, some sarcomeres stretch more than average and some less. The sarcomere length nonuniformity theory is thought to work on the descending limb of the force–length relationship exclusively (Figure 4.26), as force production and sarcomere lengths are assumed to be unstable on this part of the relationship (Hill, 1953), whereas they are

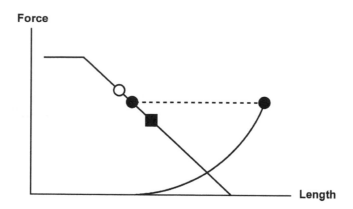

FIGURE 4.26

Illustration of force enhancement according to the sarcomere length nonuniformity theory. Imagine that the muscle is at a short length (open circle) and is stretched such that the average sarcomere length becomes that shown by the filled square. According to the sarcomere length nonuniformity theory, some sarcomeres will be stretched more than average and will fall onto the passive force curve (right filled circle), while other sarcomeres will be stretched less than average and will remain on the descending limb of the force–length relationship (left filled circle). At force equilibrium (horizontal dashed line), the force is greater (enhanced) than what one would expect based on the average sarcomere length (filled square).

assumed to be perfectly stable on the ascending limb of the force–length relationship (Allinger et al., 1996; Morgan, 1990; Zahalak, 1997).

Based on the sarcomere length nonuniformity theory, as mathematically formulated by Morgan (1990), Talbot and Morgan (1996), Allinger et al. (1996), and Zahalak (1997), as well as qualitatively derived and illustrated by Hill (1953) and Morgan et al. (2000), the following testable hypotheses can be derived.

- Force enhancement cannot occur on the ascending limb of the force–length relationship, because sarcomere length nonuniformities are not supposed to develop there.

- The steady-state isometric force following muscle stretch on the descending limb of the force–length relationship should not exceed the purely isometric force at the length from which the stretch was started, because stretching of the entire muscle (fiber) is associated with stretching of all parts of the muscle/fiber (Edman et al., 1978). More rigorously, the steady-state isometric force following muscle stretch cannot be greater than the purely isometric force at optimal muscle/fiber length.

- Muscle/fiber stiffness should be decreased in the force-enhanced state compared to the purely isometric state at the corresponding length (Morgan et al., 2000).

All three of these hypotheses have been rejected by experimental observations. Force enhancement has been observed on the ascending limb of the force–length relationship (DeRuiter et al., 2000) (Figure 4.27). The steady-state isometric force following muscle stretch has been shown to exceed the isometric force at the muscle length from which the stretch was initiated (Edman et al., 1978, 1982; Schachar et al., 2002; Wakeling et al., 2000) and the isometric force at optimal muscle length (Edman et al., 1982; Wakeling et al., 2000). Finally, fiber/muscle stiffness following stretching does not decrease, but has been reported to remain about the same (Sugi and Tsuchiya, 1988) or increase (Herzog and Leonard, 2000; Linari et al., 2000), as compared to the stiffness following a purely isometric contraction.

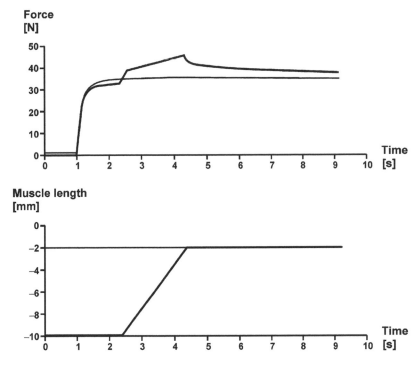

FIGURE 4.27
Illustration of an example of force enhancement following an 8-mm stretch of cat soleus (at 4 mm/s) on the ascending limb of the force–length relationship. There is a small, but consistent, force enhancement following the stretch that cannot be readily explained by the sarcomere length nonuniformity theory.

Therefore, the sarcomere length nonuniformity theory must be rejected at this point; at least, it must be rejected in the sense that it cannot explain the experimental results alone. However, it may play some role in force enhancement following muscle stretch.

A purely theoretical argument has been made (Herzog, 1998) to reconcile the observations on force enhancement with the traditional thinking of actin-myosin interaction and force production: the cross-bridge theory (Huxley, 1957; Huxley, 1969, Huxley and Simmons, 1971). The argument was based on the idea that, according to the cross-bridge theory, the steady-state isometric force in a muscle can only be enhanced in two ways: by an increase in the number of attached cross-bridges, or by an increase in the average force per cross-bridge. Some of the stiffness results suggest that the number of attached cross-bridges may be increased in the force-enhanced state (Herzog and Leonard, 2000; Linari et al., 2000). However, this idea needs further investigation.

One idea that has been associated with force enhancement for a long time, but could not be demonstrated directly, is the notion that force enhancement is associated with the "recruitment" of an elastic element during active, but not passive, muscle stretch. Edman et al. (1982) was the first to hint at such a mechanism. Others have restated it (DeRuiter et al., 2000; Edman and Tsuchiya, 1996; Noble, 1992) without direct evidence, while still others used it for the theoretical modeling of force enhancement (Forcinito et al., 1997).

Recently, we found direct evidence for passive force enhancement. We observed, first in cat semitendinosus, then in cat soleus (Herzog and Leonard, 2002), human adductor pollicis (Lee and Herzog, 2002), and single fibers of frog tibialis anterior (Wakeling et al., 2000) that the passive force following an active muscle stretch was considerably greater

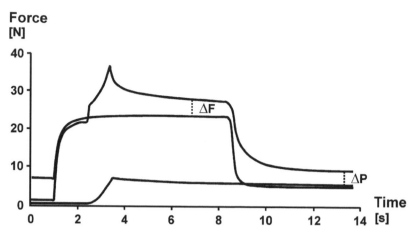

FIGURE 4.28

Illustration of passive force enhancement (ΔP) in cat soleus. Shown are the force–time histories of an isometric contraction at the final reference length (middle, steady trace), an active stretch contraction (top trace), and a passive stretch (bottom trace). After stretching, the active and passive muscles are at the final reference length. Observe the steady-state, isometric force enhancement following the active stretch (ΔF), and the passive force that remains considerably greater following deactivation of the muscle after the active stretch (passive force enhancement, ΔP) compared to the passive force following the active isometric contraction and the passive stretch.

than the passive force following a passive muscle stretch or following a purely isometric contraction (Figure 4.28). We think that the molecular spring titin might cause this passive force enhancement, as it is known that titin can change its stiffness properties depending on the mechanical (Wu et al., 2001) or physiological (Trok and Barnett, 2001) environment and, therefore, might cause a change in passive muscle force at a given length, depending on the contractile history of the muscle/fiber.

Therefore, it appears that force enhancement following muscle stretching has an active and a passive component. The passive component has not been observed at short muscle lengths, particularly not on the ascending limb of the force–length relationship. Since force enhancement occurs on the ascending limb (DeRuiter et al., 2000), but sarcomere length nonuniformities do not, it seems that the active component of force enhancement is not associated with the development of sarcomere length nonuniformities but may be associated with an increased number of attached cross-bridges (Herzog, 1998; Herzog and Leonard, 2000, 2002; Linari et al., 2000). Great passive force enhancement occurs at long muscle length and following great amplitudes of stretching, particularly on the descending limb of the force–length relationship. However, passive force enhancement was found to be smaller than the total force enhancement in all situations tested. Therefore, it appears that force enhancement following skeletal muscle stretching is composed of an active and a passive component that contribute to varying degrees to the total force enhancement, depending on the contractile history and muscle length. The origin of the active component is thought to be associated with an increase in the number of attached cross-bridges following stretching (Herzog and Leonard, 2000, 2002; Linari et al., 2000) compared to the purely isometric force at the corresponding muscle length. The origin of the passive component is not known, but the molecular spring titin seems like an ideal candidate.

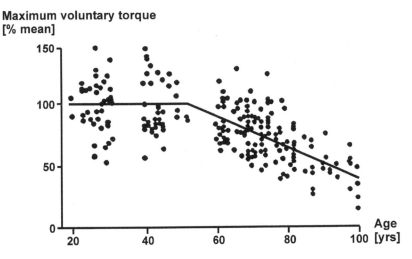

FIGURE 4.29
Maximal voluntary torque of ankle dorsiflexor and plantarflexor muscles in men and women as a function of
age. The results have been expressed as a percentage of the mean value for the youngest age group (20–30 years).
The horizontal line for subjects aged 52 years or less signifies the lack of an age effect before this time. (Reprinted
from Vandervoort, A.A. and McComas, A.J., *J. Appl. Physiol.*, 61, 361–367, 1986. With permission.)

4.7   Age and Strength

Women and men live significantly longer than ever before and often remain physically
active, at the workplace or recreationally, into old age. However, it is well documented
that muscles become smaller, strength decreases, and movement precision decreases with
increasing age.

Loss of strength with age could be caused by a decreased ability to recruit all motor
units of a muscle maximally, by a loss of contractile material, or by a combination of the
two. It has been shown in ankle muscles that the loss in force with age was not associated
with central nervous system commands (Vandervoort and McComas, 1986). Therefore,
the decrease in force must be associated with the loss of contractile material or a decrease
in the effectiveness of the excitation–contraction coupling process.

A substantial decrease in strength appears to occur in people of age > 60 years (Figure
4.29). This decrease in strength appears to be associated with a loss of muscle mass
(Inokuchi et al., 1975; Klitgaard et al., 1990; Overend et al., 1992), and a corresponding
decrease in the number of muscle fibers in a given muscle (Lexell et al., 1983). It is
important to note here that the decrease in the number of muscle fibers is greater for Type
II than Type I fibers (Klitgaard et al., 1990).

Aging is also associated with a decrease in the number of motoneurons and ventral root
axons supplying a given muscle. Decrease in functioning motor units has been found in
humans (Figure 4.30) and rats (Campbell et al., 1973) and is associated with decreased
postural stability (Rogers et al., 1992; Winter et al., 1990) and a decreased ability to control
submaximal isometric forces (Enoka, 1994). However, loss of strength in the elderly can
be prevented or offset, to a great degree, with strength-training programs (Aniansson and
Gustaffson, 1981; Grimby, 1988; Kaufman, 1985; Perkins and Kaiser, 1961).

Number of motor units

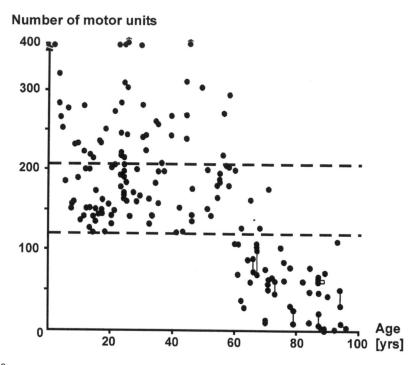

FIGURE 4.30

Loss of functioning motor units with age. The results were obtained from the extensor digitorum brevis of 207 healthy subjects aged 7 months to 97 years. The upper and lower horizontal lines show the mean value (210 units) and the lower limit of the range for subjects below 60 years (120 units), respectively. Linked values are from bilateral observations. [Reprinted from McComas, A.J., *Skeletal Muscle: Form and Function.* (Champaign, IL: Human Kinetics), 1996, with permission.]

In summary, people aged 60 years or older may have reduced muscle strength and reduced precision to perform ergonomic tasks. Systematic training of strength and precision control can largely offset the natural decline with age and may constitute an effective way to prolong a person's active living and productiveness with aging. Nevertheless, the elderly must be considered at greater risk for work-related injury associated with heavy labor.

4.8   Future Directions

Describing muscle strength and muscular properties in a well-defined, laboratory-based setup is easy compared to the challenges of identifying such properties in humans during everyday work-related or leisure-type activities.

One of the unresolved problems is the force sharing among agonistic and antagonistic muscles during voluntary movement. For example, during a lifting task, we might flex the elbow, and we can calculate the resultant elbow flexor moment. However, the elbow is crossed by numerous muscles (biceps, brachialis, brachioradialis, triceps, etc.); therefore, the resultant elbow flexor moment cannot be uniquely related to the corresponding muscle forces. This problem is typically referred to as the general distribution problem in biomechanics (Crowninshield and Brand, 1981). However, without an experimental, or theoretical, approach to determine the individual muscle forces during a given movement task, it is virtually impossible to provide good and accurate information on the control of movements, the local loading caused by a given task, and the strength and endurance capabilities of *in vivo* human muscles. Therefore, I would consider the solution of the general distribution problem as one of the great challenges in biomechanics and ergonomics.

Another issue of great controversy is the relationship of fiber (contractile element) length changes relative to the length changes in the entire muscle. For example, muscle injury is often associated with active muscle stretching; however, whether or not fibers are also stretched as the injury occurs is a largely unresolved question. Ultrasound imaging has been used quite extensively in the past decade to quantify the fiber length changes during isometric or well-controlled isokinetic contractions. These studies have provided great insight into the mechanics of *in vivo* muscle contraction, but these techniques have not been successfully applied to the investigation of normal unrestrained, voluntary movements. Such research should be performed in the future, although the necessary technology for these types of investigations does not appear to be ready yet.

Summarizing, the study of muscle forces, movement control, and loading of the human body, as well as the relationship between movement tasks and the potential for repetitive motion injury to musculoskeletal tissues presents great challenges in ergonomic situations. It is my sincere hope that some of the successful laboratory-based investigative tools are improved so that workplace biomechanical analyses can reach a new standard. Such research is important and challenging and should provide ample opportunities for ambitious and motivated young scientists.

Additional Reading

Enoka, R.M. 1994. *Neuromechanical Basis of Kinesiology*, 2nd ed. (Champaign, IL: Human Kinetics).
McComas, A.J. 1996. *Skeletal Muscle: Form and Function* (Champaign, IL: Human Kinetics).
Nigg, B.M. and Herzog, W. 1999. *Biomechanics of the Musculo-Skeletal System*, 2nd ed. (Chichester, UK: John Wiley & Sons).
Epstein, M. and Herzog, W. 1998. *Theoretical Models of Skeletal Muscle (Biological and Mathematical Considerations)* (Chichester, UK: John Wiley & Sons).

Acknowledgment

All my muscle research has been generously and continuously supported by NSERC of Canada since 1987.

References

Abbott, B.C. and Aubert, X.M. 1952. The force exerted by active striated muscle during and after change of length, *J. Physiol.*, 117, 77–86.
Abraham, L.D. and Loeb, G.E. 1985. The distal hindlimb musculature of the cat, *Exp. Brain Res.*, 58, 580–593.
Allinger, T.L., Epstein, M., and Herzog, W. 1996. Stability of muscle fibers on the descending limb of the force–length relation. A theoretical consideration, *J. Biomech.*, 29, 627–633.

Aniansson, A. and Gustaffson, E. 1981. Physical training in elderly men with specific reference to quadriceps muscle strength and morphology, *Clin. Physiol.*, 1, 87–98.

Ariano, M.A., Armstrong, R.B., and Edgerton, V.R. 1973. Hindlimb muscle fiber populations of five mammals, *J. Histochem. Cytochem.*, 21(1), 51–55.

Blix, M. 1894. Die Laenge und die Spannung des Muskels, *Skand. Arch. Physiol.*, 5, 149–206.

Bobbert, M.F. and van Soest, A.J. 1992. Effects of muscle strengthening on vertical jump height: a simulation study, *Med. Sci. Sports Exerc.*, 26, 1012–1020.

Burke, R.E., Levine, D.N., Saleman, M., and Tsairis, P. 1974. Motor units in cat soleus muscle: physiological, histochemical and morphological characteristics, *J. Physiol.*, 238, 503–514.

Burke, R.E. and Tsairis, P. 1973. Anatomy and innervation ratios in motor units of cat gastrocnemius, *J. Physiol.*, 234, 749–765.

Campbell, M.J., McComas, A.J., and Petito, F. 1973. Physiological changes in ageing muscles, *J. Neurol. Neurosurg. Psychiatry*, 36, 174–182.

Close, R. 1964. Dynamic properties of fast and slow skeletal muscles of the rat during development, *J. Physiol.*, 173, 74–95.

Close, R. 1965. Force:velocity properties of mouse muscle, *Nature*, 206, 718–719.

Close, R. and Hoh, J.F.Y. 1967. Force:velocity properties of kitten muscles, *J. Physiol.*, 192, 815–822.

Close, R.I. 1972. Dynamic properties of mammalian skeletal muscles, *Physiol. Rev.*, 52, 129–197.

Crowninshield, R.D. and Brand, R.A. 1981. The prediction of forces in joint structures: distribution of intersegmental resultants, in *Exercise and Sport Sciences Reviews* (Philadelphia, PA: Franklin Institute Press), pp. 159–181.

Daniel, T.L., Trimble, A.C., and Chase, P.B. 1998. Compliant realignment of binding sites in muscle: transient behaviour and mechanical tuning, *Biophys. J.*, 74, 1611–1621.

DeRuiter, C.J., De Haan, A., Jones, D.A., and Sargeant, A.J. 1998. Shortening–induced force depression in human adductor pollicis muscle, *J. Physiol.*, 507(2), 583–591.

DeRuiter, C.J., Didden, W.J.M., Jones, D.A., and De Haan, A. 2000. The force–velocity relationship of human adductor pollicis muscle during stretch and the effects of fatigue, *J. Physiol.*, 526.3, 671–681.

Edman, K.A.P. 1979. The velocity of unloaded shortening and its relation to sarcomere length and isometric force in vertebrate muscle fibers, *J. Physiol.*, 291, 143–159.

Edman, K.A.P., Elzinga, G., and Noble, M.I.M. 1978. Enhancement of mechanical performance by stretch during tetanic contractions of vertebrate skeletal muscle fibers, *J. Physiol.*, 281, 139–155.

Edman, K.A.P., Elzinga, G., and Noble, M.I.M. 1982. Residual force enhancement after stretch of contracting frog single muscle fibers, *J. Gen. Physiol.*, 80, 769–784.

Edman, K.A.P. and Reggiani, C. 1983. Length-tension-velocity relationships studied in short consecutive segments of intact muscle fibers in the frog, in *Contractile Mechanisms of Muscle. Mechanics, Energetics and Molecular Models, Volume II*, G.H. Pollack and H. Sugi, Eds. (New York: Plenum Press), pp. 495–510.

Edman, K.A.P. and Tsuchiya, T. 1996. Strain of passive elements during force enhancement by stretch in frog muscle fibers, *J. Physiol.*, 490.1, 191–205.

Enoka, R.M. 1994. *Neuromechanical Basis of Kinesiology* (Champaign, IL: Human Kinetics).

Epstein, M. and Herzog, W. 1998. *Theoretical Models of Skeletal Muscle: Biological and Mathematical Considerations* (New York: John Wiley & Sons).

Faulkner, J.A., Claflin, D.R., and McCully, K.K. 1986. Power output of fast and slow fibers from human skeletal muscles, in *Human Muscle Power*, N.L. Jones, N. McCartney, and A.J. McComas, Eds. (Champaign, IL: Human Kinetics), pp. 81–91.

Faulkner, J.A., Jones, D.A., Round, J.M., and Edwards, R.H.T. 1980. Dynamics of energetic processes in human muscle, in *Exercise Bioenergetics and Gas Exchange*, P. Cerretelli and B.J.Whipp, Eds. (Amsterdam: Elsevier/North-Holland Biomedical), pp. 81–90.

Forcinito, M., Epstein, M., and Herzog, W. 1997. Theoretical considerations on myofibril stiffness, *Biophys. J.*, 72, 1278–1286.

Goldman, Y.E. and Huxley, A.F. 1994. Actin Compliance: are you pulling my chain? *Biophys. J.*, 67, 2131–2136.

Gordon, A.M., Huxley, A.F., and Julian, F.J. 1966. The variation in isometric tension with sarcomere length in vertebrate muscle fibers, *J. Physiol.*, 184, 170–192.

Goslow, G.E. and Van DeGraaff, K.M. 1982. Hindlimb joint angle changes and action of the primary ankle extensor muscles during posture and locomotion in the striped skunk (*Mephitis mephitis*), *J. Zool. (London)*, 197, 405–419.

Grimby, G. 1988. Physical activity and effects of muscle training in the elderly, *Ann. Clin. Res.*, 20, 62–66.

Henneman, E. and Olson, C.B. 1965. Relations between structure and function in the design of skeletal muscles, *J. Neurophysiol.*, 28, 581–598.

Henneman, E., Somjen, G., and Carpenter, D.O. 1965. Functional significance of cell size in spinal motoneurons, *J. Neurophysiol.*, 28, 560–580.

Herzog, W. 1994. Muscle. In *Biomechanics of the Musculo-Skeletal System*, B.M. Nigg and W. Herzog, Eds. (Toronto: John Wiley & Sons), pp. 154–190.

Herzog, W. 1998. History dependence of force production in skeletal muscle: a proposal for mechanisms, *J. Electromyogr. Kinesiol.*, 8, 111–117.

Herzog, W., Guimaraes, A.C.S., Anton, M.G., and Carter-Erdman, K.A. 1991a. Moment-length relations of rectus femoris muscles of speed skaters/cyclists and runners, *Med. Sci. Sports Exerc.*, 23, 1289–1296.

Herzog, W., Kamal, S., and Clarke, H.D. 1992a. Myofilament lengths of cat skeletal muscle: theoretical considerations and functional implications, *J. Biomech.*, 25, 945–948.

Herzog, W. and Leonard, T.R. 1997. Depression of cat soleus forces following isokinetic shortening, *J. Biomech.*, 30(9), 865–872.

Herzog, W. and Leonard, T.R. 2000. The history dependence of force production in mammalian skeletal muscle following stretch-shortening and shortening-stretch cycles, *J. Biomech.*, 33, 531–542.

Herzog, W. and Leonard, T.R. 2002. Force enhancement following stretching of skeletal muscle: A new mechanism, *J. Exp. Biol.*, 205, 1275–1283.

Herzog, W., Leonard, T.R., Renaud, J.M., Wallace, J., Chaki, G., and Bornemisza, S. 1992b. Force–length properties and functional demands of cat gastrocnemius, soleus and plantaris muscles, *J. Biomech.*, 25, 1329–1335.

Herzog, W., Leonard, T.R., and Wu, J.Z. 1998. Force depression following skeletal muscle shortening is long lasting, *J. Biomech.*, 31, 1163–1168.

Herzog, W., Leonard, T.R., and Wu, J.Z. 2000. The relationship between force depression following shortening and mechanical work in skeletal muscle, *J. Biomech.*, 33, 659–668.

Herzog, W. and Read, L.J. 1993. Lines of action and moment arms of the major force-carrying structures crossing the human knee joint, *J. Anat.*, 182, 213–230.

Herzog, W., Read, L.J., and ter Keurs, H.E.D.J. 1991b. Experimental determination of force–length relations of intact human gastrocnemius muscles, *Clin. Biomech.*, 6, 230–238.

Herzog, W. and ter Keurs, H.E.D.J. 1988. Force–length relation of *in-vivo* human rectus femoris muscles, *Pflügers Arch.*, 411, 642–647.

Higuchi, H., Yanagida, T., and Goldman, Y.E. 1995. Compliance of thin filaments in skinned fibers of rabbit skeletal muscle, *Biophys. J.*, 69, 1000–1010.

Hill, A.V. 1938. The heat of shortening and the dynamic constants of muscle, *Proc. R. Soc. London*, 126, 136–195.

Hill, A.V. 1953. The mechanics of active muscle, *Proc. R. Soc. London*, 141, 104–117.

Hill, A.V. 1970. *First and Last Experiments in Muscle Mechanics* (London: Cambridge University Press).

Hodgson, J.A. 1983. The relationship between soleus and gastrocnemius muscle activity in conscious cats — a model for motor unit recruitment? *J. Physiol.*, 337, 553–562.

Horowits, R., Maruyama, K., and Podolsky, R.J. 1989. Elastic behaviour of connectin filaments during thick filament movement in activated skeletal muscle, *J. Cell Biol.*, 109, 2169–2176.

Horowits, R. and Podolsky, R.J. 1987. The positional stability of thick filaments in activated skeletal muscle depends on sarcomere length: evidence for the role of titin filaments, *J. Cell Biol.*, 105, 2217–2223.

Horowits, R. and Podolsky, R.J. 1988. Thick filament movement and isometric tension in activated skeletal muscle, *Biophys. J.* 54, 165–171.

Huxley, A.F. 1957. Muscle structure and theories of contraction, *Prog. Biophys. Biophys. Chem.*, 7, 255–318.

Huxley, A.F. and Simmons, R.M. 1971. Proposed mechanism of force generation in striated muscle, *Nature*, 233, 533–538.

Huxley, H.E. 1969. The mechanism of muscular contraction, *Science*, 164, 1356–1366.

Huxley, H.E., Stewart, A., Sosa, H., and Irving, T. 1994. X-ray diffraction measurements of the extensibility of actin and myosin filaments in contracting muscles, *Biophys. J.*, 67, 2411–2421.

Inokuchi, S., Ishikawa, H., Iwamoto, S., and Kimura, T. 1975. Age-related changes in the histological composition of the rectus abdominis muscle of the adult human, *Hum. Biol.*, 47, 231–249.

Katz, B. 1939. The relation between force and speed in muscular contraction, *J. Physiol.*, 96, 45–64.

Kaufman, T.L. 1985. Strength training effect in young and aged women, *Arch. Phys. Med. Rehabil.*, 65, 223–226.

Klitgaard, H., Zhou, M., Schiaffino, S., Betto, R., Salviati, G., and Saltin, B. 1990. Aging alters the myosin heavy chain composition of single fibers from human skeletal muscle, *Acta Physiol. Scand.*, 140, 55–62.

Kojima, H., Ishijima, A., and Yanagida, T. 1994. Direct measurement of stiffness of single actin filaments with and without tropomyosin by *in vitro* nanomanipulation, *Proc. Natl. Acad. Sci. U.S.A.*, 91, 12962–12966.

Lee, H.D. and Herzog, W. 2002. Force enhancement following muscle stretch of electrically and voluntarily activated human adductor pollicis, *J. Physiol.*, 545, 321–330.

Lexell, J., Henriksson-Larsén, K., Winblad, B., and Sjöström, M. 1983. Distribution of different fiber types in human skeletal muscles: effects of aging studied in whole muscle cross sections, *Muscle Nerve*, 6, 588–595.

Linari, M., Lucii, L., Reconditi, M., Vannicelli Casoni, M.E., Amenitsch, H., Bernstorff, S., and Piazzesi, G. 2000. A combined mechanical and x-ray diffraction study of stretch potentiation in single frog muscle fibers, *J. Physiol.*, 526.3, 589–596.

Lombardi, V. and Piazzesi, G. 1992. Force response in steady lengthening of active single muscle fibers, in *Muscular Contraction*, R.M. Simmons, Ed. (Cambridge, U.K.: Cambridge University Press), pp. 237–255.

Luff, A.R. 1981. Dynamic properties of the inferior rectus, extensor digitorum longus, diaphragm and soleus muscle of the mouse, *J. Physiol.*, 313, 161–171.

Lutz, G.J. and Rome, L.C. 1993. Built for jumping: the design of the frog muscular system, *Science*, 263, 370–372.

MacIntosh, B.R., Herzog, W., Suter, E., Wiley, J.P., and Sokolosky, J. 1993. Human skeletal muscle fiber types and force: velocity properties: model and Cybex measurements, *Eur. J. Appl. Physiol. Occup. Physiol.*, 67, 499–506.

Mai, M.T. and Lieber, R.L. 1990. A model of semitendinosus muscle sarcomere length, knee and hip joint interaction in the frog hindlimb, *J. Biomech.*, 23, 271–279.

Maréchal, G. and Plaghki, L. 1979. The deficit of the isometric tetanic tension redeveloped after a release of frog muscle at a constant velocity, *J. Gen. Physiol.*, 73, 453–467.

McComas, A.J. 1996. *Skeletal Muscle: Form and Function* (Champaign, IL: Human Kinetics).

Morgan, D.L. 1990. New insights into the behavior of muscle during active lengthening, *Biophys. J.*, 57, 209–221.

Morgan, D.L., Whitehead, N.P., Wise, A.K., Gregory, J.E., and Proske, U. 2000. Tension changes in the cat soleus muscle following slow stretch or shortening of the contracting muscle, *J. Physiol.*, 522.3, 503–513.

Nishizaka, T., Miyata, H., Yoshikawa, H., Ishiwata, S., and Kinosita, K.J. 1995. Unbinding force of a single motor molecule of muscle measured using optical tweezers, *Nature*, 377, 251–254.

Noble, M.I.M. 1992. Enhancement of mechanical performance of striated muscle by stretch during contraction, *Exp. Physiol.*, 77, 539–552.

Overend, T.J., Cunningham, D.A., Paterson, D.H., and Lefcoe, M.S. 1992. Thigh composition in young and elderly men determined by computed tomography, *Clin. Physiol.*, 12, 629–640.

Pandy, M.G. and Zajac, F.E. 1991. Optimal muscular coordination strategies for jumping, *J. Biomech.*, 24, 1–10.

Pandy, M.G., Zajac, F.E., Sim, E., and Levine, W.S. 1990. An optimal control model for maximum-height human jumping, *J. Biomech.*, 23, 1185–1198.

Perkins, L.C. and Kaiser, H.L. 1961. Results of short-term isotonic and isometric exercise programs in persons over sixty, *Phys. Ther. Rev.*, 41, 633–635.

Perrine, J.J. and Edgerton, V.R. 1978. Muscle force–velocity and power-velocity relationships under isokinetic loading, *Med. Sci. Sports Exerc.*, 10, 159–166.

Ranatunga, K.W. 1981. Temperature-dependence of shortening velocity and rate of isometric tension development in rat skeletal muscle, *J. Physiol.*, 329, 465–483.

Rayment, I., Holden, H.M., Whittaker, M., Yohn, C.B., Lorenz, M., Holmes, K.C., and Milligan, R.A. 1993. Structure of the actin-myosin complex and its implications for muscle contraction, *Science*, 261, 58–65.

Rogers, M.W., Kukulka, C.G., and Soderberg, G.L. 1992. Age-related changes in postural responses preceding rapid self-paced and reaction time arm movements, *J. Gerontol.*, 47, M159–M165.

Sacks, R.D. and Roy, R.R. 1982. Architecture of the hind limb muscles of cats: functional significance, *J. Morphol.*, 173, 185–195.

Schachar, R., Herzog, W., and Leonard, T.R. 2002. Force enhancement above the initial isometric force on the descending limb of the force–length relationship, *J. Biomech.*, 35, 1299–1306.

Smith, J.L., Betts, B., Edgerton, V.R., and Zernicke, R.F. 1980. Rapid ankle extension during paw shakes: selective recruitment of fast ankle extensors, *J. Neurophysiol.*, 43, 612–620.

Spector, S.A., Gardiner, P.F., Zernicke, R.F., Roy, R.R., and Edgerton, V.R. 1980. Muscle architecture and force–velocity characteristics of cat soleus and medial gastrocnemius: implications for motor control, *J. Neurophysiol.*, 44, 951–960.

Sugi, H. and Tsuchiya, T. 1988. Stiffness changes during enhancement and deficit of isometric force by slow length changes in frog skeletal muscle fibers, *J. Physiol.*, 407, 215–229.

Suter, E., Herzog, W., Sokolosky, J., Wiley, J.P., and MacIntosh, B.R. 1993. Muscle fiber type distribution as estimated by Cybex testing and by muscle biopsy, *Med. Sci. Sports Exer.*, 25, 363–370.

Talbot, J.A. and Morgan, D.L. 1996. Quantitative analysis of sarcomere non-uniformities in active muscle following a stretch, *J. Muscle Res. Cell Motil.*, 17, 261–268.

ter Keurs, H.E.D.J., Iwazumi, T., and Pollack, G.H. 1978. The sarcomere length-tension relation in skeletal muscle, *J. Gen. Physiol.*, 72, 565–592.

Thorstensson, A., Grimby, G., and Karlsson, J. 1976. Force–velocity relations and fiber composition in human knee extensor muscles, *J. Appl. Physiol.*, 40, 12–15.

Trok, B.M. and Barnett, V.A. 2001. Oxidative modulation of skeletal muscle titin's elasticity *in situ*. *Biophys. J. Book. Abstr.* 80(1), 275a (Abstr.).

van Leeuwen, J.L. and Spoor, C.W. 1992. On the role of biarticular muscles in human jumping, *J. Biomech.*, 25, 207–209.

van Soest, A.J., Schwab, A.L., Bobbert, M.F., and Van Ingen Schenau, G.J. 1993. The influence of the biarticularity of the gastrocnemius muscle on vertical-jumping achievement, *J. Biomech.*, 26, 1–8.

Vandervoort, A.A. and McComas, A.J. 1986. Contractile changes in opposing muscles of the human ankle joint with aging, *J. Appl. Physiol.*, 61, 361–367.

Wakabayashi, K., Sugimoto, Y., Tanaka, H., Ueno, Y., Takezawa, Y., and Amemiya, Y. 1994. X-ray diffraction evidence for the extensibility of actin and myosin filaments during muscle contraction, *Biophys. J.*, 67, 2422–2435.

Wakeling, J., Herzog, W., and Syme, D. 2000. Force enhancement and stability in skeletal muscle fibers. *XIth Congress of the Canadian Society for Biomechanics*, p. 145 (Abstr.).

Walker, S.M. and Schrodt, G.R. 1973. I segment lengths and thin filament periods in skeletal muscle fibers of the Rhesus monkey and the human, *Anat. Rec.*, 178, 63–81.

Walmsley, B., Hodgson, J.A., and Burke, R.E. 1978. Forces produced by medial gastrocnemius and soleus muscles during locomotion in freely moving cats, *J. Neurophysiol.*, 41, 1203–1215.

Wickiewicz, T.L., Roy, R.R., Powell, P.L., and Edgerton, V.R. 1983. Muscle architecture of the human lower limb, *Clin. Orthop.*, 179, 275–283.

Winter, D.A., Patla, A.E., and Frank, J.S. 1990. Assessment of balance control in humans, *Med. Prog. Technol.*, 16, 31–51.

Woittiez, R.D., Huijing, P.A., Boom, H.B.K., and Rozendal, R.H. 1984. A three-dimensional muscle model: a quantified relation between form and function of skeletal muscles, *J. Morphol.*, 182, 95–113.

Wu, Y., Bell, S.P., Lewinter, M.M., and Granzier, H.L.M. 2001. Alterations in titin and collagen underlie elevated passive muscle stiffness in rapid-pacing induced dilated cardiac myopathy of dogs. *Biophys. J. Book Abstr.* 80(1), 262a (Abstr.).

Zahalak, G.I. 1997. Can muscle fibers be stable on the descending limbs of their sarcomere length-tension relations? *J. Biomech.*, 30, 1179–1182.

5

The Ability to Persist in a Physical Task

Brian R. MacIntosh and Andrea N. Devrome

CONTENTS

ABSTRACT Endurance can be considered to be the ability to persist with a physical task. This applies whether the task is by a small muscle group or is considered to be whole-body exercise. Endurance is important in the workplace when physical exertion is part of the job expectation, or when repetitive movements are performed by small muscle groups (i.e., hand movements). Endurance is typically measured by performing a constant muscular effort at some percent of maximal voluntary force, or by repetitive voluntary contractions until a specific task can no longer be performed. Females tend to have better endurance than males when asked to perform constant or repetitive contractions at some fixed percent of maximum. There is no gender difference when maximal effort contractions are considered. Aging does not have a specific effect on endurance, but performance of a fixed task, such as repetitively lifting a given object, will be impaired in proportion to the decrease in maximal voluntary strength. Endurance is improved by training or practice and results in performance of the task with less disturbance to homeostasis.

5.1 Outline

Endurance is one of those terms that creeps into everyday language, and we all seem to understand its meaning. Someone with endurance can keep going longer than someone without. Although it is tempting to think that a person with more strength would have an advantage in such a challenge, this is apparently not the case. In this chapter, we will provide a more explicit definition of endurance and outline the importance of endurance in everyday life and athletic events. We will then discuss some of the ways endurance is measured at the muscle and whole-body levels. This will be followed by a discussion of some of the factors that influence endurance: gender, aging, and training. Endurance is clearly an important feature of our movement capabilities, and understanding endurance will enhance our comprehension of specific physical limitations and ways to optimize physical performance.

5.2 Definition of Endurance

For the purpose of this chapter, endurance is defined generally as the ability to persist in a physical task. Endurance is typically measured in time (duration). The specific task and the physical capabilities of the individual with respect to that task will determine the actual duration. In general, the lower the relative intensity of the task, the longer that duration will be.

The specific task may be as simple as a maintained submaximal muscle contraction or intermittent contractions where the effort is alternated with a rest period. Intermittent contractions can be isometric (no movement) or dynamic (movement involved). Strictly speaking, an honest comparison of endurance between two individuals should require them to perform the same physical task and measure how long each can continue performing that task. However, scientists recognize that someone with greater strength will have an advantage in such a comparison, so they make comparisons of endurance where the physical task is normalized with respect to strength of the participants. In some cases, the intensity of contraction can be designated relative to a maximal voluntary contraction

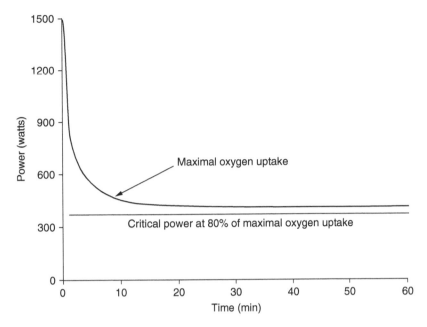

FIGURE 5.1
Intenstity–duration relation for whole-body exercise, cycling. This relationship is determined from a series of constant power trials to exhaustion, lasting 3–12 min.

(MVC). In the case of dynamic contractions, intensity can be expressed relative to the highest load that can be moved through the designated range of motion. This is also known as the one repetition maximum (1 RM). Alternatively, dynamic contractions can be quantified relative to maximal power output for the prescribed movement.

Endurance can also be used to describe the ability to persist with whole-body exercise such as walking, running, swimming, cycling, etc. With these activities, endurance can be related to the intensity–duration relation, where special markers of intensity include values associated with one of the following: peak power, peak speed, maximal oxygen uptake, lactate threshold, or critical power (see Figure 5.1).

It is important to realize that endurance and fatigue are reciprocal concepts. At the limit of endurance, fatigue (central or peripheral) has accumulated to the point of preventing the continuation of the physical task at the desired intensity. When the task requires a relatively high effort, a little bit of fatigue can prevent continuing. When the intensity of the physical task is low, considerably more fatigue would be required to prevent continuing with the task. At the limit of endurance, two people can have similar fatigue yet very different endurance. They have a similar reduction in contractile capability, but it took them different amounts of time to get there.

5.3 Ergonomic Implications for Endurance

In spite of automation and the proliferation of laborsaving devices, physical effort is still an essential part of our daily living. The major difference in this respect is that the physical effort is often quite repetitive. This applies to workers on an assembly line or conveyor belt as well as to those who work much of the day at a keyboard.

5.3.1   Endurance in Our Daily Lives

Keep in mind that endurance is the ability to persist in a physical task. When the physical task is keyboard or mouse operation, the limiting factor could be shoulder or arm discomfort from the relatively persistent contractile effort needed in these regions to permit the digital or manual dexterity required of this particular physical task. Appropriate body position and posture are important features to ensure the ability to persist with such a task. Inappropriate ergonomics can lead to chronic injury in situations that involve repetitious activations of muscle groups, even when the involved muscles are relatively small.

In special circumstances (i.e., aging or unique physical impairments), the endurance associated with the simple task of walking or supporting the body weight may have important health implications. It has been demonstrated that elderly women with a history of falls have less endurance than age-matched controls (Schwendner et al., 1997). The implication here is that the impaired ability to persist with this physical task may have been a factor contributing to the increased incidence of falls. It seems likely that a multitude of daily tasks associated with occupation, recreation, or daily living can lead to reduced contractile performance of specific muscle groups (due to limited endurance), resulting in impaired performance or accidents causing injury. More specifically, muscle fatigue can affect skill in physical tasks. For example, performance of carpenters has been reported to be impaired after a task that caused fatigue in shoulder muscles (Hammarskjold and Harmsringdahl, 1992).

The ability to persist in a physical task may also be a factor that hinders voluntary participation in an active lifestyle. Regular physical exercise is known to reduce the risk for several chronic diseases (Booth et al., 2002; Katzmarzyk et al., 2000). Lack of endurance, then, can have a serious impact on quality of life as well as health care costs.

5.3.2   Endurance Is Also Important to Athletes

We do not have contests to see how long someone can persist with a physical task (arm wrestling and some of the "strong-man" competitions appear to be exceptions). Many events that are considered "endurance" events are actually events of a fixed distance, and the winner is the one who can complete the distance in the shortest amount of time. This relates to endurance in that the winner is the one who can maintain the highest speed for sufficient duration to complete the distance. When competing in such events, athletes typically maintain a constant pace that is as high as they perceive they can sustain just long enough to complete the distance. This concept is nicely illustrated in a graph that shows world records for different distances of a given athletic event. Swimming world records are shown in Figure 5.2. This graph looks very similar to the intensity–duration relation shown in Figure 5.1. The longer the distance of the race, the lower will be the intensity that can be sustained. Although we typically think of endurance events as events of long duration, the definition of endurance that we have chosen does not have a time restriction. It is just as relevant to speak of endurance with respect to maintaining an intensity of exercise associated with maximal oxygen uptake (4–8 min) as it is an exercise associated with critical power (30–60 min).

5.4   How Is Endurance Quantified?

There are several circumstances where measurement of endurance is desirable. In the research literature, the majority of such measurements of endurance are done either to

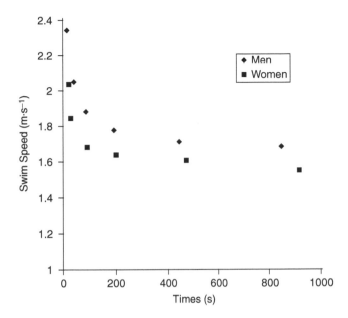

FIGURE 5.2
World longcourse swimming records for 50, 100, 200, 400, 800 and 1500 m freestyle. Data obtained from: http://www.fina.org/recordshome.html.

evaluate the impact of specific conditions or treatments on the ability to persist with a physical task, or to compare the endurance capabilities of different groups. Endurance may also be measured to evaluate the specific physical capabilities of an individual, for example, as an applicant for a physically demanding job. Endurance may also be measured to evaluate the impact of a training program. Regardless of the reason for the measurement, it is important that the measurement is chosen with an appropriate rationale, and that the measurement gives a valid estimate of endurance.

5.4.1 Endurance Relates to the Intensity–Duration Relation

Any single measurement of endurance can be depicted on a graph of the relationship between intensity of exercise and the duration that the exercise can be sustained. The intensity–duration relation has been used to depict endurance for a broad range of tasks, regardless of the mass of muscle involved. This can be said for an isolated single muscle fiber, or whole muscle, as well as for a muscle or muscle group activated voluntarily in the body, or for whole-body exercise.

The equation for a hyperbola has typically been used to fit to data relating the duration of persistence for a given task at different intensities. This approach was first used to demonstrate properties of a synergistic muscle group performing repetitive dynamic muscle contractions (Monod and Scherrer, 1965), but has been most critically advanced for whole-body exercise (Billat et al., 2000; Bishop et al., 1998; Brickley et al., 2002; Hill, 1993).

The hyperbola that has most often been used to describe the intensity–duration relation is: $W' = T_{lim} \times (P - CP)$, where P is any power output, T_{lim} is the duration that P can be sustained, CP (critical power) is the asymptote, and W' is a constant representing the absolute amount of work that can be performed at exercise intensities above CP. This equation fits the data reasonably well when the data are obtained in a series of time trials lasting across a range of durations from 3 to 12 min (Bishop et al., 1998). The asymptote

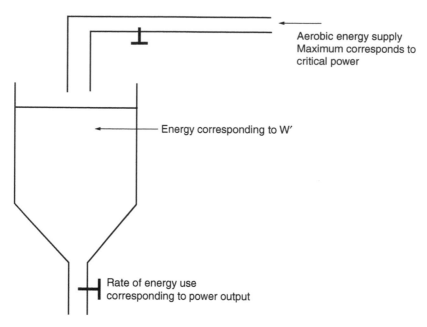

FIGURE 5.3

Illustration of the energy concepts related to critical power and W'. When the rate of energy use does not exceed the rate of supply of energy from aerobic sources, then there will be no net use of the energy corresponding to W'. Note that W' is a measure of work, so this energy relies on a known efficiency. When the rate of energy use exceeds the rate supplied by aerobic metabolism, the energy corresponding to W' will be used, and the duration of exercise will be limited by the rate of use of this energy and the magnitude of energy available. (Adapted from Bishop, D., D.G. Jenkins, and A. Howard, *Int. J. Sports Med.*, 19, 125–129, 1998.)

of this relation represents a power output (CP) that can theoretically be sustained indefinitely. For whole-body exercise, this intensity is assumed to be the highest intensity of exercise that can be sustained with energy provided only by oxidative metabolism (no net contribution from glycolysis leading to the accumulation of lactate). This should represent the anaerobic threshold (Svedahl and MacIntosh, 2003). Of course, the intensity of exercise representing the anaerobic threshold cannot be sustained indefinitely, so this remains a hypothetical concept.

A simple way to consider the concepts represented in this apparently hyperbolic relation has been proposed by Bishop et al. (1998). In their model, the W' can be achieved by using a finite amount of energy that can be obtained from nonaerobic processes, and CP represents the highest rate of energy use that can be sustained without contribution from these nonaerobic sources (see Figure 5.3). The nonaerobic processes would represent phosphocreatine stores and glycolysis leading to accumulation of lactic acid. According to this model, exercise could be sustained indefinitely at intensities up to CP. Once CP was exceeded, there would be a requirement for energy to come from the finite (nonaerobic) energy stores. Exercise could continue until these stores were depleted. The duration of exercise is therefore dependent on the magnitude of the energy store associated with W' and the rate of use of this energy.

5.4.2 Specific Tests of Endurance

There are typically three forms of endurance test: (a) sustained isometric contraction of a muscle or synergistic muscle group, (b) intermittent contractions of a muscle or muscle group, and (c) whole-body exercise. Regardless of which of these is under consideration,

TABLE 5.1

Examples of Endurance Tests

Muscle	Sustained/ Intermittent	Isometric/ Dynamic	Duration	Source
Elbow flexor	15% MVC	Isometric	900 s	(Semmler et al., 1999)
Elbow flexor	15–80% MVC	Isometric	<1–50 min	(Hagberg, 1981)
Elbow flexor	15–50% MVC	30° · s⁻¹	<1–45 min	(Hagberg, 1981)
Elbow flexor	80% MVC	Isometric	25 s	(Felici et al., 2001)
Elbow flexor	20% MVC	Isometric	500–2000 s	(Hunter and Enoka, 2001)
Knee extensor	$180° · s^{-1}$	Dynamic	30 reps	(Luo et al., 2001)
Knee extensor	$180° · s^{-1}$	Dynamic	50 reps	(Grimby et al., 1992)

the general relationship illustrated in Figure 5.1 applies. The higher the intensity of effort, the shorter the duration of time that intensity can be maintained, but there is considerable variability in the relationship, as seen in Table 5.1, which presents examples of endurance tests from the literature. However, the mechanisms that limit the exercise duration may vary. Each of these types of endurance test will be considered further below.

5.4.3 Sustained Submaximal Isometric Contractions

It is common to measure endurance by asking the participant to maintain a constant submaximal isometric force. This approach has been used for several muscle groups in the human body: elbow flexors (Hagberg, 1981; Proske and Morgan, 2001), adductor pollicis (Fulco et al., 1999), knee extensors (Ng et al., 1994; Wretling and Henriksson-Larsén, 1998), and other muscles. Before the test can begin, the MVC is assessed. Once the force or torque associated with maximal effort is known, then the target force can be calculated. The actual target force has varied from less than 20% to greater than 80% of maximum voluntary force. Clearly, the actual duration of the test is dependent on the relative force sustained, and this can vary from seconds to several minutes. This approach for the measurement of endurance is frequently used in studies that compare various groups or that evaluate interventions to alter endurance, but certain problems have become apparent where such comparisons are made.

It is generally thought that normalizing the contraction to the maximal voluntary force is a fair way to compare groups. It seems reasonable to expect that endurance associated with maintaining the same absolute force would favor the individual who is stronger. However, there are some observations that shed doubt on the fairness of comparing similar relative contractions.

5.4.4 Problems with Sustained Submaximal Isometric Contractions

There are two key problems with sustained relative submaximal isometric contractions that make them less than ideal for evaluation of endurance. One of these problems has to do with the impact of internal pressure on blood flow, and the second one is the concern that absolute strength may alter the expected endurance, negating the attempt to normalize the testing procedure to compare subjects of different physical stature. In most cases of assessment of endurance, there is a necessary comparison between candidates and/or between before and after a treatment that may include training, nutritional supplementation, or other ergogenic intervention. Because of the nature of comparison, there must be reasonable methods of normalization. However, normalization according to MVC may not be appropriate.

5.4.5 Blood Flow Is Compromised during Isometric Contractions

When a muscle contracts, the internal pressure can exceed arterial pressure, and blood flow can be compromised or even stopped. During intermittent contractions, blood flow can occur during the relaxation phase, like in the heart. The relative force at which blood flow is entirely occluded is variable, but is thought to be around 50%. There is not much difference in endurance between occluded and free circulating conditions for sustained isometric contractions of knee extensors at 50% of MVC (Hisaeda et al., 2001). When the sustained contractile response is less than 50%, then variability in blood flow response may be a factor contributing to variability in endurance. Very little research has been done on this topic. It is likely that absolute force affects the internal pressure, making it unreasonable or unfair to make a comparison at a given percent of MVC in conditions where MVC may be different.

5.4.6 Do the Weak Have More Endurance or Less?

An important question that still needs to be addressed in the scientific literature is whether endurance should be quantified under identical absolute conditions or if conditions should be normalized with respect to strength. Figure 5.4 illustrates the torque–duration relation for three hypothetical subjects with different strength. These curves are typically obtained by a series of trials to exhaustion, with different constant force. Can you tell which one has the best endurance? This situation is complicated in research where comparison is made at different times, when the MVC would have changed.

Immobilization is known to cause muscle atrophy and a corresponding decrease in strength. Intuitively, one would also expect this intervention to cause a decrease in endurance. This is apparently not the case. Semmler et al. (2000) immobilized one arm of a group of otherwise healthy individuals for 4 weeks. These individuals experienced decreases in the maximum voluntary force, but endurance did not decrease. In fact, in a

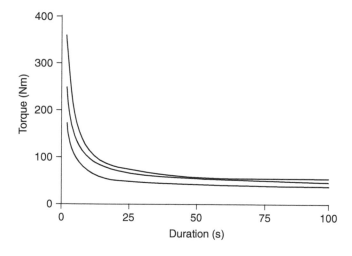

FIGURE 5.4

Intensity–duration relation for isometric knee extensor effort for three individuals with different strength (maximal torque). Clearly, when measured in absolute terms, the strongest individual has greater endurance, when measured for any sustained torque less than 65 N·m, and always has greater endurance than the weakest subject. When the required effort is less than 65 N·m, then the subject with the middle value for strength has the greatest endurance. Although the weakest subject appears to have the lowest value for endurance across the full range of effort illustrate, the shape of the curve is very similar to that for the strongest person, and this individual would have similar endurance to the strongest person if asked to sustain the same percent of maximum torque.

subgroup of these subjects, endurance time increased. Members of this subgroup were able to maintain 15% of MVC much longer after the period of immobilization. Would it have been more appropriate to compare endurance before and after immobilization by having them perform the same absolute task?

It is interesting to note that the authors of this paper detected a neural adaptation in their subjects that may have contributed to the improved endurance. This adaptation has been thought to permit occasional rest for some motor units, thereby avoiding fatigue. This observation raises the question of whether it is appropriate to compare endurance within a group of subjects under different conditions when MVC has changed. Certainly in situations where endurance is evaluated for occupational requirements, the physical task may not relate to individual MVC. A given absolute force may be required. In this particular study (Semmler et al., 2000), the ability to hold the same absolute force before and after the immobilization was not evaluated.

The neural adaptation that was observed by Semmler et al. (2000) may not have been the only factor contributing to the enhanced endurance reported in this study. In fact, some subjects who did not appear to have the adaptation also had prolonged endurance. These observations lead to the question of whether or not it is fair to use a load relative to MVC to compare endurance between conditions or groups.

5.4.7 Intermittent Contractions to Measure Endurance

Intermittent contractions are also used to measure endurance. These contractions can be isometric or dynamic, and the effort-to-rest ratio is usually prescribed. The rest may be occupied by activation of the antagonist muscle group in cases of dynamic contractions, to permit return to the starting position. One advantage of using intermittent contractions to measure endurance is that blood flow can occur during the rest period. Therefore, one of the concerns with continuous contraction can be averted. A clear disadvantage of dynamic contractions is that it is not as simple to normalize with respect to the load that is applied.

Different approaches have been used to evaluate endurance of muscle groups using intermittent contractions. Intermittent isometric contractions can be evaluated in a manner similar to the approach with continuous contractions. This begins with determination of the MVC. Subsequently, visual feedback can be used to provide a target force that the subject is to maintain, and audio or visual cues can be used to regulate the times of effort and rest. Intermittent dynamic contractions can be performed with a fixed percent of 1 RM. This exercise would continue until the participant was unable to complete a repetition.

Intermittent isokinetic contractions are also used to evaluate endurance (Grimby et al., 1992; Wretling and Henricksson-Larsén, 1998). In this case, the common approach is to have the participant make repeated maximal effort with a fixed angular velocity, and measure the time or number of contractions until peak torque or work per contraction reaches 50% of the initial value. Alternatively, a fixed number of contractions can be executed, and the measurement is an index of fatigue (relative decline in peak torque).

5.4.8 Whole-Body Exercise to Measure Endurance

Measurement of endurance with whole-body exercise is typically done with a treadmill or some form of ergometer. Although it may be useful to compare endurance between individuals at a given absolute power output, a more common approach is to measure maximal oxygen uptake and ask participants to maintain a constant fixed percent of their maximal oxygen uptake. An alternative approach is to measure endurance in whole-body exercise during a time trial. With this approach, work or distance traveled can be the outcome measurement. The test is to perform the greater amount of work in a fixed period of time or cover a given distance in as short a time as possible. All of the characteristic features that influence endurance in the usual tests of endurance will play a role in performance of these tests, and additional factors may be important.

5.5 A Physiological View of Muscular Endurance

What is it about a muscle that permits the avoidance (or delay) of muscle fatigue? The easiest way to think about this is to consider the energetics of the physical task and the disturbance to the homeostatic condition that is brought about by participating in the physical task. Endurance can be prolonged if the energy cost of the task can be reduced and/or the provision of energy can be achieved with less disturbance to the homeostatic condition. The relative energy cost of performing a task is often expressed as the efficiency or the economy.

5.5.1 What Is the Difference between Efficiency and Economy?

"Efficiency" is a term that is misused far too often. In energetic considerations, the efficiency of a process relates the proportion of energy that is conserved, relative to the total energy used in the task. Chemical efficiency considers the amount of energy in the bonds of a specific chemical product of a reaction relative to the amount of energy in the substrate. The difference is lost as heat during the chemical reaction. For example, when the metabolic substrates (carbohydrates, fats, and proteins) are converted to CO_2 and H_2O in our muscle cells, some of the chemical energy from these compounds will be conserved as chemical energy in the terminal phosphate bond of adenosine triphosphate (ATP). The proportion of the chemical energy conserved is approximately 60%. Therefore, the efficiency of this energy transduction is 60%. The remainder of the chemical energy that was available in the metabolic substrates will be lost as heat.

Mechanical efficiency is the proportion of energy conserved as potential energy or mechanical work relative to the total energy that was supplied to perform the movement under consideration. Efficiency is: work × energy cost^{-1}. Considerable research has been undertaken to understand the energetics of muscle contractions, and it is known that mechanical efficiency of a skeletal muscle can be no greater than 50% when the energy source is considered to be ATP. Therefore, if both steps are considered, the best efficiency of energy transduction from metabolic substrates to mechanical work is half of 60%, or 30% (Barclay et al., 1993). This optimal value would only occur in ideal circumstances.

The efficiency of whole-body exercise typically ranges from zero to about 25%. Cycle ergometry is an exercise that the human body can perform with a relatively high efficiency, while running has a very low efficiency because there is very little net mechanical work accomplished.

Elite cyclists have efficiencies in cycle ergometry ranging from 18 to 24% (Coyle et al., 1992). The efficiency of cyclists is related to the proportion of Type I myosin isoform that is present in their vastus lateralis muscle (Coyle et al., 1992). It is important to point out that this observation should not be construed as evidence that a preponderance of Type I myosin isoform in a muscle gives it better efficiency than muscles with more Type II myosin. Efficiency of muscle contraction is tightly linked to velocity of shortening. Cycling

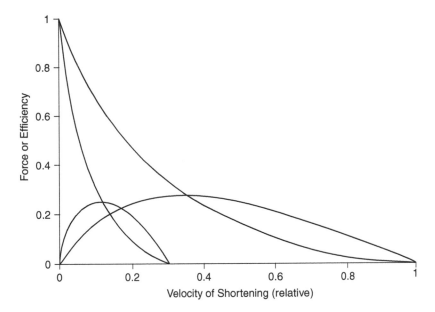

FIGURE 5.5
Force–velocity and efficiency–velocity relationships. The shape and relative magnitudes of the efficiency–velocity relationships are similar between muscles with different myosin heavy chain isoforms, but they are scaled differently.

is performed at relatively slow muscle shortening velocities. When the velocity of shortening is much higher, muscles with more Type II myosin have greater efficiency (see Figure 5.5). In many cases, such as isometric contractions, work is not done during a muscle contraction, and mechanical efficiency is zero, regardless of myosin isoform composition of the active muscles.

There are many conditions or circumstances where mechanical work is not the outcome of interest, such as isometric contractions or repeated motion where net work is zero. Under these conditions, mechanical efficiency is not informative. Economy may be a more appropriate term to use in describing the energetics of muscle contraction or exercise in these circumstances. Economy can be considered to be the energy cost to perform a specified task. For example the economy of maintaining an isometric force may be of interest. In this case, the energy cost (in kcal or kJ) can be expressed relative to the tension–time integral. Alternatively, economy might be of interest for various forms of locomotion (walking, cycling, rowing, etc.). In this case the economy can be expressed as an energy cost per unit distance traveled (kcal·km^{-1}). In cases where the energy cost of locomotion is considered to be body mass dependent, like walking and running, economy can be expressed relative to body mass (kcal·km^{-1}·kg^{-1}). Economy of locomotion is one of the important determinants of success in races, particularly when there is a considerable variation in economy, as there is for race-walking (Hagberg and Coyle, 1983).

5.5.2 What Determines the Energy Cost of a Muscle Contraction?

Endurance relates to the ability to perform a task with less energy, and less disturbance to homeostasis. It is therefore important to have knowledge of the factors that influence the energy cost of a muscle contraction.

There are two primary components to the energy cost of a muscle contraction: crossbridge turnover and ion pumping. Ion pumping can also be considered the energy cost of activation, and involves the Na$^+$–K$^+$ ATPase and the Ca^{2+} ATPase. These ion pumps work to restore concentration gradients for Na$^+$, K$^+$, and Ca^{2+}. The total amount of energy required by these ion pumps is dependent on the number of activations (action potentials) per motor unit and the size and number of motor units activated. Cross-bridge turnover has to do with the ATP required to dissociate actin from myosin during each cross-bridge cycle. The total amount of ATP needed for this process will depend on the number of cross-bridges cycling, the individual rate at which cross-bridges hydrolyze ATP, and the duration of the contraction. There are several factors that impact the amount of ATP required for this aspect of muscle contraction: frequency of activation, number of motor units activated, muscle length, time-dependent cross-bridge turnover, and rate of length change. These determinants of the energy cost of muscle contraction have been reviewed by others (MacIntosh et al., 2000; Stainsby and Lambert, 1979) and will be briefly described here.

When a motor nerve is activated by an action potential transmitted down a motoneuron, Ca^{2+} is released into the sarcoplasm of each muscle fiber associated with that motor unit. Free Ca^{2+} concentration will transiently increase, and Ca^{2+} will bind to troponin, permitting cross-bridge interaction. The number of cross-bridges engaged in a muscle cell depends primarily on the proportion of troponin that binds Ca^{2+}, and that depends on the free Ca^{2+} concentration in the sarcoplasm. Generally, a higher free Ca^{2+} concentration will be reached when the frequency of action potentials is higher. According to the length–tension relation, muscle length will also affect the number of cross-bridges that are available for interaction with actin.

During an isometric contraction, any cross-bridge apposed by activated actin can bind to that actin and undergo the sequence of chemical events we refer to as a cross-bridge cycle. When the cycle occurs in a very short period of time, the cross-bridge can use ATP very quickly. Under isometric conditions, the time-dependent turnover of cross-bridges is related to the predominant myosin isoform. In human muscles, three myosin heavy chain isoforms are found: Type I, Type IIa, and Type IIx (Pette, 2002). The time-dependent turnover of a cross-bridge is slowest for Type I, faster for Type IIa, and fastest for Type IIx. It is generally thought that the size of motor units corresponds to the myosin isoform; Type I motor units are smallest, and Type IIx are the largest. However, if we consider three hypothetical motor units of identical length and myofibrillar cross-sectional area, the acto–myosin ATP turnover during an isometric contraction of any given force would be lowest in the Type I motor unit, highest in the Type IIx motor unit, and intermediate in the Type IIa motor unit.

The energy cost of myosin ATPase interaction changes when muscle contraction results in length change. At a relatively slow velocity, the cross-bridge will reach the limits of its translation in less time than it would normally have remained attached. This reduces the time required for a cross-bridge cycle. This time gets progressively shorter as velocity of shortening increases, so the rate of ATP use by a cross-bridge will tend to increase as velocity of shortening increases. This tendency is countered by another velocity-dependent factor. As velocity increases, the probability for a given cross-bridge binding to an actin active site is reduced. For this reason, fewer cross-bridges are bound at any instant during a high-velocity contraction. This effect contributes to the decrease in force at high velocities that relates to the force–velocity property of muscle. The influence of velocity of shortening on the energy cost of muscle contraction is such that, at slow velocities the rate of ATP use is greater than for an isometric contraction, and at higher velocities the rate of ATP use can decrease to below that for an isometric contraction with the same level of activation. The effect of velocity of shortening on a muscle contraction can be seen in Figure 5.6.

These factors that determine the energy cost of muscle contraction should be kept in mind as we further consider the factors that influence endurance. Endurance will be

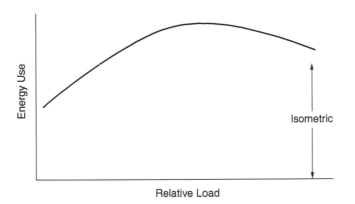

FIGURE 5.6
The relationship between energy per contraction and relative load. As load decreases from the isometric condition, velocity will increase according to the force–velocity relation. As load decreases (and velocity increases), energy use per contraction first increases then decreases. (Adapted from Stainsby, W.N. and J.K. Barclay, *Am. J. Physiol.*, 230, 1008–1012, 1976.)

determined by the rate of energy use in a given muscle fiber and the ability of that fiber to maintain homeostasis in the face of high ATP demand.

5.5.3 What Does Homeostasis Have to Do with Endurance?

Homeostasis is the tendency to maintain a constant internal environment. In the resting condition, the temperature and concentration of ions and various molecules are regulated to remain at specific levels in the various fluid compartments of the body. These compartments include: intracellular, extracellular, and vascular. Exercise tends to disturb the condition of homeostasis, and it is quite likely that a substantial disturbance to homeostasis is an important factor in limiting endurance. The ability to undertake exercise of a given intensity with less disturbance to this homeostasis is a feature that permits better endurance.

There are specific ions and molecules that are more likely to be disturbed during exercise. These include: H^+, K^+, Ca^{2+}, Na^+, glucose, ATP, adenosine diphosphate (ADP), phosphocreatine (PCr), inorganic phosphate (P_i), NH_3, and lactate. The partial pressures of O_2 and CO_2 are also altered during exercise, and this is particularly evident in the venous blood during exercise with a substantial portion of the body's muscle mass.

5.5.4 Exercise of Single Muscles or Small Muscle Groups

Low-intensity exercise of a small muscle group will result in some changes within the muscle fibers associated with the active motor units. These changes include the following: increased temperature, cytoplasmic [ADP] (molar ADP concentration), $[P_i]$, and $[Na^+]$, as well as decreased cytoplasmic $[NAD^+]$, [PCr], [ATP], $[K^+]$, etc. When that low intensity of exercise can be sustained for a long time, the rate of these changes is usually slow. As the intensity of exercise increases, the magnitude of change in concentration of these substances is likely to increase. Furthermore, with increased motor unit recruitment, additional volume of muscle will be involved in these changes, and with more motor units involved, the likelihood of change (disturbance to homeostasis) in the extracellular and vascular compartments.

5.5.5   The Importance of Calcium Homeostasis

Calcium is very important in intracellular signaling processes. At rest, cytoplasmic $[Ca^{2+}]$ is about 70 nM, and there is a high concentration of free Ca^{2+} in the sarcoplasmic reticulum. During each activation of a given motor unit, the cytoplasmic $[Ca^{2+}]$ increases transiently and then falls toward the resting level. The transient rise in cytoplasmic $[Ca^{2+}]$ activates the contractile process and also signals activation of several enzymes. The magnitude and frequency of the Ca^{2+} transients are very important to the magnitude of contractile response and activation of processes that maintain the homeostatic condition, especially with respect to maintaining ATP concentration. Inability to maintain a constant supply of ATP can cause disturbance of Ca^{2+} homeostasis and impaired muscle activation. In a fatiguing muscle fiber, resting $[Ca^{2+}]$ increases, free $[Ca^{2+}]$ in the terminal cisternae decreases (Allen and Westerblad, 2001), and the calcium transients associated with activation become smaller (Westerblad and Allen, 1991). These changes contribute to fatigue or the approach of the limits of endurance.

5.5.6   Factors That Prevent Radical Disturbance to the Homeostatic Condition

It can be considered that the goal in exercise is to replenish ATP at the rate that it is being used, and to do so with the smallest possible change in the distribution of ions, accumulation of heat, and concentration of metabolic products. The key features that are important in this task are the enzymes of the metabolic processes that replenish ATP, as well as the pumps and transporters that are embedded in the membranes. An increase in the metabolic enzyme capacity permits replenishment of ATP in the active muscle cells with less decrease in [ATP] and less corresponding increase in [ADP]. The increased numbers of membrane proteins help prevent changes in heat and cytoplasmic concentration of ions and of substances used in or produced in the metabolic processes. Training adaptations contribute to minimizing this disturbance of homeostasis during physical effort by modifications in enzyme, transporter, and vascular networks. Disturbance to homeostasis is decreased by several training adaptations, including modifications to enzymes, transporters, and vascular networks.

5.6   Factors That Influence Endurance

5.6.1   Do Women Have More Endurance Than Men?

Research has established that certain differences exist in muscular endurance between men and women. It has been shown that women have significantly longer endurance when performing fatiguing submaximal contractions at a given percent of their MVC. This has been demonstrated consistently in several different muscle groups, including adductor pollicis, elbow flexors, extrinsic finger flexors, and knee extensors. The exact mechanism for the greater muscular endurance in women is not clearly understood, though scientists have suggested that several factors are involved. These contributing factors include: fiber-type composition, substrate use, hormones, and muscle activation. Though experimentation has clearly demonstrated that these factors are involved, relative contributions and specific effects remain controversial.

Studies have shown that women have a greater percentage of Type I (slow twitch), fatigue-resistant muscle fibers than men, and this is consistently the case in different muscle groups (Fulco et al., 1999; Simoneau and Bouchard, 1989). It has been suggested that glycogen depletion would be more rapid in males than females because of the fiber-type differences. However, many of the endurance tests are too short to rely on glycogen depletion as an end point. It is known that slow twitch fibers fatigue more slowly than fast twitch fibers, so this factor alone could explain the gender difference in endurance. Fiber-type differences in endurance may be related to differences in ATP turnover associated with myosin heavy chain composition. Type I fibers have slower ATP turnover, resulting in less energy use (Crow and Kushmerick, 1982). Furthermore, slow twitch muscle is associated with a higher density of $Na^+ñK^+$ ATPase (Chin and Green, 1993), so it is possible that it is better capable of preserving ion gradients than fast twitch muscle.

Another possible mechanism for the decreased muscular endurance observed in men is the fact that, in many studies, subjects perform exercises at an intensity that is relative to their maximum. Because men are stronger than women, a given percentage of their maximum will generally be at a higher absolute force compared to their women counterparts. It has been suggested that the men, who are working at higher loads, might experience a greater degree of local vascular occlusion as a result of the isometric contractions (Hunter and Enoka, 2001). This mechanism may indeed exist; however, a recent study matched men and women for absolute strength of the adductor pollicis muscle, such that the subjects were working at the same relative and absolute loads. This study still observed a consistently longer endurance time for women than men (Fulco et al., 1999). This observation disputes the mechanism of differential vascular occlusion.

Gender-dependent differences in endurance may also be due to the presence of estrogen. Premenopausal women also have up to 100 times greater circulating levels of estrogens than men (Tiidus, 1995). Studies have shown that estrogen has strong antioxidant properties and may contribute to membrane stability (Tiidus, 2000). Estrogen has been shown to protect against enzyme leakage, which is a symptom of membrane disruption that may occur with fatiguing activity. If muscle damage is a factor that limits the duration of a given activity, then estrogen may be responsible, in part, for the greater endurance observed in females. However, estrogen is not likely the mechanism for the gender-dependent difference in endurance for elderly subjects (> 70 years) (Hicks and McCartney, 1996).

A final possible cause for the observed difference between muscular endurance in men and women is the degree of muscle activation. It has been shown that men have a greater loss of maximal voluntary neural activation than women when performing repeated maximal isometric contractions (Hakkinen, 1993). There is some contradictory evidence regarding this issue. Other studies have shown that there is no difference in activation between men and women after performing repetitive fatiguing submaximal contractions (Hunter and Enoka, 2001). It is possible that this difference is due to the type of fatigue protocol that is implemented and whether the loads are low or high. Hakkinen (1993) suggests that women may have less "aggressiveness" than men and might be exercising at lower intensities, thus experiencing less fatigue. The impact of such a difference is difficult to rationalize with respect to differences in endurance. As long as the scientists conducting the test make sure the subjects maintain the target force adequately, then aggression does not enter the picture during the fatigue test. Aggression could influence the determination of maximum voluntary force production. If the subjects performed a less-than-maximal voluntary contraction, then the relative force that would be used during endurance measurements would be a lower percentage of their true maximum. Researchers could use twitch interpolation to determine whether subjects were contracting maximally or not. As a result, it would be easy to confirm the participation of this mechanism.

5.6.2   Does Endurance Decrease with Age?

Most research in age-related changes in skeletal muscle function has focused on muscular strength and structure. As a result, relatively little is known of the changes that occur with muscular endurance as we get older. Though alterations in the ability to exert maximal force are important to identify, some suggest that it is not as relevant to the performance of daily activities as is muscular endurance (Bemben, 1998). Some might even suggest that muscular endurance is a more practical measure of neuromuscular function. It is more likely that an understanding of both muscular strength and endurance are important, because one undoubtedly affects the other. However, in this section, we will look at the changes in muscular endurance and possible causes of these changes in aging individuals.

As described earlier, a variety of tests is used to measure muscular endurance. Of particular interest are studies that use a relative percent of maximal strength to measure muscle endurance vs. tests that use absolute strength to compare muscular fitness between young and older individuals. As will be described below, there are many controversial findings when both methods are used. It remains to be seen which type of test has greater relevance to daily functioning. In an occupational situation such as an assembly line, older individuals would be expected to perform the same physical tasks as their younger counterparts. In this case, absolute endurance is much more relevant. If the older worker has less strength, they will have to work at a higher relative intensity to be able to perform the task. In fact, it is difficult to identify any daily activities where relative endurance might be important.

There are many conflicting observations regarding changes to muscular endurance with aging, in both human and animal models. One study in rats demonstrated no difference in the fatigability in the soleus muscles of young vs. old rats; however there was an increase in glycolysis and glycogen utilization in the older rats (Fitts et al., 1984). The increased carbohydrate use in the older rats suggests that they have a different substrate mix, and higher carbohydrate use is generally considered to be associated with a higher metabolic crossover intensity (Brooks and Mercier, 1994), indicating a lower tolerance for continued exercise. The crossover intensity is the intensity of exercise above which there is a higher proportion of carbohydrate than fat used in the exercise. Additional studies have confirmed that sustained power is reduced in the extensor digitorum longus (EDL) muscle of old mice when compared to young animals (Faulkner and Brooks, 1995). This controversy extends to human studies. In some studies, muscular endurance has been shown to be similar in old and young individuals when the contractions are performed at the same relative intensities (Larsson, 1978). In other studies, where electrically evoked contractions are measured, older subjects display a greater relative loss of force (Davies et al., 1986). In general, most authors appear to agree that endurance based on absolute load decreases with age, but endurance measured at a percent of maximum effort is observed.

Researchers suggest that observed declines in muscular endurance are due in part to the well-documented age-related atrophy that occurs in older individuals. Maximal strength declines with muscle atrophy. If an aged individual who has experienced muscle atrophy performs any given task at the same absolute load as a control subject, the older person will likely be working at a higher percent of his or her maximum and will fatigue faster. Again, there is some controversy surrounding this issue. It has been shown that, when an activity is performed at the same percent of maximum, atrophied muscles display better endurance capabilities (Larsson, 1978).

Studies examining muscle biopsies consistently show a decrease in Type II muscle fibers with increasing age (Larsson et al., 1979). It is possible that the maintenance of muscular endurance observed in aging muscle occurs as a result of the greater proportion of Type I fibers that remain.

5.6.3 Several Other Factors Can Influence Endurance

Endurance associated with any physical task can be altered (improved or decreased) by a variety of circumstances. For example, high temperature and dehydration impair endurance (Doherty et al., 1994). Procedures and supplements that enhance performance of a physical task are called ergogenic aids. A few examples of effective ergogenic aids include the following: caffeine, water, sports drinks, proper nutrition, training, anabolic/androgenic steroids, etc. Although there are many ergogenic aids that actually do enhance performance, there are many more pseudo-ergogenics that purport to enhance performance, but for which there is no evidence to substantiate such a claim. In other cases, evidence is conflicting. An example of this latter possibility is creatine monohydrate. The study of ergogenic aids is very important. The ability to sustain effort in a physical task has important implications in recreation, occupations, and daily living. Understanding how to optimize such performance will improve productivity and quality of life.

5.7 Endurance Can Be Enhanced by Training

The human body tends to adapt to stress placed on it such that subsequent similar stress is more readily tolerated. Training or regular practice of specific movements leads to adaptations that permit longer endurance. This makes training one of the best ergogenic aids known. The training-induced adaptations that occur can generally be placed in three categories: (a) adaptations that reduce the energy cost, (b) adaptations that enhance the provision of energy for the task, and (c) adaptations that reduce the disturbance to homeostasis. The magnitude of adaptation that can be expected is dependent on several factors: initial level, intensity of training, duration of training, frequency of training, and progression. This chapter will not deal further with the details of training programs. However, the specific adaptations that occur as a result of training will be presented.

5.7.1 Adaptations That Reduce Energy Cost

There appears to be some disagreement in the research literature over whether or not specific endurance training will result in a different energy cost for the same physical task. There are probably several reasons for this disparity. Such reasons would include: some training activities may not induce a change in energy cost; the duration of the study may not have been sufficient to allow this change; or the measurement may not have been sensitive enough to detect a small change.

There are two mechanisms by which a change in energy cost of performing a task would be likely to occur. These two mechanisms can be gleaned from the earlier discussion of the determinants of the energy cost of muscle contraction: altered activation and decreased myosin ATPase activity. In this case, activation can be simply thought of in terms of more refined activation of motor units, including less activation of antagonists. Practice of a task, particularly a more complex one, will lead to more effective motor control patterns, allowing a task to be accomplished with less activation and hence less energy cost. Slower myosin ATP turnover can be achieved by increased expression of Type I or IIa myosin ATPase. This can occur due to a relative hypertrophy of fibers expressing these myosin isoforms or by altered myosin isoform expression; for example, fibers formerly expressing Type IIx myosin heavy chain now express Type IIa (Howald et al., 1985). Hybrid fibers (individual fibers expressing more than one isoform) are more prevalent during a training program (Pette, 2001). Type I myosin is more economical when maintaining a constant (isometric) force or when the movement of concern is relatively slow (Barclay et al., 1993). Similarly, it would be expected that Type IIa myosin would be more economical than Type IIx. It has often been observed that endurance athletes have very few Type IIx (formerly considered to be IIb) motor units. This adaptation relates to improved endurance.

5.7.2   Adaptations That Increase Provision of Energy

The key adaptation to endurance training that contributes to increased provision of energy is an increase in total mitochondrial protein content (Alway, 1991; Wibom et al., 1992). This increase corresponds to an increase in maximal oxygen uptake (in the case of whole-body exercise), but the main impact of this adaptation is that a given rate of ATP synthesis can be sustained with less disturbance to homeostasis.

5.7.3   Other Cellular Changes That Help with Homeostasis

When phosphocreatine is hydrolyzed by creatine kinase to rephosphorylate cytoplasmic ADP, forming ATP, creatine concentration increases. Creatine stimulates oxidative metabolism in the mitochondria. If a given concentration of creatine in the cytoplasm is associated with a certain rate of ATP generation per volume of mitochondria, then more mitochondria can generate that amount of ATP with less change in the creatine concentration. This means that a given task can be accomplished with less change in PCr concentration, i.e., less departure from homeostasis. A similar situation occurs for other key metabolic intermediates.

Vascular changes can also contribute to the improved aerobic metabolism. Training has been shown to result in increased capillary density and more homogeneous perfusion (Kalliokoski et al., 2001). These adaptations will contribute by more effective delivery of substrates and O_2, as well as improved removal of substances that might otherwise accumulate in the extracellular space, including lactate, K^+, H^+, and heat. These changes in vascularization contribute to minimizing the impact of exercise on the disturbance to cellular homeostasis.

In addition to the changes described above, improved $Na^+ñK^+$ ATPase activity has been reported with chronic electrical stimulation in rats (Hicks et al., 1997) and following a brief training program in human subjects (Green et al., 1993). This increased pump activity would be expected to contribute to better maintenance of ion gradients across the sarcolemma.

5.8   Future Directions in Endurance Research

Clearly, there is a need for further research into endurance. The majority of research on this topic to date has been concerned with identifying general factors that might influence endurance and quantifying differences between identifiable populations. This research needs to be supplemented by investigation into the relevant method of normalizing the load that should be used for such research. Furthermore, there is a need to study specific factors that may limit endurance. This work should consider possible limits to endurance for work of muscle groups of varying sizes, while exercising over a range of intensities.

5.9   Summary

Endurance, the ability to persist in a physical task, is important in our daily lives. Endurance permits us to perform physical tasks without undue stress and can be measured across the intensity spectrum with isometric or dynamic contractions. There are many factors that impact endurance, including gender, aging, training, and nutrition. Enhanced endurance is associated with economical or efficient performance and minimal disturbance to homeostasis. Males tend to have less endurance than females, but this difference is diminished or lost when maximal effort is considered. Aging does not appear to have a major impact on endurance when it is measured relative to MVC. Training, appropriate nutrition, and certain ergogenic aids can optimize endurance.

References

Allen, D.G. and H. Westerblad. 2001. Role of phosphate and calcium stores in muscle fatigue, *J. Physiol.*, 536(3), 657–665.

Alway, S.E. 1991. Is fiber mitochondrial volume density a good indicator of muscle fatigability to isometric exercise? *J. Appl. Physiol.*, 70(5), 2111–2119.

Barclay, C.J., J.K. Constable, and C.L. Gibbs. 1993. Energetics of fast- and slow-twitch muscles of the mouse, *J. Physiol. (London)*, 472, 61–80.

Bemben, M.G. 1998. Age-related alterations in muscular endurance, *Sports Med.*, 25(4), 259–269.

Billat, V.L., R.H. Morton, N. Blondel, S. Berthoin, V. Bocquet, J.P. Koralsztein, and T.J. Barstow. 2000. Oxygen kinetics and modelling of time to exhaustion whilst running at various velocities at maximal oxygen uptake, *Eur. J. Appl. Physiol. Occup. Physiol.*, 82(3), 178–187.

Bishop, D., D.G. Jenkins, and A. Howard. 1998. The critical power function is dependent on the duration of the predictive exercise tests chosen, *Int. J. Sports Med.*, 19, 125–129.

Booth, F.W., M.V. Chakravarthy, and E.E. Spangenburg. 2002. Exercise and gene expression: physiological regulation of the human genome through physical activity, *J. Physiol.*, 543(2), 399–411.

Brickley, G., J. Doust, and C.A. Williams. 2002. Physiological responses during exercise to exhaustion at critical power, *Eur. J. Appl. Physiol. Occup. Physiol.*, 88, 146–151.

Brooks, G.A. and J. Mercier. 1994. Balance of carbohydrate and lipid utilization during exercise: the "crossover" concept, *J. Appl. Physiol.*, 76(6), 2253–2261.

Chin, E.R. and H.J. Green. 1993. Na$^+$ñK$^+$ ATPase concentration in different adult rat skeletal muscles is related to oxidative potential, *Can. J. Physiol. Pharmacol.*, 71, 615–618.

Coyle, E.F., L.S. Sidossis, J.F. Horowitz, and J.D. Beltz. 1992. Cycling efficiency is related to the percentage of Type I muscle fibers, *Med. Sci. Sports Exerc.*, 24(7), 782–788.

Crow, M.T. and M.J. Kushmerick. 1982. Chemical energetics of slow- and fast-twitch muscles of the mouse, *J. Gen. Physiol.*, 79, 147–166.

Davies, C.T., D.O. Thomas, and M.J. White. 1986. Mechanical properties of young and elderly human muscle, *Acta Medica Scand. Suppl.*, 711, 219–226.

Doherty, T.J., T. Komori, D.W. Stashuk, A. Kassam, and W.F. Brown. 1994. Physiological properties of single thenar motor units in the F-response of younger and older adults, *Muscle Nerve*, 17(8), 860–872.

Faulkner, J.A. and S.V. Brooks. 1995. Muscle fatigue in old animals. Unique aspects of fatigue in elderly humans, *Adv. Exp. Med. Biol.*, 384, 471–480.

Felici, F., A. Rosponi, P. Sbriccoli, M. Scarcia, I. Bazzucchi, and M. Iannattone. 2001. Effect of human exposure to altitude on muscle endurance during isometric contractions, *Eur. J. Appl. Physiol. Occup. Physiol.*, 85(6), 507–512.

Fitts, R.H., J.P. Troup, F.A. Witzmann, and J.O. Holloszy. 1984. The effect of ageing and exercise on skeletal muscle function, *Mech. Ageing Dev.*, 27(2), 161–172.

Fulco, C.S., P.B. Rock, S.R. Muza, E. Lammi, A. Cymerman, G. Butterfield, L.G. Moore, B. Braun, and S.F. Lewis. 1999. Slower fatigue and faster recovery of the adductor pollicis muscle in women matched for strength with men, *Acta Physiol. Scand.*, 167(3), 233–239.

Green, H.J., E.R. Chin, M. Ball-Burnett, and D. Ranney. 1993. Increases in human skeletal muscle $Na^+ñK^+$-ATPase concentration with short-term training, *Am. J. Physiol. Cell Physiol.*, 264, C1538–C1541.

Grimby, G., A. Aniansson, M. Hedberg, G.-B. Henning, U. Grangård, and H. Kvist. 1992. Training can improve muscle strength and endurance in 78- to 84-yr-old men, *J. Appl. Physiol.*, 73(6), 2517–2523.

Hagberg, J.M. and E.F. Coyle. 1983. Physiological determinants of endurance performance as studies in competitive racewalkers, *Med. Sci. Sports Exerc.*, 15(4), 287–289.

Hagberg, M. 1981. Muscular endurance and surface electromyogram in isometric and dynamic exercise, *J. Appl. Physiol.*, 51(1), 1–7.

Hakkinen, K. 1993. Neuromuscular fatigue and recovery in male and female athletes during heavy resistance exercise, *Int. J. Sports Med.*, 14(2), 53–59.

Hammarskjold, E. and K. Harmsringdahl. 1992. Effect of arm-shoulder fatigue on carpenters at work, *Eur. J. Appl. Physiol. Occup. Physiol.*, 64(5), 402–409.

Hicks, A., K. Ohlendieck, S.O. Göpel, and D. Pette. 1997. Early functional and biochemical adaptations to low-frequency stimulation of rabbit fast-twitch muscle, *Am. J. Physiol. Cell Physiol.*, 273(1), C297–C305.

Hicks, A.L. and N. McCartney. 1996. Gender differences in isometric contractile properties and fatigability in elderly human muscle, *Can. J. Appl. Physiol.*, 21, 441–454.

Hill, D.W. 1993. The critical power concept: a review, *Sports Med.*, 16(4), 237–254.

Hisaeda, H.O., M. Shinohara, M. Kouzaki, and T. Fukunaga. 2001. Effect of local blood circulation and absolute torque on muscle endurance at two different knee-joint angles in humans, *Eur. J. Appl. Physiol. Occup. Physiol.*, 86(1), 17–23.

Howald, H., H. Hoppeler, H. Claassen, O. Mathieu, and R. Straub. 1985. Influences of endurance training on the ultrastructural composition of the different muscle fiber types in humans, *Pflügers Arch.*, 403, 369–376.

Hunter, S.K. and R.M. Enoka. 2001. Sex differences in the fatigability of arm muscles depends on absolute force during isometric contractions, *J. Appl. Physiol.*, 91(6), 2686–2694.

Kalliokoski, K.K., V. Oikonen, T.O. Takala, H. Sipila, J. Knuuti, and P. Nuutila. 2001. Enhanced oxygen extraction and reduced flow heterogeneity in exercising muscle in endurance-trained men, *Am. J. Physiol. Endocrinol. Metab.*, 280, E1015–E1021.

Katzmarzyk, P.T., N. Gledhill, and R.J. Shephard. 2000. The economic burden of physical inactivity in Canada, *CMAJ*, 163(11), 1435–1440.

Larsson, L. 1978. Morphological and functional characteristics of the aging skeletal muscle in man. A cross-sectional study, *Acta Physiol. Scand., Suppl.*. 457, 1–36.

Larsson, L., G. Grimby, and J. Karlsson. 1979. Muscle strength and speed of movement in relation to age and muscle morphology, *J. Appl. Physiol.*, 46(3), 451–456.

Luo, Y.M., N. Hart, N. Mustfa, R.A. Lyall, M.I. Polkey, and J. Moxham. 2001. Effect of diaphragm fatigue on neural respiratory drive, *J. Appl. Physiol.*, 90, 1691–1699.

MacIntosh, B.R., R.R. Neptune, and A.J. Van den Bogert 2000. Intensity of cycling and cycle ergometry: power output and energy cost, in *Biomechanics and Biology of Movement*, B.M. Nigg, B.R. MacIntosh, and J. Mester, Eds. (Champaign, IL: Human Kinetics), pp. 129–148.

Monod, H. and J. Scherrer. 1965. The work capacity of a synergistic muscle group, *Ergonomics*, 8, 329–338.

Ng, A.V., J.C. Agre, P. Hanson, M.S. Harrington, and F.J. Nagle. 1994. Influence of muscle length and force on endurance and pressor responses to isometric exercise, *J. Appl. Physiol.*, 76(6), 2561–2569.

Pette, D. 2001. Historical perspectives: plasticity of mammalian skeletal muscle, *J. Applied Physiol.*, 90(3), 1119–1124.

Pette, D. 2002. The adaptive potential of skeletal muscle fibers, *Can. J. Appl. Physiol.*, 27(4), 423–448.

Proske, U. and D.L. Morgan. 2001. Muscle damage from eccentric exercise: mechanism, mechanical signs, adaptation and clinical applications, *J. Physiol.*, 537(2), 333–345.

Schwendner, K.I., A.E. Mikesky, W.S. Holt, Jr., M. Peacock, and D.B. Burr. 1997. Differences in muscle endurance and recovery between fallers and nonfallers, and between young and older women, *J. Gerontol. Biol. Sci. Med. Sci.*, 52(3), M155–M160.

Semmler, J.G., D.V. Kutzscher, and R.M. Enoka. 1999. Gender differences in the fatigability of human skeletal muscle, *J. Neurophysiol.*, 82, 3590–3593.

Semmler, J.G., D.V. Kutzscher, and R.M. Enoka. 2000. Limb immobilization alters muscle activation patterns during a fatiguing isometric contraction, *Muscle Nerve*, 23(9), 1381–1392.

Simoneau, J.A. and C. Bouchard. 1989. Human variation in skeletal muscle fiber-type proportion and enzyme activities, *Am. J. Physiol.*, 257(4 Pt 1), E567–E572.

Stainsby, W.N. and J.K. Barclay. 1976. Effect of initial length on relations between oxygen uptake and load in dog muscle, *Am. J. Physiol.*, 230, 1008–1012.

Stainsby, W.N. and C.R. Lambert 1979. Determinants of oxygen uptake in skeletal muscle, *Exerc. Sport. Sci. Rev.*, 7, 125–152.

Svedahl, K. and B.R. MacIntosh 2003. Anaerobic threshold: the concept and methods of measurement, *Can. J. Appl. Physiol.*, 28(2), 299–323.

Tiidus, P.M. 1995. Can estrogens diminish exercise induced muscle damage? *Can. J. Appl. Physiol.*, 20(1), 26–38.

Tiidus, P.M. 2000. Estrogen and gender effects on muscle damage, inflammation, and oxidative stress, *Can. J. Appl. Physiol.*, 25(4), 274–287.

Westerblad, H. and D.G. Allen. 1991. Changes of myoplasmic calcium concentration during fatigue in single mouse muscle fibers, *J. Gen. Physiol.*, 98, 615–635.

Wibom, R., E. Hultman, M. Johansson, K. Matherei, D. Constantin-Teodosiu, and P.G. Schantz. 1992. Adaptation of mitochondrial ATP production in human skeletal muscle to endurance training and detraining, *J. Appl. Physiol.*, 73(5), 2004–2010.

Wretling, M.L. and K. Henricksson-Larsén. 1998. Mechanical output and electromyographic parameters in males and females during fatiguing knee-extensions, *Int. J. Sports Med.*, 19, 401–407.

6

Measurement of Localized Muscle Fatigue in Biceps Brachii Using Objective and Subjective Measures

Shrawan Kumar, Tyler Amell, Yogesh Narayan, and Narasimha Prasad

CONTENTS

ABSTRACT   Localized muscle fatigue (LMF) is of significant interest to many professionals, including ergonomists, for safety interventions. In the literature, many indices have been used to indicate LMF. However, no gold standard could be identified. For determining a standard indicator for LMF, nine subjects were required to exert their previously measured maximal voluntary contraction (MVC) and 40% MVC in elbow flexion for as long as they could. Their magnitude of force of exertion, electromyograph (EMG) amplitude, median frequency (MF), muscle bed blood volume (BV), and muscle oxygenation were measured for MVC. Similarly, the muscle force exertion, EMG amplitude, MF, BV, oxygenation, oxygen uptake, ventilation volume, and heart rate were measured for 40% MVC. Rate of perceived exertion (RPE), visual analog score (VAS), and body part discomfort rating (BPDR) were measured for both contractions. The obtained data were subjected to analysis of variance (ANOVA) with repeated measures correlation analysis to determine correlation between force of contraction and independent variables. A regression analysis was carried out to predict the force from independent variables. The mean MVC and durations of hold scores were (with standard deviations) 197.6 N (25.9 N) and 58.15 s (17.6 s), respectively. The corresponding 40% MVC values were 79 N (10.4 N) and 275 s (35.8 s). The ANOVA revealed that the values at different percentiles of the task cycles were significantly different for both contraction levels ($p < .001$). The MF was the strongest indicator of LMF in MVC ($r = 0.91$, $p < .001$), but in the 40% MVC contraction, the VAS score was better. None of the variables were found to consistently represent LMF in different levels of muscle contraction. A different grouping of objective and subjective variables for MVC and 40% MVC increased the predictability of force over the single variables mentioned before. A gold standard for representing LMF still eludes us.

6.1 Introduction

Fatigue, though a word of common and everyday language, is a diverse and complex phenomenon in the scientific domain. Fatigue as it relates to human performance is important for both the cognitive and physical domains. In the physical domain, the nature of fatigue is different when it relates to the individual as a whole or a specific muscle tissue. Generally, a person's fatigue may be measured in the rate as well as total physiological energy spent. The objective measures used for these are oxygen uptake and heart rate. The subjective measure most suitable for fatigue has been Borg's scale (Borg, 1982). Among the tissues affected by prolonged and/or repetitive work are connective tissues, which undergo mechanical fatigue, and muscles, which undergo physiological and mechanical fatigue. Whereas the feeling of muscle fatigue is a common experience, its quantitative measurement has not been entirely satisfactory.

A variety of methods have been used for measuring localized muscle fatigue. These fall under categories of objective and subjective methods. Objectively, the clearest indication of fatigue is the duration for which a load can be held or decay in the force over time. Inability to maintain a level of force over a designated time is an indication of force fatigue. The magnitude of force (strength) and duration it can be held (endurance) were initially demonstrated by Rohmert (1960). However, the ability to sustain a load or enhance the level of strength is subjectively modifiable. Also, despite the discomfort, people can maintain their contraction for a period longer than the normal endurance time, if motivated to do so. Thus the strength and endurance curve is obscured, making it difficult to measure even force fatigue quantitatively.

Electromyography has been considered a reliable tool for an indication of localized muscle fatigue (Chaffin, 1973; De Luca, 1985). In recent years, other authors have stated that the median frequency (MF) of the EMG power spectrum is sensitive to the physiological manifestation of fatigue (Kumar and Narayan, 1998; Kumar et al., 2001; Mannion and Dolan, 1994; Roy and De Luca, 1996). The rate of decline of the MF during a sustained contraction has commonly been used as an index of muscle fatigue (Kumar and Narayan, 1998; Kumar et al., 2001; Mannion and Dolan, 1994; Roy and De Luca, 1996). An accompanying increase in EMG amplitude has also been considered a quantitative measure of fatigue (Kumar and Narayan, 1998; Kumar et al., 2001; Ng et al., 1997; van Dieen et al., 1993, 1996). Other spectral parameters of the EMG signal, such as mean power frequency (Kwanty et al., 1970), the ratio of low frequency to high frequency (Bigland-Ritchie et al., 1981), and the number of zero crossings (Hagg, 1981), have also been used in some studies as indices of fatigue but were found to be more affected by noise than the measure of median frequency (Stuten and De Luca, 1981). Although, among EMG variables, the slope of decline of the median frequency in sustained contractions has been considered as an index of fatigue, when it is compared against the standard of inability to hold or maintain a force, the slope of MF decline does not mirror it (Kumar and Narayan, 1998; Kumar et al., 2001). On the other hand, the EMG amplitude increases with the decrease in force, but this inverse relationship is variable. Therefore the EMG variables, though helpful indicators, are far from being a gold standard index of fatigue.

Among other local factors that affect muscle fatigue are blood perfusion of the muscle and level of its oxygenation (Grandjean, 1988). Until recently these measurements could not be made except invasively; hence, they were largely ignored. With the advent of near-infrared spectroscopy (NIRS), such measurements can now be made noninvasively (Hamoaka et al., 1996; McCully and Hamoaka, 2000). This is a noninvasive technique that depends on detecting a continuous stream of reflected light from the light source through muscle tissue. The light source of the device emits 700–900 nm wavelength light into the muscle. Heme and myoglobin are the primary absorbing compounds of these frequencies, and the absorption of light by the heme group is detected as optical density. This is a semiquantitative measure and is expressed in percent change in optical density, allowing us to infer the oxygenation/deoxygenation of the blood in the muscle and the blood volume of the muscle. Using this device, several authors have presented their data. De Blasi et al. (1993, 1996), Murthy et al. (1997), and Hicks et al. (1999) have studied these variables in the forearm muscles. De Blasi et al. (1993) studied oxygen consumption of the brachioradialis muscle and its recovery time. They did not find any difference in the oxygen uptake of the muscle in isometric maximal voluntary contraction (MVC) with and without vascular occlusion. However, vascular occlusion resulted in complete desaturation of the hemoglobin and myoglobin in 15 to 20 seconds. De Blasi et al. (1996), in a separate study, reported that the oxygen desaturation rate occurring during incremental levels of isometric exercise without vascular occlusion was not proportional to the strength of the contraction. Murthy et al. (1997) reported that the tissue oxygenation at low levels of isometric contraction (5, 10, 15, and 50% MVC) of the extensor carpi radialis brevis muscle resulted in deoxygenation to 89, 81, 78, and 48% of the resting value, which was assigned 100%. They stated that this oxygen depletion may be directly linked to the muscle fatigue. Hicks et al. (1999), after a comparative study of submaximal isometric contractions (10 and 30% MVC) measured by NIRS and direct venous blood sampling, suggested caution in application of NIRS technology. However, a host of other studies have reported oxygenation and blood volume in quadriceps and erector spinae. Nonetheless, a direct comparison between the force fatigue and deoxygenation and blood perfusion has not been reported.

It is generally understood that there is no direct reliable quantitative measure of localized muscle fatigue. Therefore, different authors have chosen from a variety of indicators of fatigue, such as EMG median frequency, muscle deoxygenation, oxygen uptake, heart rate, rate of perceived exertion (RPE), and visual analog scale (VAS) scores. The objective of this study was to measure EMG amplitude, median frequency, oxygen uptake, ventilation, heart rate, deoxygenation, blood volume, RPE, VAS, and Body Part Discomfort Rating (BPDR) in one study and examine their relationship with the force fatigue and their interrelationship in isometric MVC and 40% of MVC of the biceps brachii in sustained contractions.

6.2   Method

6.2.1   Subjects

Nine healthy male subjects with no self-reported cardiovascular, musculoskeletal, neuro-muscular, or psychiatric disorders were recruited from the university student population through a campus-wide advertisement. The mean (standard deviation) age, weight, and height of the experimental sample were 25.7 (2.1) years, 78.9 (9.5) kg, and 180.1 (4.3) cm, respectively. The study was approved by the Health Research Ethics Board, and all subjects provided an informed consent by signing the consent form. The subjects were remunerated at the rate of $10.00 per hour.

6.2.2   Task

The subjects stood upright in front of an iron chain anchored to the floor with an intervening load cell and a round handle at the upper end. The subjects flexed their right (dominant) arms by 90° at the elbow while the upper arms hung vertically down from the shoulders. The level of the handle was adjusted individually to the height of the subject's hand. In this position, the subjects were required to give their isometric MVC for elbow flexion without changing the preset angle at the wrist, elbow, or shoulder. The subjects were instructed to build their strength to maximum gradually within the first 2 seconds and maintain the maximal level of contraction for another 3 seconds, at which time the contraction was terminated. Subsequent to this contraction the subjects were given a 2-minute rest. Strength values were taken at the subjects' 100% effort.

   The subjects, after a minimum 2-minute rest, acquired the task position described above. In front of the subject, a visual feedback mechanism was adjusted to show 40% of isometric MVC. The subjects were instructed to exert elbow flexion force in isometric mode such that their contraction indicator reached one displayed on the screen, and they maintained it as long as they could. Even when the force declined below this level, they were asked to keep exerting until they could no longer exert. Subsequent to this exertion the subjects were provided a rest of 20 minutes, at which time they were asked to exert their MVC and hold it in maximal effort as long as they could.

6.2.3   Equipment and Tools

6.2.3.1   *EMG Devices*

The EMG electrodes were an active-surface electrode (Model DE-2.3, Delsys Inc., Wellesley, MA). They were made of two silver–silver chloride bands (10 mm long and 1 mm wide) mounted in parallel 1 cm apart in a molded block. The block also housed a miniature preamplifier with a gain of 100 (to amplify the signal on site) and a high-pass filter at 8 Hz. These electrodes were connected to amplifiers, worn around waist by means of a short cable, that had multiple gain. The amplifiers had a time constant of 20 ms, a low nonlinearity, and 120-dB signal-to-noise ratio.

6.2.3.2 Oxygen Uptake Measuring Devices

6.2.3.2.1 Morgan Oxylog 2

This is a portable oxygen consumption–measuring device (Morgan Medical Ltd., Kent, UK). This device could measure 0.1 to 10 liters oxygen consumption per minute. The flow transducer was a turbine-type flow meter attached to a facemask to enable measurement of inspiratory volumes in the range of 5–500 liters per minute (lpm). The Oxylog 2 contained Figaro KE-25 oxygen fuel type oxygen sensors and operated with 5-volt power supply from four NiCd cells. The device, among other variables, measured minute and total oxygen consumption and minute and total inspired volume of the air. The airflow was directed to the Oxylog through an air hose connected at one end to the mouthpiece or facemask and at the other end to the gas inlet of the device.

6.2.3.2.2 Polar Vantage XL Heart Rate Monitor

The device consisted of a transmitter incorporated in a chest band worn by the subject. The transmitter relayed the heart rate to a wristwatch receiver, which could collect and hold data for up to 33 hours. This heart rate data was subsequently downloaded into a data logger.

6.2.3.3 Near Infrared Spectrascope

The model MRM-96 and the accompanying software NIRCOM (NIM, Inc., Philadelphia) were used in this study. This was a noninvasive oxygen and blood volume trend monitor designed to measure data from the muscle bed. It was a continuous wave spectrometer that operated by transmitting light into the skin below the probe, collecting reflected light, and processing the signals by computing optical density from the measurements. The light that was not absorbed or scattered away from the detector was collected by the photodiode detectors in the probe and converted to digital data. The MRM-96 emitted lights of wavelengths 760 and 850 nm, which were indicators of the relative quantities of oxy- and deoxyhemoglobin. With the relative change in quantities of these compounds, the light absorbed by them changed as well. The oxygenation level was indicated by the difference between two lights, and the blood volume in the muscle bed was indicated by the sum of the two lights.

6.2.3.4 Load Cell and Force Monitor

A load cell (LCCB-IK from Omega, Laval, PQ), was connected to the force monitor (Prototype Design, Ann Arbor, MI). The tension load cell fed to the force monitor, which conditioned and amplified the signal and displayed the reading through a LED display. The output of the force monitor was fed to the Data Translation A to D card (DT2801A) of the computer for continuous sampling of the signals.

6.2.3.5 Subjective Measures

6.2.3.5.1 Visual Analog Scale

The subjects were provided a 10-cm horizontal line with anchors at the two ends reading "not fatigued at all" and "worst imaginable fatigue." The subjects crossed the horizontal line according to their feeling at the time of measurement. With maximal contraction, the subjects were given this test twice, once before starting the test and the other after completing the test. For the 40% MVC contraction the subjects were presented with a new scale every minute for marking.

6.2.3.5.2   *Body Part Discomfort Rating (BPDR)*

The subjects were provided with a blank line diagram of the entire body and asked to place an "X" on the part of the body where they felt discomfort due to fatigue. Subsequent to placing an "X" they were asked to rate their fatigue on a scale of 1 to 10, 1 being no fatigue and 10 being the maximal fatigue. Like the VAS, this test was done only twice with maximal contraction, but with 40% MVC contraction the rating was completed every minute.

6.2.3.5.3   *Rate of Perceived Exertion*

The rate of perceived exertion (RPE) was filled out by the subjects according to Borg (1982). The RPE was scored on a 20-point scale. Like other subjective measures, the subjects determined their RPE only twice with maximal contraction but every minute for 40% MVC.

6.2.4   Procedures

After the subjects signed the informed consent form, they were familiarized with the equipment and procedures. They were asked to come another day, when they were first weighed and measured. Their biceps were palpated, and the precalibrated NIRS probe was applied on the muscle. Adjacent to the NIRS probe, a single bipolar surface electrode was applied to skin over bicep brachii after abrading it with a paper towel soaked with alcohol–acetone. After application of the probe and the electrode, the subjects were grounded, and it was ensured that the two precalibrated measuring devices were working as desired. At this stage, the Polar chest band was applied after slightly wetting the contact of the device. The wristband receiver was tied around the wrist and checked to ensure that the device was working properly. Finally, the subjects were fitted with the facemask, with mouthpiece and airflow valve, and connected to the Oxylog by air hose. Such prepared subjects were allowed to rest for a period of 5 minutes to obtain the resting values of oxygen uptake, ventilation, heart rate, blood oxygenation, blood volume, and the EMG signal. Subsequently, the subjects were asked to perform their isometric MVC. After 5 minutes of rest, the subjects were again placed in the experimental setting to perform their 40% MVC, while all variables were measured as described above. After 20 minutes of rest, the trial for maximal contraction was conducted.

6.2.5   Data Acquisition

The suitably conditioned and amplified data from the load cell, Oxylog, NIR spectroscope, and EMG amplifier were sampled at 1 kHz for the entire duration of isometric MVC, 40% MVC fatiguing contraction, and maximal fatiguing contraction. These data were fed to the Data Translation (DT 2801A) A to D board for conversion into digital values. These were then stored in the computer memory. A Pentium computer was used as the controller and storage device. The data acquisition was carried out using specially written software for data collection.

6.2.6 Data Analysis

The force trace was smoothed using a 7-point smoothing routine, repeated once. From this trace, the peak value was read and the average value calculated. Each task duration for each individual was divided into 100 equal parts, and values corresponding to each percentile and each 10-percentile of the task durations were extracted. The percentile values of all subjects were pooled, and descriptive statistics were calculated for further statistical analysis. A similar processing was done for oxygen uptake, heart rate, ventilation volume, muscle blood oxygenation, muscle-bed blood volume, and EMG amplitude. Similar values for VAS, RPE, and BPDR scores were obtained by interpolating between the measured points and extracting the relevant percentile values. The EMG signals were subjected to fast Fourier transform analysis to obtain the median frequency according to methods described in Kumar and Narayan (1998) and Kumar et al. (2001). These quantitative values were used to develop the magnitude of the signal and the task percentile plot.

The statistical analysis consisted of repeated measure univariate analysis of variance for each variable separately, followed by a repeated measure multivariate analysis of variance combining all variables for MVC. For 40% MVC, contraction a Pearson Product Moment Correlation was calculated between the magnitude of force and the magnitude of each of the measured variables to determine the strength of correlation of each independent variable with force. A multiple correlation matrix was developed for MVC and 40% MVC variables to glean the interrelationship between all independent variables. Finally, to examine the predictability of force from other variables, a multiple regression equation was developed, with force as dependent variable and conditioning on other variables. The repeated measure considered for the feature of the design.

6.3 Results

6.3.1 Force and Time

The mean maximal voluntary contraction force of the experimental sample for elbow flexion was 197.6 N (SD 25.9 N). The mean duration for which the maximal contraction effort could be sustained was 58.1 s (SD 17.6 s), though the force exerted kept declining through the entire duration. Similarly the mean 40% force recorded was 79.0 N (SD 10.4 N), and the mean duration for sustaining this load was 275 s (SD 35.8 s).

6.3.2 Variables in Maximal Voluntary Fatiguing Contraction

At maximal voluntary contraction, the continuous monitoring at 1 kHz of the force, EMG, blood oxygenation, and blood volume provided continuous data presented in Figure 6.1. The means and standard deviations are presented at intervals corresponding to 10% of each task. A steady decline in force was mirrored most closely by the median frequency of the recorded EMG. A spectral plot of the bicep in MVC fatiguing contraction demonstrates this (Figure 6.2). The EMG amplitude also declined steadily and progressively for up to 80% of the task cycle, beyond which it dropped more rapidly.

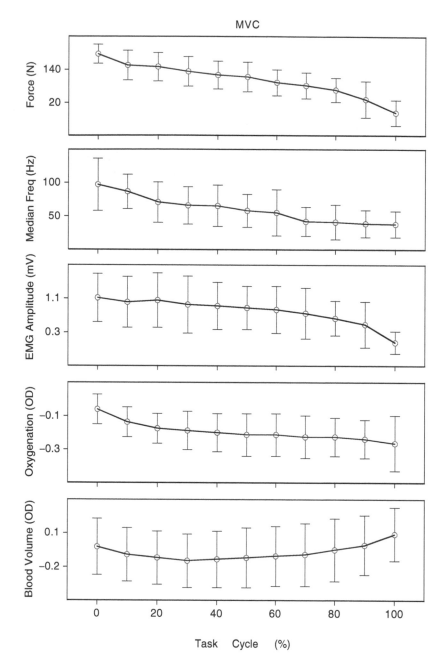

FIGURE 6.1
The pattern of variables in maximal fatiguing contraction.

6.3.3 40% MVC Fatiguing Contraction

In this contraction the force, optical densities for blood oxygen and blood volume, EMG amplitude, and EMG median frequency were continuously monitored at 1 kHz. The oxygen uptake, ventilation volume, heart rate, visual analog scale score, and the RPE score were obtained at one-minute intervals. In the intervals between readings, the data were interpolated, and the values corresponding to appropriate task percentile value were

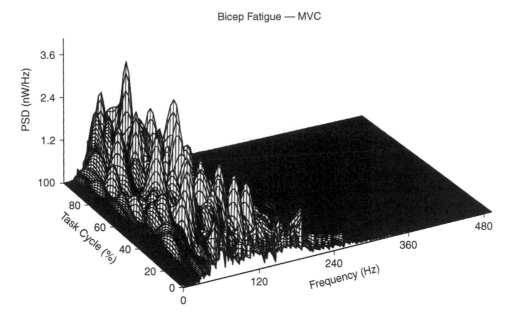

FIGURE 6.2
A spectral plot of the bicep fatigue in maximal fatiguing contraction.

extracted from the interpolation. The patterns in variation of measures are presented in Figure 6.3. With a progressive and gradual decline in force, there was a gradual and progressive decline in the MF of the EMG, but its amplitude remained relatively stable (Figure 6.3). The spectral plot of the EMG of the biceps brachii in this contraction did not demonstrate the same progressive shift of MF as in MVC contractions (Figure 6.4). The optical density of the blood oxygen and the blood volume continued to increase, indicating a progressive increase in blood volume as well as blood oxygen levels (Figure 6.3). The oxygen uptake, ventilation volume, heart rate, VAS score, and RPE score all progressively increased with a progressive decrease in force (Figure 6.3).

6.3.4 Correlation Analysis

The correlation coefficients between the force and other measured variables for MVC and 40% MVC fatiguing trials are presented in Table 6.1 and Table 6.2. In the MVC fatiguing contraction, the MF of EMG was strongly correlated with the force ($r = 0.91$; $p < .001$). Blood oxygenation and blood volume also were correlated with the force, albeit moderately (Table 6.1). Intercorrelation coefficients between the independent variables largely remained at low to moderate levels. Only the blood volume optical density and EMG amplitude reached a moderate correlation coefficient value ($r = 0.61$; $p < .001$). Most correlation coefficient values were positive, except the correlation between blood oxygenation level and blood volume and EMG amplitude.

The correlation matrix for the 40% MVC fatiguing contraction presented a different picture (Table 6.2). With progressively decreasing force during this contraction, only EMG MF and amplitude were positively correlated ($r = 0.34$ and 0.11, respectively; $p < .001$). All other variables were negatively and significantly correlated with the force (Table 6.2). On average, VAS and RPE scores showed modest correlation with force ($r = -0.67$ and -0.54, respectively; $p < .001$) (Table 6.2). They consistently demonstrated good correlation with all other independent variables as well, except blood oxygen optical density and

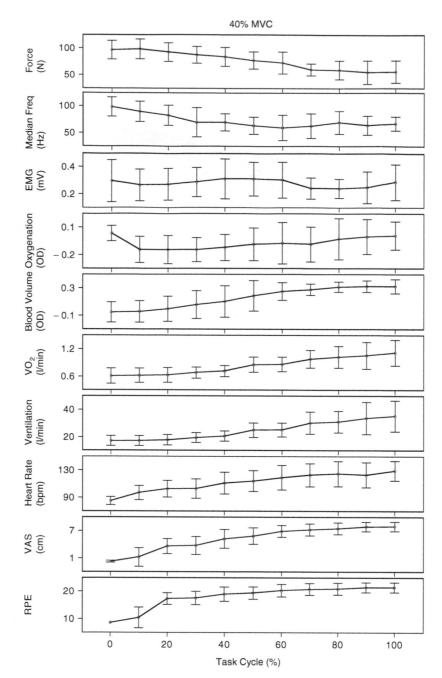

FIGURE 6.3
The pattern of variables in 40% MVC fatiguing contraction.

EMG amplitude. The only variable that had better correlation with force was blood volume optical density ($r = -0.78$; $p < .001$). The blood oxygenation optical density and MF of the EMG showed only a modest correlation with force ($r = -0.39$ and 0.39, respectively; $p < .001$). Metabolic variables were highly intercorrelated, as expected.

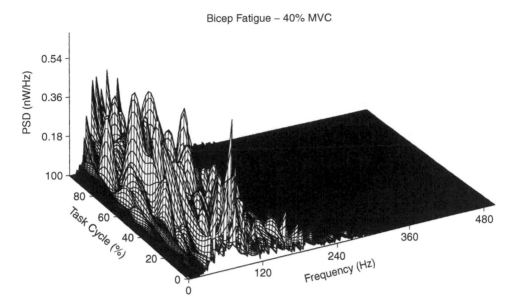

FIGURE 6.4
A power spectral plot of bicep fatigue in a fatiguing contraction at 40% MVC.

TABLE 6.1

Correlation between Force and Measured Variables for MVC Fatiguing Contractions

		Force (N)	Oxygenation (OD)	Blood Volume (OD)	Me (Hz)	EMG Amp. (mV)
Force (N)	Correlation	1.00	0.46	0.30	0.91	0.39
	Sig		0.001	0.001	0.001	0.001
Oxygenation (OD)	Correlation		1.00	−0.35	0.45	−0.19
	Sig			0.001	0.001	0.001
Blood Volume (OD)	Correlation			1.00	0.18	0.61
	Sig				0.001	0.001
Med. freq. (Hz)	Correlation				1.00	0.36
	Sig					0.001
EMG amp. (mV)	Correlation					1.00
	Sig					

6.3.5 Analysis of Variance with Repeated Measures Results

The summary tables of analysis of variance with repeated measures for the MVC fatiguing and 40% fatiguing contraction for all independent and dependent variables are presented in Table 6.3 and Table 6.4. The dependent variable of force and all independent variables (blood oxygenation, blood volume, EMG MF, and EMG amplitude) were significantly different at various task percentile levels within the subject ($p < .001$) and between subjects ($p < .001$) in the MVC trial. The latter observation indicates significant and progressive change with increasing task duration despite the effort of the subjects to maintain at a constant level. Identical results were obtained in 40% MVC fatiguing trials, where, in addition to the dependent variable of force, the independent variables blood oxygenation, blood volume, EMG amplitude, MF, oxygen uptake, ventilation, heart rate, RPE score,

TABLE 6.2

Correlation between Force and Measured Variables for 40% MVC Fatiguing Contractions

		Force (N)	Oxygenation (OD)	Blood Volume (OD)	Med.Freq (Hz)	EMG Amp (mV)	Heart Rate (bpm)	VO_2 (l/min)	Ventilation (l/min)	VAS (cm)	RPE	BPD Upper Arm
Force	Correlation	1.00	-0.41	-0.72	0.33	0.15	-0.33	-0.44	-0.39	-0.57	-0.42	-0.47
	Sig.		0.001	0.001	0.001	0.001	0.001	0.001	0.001	0.001	0.001	0.001
Oxygenation (OD)	Correlation		1.00	0.41	0.266	-0.03	-0.28	-0.20	-0.08	0.31	0.1	0.1
	Sig.			0.001	0.001	0.47	0.001	0.001	0.028	0.001	0.1	0.011
Blood volume (OD)	Correlation			1.00	-0.09	-0.3	0.32	0.56	0.54	0.63	0.47	0.56
	Sig.				0.014	0.001	0.001	0.001	0.001	0.001	0.001	0.001
Med. freq (Hz)	Correlation				1.00	-0.17	-0.56	-0.35	-0.27	-0.45	-0.5	-0.51
	Sig.					0.001	0.001	0.001	0.001	0.001	0.001	0.001
EMG amp (mV)	Correlation					1.00	0.04	-0.21	-0.27	-0.14	-0.14	-0.19
	Sig.						0.229	0.001	0.001	0.001	0.001	0.001
Heart rate (bpm)	Correlation						1.00	0.66	0.71	0.60	0.65	0.67
	Sig.							0.001	0.001	0.001	0.001	0.001
VO_2 (l/min)	Correlation							1.00	0.91	0.53	0.51	0.58
	Sig.								0.001	0.001	0.001	0.001
Ventilation (l/min)	Correlation								1.00	0.59	0.56	0.64
	Sig.									0.001	0.001	0.001
VAS (cm)	Correlation									1.00	0.923	0.87
	Sig.										0.001	0.001
RPE	Correlation										1.00	0.85
	Sig.											0.001
BPD upper arm	Correlation											1.00
	Sig.											

TABLE 6.3

Univariate ANOVA (with Repeated Measures)
Summary Table for MVC Fatiguing Contraction

S	F	Hypo-dF	Sig	
Force	2133.0	6	0.001	Within sub
	2896.7	1	0.001	Between sub
O2	759222.1	6	0.001	Within sub
	2831.1	1	0.001	Between sub
BV	1871.1	6	0.001	Within sub
	4133.4	1	0.001	Between sub
MF	5093.4	6	0.001	Within sub
	2799.7	1	0.001	Between sub
Ampl.	22266.2	6	0.001	Within sub
	1485.7	1	0.001	Between sub

TABLE 6.4

Univariate ANOVA (with Repeated Measures)
Summary Table for 40% MVC Fatiguing Contraction
Univariate for Subjects

	F	Hypo-dF	Sig	
Force	36187.2	6	0.001	Within
	793.6	1	0.001	Between
O2	7467.7	6	0.001	Within
	849.7	1	0.001	Between
Blood Volume	8587.0	6	0.001	Within
	44.5	1	0.001	Between
EMG ampl.	13019.9	6	0.001	Within
	8453.1	1	0.001	Between
MF	4054.4	6	0.001	Within
	5743.9	1	0.001	Between
VO2	1511.9	6	0.001	Within
	1650.1	1	0.001	Between
HR	31851.9	6	0.001	Within
	14618.7	1	0.001	Between
Vent.	22983.2	6	0.001	Within
	1023.2	1	0.001	Between
RPE	30255.0	6	0.001	Within
	1079.7	1	0.001	Between
VAS	611.9	6	0.001	Within
	559.7	1	0.001	Between

and VAS score showed significantly different magnitudes of these variables at different task cycles within the subject ($p < .001$) and between the subjects ($p < .001$). Thus, in both contractions, as the time progressed the fatigue clearly progressed.

6.3.6 MVC Force

With force as the outcome variable and EMG, MF, BV, and oxygen as independent variables, a linear model was fitted under repeated measure framework using SAS software. The fitted model is given by:

Force = −34.36 + (63.56 EMG) + (1.022 MF) − (167.12 BV) + (0.33 oxygen) with the Akaike Information Criterion (AIC) measure of goodness-of-fit being 496.9 and residual variance estimate being 478.35. All the independent variables were significant. Based on the absolute values of t-ratios, we can order MF, EMG amplitude, oxygen, and BV in terms of statistical significance of these variables in describing force. Under the presence of EMG amplitude, MF, oxygen, and the variable BV had a negative association with the force, while all other variables had positive association with force. To examine significance of each of the independent variables on force, a similar model was fitted with one independent variable at a time. The residual error variance estimates for EMG amplitude, median frequency, blood volume, and muscle oxygenation were 1125.74, 1707.92, 2748.59, and 2156.99, respectively, compared to 472.35. Based on the above analysis corresponding to models, each with one independent variable, one can rank these independent variables in terms of most significant variable in explaining force as EMG amplitude, MF, oxygenation, and blood volume using smaller residual variance criteria. The above ranking is also true based on correlation coefficients.

6.3.7   40% MVC Force

The set of independent variables for this case consists of blood volume, blood oxygenation, VO_2, ventilation, heart rate, EMG amplitude, EMG median frequency, visual analog scale, rate of perceived exertion, and body part discomfort rating. The final frequency model is given below.

Force = 140.92 + (53.17 EMG amplitude) – (41.9 BV) + (0.045 blood oxygenation) –
(21.15 VO_2) – (0.64 heart rate) + (0.68 RPE) +
(1.39 discomfort rating of the upper arm).

The order of importance of significant predictors was determined based on absolute values of t-statistics. The goodness-of-fit of the final model was evaluated using reduction in mean squared error. An 80% reduction in mean squared error with significant predictors was observed compared to the model without any significant predictors.

6.4   Discussion

Localized muscle fatigue has been considered to be of great significance in possible injury mechanisms (Kumar, 2001; Kumar and Narayan, 1998; Kumar et al., 2001). It is common knowledge and observation that the fatiguing muscles progressively deteriorate in performance. There are many possible sites where fatigue may set in and many possible mechanisms through which it may manifest itself. With respect to the possible sites, there may be fatigue at the neuron level, when it is simply unable to maintain the train of stimuli that carry the contraction command, or at the neuromuscular junction, due to depletion of neurotransmitters such as acetylcholine, or at the muscle membrane, which may show lack of responsiveness to the central stimuli. While each of these factors may possibly contribute to the phenomenon, the EMG recordings in this experiment do not tend to support this argument. With the onset and progression of fatigue, there is an initial reduction in signals, followed by an increment in amplitude. This clearly indicates that the electrophysiological mechanism of muscle contraction remains intact till the end. It is due to the significant decline in force that the synchronization of motor units occurs

TABLE 6.5

Regression Analysis for MVC Fatiguing Contraction

Variable	R	R2	DF	F	Beta	Sig
Oxygenation	0.458	0.209	1	241.26	58.7	0.001
Blood Volume	0.303	0.091	1	91.49	9.24	0.001
MF	0.910	0.828	1	4361.43	0.826	0.001
EMG	0.393	0.153	1	165.34	0.180	0.001

TABLE 6.6

Regression Analysis for 40% MVC Fatiguing Contractions

Variable	R	R2	dF	F	Beta	Sig.
Oxygenation	0.395	0.155	1	130.37	−60.365	0.001
Blood Volume	0.783	0.614	1	1119.93	−77.052	0.001
VO$_2$	0.518	0.268	1	332.90	−40.035	0.001
HR	0.422	0.177	1	196.25	−0.596	0.001
Vent	0.463	0.213	1	247.01	−1.141	0.001
MF	0.397	0.156	1	169.41	0.504	0.001
EMG	0.119	0.013	1	13.10	18.752	0.001
VAS	0.671	0.449	1	741.79	−4.850	0.001
RPE	0.542	0.293	1	376.63	−2.361	0.001

(Bigland-Ritchie et al., 1981) to maximize the electrical impact on the contractile phenomenon. With the low-level (40%) prolonged contraction presented in this study, there did not appear to be a precipitous decline in the EMG amplitude. This is clearly seen in the data presented here that the EMG amplitude had a correlation of only 0.39 and 0.11 in MVC and 40% fatiguing contractions, respectively. In these two conditions, the EMG amplitude accounted for only 15.3 and 1.3% of variance in predicting force in maximal and 40% MVC contractions (Table 6.5 and Table 6.6). Spectral compression, as measured by the decline in MF and the EMG signals, showed a high correlation with the force in the MVC fatiguing contraction ($r = 0.910$; $p < .001$), explaining 82.8% of variance in regression. It is unclear from this study whether the spectral compression is an adaptation on the part of the nervous control mechanism to maximize or maintain the phenomenon of contraction, or whether it is an adaptation to the declining force. In any event, the MF does not behave in the same manner with 40% fatiguing contraction as with MVC fatiguing contraction, showing a correlation coefficient of 0.397 ($p < .001$) and explaining only 15.6% of the variance in prediction of the force. This divergent behavior of spectral compression with different magnitudes of contraction remains enigmatic.

The physiology of muscle contraction is complex and is the resultant expression of nervous, metabolic, mechanical, and subjective factors. The primary energy source for muscles to contract is ATP, which breaks down and releases energy required for contraction. Since the muscles have only a finite amount of ATP stored, in any prolonged contraction it needs to be replenished by reconstitution of the compound in the presence of oxygen brought to the muscle by blood. Without the energy supply, the muscle will grind to a halt. It would, therefore, be logical that muscle blood volume and its level of oxygenation would adequately and faithfully mirror the degree of fatigue and decline in force. While these variables have shown a modest correlation ($r = 0.458$ and 0.303, respectively) in the MVC fatiguing contraction, they have accounted for only 20.9 and 9.1% of variance in force, respectively. In the 40% MVC fatiguing contraction, the blood oxygenation and blood volume, as measured by the NIRS optical density, demonstrate correlation coefficients of 0.395 and 0.783, explaining 15.5 and 61.4% of variance in force, respectively. Such varied responses for MVC and 40% MVC fatiguing contractions are puzzling. In either case, given such direct relationship between muscle contraction and the biochemical demands (e.g., ATP), the lack of better association with muscle force and muscle blood oxygenation and blood volume level remains unsatisfactory. The possibility that it is due to the experimental measuring device cannot be ruled out, in light of lack of concordance reported between Doppler and NIRS measurements (Hicks et al., 1999).

The measures of oxygen uptake, ventilation, and heart rate changes, which could be done only during 40% MVC fatiguing contractions, also revealed only modest correlations, ranging between $r = 0.42$ and 0.51, and explained between 17.7 and 26.8% of variance in force. This modest correlation and low level of predictability may be more acceptable for these variables, due to the fact that a single bicep brachii is a relatively small component in proportion to the whole body. The physiological demand changes in a small part of the body are likely to influence the whole-organism measurements only mildly. Perhaps it is for this reason that only a modest correlation between the central measure of RPE and the force ($r = 0.54$) and only 29.3% of the variance in force were accounted for by this variable. The RPE scale is based on the physiological cost and the heart rate (Borg, 1982). Interestingly, VAS score, a local indicator, had stronger correlation ($r = 0.67$) and accounted for 44.9% of variance in the force. This score was based on the difficulty individuals felt in maintaining the level of contraction. This variable may have had an integrative and summative effect on local factors.

The issue of considering force decline as the gold standard for fatigue may be questioned as well. The observation of decline in force with prolonged contraction is unmistakable. Similarly, local discomfort and soreness with such prolonged contraction is also an invariable experience. Despite electrophysiological, metabolic, mechanical, and subjective linkages with fatigue, none of the variables alone have proved to be a reliable index of the phenomenon. For this study, it would appear that the decline in MF and the EMG signals in MVCs and perhaps high-level contractions may be one of the better predictors. In low-level contraction, however, blood volume and VAS scores may be more closely related to fatigue. In the ergonomic literature various individual variables, as measured in the current experiment, have been used as indicators of fatigue. However, the current study does not support the use of any single variable as a valid and reliable index of fatigue. Instead, groups of variables, as described before, account for greater amounts of variance. This common but complex phenomenon eludes convenient indexing even today.

References

Bigland-Ritchie, B., Donovan, E.F., and Roussos, C.S. 1981. Conduction velocity and EMG power spectrum changes in fatigue of sustained maximal efforts, *J. Appl. Physiol.*, 51, 1300–1305.

Borg, G. 1982. Physiological bases of perceived exertion, *Med. Sci. Sports Exerc.*, 14, 377–381.

Chaffin, D. 1973. Localized muscle fatigue definition and measurement, *J. Occup. Med.*, 15, 346–354.

De Blasi, R.A., Cope, M., Elwell, C., Safoue, F., and Ferrasi, M. 1993. Noninvasive measurement of human forearm oxygen consumption by near infrared spectroscopy, *Eur. J. Appl. Physiol.*, 67, 20–25.

De Blasi, R.A., Sfareni, R., Pietranico, B., Mega, A.M., and Ferrari, M. 1996. Noninvasive measurement of brachioradial muscle VO_2–blood flow relationship during graded isometric exercise, *Adv. Exp. Med. Biol.*, 388, 293–298.

De Luca, C. 1985. Myoelectric manifestations of localized muscle fatigue in humans, *Crit. Rev. Biomed. Eng.*, 11, 251–279.

Grandjean, E. 1988. *Fitting the Task to the Man*, 4th ed. (London: Taylor and Francis).

Hagg, G. 1981. Electromyographic fatigue analysis based on the number of zero crossings, *Pflugers Arch.*, 391, 78–80.

Hamoaka, T., Iwane, H., Shimomitsu, T., Katsumura, T., Murase, N., Nishio, S., Osada, T., Kurosawa, Y., and Chance, B. 1996. Noninvasive measures of oxidative metabolism on working human muscles by near infrared spectroscopy, *J. Appl. Physiol.*, 81, 1410–1417.

Hicks, A., McGill, S., and Hughson, R.L. 1999. Tissue oxygenation by near infrared spectroscopy and muscle blood flow during isometric contractions of the forearm, *Can. J. Appl. Physiol.*, 24, 216–230.

Kumar, S. 2001. Theories of musculoskeletal injury causation, *Ergonomics*, 44, 17–47.

Kumar, S. and Narayan, Y. 1998. Spectral parameters of trunk muscles during fatiguing isometric axial rotation in neutral posture, *J. Electromyogr. Kinesiol.*, 8, 257–267.

Kumar, S., Narayan, Y., Stein, R.B., and Snyders, C. 2001. Muscle fatigue in axial rotation of the trunk, *Int. J. Ind. Ergon.*, 28, 113–125.

Kwanty, E., Thomas, D.H., and Kwanty, H.G. 1970. An application of signal processing techniques to the study of myoelectric signals, *IEEE Trans. Biomed. Eng.*, 17, 303–312.

Mannion, A.F. and Dolan, P. 1994. Electromyographic median frequency changes during isometric contraction of the back extension to fatigue, *Spine*, 19, 1223–1229.

McCully, K.K. and Hamoaka, T. 2000. Near infrared spectroscopy: what can it tell us about oxygen saturation in skeletal muscle? *Exerc. Sport Sci. Rev.*, 28, 123–127.

Murthy, G., Kahan, N.J., Hargens, A.R., and Rempel, D. 1997. Forearm muscle oxygenation decreases with low levels of voluntary contraction, *J. Orthop. Res.*, 15, 507–511.

Ng, J.K.F., Richardson, C.A., and Jull, G.A. 1997. Electromyographic amplitude and frequency changes in the iliocostalis lumborum and multifidus muscles during a trunk holding test, *Phys. Ther.*, 77, 954–961.

Rohmert, W. 1960. Ermittlung von Erholungspausen fur Statistiche Arbeit des Menschen, *Int. Z. Angew. Physiol.*, 18, 123–124.

Rohmert, W. 1960. Arbeitsmedizin, sozial medizin, *Arbeitshygiene*, 22, 118–123.

Roy, S. and De Luca, C. 1996. Surface electromyographic assessment of low back pain, in *Electromyography in Ergonomics*, S. Kumar and A. Mital, Eds. (London: Taylor and Francis), pp. 259–295.

Stuten, F.B. and De Luca, C.J. 1981. Frequency parameters of the myoelectric signal as a measure of muscle conduction velocity, *IEEE Trans. Biomed. Eng.*, 28, 515–523.

Van Dieen, J., Toussaint, H., Thissen, C., and Van De Van, A. 1993. Spectral analysis of erector spinae EMG during intermittent isometric fatiguing exercise, *Ergonomics*, 36, 407–414.

Van Dieen, J., Boke, B., Oosterhuis, W., and Toussaint, H.M. 1996. The influence of torque and velocity on erector spinae muscle fatigue and its relationship to changes of electromyogram spectrum density, *Eur. J. Appl. Physiol. Occup. Physiol.*, 72, 310–315.

7

Psychophysical Aspects of Muscle Strength

Patrick G. Dempsey

CONTENTS

ABSTRACT The measurement of human muscle strength is of fundamental importance to ergonomic research and practice. Psychophysics has been utilized as one approach to measuring muscle strength, because the strength of sensation of exerting muscle force increases with increasing force according to the psychophysical power law. One of the most widely used applications of psychophysics to ergonomics is the collection of maximum voluntary contraction data, which is often used to design workplaces, tools, and equipment. Another methodology with increasing application is using psychophysical methods to have workers estimate the magnitude of force production while performing tasks at the workplace. Further areas of suggested research are also discussed.

7.1 Introduction

One of the fundamental ergonomics job design and assessment concepts is that of comparing the demands of work with the respective capabilities and limitations of the human body. A comprehensive assessment can include the capabilities and limitations of body subsystems, including the musculoskeletal, cardiovascular, and central nervous systems. In the case of muscular exertion, assessing maximal exertion capabilities and the magnitude of exertions while performing tasks have been key components of assessing the degree of compatibility between strength capabilities and job demands.

Databases of maximal strength typically provide the human capabilities information for practitioners, although these data are sometimes collected from worker populations. Various techniques are used to assess exertion levels while performing tasks. Oftentimes, ergonomic guidelines for muscular exertion are expressed as some percentage of maximum strength, depending on the frequency and duration of the activity. Psychophysics has contributed to methods both of collecting strength data and of assessing exertions at the workplace.

This chapter will begin with an overview of the basic theories and principles of psychophysics and how these principles are related to generating and measuring muscular force production. Inherent in the term psychophysics are both psychological and physical components. The applications of psychophysics to ergonomic problems will be discussed, particularly with respect to collecting strength data and using psychophysical estimates of force production to assess upper-extremity-intensive tasks. Finally, future directions for research of the psychophysics of muscular exertion and assessment will be suggested.

7.2   Psychophysics and Force Production

Psychophysics is a branch of psychology that deals with the relationships between stimuli and sensations. These relationships have been described by the psychophysical power law developed by Stevens (1960), where the psychological magnitude (sensation) ψ grows as a power function of the physical magnitude ϕ (stimulus) in the following manner:

$$\psi = k\phi^n$$

The constant k depends on the units of measure, and the exponent n has a value that varies for different sensations. The value of n has been found to range from less than 1 for stimuli such as smell and brightness to as high as 3.5 for electric shock (Stevens, 1960).

When considering muscle exertion, there are several physiologic entities that can generate a sensory stimulus. The exertion of force excites the kinesthetic, touch, and pressure senses (Stevens and Mack, 1959). Several of these mechanisms will be briefly discussed.

Both the Golgi tendon organs and muscle spindles are capable of providing feedback when muscles are contracted. The Golgi tendon organs are located in tendons near the point of attachment of the muscle and are arranged in series with the muscle fibers. These organs primarily react to tension in the muscle (as opposed to passive stretching) by discharging nerve impulses that excite inhibitory interneurons. The interneurons then inhibit the motor neurons of the contracting muscle, essentially limiting the force production.

The muscle spindles, another type of proprioceptor, are most abundant in the muscles of the arms and legs. The primary role of the muscle spindles is to sense stretch of the muscle. If the muscle is stretched sufficiently, the spindles can initiate a stronger contraction to reduce the degree of stretch.

Oftentimes, muscle exertions are carried out for the purpose of exerting force on something in the external environment. The touch sense will be activated by mechanical stimulation of the skin. If the force is high enough or the surface force exerted on the skin is irregular or results in high pressure, the pain receptors (nociceptors) may be activated.

Some of the early studies of the psychophysics of force exertion were concerning grip force exertion. Stevens and Mack (1959) used ratio production, magnitude production, and magnitude estimation to investigate grip force production. For the ratio production experiment, subjects were asked to exert a light, moderate, or heavy grip force and then to exert a force that they believed was either half or twice as large. For the magnitude production experiment, subjects were given a number and asked to exert grip force proportional to the number. Subjects were "calibrated" by exerting a "moderate" force and being told that this force was a 10. Subjects were then given the numbers 3, 6, 10, 20, and 30 and asked to produce forces in proportion relative to the 10. For magnitude estimation, subjects exerted a 17-lb force that was indicated as being a "10." Subjects then estimated the magnitude of pulls between 5 and 35 lb. The experiment was replicated using 15 lb as the calibration weight, and forces between 4 and 40 lb for the magnitude estimation trials.

For the ratio production study, Stevens and Mack (1959) found that the average ratio of the larger force to the lower force that was supposed to be a 2 to 1 ratio was, on average, a ratio of 1.4 to 1. For the magnitude production study, apparent force grew as approximately the square of the physical force. Similar results were obtained for magnitude estimation, but rather than the function growing as a square, the exponent was about 1.6.

Two additional experiments were performed by Stevens and Mack (1959) where subjects matched the intensity of grip force to the intensity of stimuli of different magnitudes involving weight lifted and pressure on the palm. The results are discussed below, as additional studies using the same protocol were conducted by Stevens et al. (1960).

Stevens et al. (1960) investigated the relationships between the intensity of five criterion stimuli (electric shock, vibration, white noise loudness, 1000-Hz tone loudness, and white light) and force of handgrip intended to match the magnitude of the criterion stimuli presented. The exponents of the power function ranged from 0.33 for brightness of white light to 3.5 for electrical shock. For weight lifted, the exponent was 1.45.

7.3 Maximum Voluntary Contraction

One of the most basic uses of psychophysical techniques by ergonomists has been that of collecting maximum voluntary contraction (MVC) data for different types of exertions ranging from individual joint strengths to more complex strengths such as lifting. These MVC data are often used to define human capabilities and limitations for tasks with specific strength requirements. For example, in emergency situations (such as opening an emergency door release), it is necessary to know the maximum strength capabilities of the relevant population. For repetitive tasks, knowledge of strength distributions is used so that tasks do not exceed a certain percentage of MVC to avoid fatigue, etc. MVC data can be collected for both static and dynamic tasks.

Like the psychophysical approach to manual materials handling (e.g., Snook and Ciriello, 1991; Snook and Irvine, 1967), collecting MVC data is a variant of the psychophysical technique called magnitude production. Originally, this technique was used by giving subjects a number and having them produce a sensation they felt matched the magnitude of the number (Stevens, 1975). In many materials-handling studies that utilize psychophysics, subjects are asked to pick a load or force (stimulus) that corresponds to them not being overheated, tired, or similar nomenclature (Ayoub and Dempsey, 1999). Thus, the "magnitude" they are presented is related to the level of exertion. When collecting MVC data, subjects are asked to produce the maximum muscular exertion they are willing to produce. For those interested in more details about the application of psychophysics to materials handling, see the chapter by Gallagher et al. (Chapter 8).

Because of the "voluntary" nature of this type of test, care is warranted when utilizing the procedure. Some of the early ergonomic guidelines for this type of test were presented by Caldwell et al. (1974) and Chaffin (1975). Some of the factors related to psychophysics, and specifically the expression or interpretation of "maximum," will be discussed briefly.

Chaffin (1975) recommended that subjects should be given instructions in an objective manner that does not involve emotional appeal. Monetary incentives, fear, noise, spectators, or other factors that could emotionally influence subjects should be avoided. It is not difficult to imagine how these factors could introduce variation in how subjects interpret "maximum." Kroemer (1999) lists additional sources of variation in maximal exertion, including arousal of ego involvement, drugs, hypnosis, and verbal encouragement. Interestingly, the author has observed considerable variation in the amount of verbal encouragement different experimenters utilize. Because of the inherently subjective nature of these tests, care must be taken so that the results are reliable.

One issue relevant to psychophysical applications to collecting MVC data is whether or not the apparatus is designed to eliminate pain or discomfort during the data collection. If the apparatus is designed in a manner such that the nociceptors are excited during data collection, the pain may lead to less-than-optimal strength values. Thus, handles and surfaces should be designed to minimize pain and discomfort during data collection. For example, a small diameter handle is not suggested for collecting pulling force data, as the discomfort may lead to less-than-maximal values.

7.4   Psychophysical Estimates of Exertion Magnitude

One application of psychophysics that has been popular, particularly among ergonomics practitioners, is that of using psychophysical estimates of force production to assess the force requirements of upper-extremity-intensive tasks. It remains difficult to obtain accurate measurements of force exertion magnitudes while workers perform tasks. Direct measurement techniques, such as electromyography, are difficult to apply in the workplace. Electromyography, as an example, requires a complex calibration to relate the electromyography signals to force magnitude. Although this can be performed for static postures with reasonable accuracy, most occupational tasks are dynamic in nature and are not readily analyzed. This has led to practitioners asking workers to exert a force (often grip or pinch force) on a dynamometer that they believe is equal to the force required by the task.

The problems, or perhaps difficulty, associated with direct measurement led Drury (1987) to recommend that, when electromyography was impractical due to work conditions or the lack of expertise, workers could be asked to replicate forces on a dynamometer. This recommendation was based upon the results of Yonda (1985) that people can "reproduce forces very accurately" (Drury, 1987). This technique has widespread use amongst practitioners, due the practicality and the lack of the requirement for expensive equipment.

Recently, Casey et al. (2002) performed an investigation of the ability of subjects to provide psychophysical estimates of grip force production during three simulated tasks. A screwdriver task, ratchet task, and lift-and-carry task were simulated for four conditions representing graded force requirements. This was accomplished by increasing the torque requirements for the screwdriver and ratchet tasks and increasing weight for the carry task. The weights were attached to a cylindrical handle so that grip force was required. The results indicated that subjects were quite accurate, on average, in replicating average grip forces. For different task conditions and subject experience level, the average errors were between −15.8 and 6.7%. Peak forces were largely underestimated, with average errors between −54 and −39%. Subjects were asked to "grip the handle with the same force you just used while you performed the task," and different instructions could have resulted in more accurate peak assessments. There was large between-subject variation in accuracy; thus it is recommended that practitioners use this approach cautiously when estimates are obtained from one or a few workers. Future studies may examine the effects of different instructions to see if peak forces can be accurately assessed.

Recently, the American Conference of Governmental Industrial Hygienists (2001) has published a threshold limit value for repetitive upper-extremity-intensive tasks, called the Hand Activity Level (HAL). The methodology requires assessment of Hand Activity Level, which is based on the distribution of work and recovery periods and on normalized peak force exertion. Peak force is defined as the peak force required by the task or job divided by the individual's or population's peak force capability.

The American Conference of Governmental Industrial Hygienists (2001) suggested several methodologies for assessing hand force, including worker ratings, observer ratings, biomechanical analyses, force gauges, and electromyography. Of these, the worker ratings rely on psychophysical techniques. The method suggested was using the Borg CR-10 scale [see Borg (1998) for a comprehensive review of perceived exertion and pain scales]. This scale can be used to rate pain and exertion with verbal anchors ranging from 0 ("Nothing at all") to 10 ("Extremely strong 'Max P'"). In the case of HAL, the worker is asked to rate the peak force exerted while performing a job. The rating scale is assumed to parallel the percentage of maximum muscle strength exerted by a scale of 1:10. Although Borg's scales have been widely used in ergonomics research and practice, the application to assessing percentage of maximum strength a task requires has not been validated or studied extensively.

7.5 Future Directions

There are several areas of research related to the psychophysics of muscle exertion and assessment that would be valuable, particularly with respect to ergonomics practice. Research that has been conducted on psychophysical ratings of perceived exertion and pain has shown that, with proper application, fairly strong relationships between workload and pain, and the perception of the magnitude of the workload or pain can be established (e.g., Borg, 1998). Basic research on the ability to rate the level of exertion could lead to very useful tools for the practitioner. As was mentioned earlier, direct measurement of force exertion continues to be difficult in the workplace and is often expensive and requires substantial expertise. The work of Casey et al. (2002) and the work of Yonda (1985), cited by Drury (1987), suggest that fairly accurate estimates of the magnitude of muscular exertion can be obtained through psychophysical estimates. Advancing these approaches and better defining how well the techniques work in different situations is warranted, however, as these studies were not extensive. The effect of instruction is likely to be very important, and basic issues associated with the way these techniques are used need to be resolved. The author is aware of many practitioners who use these techniques, and more knowledge concerning the reliability and validity for different populations and tasks is needed.

Another area of research that has potential for substantial impact in the workplace is the use of psychophysical techniques to assess the strength of injured workers, particularly those with musculoskeletal disorders. There is little knowledge of the strength decrements in injured populations, and few methods exist for assessing strength capacity of injured workers. One example of this type of research is the study of acceptable maximum effort by Khalil et al. (1987). Acceptable maximum effort was defined as the highest level of voluntary effort, solicited through various strength measurement protocols, that back pain patients were able to achieve without unacceptable levels of pain. The method was found to be reliable, and the authors suggested the use of the technique as a means to assess functional abilities. Such research could help match the reduced capacities of injured workers to jobs that do not exceed these capacities.

References

American Conference of Governmental Industrial Hygienists. 2001. *Threshold Limit Values for Chemical Substances and Physical Agents: Biological Exposure Indices* (Cincinnati, OH: ACGIH).

Ayoub, M.M. and Dempsey, P.G. 1999. The psychophysical approach to manual materials handling task design, *Ergonomics*, 42, 17–31.

Borg, G. 1998. *Borg's Perceived Exertion and Pain Scales* (Champaign, IL: Human Kinetics).

Caldwell, L.S., Chaffin, D.B., Dukes-Dobos, F.N., Kroemer, K.H., Laubach, L.L., Snook, S.H., and Wasserman, D.E. 1974. A proposed standard procedure for static muscle strength testing, *Am. Ind. Hyg. Assoc. J.*, 34, 201–206.

Casey, J.S., McGorry, R.W., and Dempsey, P.G. 2002. Getting a grip on grip force estimates: a valuable tool for ergonomic evaluations, *Prof. Saf.*, 47, 18–24.

Chaffin, D.B. 1975. Ergonomics guide for the assessment of human static strength, *Am. Ind. Hyg. Assoc. J.*, 35, 505–511.

Drury, C.G. 1987. A biomechanical evaluation of the repetitive motion injury potential of industrial jobs, *Sem. Occup. Med.*, 2, 41–49.

Khalil, T.M., Goldberg, M.L., Asfour, S.S., Moty, E.A., Rosomoff, R.S., and Rosomoff, H.L. 1987. Acceptable maximum effort (AME): A psychophysical measure of strength in back pain patients, *Spine*, 12, 372–376.

Kroemer, K.H.E. 1999. Assessment of human muscle strength for engineering purposes: a review of the basics, *Ergonomics*, 42, 74–93.

Snook, S.H. and Ciriello, V.M. 1991. The design of manual handling tasks: revised tables of maximum acceptable weights and forces, *Ergonomics*, 34, 1197–1213.

Snook, S.H. and Irvine, C.H. 1967. Maximum acceptable weight of lift, *Am. Ind. Hyg. Assoc. J.*, 28, 322–329.

Stevens, J.C. and Mack, J.D. 1959. Scales of apparent force, *J. Exp. Psychol.*, 58, 405–413.

Stevens, J.C., Mack, J.D., and Stevens, S.S. 1960. Growth of sensation on seven continua as measured by force of handgrip, *J. Exp. Psychol.*, 59, 60–67.

Stevens, S.S. 1960. The psychophysics of sensory function, *Am. Sci.*, 48, 226–253.

Stevens, S.S. 1975. *Psychophysics: Introduction to Its Perceptual, Neural, and Social Prospects* (New York: John Wiley & Sons).

Yonda, R.A. 1985. An Investigation of the Human Ability to Replicate Task-Produced Forces on a Load Cell Apparatus, unpublished M.S. thesis, State University of New York at Buffalo.

8

Isometric, Isoinertial, and Psychophysical Strength Testing: Devices and Protocols

Sean Gallagher, J. Steven Moore, and Terrence J. Stobbe

CONTENTS

ABSTRACT Many jobs in industry place severe demands on the worker's musculoskeletal system — demands that may approach or exceed worker voluntary strength capabilities. There is evidence to suggest that such jobs increase the likelihood that the worker will experience a musculoskeletal disorder. For this reason, a great deal of effort has recently been focused on the development of methods to evaluate muscular strength capabilities of workers, both for purposes of ergonomic job design and for the development of worker selection procedures. However, the necessity of using indirect measures of muscular strength makes its assessment quite complex, which has sometimes led to confusion and misunderstanding regarding appropriate uses of strength measurement techniques. The purpose of this chapter is to provide information regarding the appropriate procedures for the measurement and reporting of strength test results for three common measurement techniques used in ergonomics (isometric, isoinertial, and psychophysical). It is hoped that the information contained in this chapter will provide the reader with a better understanding of the advantages, disadvantages, caveats, and limitations associated with the use of these strength assessment techniques.

8.1 Ergonomic Relevance

An understanding of the muscular strength characteristics of both individuals and populations can assist the ergonomist in developing appropriate interventions to reduce musculoskeletal disorder risk. Numerous strength assessment techniques exist for this purpose. However, muscular strength is a complicated function, and failure to adhere to established principles of measurement and analysis can lead to gross misinterpretation of test results. This chapter describes three strength assessment techniques commonly used by ergonomists (i.e., isometric, isoinertial, and psychophysical) and discusses issues important to proper assessment of muscular strength for each of these techniques.

8.2 Introduction

Muscular strength is a complicated function that can vary greatly depending on the methods of assessment. As a result, there is often a great deal of confusion and misunderstanding of the appropriate uses of strength testing in ergonomics. It is not uncommon to see these techniques misapplied by persons who are not thoroughly familiar with the caveats and limitations inherent with various strength assessment procedures. The purposes of this chapter are: (a) to familiarize the reader with three common techniques of strength assessment used in ergonomics (isometric, isoinertial, and psychophysical) and (b) to describe the proper applications of these techniques in the attempt to control work-related musculoskeletal disorders (WMSDs) in the workplace.

This chapter contains three sections, one for each of the strength measurement techniques listed above. Each section describes the strength measurement technique and reviews the relevant published data. Equipment considerations and testing protocols are described, and the utility of the tests in the context of ergonomics is also evaluated. Finally, each section concludes with a discussion of the measurement technique with regard to the Criteria for Physical Assessment in Worker Selection (Chaffin and Andersson, 1991). In this discussion, each measurement technique is subjected to the following set of questions:

1. Is it safe to administer?
2. Does it give reliable quantitative values?
3. Is it related to specific job requirements?
4. Is it practical?
5. Does it predict risk of future injury or illness?

It is hoped that this chapter will provide a resource that can be used to better understand and properly apply these strength assessment techniques in the effort to reduce the risk of WMSDs.

8.3 Isometric Strength

Isometric strength is defined as the capacity to produce force or torque with a voluntary isometric [muscle(s) maintain(s) a constant length] contraction. The key thing to understand about this type of contraction and strength measurement is that there is no body movement during the measurement period. The tested person's body angles and posture remain the same throughout the test.

Isometric strength has historically been the one most studied and measured. It is probably the easiest to measure and the easiest to understand. Some strength researchers feel that isometric strength data may be difficult to apply to some "real life" situations, because in most real circumstances people are moving — they are not static. Other researchers counter that it is equally difficult to determine the speed of movement of a person or group of persons doing a job (they all move in their own unique manners and at their own speed across the links and joints of the body). Thus, dynamic strength test data collected on persons moving at a different speed and/or in a different posture from the "real world" condition will be just as hard to apply. In truth, neither is better — they are different measurements, and both researchers and users should collect/use data which they understand and which fits their application.

8.3.1   Workplace Assessment

When a worker is called upon to perform a physically demanding lifting task, moments (or torques) are produced about various joints of the body by the external load (Chaffin and Andersson, 1991). Often these moments are augmented by the force of gravity acting on the mass of various body segments. For example, in a biceps curl exercise, the moment produced by the forearm flexors must counteract the moment of the weight held in the hands, as well as the moment caused by gravity acting on the center of mass of the forearm. In order to successfully perform the task, the muscles responsible for moving the joint must develop a greater moment than that imposed by the combined moment of the external load and body segment. It should be clear that for each joint of the body, there exists a limit to the strength that can be produced by the muscle to move ever-increasing external loads. This concept has formed the basis of isometric muscle strength prediction modeling (Chaffin and Andersson, 1991).

The following procedures are generally used in this biomechanical analysis technique. First, workers are observed (and usually photographed or videotaped) during the performance of physically demanding tasks. For each task, the posture of the torso and the extremities are documented at the time of peak exertion. The postures are then re-created using a computerized software package, which calculates the load moments produced at various joints of the body during the performance of the task. The values obtained during this analysis are then compared to population norms for isometric strength obtained from a population of industrial workers. In this manner, the model can estimate the proportion of the population capable of performing the exertion, as well as the predicted compression forces acting on the lumbar disks resulting from the task.

Figure 8.1 shows an example of the workplace analysis necessary for this type of approach. Direct observations of the worker performing the task provide the necessary data. For example, the load magnitude and direction must be known (in this case, a 200-N load acting downward), as well as the size of the worker, the postural angles of the body (obtained from photographs or videotape), and whether the task requires one or two hands. Furthermore, the analysis requires accurate measurement of the load center relative to the ankles and the low back. A computer analysis program can be used to calculate the strength requirements for the task and the percentage of workers who would be likely to have sufficient strength capabilities to perform it. Results of this particular analysis indicate that the muscles at the hip are most stressed, with 83% of men having the necessary capabilities, but only slightly more than half of women would have the necessary strength in this region. These results can then be used as the basis for determining those workers who have adequate strength for the job. However, such results can also be used as ammunition for recommending changes in job design (Chaffin and Andersson, 1991).

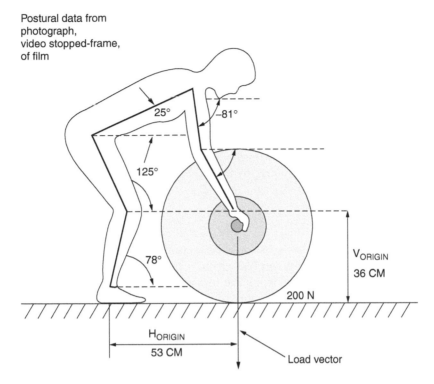

FIGURE 8.1
Postural data required for analysis of joint moment strengths using the isometric technique. (With permission of The Regents of The University of Michigan.)

8.3.2 The Isometric Testing Protocol

The basic testing protocol for isometric strength testing was developed by Caldwell et al. (1974) and published in an AIHA Ergonomics Guide by Chaffin (1975). The protocol outlined herein includes additional information determined by researchers since that time. When conducting isometric testing, there are a number of factors that must be considered and controlled (if possible) to avoid biased results. These factors include the equipment used to make the measurements, the instructions given to the person tested, the duration of the measurement period, the person's posture during the test, the length of the rest period between trials, the number of trials a person is given for each test, the tested person's physical state at the time of testing, the type of postural control used during the tests, and the environmental conditions during the test.

8.3.2.1 Test Duration

The length of an isometric strength test can impact the result in two ways. If it is too long, the subjects will fatigue and their strength scores will decline. If it is too short, the subjects will not reach their maximum force level before the test is terminated. The existing AIHA Guide suggests a four-second test, with the score being the average strength displayed during the second through fourth seconds. The appropriate three-second period can be determined as follows.

If the measuring equipment has the capability, collect strength data by having the subjects begin their contraction with the equipment monitoring the force until some preselected threshold is reached (usually 20 to 30% below the expected maximum force for the person and posture). Have equipment wait one second, and then have the equipment average the displayed force for the next three seconds. This is easily done with computerized systems.

If the equipment does not have the above capability, then have the person tested begin the test and gradually increase force over a one-second period. The force should be measured and averaged over the next three seconds. In complex whole-body tests, where multiple functional muscle groups are involved, it may take individuals a few seconds to reach their maximum. Under these conditions, the data collectors must adjust the premeasurement time interval accordingly, and they must carefully monitor the progress of the testing to insure that they are, in fact, measuring the maximal force during the three-second period.

8.3.2.2 Instructions

The instructions to the person tested should be factual, include no emotional appeals, and be the same for all persons in a given test group. This is most reliably accomplished with standardized written instructions, since the test administrator's feelings about the testee or the desired outcome may become evident during verbal instruction.

The following additional factors should also be considered. The purpose of the test, the use of the test results, the test procedures, and the test equipment should be thoroughly explained to the persons tested. Generally, the anonymity of the persons tested is maintained, but if names may be released, the tested person's written permission must be obtained. Any risks inherent to the testing procedure should be explained to the persons tested, and an informed consent document should be provided to, and signed by, all participating persons. All test participants should be volunteers.

Rewards, performance goals, encouragement during the test (for example "pull, pull, pull, you can do it, etc."), spectators, between-person competition, and unusual noises will all affect the outcome of the tests and must be avoided. Feedback to the tested person should be positive and qualitative. Feedback should not be provided during the test exertion, but may be provided after a trial or test is complete. No quantitative results should be provided during the testing period, because they may change the person's incentive and, thus, the test result.

To the tested person, a four-second maximal exertion seems to take a long time. During the test, feedback in the form of a slow four count or some other tester–testee agreed upon manner should be provided so the tested person knows how much longer a test will last.

8.3.2.3 Rest Period Length

Persons undergoing isometric strength testing will generally be performing a series of tests, with a number of trials for each test. Under these conditions, a person could develop localized muscle fatigue, and this must be avoided, since it will result in underestimating strength. Studies by Schanne (1972) and Stobbe (1982) have shown that a minimum rest period of two minutes between trials of a given test or between tests is adequate to prevent localized muscle fatigue. The data collector must be alert for signs of fatigue, such as a drop in strength scores as a test progresses. The person tested must be encouraged to report any symptoms of fatigue, and the rest periods should be adjusted accordingly. Whenever possible, successive tests should not stress the same muscle groups.

8.3.2.4   Number of Trials for Each Test

The test-retest variability for this type of testing is about 10%. It is higher for people with limited experience either with isometric testing or with forceful physical exertion in general. In addition, these people will often require a series of trials of a test to reach their maximum. The use of a single trial of a test will generally underestimate a person's maximum strength, and may underestimate it by more than 50%. A two-trial protocol results in less of an underestimate, but it may still exceed 30% (Stobbe and Plummer, 1984).

For this reason, the preferred approach to determining the number of trials for each test is to make the choice on the basis of performance. Begin by having the subject perform two trials of the test. The two scores are then compared, and if they are within 10% of each other, the highest of the two values is used as the estimate of the person's maximal strength, and you proceed to the next test. If the two values differ by more than 10%, additional trials of the same test are performed until the two largest values are within 10% of each other. Using this approach, Stobbe and Plummer (1984) averaged 2.43 trials per test across 67 subjects performing an average of 30 different strength tests. In any case, a minimum of two trials is needed for each test.

8.3.2.5   When to Give Tests

A person's measured strength is, for a variety of reasons, somewhat variable. It will not be constant over time, or over a workday. However, in the absence of specific muscle strength training, it should remain within a relatively narrow range. It is generally higher at the beginning of a workday than at the end. The fatigue-induced strength decrement will vary from person to person and will depend on the nature of the work done during the day. A person who performs repetitive lifting tasks all day can be expected to have a large lifting strength decrement over a workday, whereas a sedentary worker should have little or no decrement. Based on these results, the fairest evaluation of a person's maximum strength can be done at the beginning of, or at least early in, a workday.

8.3.2.6   Test Posture

Measured strength is highly posture dependent. Even small changes in the body angles of persons being tested and/or changes in the direction of force application can result in large changes in measured strength. When collecting strength data, a researcher should first determine what type of data is sought, and then one or more strength tests that will provide that specific type of data should be designed. If, for example, the test is being done to determine whether people are physically fit for a job, the test posture should emulate, to the extent possible, the posture required on the job.

Once the test posture has been determined, the researcher must insure that the same posture is used on each trial of the test. The researcher must monitor the test to insure that the person's posture does not change during the test. If these things are not done, the test results will be erratic and may seriously overestimate or underestimate the person's actual maximal strength.

8.3.2.7   Restraint Systems

Restraint systems are generally used either to confine a person to the desired test posture, or to isolate some part of the tested person's body so that a specific muscle group (or groups) can be tested (see Figure 8.2). In addition, restraint systems help to assure that all persons participating in a given study will be performing the same test. The type and

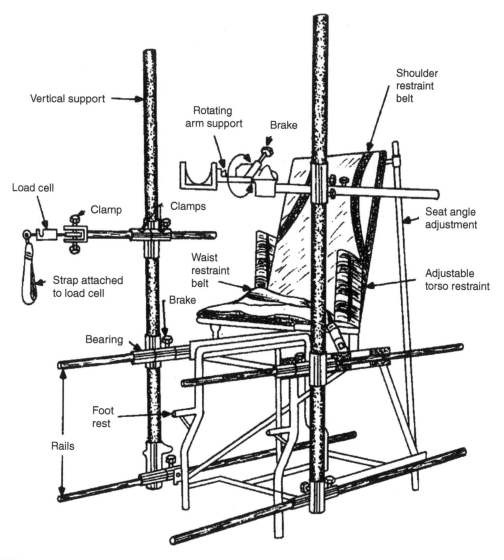

FIGURE 8.2
Example of a test fixture designed to restrain various body segments during isometric strength testing. (From *Occupational Biomechanics*, Chaffin, D.B. and G.B.J. Andersson, © 1991 by John Wiley & Sons. Reprinted by permission of John Wiley & Sons.)

location of restraint system used can have a major impact on test results. Similarly, the lack of a restraint system can allow the posture to vary or allow the use of the wrong or additional muscle groups, both of which will impact test results.

Any restraint system used should be comfortable. It should be padded in a manner that prevents local tissue stress concentrations during the test. It should be positioned so that the correct muscle group(s) and posture(s) are used and maintained. This latter item often requires some experimentation to correctly achieve.

For many strength tests, the use of a restraint system will be necessary if consistent and meaningful results are to be achieved. Researchers reporting strength testing results should describe the restraints used and their location in detail, so that other researchers and persons applying their data will be able to interpret it correctly. The nonuse of restraints should also be reported.

8.3.2.8   Environmental Conditions

The environmental conditions selected for the testing periods should be appropriate to the purpose of the test. For most testing, the environmental conditions found in a typical office building or medical department will be acceptable. In some cases, the effects of the environment on measured strength or physical performance must be determined, and then appropriate conditions can be established (e.g., worksites requiring exposure to hot or cold temperature extremes).

8.3.2.9   Equipment

Isometric strength testing equipment has not been standardized. Any equipment that has the capability to perform the necessary timing and averaging described above under "test duration" is probably acceptable. Today, this varies from dedicated force measurement devices such as the force monitor developed in the 1970s at University of Michigan, to a force transducer coupled to a PC via an A to D converter and managed by appropriate software, to complex multiple-mode strength-measuring devices manufactured by companies like Cybex, Chattex, Loredan, and Isotechnologies. The associated prices vary from one thousand plus dollars to as high as fifty to one hundred thousand dollars.

Equipment price is not the issue. Rather, it is equipment function that is at issue. Researchers should select or build equipment suited to their needs. Researchers must also understand what is happening inside the device (and its associated software) that they are using, so that the data they collect can be properly interpreted.

The human–equipment interface is another matter that can impact the test results. The interface must be appropriate to the task measured, it should be comfortable (unless discomfort effects are being studied), and it should give the person tested a sense of security about the test. Persons will generally be providing a maximal exertion in a situation where there is no movement. If they fear that the testing system may fail or move unexpectedly, they will not give a maximal performance. Similarly, the equipment must be strong enough to remain intact under the maximum load placed on it. If it fails unexpectedly, someone is going to be injured — perhaps severely.

8.3.2.10   Subjects

The subjects selected for strength testing will determine the results obtained. This means that when strength data are collected, the selection of subjects must appropriately represent the population it claims to describe (for example, design data for retired persons should be collected on retired persons, and design data for entry-level construction workers should be collected on young healthy adults).

For general research purposes, persons participating in a strength-testing project should not have a history of musculoskeletal injuries. There are other medical conditions, including hypertension, which may pose a threat of harm to a participant. Whenever possible, prospective participants should be medically evaluated and approved before participating in a strength-testing project.

The following data should be provided about the subject population when reporting strength testing results:

1. Gender
2. Age distribution
3. Relevant anthropometry (height, weight, etc.)
4. Sample size

5. Method by which sample was selected and who it is intended to represent
6. Extent of strength training done by participants, and their experience with isometric testing
7. Health status of participants (medical exam and/or questionnaire recommended)

8.3.2.11   Isometric Strength Data Reporting

The minimum data which should be reported for strength testing projects are:

1. Mean, median, and mode of data set
2. Standard deviation of data set
3. Skewness of data set (or histogram describing data set)
4. Minimum and maximum values

8.4   Evaluation of Isometric Strength Testing According to Physical Assessment Criteria

A set of five criteria have been purposed to evaluate the utility of all forms of strength testing. Isometric strength testing is evaluated with respect to these criteria in the following sections.

8.4.1   Is It Safe to Administer?

Any form of physical exertion carries with it some risk. The directions for the person undergoing an isometric test specifically state that participants should slowly increase the force until they reach what they feel is a maximum, and to stop if at any time during the exertion they feel discomfort or pain. The directions also expressly forbid jerking on the equipment. When isometric testing is performed in this manner it is quite safe to administer, because the tested person is deciding how much force to apply, over what time interval, and how long to apply it. The only known complaints relating to participation in isometric testing are some residual soreness in the muscles that were active in the test(s), and this is rarely reported.

8.4.2   Does the Method Provide Reliable Quantitative Values?

The test-retest variability for isometric testing is 5 to 10%. In the absence of a specific strength-training program, individual isometric strength remains relatively stable over time. When the number of trials is based on the 10% criterion discussed earlier, the recorded strength is near or at the tested person's maximum voluntary strength. Assuming the above factors, if test postures are properly controlled, isometric strength testing is highly reliable and quantitative.

8.4.3   Is the Method Practical?

Isometric strength testing has already been used successfully in industry for employee placement, in laboratories for the collection of design data, and in rehabilitation facilities for patient progress assessment.

8.4.4   Is the Method Related to Specific Job Requirements (Content Validity)?

Isometric strength testing can be performed in any posture. When it is conducted for employee placement purposes, the test postures should be as similar as possible to the postures that will be used on the job. The force vector applied by the tested person should also be similar to the force vector that will be applied on the job. When these two criteria are met, isometric strength testing is closely related to job requirements. However, it should be noted that results obtained using isometric strength testing loses both content and criterion-related validity as job demands become more dynamic.

8.4.5   Does the Method Predict the Risk of Future Injury or Illness?

A number of researchers have demonstrated that isometric strength testing does predict risk of future injury or illness for people on physically stressful jobs (Chaffin et al., 1978; Keyserling et al., 1980). The accuracy of this prediction is dependent on the quality of the job evaluation on which the strength tests are based and the care with which the tests are administered.

8.5   Isoinertial Strength Testing

Kroemer (1982, 1983, 1985) and Kroemer et al. (1990) define the isoinertial technique of strength assessment as one in which *mass properties of an object are held constant*, as in lifting a given weight over a predetermined distance. Several strength assessment procedures possess the attribute in this definition. Most commonly associated with the term is a specific test developed to provide a relatively quick assessment of a subject's maximal lifting capacity using a modified weight lifting device (Kroemer, 1982; McDaniel et al., 1983). The classic psychophysical methodology of assessing maximum acceptable weights of lift is also an isoinertial technique under this definition (Snook, 1978).

   While the definition provided by Kroemer (1982) and Kroemer et al. (1990) has been most widely accepted in the literature, some have applied the term "isoinertial" to techniques that differ somewhat from the definition given above, such as in a description of the Isotechnologies B-200 strength testing device (Parnianpour et al., 1988). Rather than lifting a constant mass, the B-200 applies a constant force against which the subject performs an exertion. The isoinertial tests described in this paper apply to situations in which the mass to be moved by a musculoskeletal effort is set to a constant.

8.5.1   Is Isoinertial Testing Psychophysical, or Is Psychophysical Testing Isoinertial?

As various types of strength tests have evolved over the past few decades, there have been some unfortunate developments in the terminology that have arisen to describe and/ or classify different strength assessment procedures. This is particularly evident when one tries to sort out the various tests that have been labeled "isoinertial." One example was cited above. Another problem that has evolved is that the term "isoinertial strength" has developed two different connotations. The first connotation is the conceptual definition — isoinertial strength tests describe any strength test where a constant mass is handled. However, in practice, the term is often used to denote a *specific* strength test where subjects' maximal lifting capacity is determined using a machine where a constant mass is lifted

(Kroemer, 1982; McDaniel et al., 1983). Partially as a result of this dual connotation, the literature contains references both to the "isoinertial strength test" as a psychophysical variant (Ayoub and Mital (1989) and to the psychophysical method as an "isoinertial strength test"(Chaffin and Andersson, 1991; Kroemer et al., 1990). In order to lay the framework for the next two sections, the authors would like to briefly discuss some operational definitions of tests of isoinertial and psychophysical strength.

When Ayoub and Mital (1989) state that the isoinertial strength test is a variant of the psychophysical method, they refer to the specific strength test developed by Kroemer (1982) and McDaniel et al. (1983). Clearly, this isoinertial protocol has many similarities to the psychophysical method: both are dynamic; weight is adjusted in both; both measure the load a subject is willing to endure under specified circumstances, etc. However, while both deal with lifting and adjusting loads, there are significant differences between the psychophysical (isoinertial) technique and the Kroemer/McDaniel (isoinertial) protocol, both procedurally and in use of the data collected in these tests. For purposes of this paper we will designate the Kroemer/McDaniel protocol Maximal Isoinertial Strength Tests (MIST). This section deals with the latter isoinertial technique, which differs from the psychophysical technique on the following counts:

1. In Maximal Isoinertial Strength Tests, the amount of weight lifted by the subject is *systematically adjusted by the experimenter, primarily through increasing the load to the subject's maximum*. In contrast, in psychophysical tests, *weight adjustment is freely controlled by the subject, and may be upwards or downwards*.

2. The Maximal Isoinertial Strength Tests discussed in this section are designed to quickly establish an individual's *maximal strength* using a *limited number of lifting repetitions*; whereas psychophysical strength assessments are typically performed over a *longer duration of time* (usually at least 20 minutes), and instructions are that the subject select an *acceptable (submaximal) weight of lift*, not a maximal one. Due to the typically longer duration of psychophysical assessments, greater aerobic and cardiovascular components are usually involved in the acceptable workload chosen.

3. Isoinertial strength tests have traditionally been used as a *worker selection tool* (a method of matching physically capable individuals to demanding tasks). While psychophysical tests can and have been used for worker selection, the principle focus of psychophysical methods has been to establish data that can be used for the purpose of *ergonomic job design* (Snook, 1978).

8.5.2 Published Isoinertial Strength Data

The LIFTEST and SAT procedures are isoinertial techniques of strength testing that attempt to establish the maximal amount of weight that a person can safely lift (Kroemer, 1982). In these techniques a preselected mass, constant in each test, is lifted by the subject (typically from knee height to knuckle height, elbow height, or to overhead reach height). The amount of weight to be lifted is at first relatively light, but the amount of mass is continually increased in succeeding tests until it reaches the maximal amount that the subject voluntarily indicates he/she can handle. This technique has been used extensively by the U.S. Air Force (McDaniel et al., 1983) and is applicable to dynamic lifting tasks in industry as well (Jiang et al., 1986; Kroemer, 1982).

Since a constant mass is lifted in LIFTEST, the acceleration of the load during a test is dependent on the force applied to the load during the test (in accordance with Newton's Second Law: $F = ma$). The dynamic nature of this procedure, the fact that a constant mass

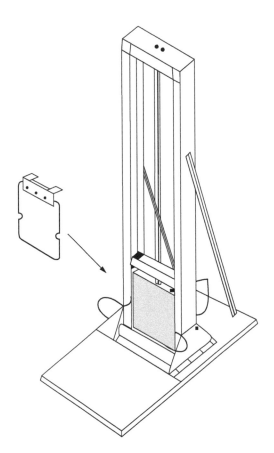

FIGURE 8.3
The Incremental Weight Lift Machine. The barrier has been removed to expose the stack of weights.

is being lifted, and the subject's freedom to choose the preferred lifting technique, all give the LIFTEST a general similarity to certain types of industrial lifting tasks. A unique aspect of the LIFTEST technique is that it is the only strength measurement procedure discussed in this document where results are based on the success or failure to perform a prescribed criterion task. The criterion tasks studied have typically included lifting to shoulder height (Jiang et al., 1986a; Kroemer, 1985; McDaniel et al., 1983; Ostrom et al., 1990), elbow height (Jiang et al., 1986a; Kroemer, 1985), or knuckle height (Jiang et al., 1986a; Kroemer, 1983). The USAF also developed a muscular endurance test using an Incremental Lift Machine (see Figure 8.3), or ILM (McDaniel et al., 1983).

The LIFTEST shoulder height maximal strength test has demonstrated the highest correlation with manual materials-handling activities (Jiang et al., 1986a), and has been subjected to a biomechanical analysis by Stevenson et al. (1990). Stevenson et al. (1990) demonstrated that this criterion task could be separated into three distinct phases: (1) a powerful upward pulling phase, where maximal acceleration, velocity, and power values are observed, (2) a wrist changeover maneuver (at approximately elbow height), where momentum is required to compensate for low force and acceleration, and (3) a pushing phase (at or above chest height), characterized by a secondary (lower) maximal force and acceleration profile.

The analysis by Stevenson et al. (1990) suggested that successful performance of the criterion shoulder-height lift requires a technique quite different from the concept of slow, smooth lifting usually recommended for submaximal lifting tasks. On the contrary, lifting

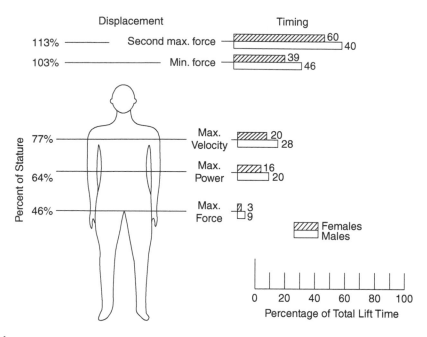

FIGURE 8.4

Displacement and timing parameters for a 1.83 m maximal isoinertial lift. Figure illustrates anatomical landmarks for the location of key events, found to be consistent for both genders. (From Stevenson, J.M., et al., *Ergonomics*, 33(2), 161–172, 1990. Reprinted with permission of Taylor and Francis.)

of a maximal load requires a rapid and powerful lifting motion. This is due in large part to the need to develop sufficient momentum to allow successful completion of the wrist changeover portion of the lift. Most lift failures occur during the wrist changeover procedure, probably the result of poor mechanical advantage of the upper limb to apply force to the load at this point in the lift. Stevenson et al. (1990) found that certain anatomical landmarks were associated with maximal force, velocity, and power readings (see Figure 8.4). Maximal force readings were found to occur at mid-thigh, maximal velocity at chest height, minimum force was recorded at head height, and the second maximal acceleration (pushing phase) was observed at 113% of the subject's stature.

8.5.2.1 *The Strength Aptitude Test (McDaniel et al., 1983)*

The Strength Aptitude Test (SAT) is a classification tool for matching the physical strength abilities of individuals with the physical strength requirements of jobs in the Air Force (McDaniel, 1994). The SAT is given to all Air Force recruits as part of their preinduction examinations. Results of the SAT are used to determine whether the individual tested possesses the minimum strength criterion that is a prerequisite for admission to various Air Force Specialties (AFSs). The physical demands of each AFS are objectively computed from an average physical demand weighted by the frequency of performance and the percent of the AFS members performing the task. Objects weighing less than 10 pounds are not considered physically demanding and are not considered in the job analysis. Prior to averaging the physical demands of the AFS, the actual weights of objects handled are converted into equivalent performance on the incremental weight-lift test using regression equations developed over years of testing. These relationships consider the type of task (lifting, carrying, pushing, etc.), the size and weight of the object handled, as well as the type and height of the lift. Thus, the physical job demands are related to, but are not

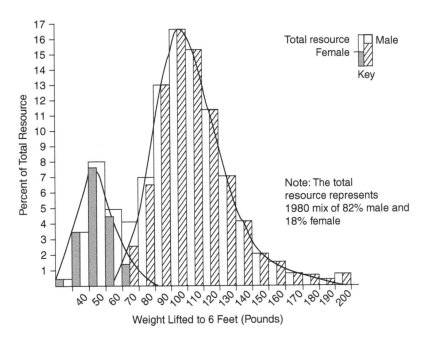

FIGURE 8.5
Distribution of weight lifted in a 1.83 m maximal isoinertial lift for male and female United States Air Force recruits. [Reprinted from McDaniel, J.W., R.J. Shandis, and S.W. Madole: *Weight Lifting Capabilities of Air Force Basic Trainees* (AFAMRL–TR–83–0001). Dayton, Ohio: Wright–Patterson AFBDH, Air Force Aerospace Medical Research Laboratory, 1983.]

identical to, the ability to lift an object to a certain height. Job demands for various AFSs are reanalyzed periodically for purposes of updating the SAT.

The first major report describing this classification tool was a study of 1671 basic trainees, including 1066 males and 605 females (McDaniel et al., 1983). The incremental weight-lift tests started with an 18.1-kg weight, which was to be raised to 1.83 m or more above the floor. This initial weight was increased in 4.5-kg increments until subjects were unable to raise the weight to 1.83 m. Maximal weight lift to elbow height was then tested as a continuation of the incremental weight-lift test. In the test of lifting the weight to 1.83 m, males averaged 51.8 kg (± 10.5 SD), while females averaged 25.8 kg (± 5.3). The respective weights lifted to elbow height were 58.6 kg (± 11.2) and 30.7 kg (± 6.3). The distributions of weight lifting capabilities for both male and female basic trainees in lifts to 6 feet are provided in Figure 8.5. Results of the elbow height lift are presented in Table 8.1. McDaniel et al. (1983) also performed a test of isoinertial endurance. This involved holding a 31.8-kg weight at elbow height for the duration the subject could perform the task. Male basic trainees were able to hold the weight for an average of 53.3 seconds (± 22.11), while female basic trainees managed to hold the weight an average of 10.3 seconds (± 10.5).

When developing the SAT, the Air Force examined more than 60 candidate tests in an extensive, four-year research program and found the incremental weight lift to 1.83 m to be the single best test of overall dynamic strength capability, which was both safe and reliable (McDaniel, 1994). This finding was confirmed by an independent study funded by the U.S. Army (Myers et al., 1984). This study compared the SAT to a battery of tests developed by the Army (including isometric and dynamic tests) and compared these with representative heavy demand tasks performed within the Army. Results showed the SAT to be superior to all others in predicting performance on the criterion tasks.

TABLE 8.1

Weight Lifted by Male and Female U.S. Air Force
Recruits Using Maximal Isoinertial Lift to Elbow Height

Percentile	Males Pounds	Kilograms	Females Pounds	Kilograms
1	80	36.3	40	18.1
5	93	42.2	48	21.8
10	100	45.4	52	23.6
20	109	49.5	58	26.3
30	116	52.6	61	27.7
40	122	55.4	65	29.5
50	127	57.6	68	30.9
60	133	60.3	71	32.2
70	140	63.5	75	34.0
80	150	68.1	78	35.4
90	160	72.6	85	38.6
95	171	77.6	90	40.8
99	197	89.4	100	45.4
Mean	129	58.6	68	30.7
S.D.	25	11.2	14	6.3
Minimum	50	22.7	< 40	< 18.1
Maximum	> 200	> 90.7	100	49.9
Number	1066		605	

8.5.2.2 *LIFTEST*

Kroemer (1982, 1985) described results of a study using a similar apparatus to the one used by the U.S. Air Force. The sample consisted of 39 subjects (25 male) recruited from a university student population. The procedures were similar to McDaniel et al. (1983), with the exception that the minimum starting weight was 11.4 kg, and that maximal lifting limits were established to prevent overexertion. These were 77.1 kg for floor-to-knuckle-height tests, and 45.4 for floor-to-overhead-reach tests. The following procedure was used for establishing the maximal load: if the initial 11.4-kg weight was successfully lifted, the weight was doubled to 22.7 kg. Additional 11.4-kg increments were added, until an attempt failed or the maximal lifting limit was reached. If an attempt failed, the load was reduced by 6.8 kg. If this test weight was lifted, 4.5 kg was added; if not, 2.3 kg was subtracted. This scheme allowed quick determination of the maximal load the subject could lift.

In Kroemer's studies, 6 of 25 male subjects exceeded the cut-off load of 100 pounds (45.4 kg) in overhead reach lifts (Kroemer, 1982, 1985). All 14 females stayed below this limit. The 19 remaining male subjects lifted an average of 27 kg. The female subjects lifted an average of 16 kg. In lifts to knuckle height, 17 of the 25 male (but none of the female) subjects exceeded the 77.1 kg cut-off limit. The remaining subjects lifted an average of about 54 kg, with males averaging 62 kg and females 49 kg. The coefficients of variation for all tests were less than 8%. Summary data for this study are given in Table 8.2.

TABLE 8.2

Results of Maximal Isoinertial Strength Tests for 25 Male and 14 Female Subjects

	All				Male				Female			
	Mean	SD	CV	N	Mean	SD	CV	N	Mean	SD	CV	N
Overhead												
Liftest (kg)	27.0	10.3	3.5%	33	34.8	5.2	3.2%	19	16.3	3.7	3.9%	14
Lift > 45.5 kg	—	—	—	6	—	—	—	6	—	—	—	0
Knuckle												
Liftest (kg)	53.9	13.4	6.9%	22	62.2	7.8	5.2%	8	49.1	13.7	7.8%	14
Lift > 77 kg	—	—	—	17	—	—	—	17	—	—	—	0

8.5.3   Evaluation of Isoinertial Strength Tests According to Physical Assessment Criteria

8.5.3.1   Is It Safe to Administer?

The MIST procedures described above appear to have been remarkably free of injury. Isoinertial procedures have now been performed many thousands of times without report of verifiable injury. However, reports of transitory muscle soreness have been noted. The temporary muscle soreness associated with isoinertial testing has been similar to that experienced in isokinetic tests, but has been reported less frequently than that experienced with isometric strength tests.

McDaniel et al. (1983) present some useful recommendations for design of safe isoinertial weight-lift testing procedures. The following list summarizes the recommendations made by these authors.

1. Weight lifting equipment should be designed so that the weights and handle move only in a vertical direction.
2. Sturdy shoes should be worn, or the subject may be tested barefoot. Encumbering clothing should not be worn during the test.
3. The initial weight lifted should be low — 20 to 40 pounds. Weights in this range are within the capability of almost everyone. Weight increments should be small.
4. The upper limit should not exceed the largest job-related requirement or 160 pounds, whichever is less.
5. The starting handle position should be one to two feet above the standing surface. If the handle is lower, the knees may cause obstruction. If the handle is too high, the subjects will squat to get their shoulders under it prior to lifting. A gap between the handles will allow them to pass outside the subject's knees when lifting, allowing a more erect back and encouraging the use of leg strength.
6. The recommended body orientation prior to lifting should be (a) arms straight at the elbow, (b) knees bent to keep the trunk as erect as possible, and (c) head aligned with the trunk. The lift should be performed smoothly, without jerk.
7. A medical history of the subject should be obtained. If suspicious physical conditions are identified, a full physical examination should be performed prior to testing. Subjects over 50 years of age or pregnant should always have a physical prior to testing.

8. All sources of overmotivation should be minimized. Testing should be done in private and results kept confidential. Even the test subject should not be informed until the testing is completed.

9. If the subject pauses during a lift, the strength limit has been reached, and the test should be terminated. Multiple attempts at any single weight level should not be allowed.

10. The testing should always be voluntary. The subject should be allowed to stop the test at any time. The subject should not be informed of the criteria prior to or during the test.

It is noteworthy that, as of 1994, over 2 million subjects have been tested on the SAT without any back injury or overexertion injury (McDaniel, 1994).

8.5.3.2   Does the Method Provide Reliable, Quantitative Values?

Kroemer et al. (1985) reported LIFTEST coefficients of variation (measures of intra-individual variability in repeated exertions) of 3.5 for all subjects in overhead lifts and 6.9 in lifts to knuckle height. The same study showed somewhat higher variability in tests of isometric strength (coefficient of variations ranging from 11.6 to 15.4). Test-retest reliability was not reported by McDaniel et al. (1983).

8.5.3.3   Is It Practical?

Isoinertial techniques generally appear practical in terms of providing a test procedure that requires minimal administration time and minimal time for instruction and learning. Even in a worst-case scenario, the isoinertial procedures used by Kroemer (1983) would take only a few minutes to determine the maximal weight lifting capability of the subject for a particular condition. The McDaniel et al. (1983) procedure can be performed in approximately 3–5 minutes.

   Practicality is determined in part by cost of the equipment required, and on this account, the cost of isoinertial techniques is quite modest. The equipment needed to develop the LIFTEST devices used by McDaniel et al. (1983) and Kroemer (1982) would not be prohibitive for most applications. In fact, Kroemer (1983) states that the device is easily dismantled and could easily be transported to different sites in a small truck or station wagon, or perhaps in a mobile laboratory vehicle.

8.5.3.4   Is It Related to Specific Job Requirements?

Since industrial lifting tasks are performed dynamically, isoinertial strength tests do appear to provide some useful information related to an individual's ability to cope with the dynamic demands of industrial lifting. McDaniel (1994) reported that these tests are predictive of performance on a wide range of dynamic tasks, including asymmetric tasks, carrying, and pushing tasks. Furthermore, Jiang et al. (1986) demonstrated that the isoinertial lifting test to 6 feet was more highly correlated with psychophysical tests of lifting than isometric tests.

8.5.3.5   Does It Predict Risk of Future Injury or Illness?

The ability of a strength test to predict risk of future injury or illness is dependent upon performance of prospective epidemiological studies. As of this writing, no such studies have been conducted on the isoinertial techniques described above.

8.6 Psychophysical Strength

According to contemporary psychophysical theory, the relationship between the strength of a perceived sensation (S) and the intensity of a physical stimulus (I) is best expressed by a power relationship (Stevens, 1957).

$$S = kI^n \tag{1}$$

This psychophysical principle has been applied to many practical problems, including the development of scales or guidelines for effective temperature, loudness, brightness, and ratings of perceived exertion. Based on the results of a number of experiments using a variety of scaling methods and a number of different muscle groups, the pooled estimate of the exponent for muscular effort and force is 1.7 (Jones, 1986).

When applying this principle to work situations, it is assumed that individuals are capable and willing to consistently identify a specified level of perceived sensation (S). For manual materials-handling tasks, this specified level is usually the *maximum acceptable weight* or *maximum acceptable force*. The meaning of these phrases are defined by the instructions given to the test subject (Snook, 1985a):

> You are to work on an incentive basis, working as hard as you can without straining yourself, or becoming unusually tired, weakened, overheated, or out of breath.

If the task involves *lifting*, the experiment measures the maximum acceptable weight of lift. Similarly, there are maximum acceptable weights for *lowering* and *carrying*. Such tests are isoinertial in nature; however, in contrast to the tests described above, they are typically used to test submaximal, repetitive handling capabilities. Data are also available for *pushing* and *pulling*. These are reported as maximum acceptable forces and include data for initial as well as sustained pulling or pushing.

8.6.1 Why Use Psychophysical Methods?

Snook identified several advantages and disadvantages to using psychophysical methods for determining maximum acceptable weights (Snook, 1985b). The advantages include:

1. The method is a realistic simulation of industrial work (face validity).
2. It is possible to study intermittent tasks (physiological steady state not required).
3. The results are consistent with the industrial engineering concept of "a fair day's work for a fair day's pay."
4. The results are reproducible.
5. The results appear to be related to low back pain (content validity).

Disadvantages include:

1. The tests are performed in a laboratory.
2. It is a subjective method that relies on self-reporting by the subject.
3. The results for very high frequency tasks may exceed recommendations for energy expenditure.
4. The results are insensitive to bending and twisting.

In terms of the application of the data derived from these studies, Liberty Mutual preferred to use it to design a job to fit the worker, since this application represented a more permanent engineering solution to the problem of low back pain in industry (Snook, 1978). This approach not only reduces the worker's exposure to potential low back pain risk factors, but also reduces liability associated with worker selection (Snook, 1978).

8.6.2   Published Data

8.6.2.1   Liberty Mutual

Snook and Ciriello at the Liberty Mutual Insurance Company have published the most comprehensive tables for this type of strength assessment (Snook and Ciriello, 1991). The most recent data is summarized in nine tables, organized as follows (Snook and Ciriello, 1991):

1. Maximum acceptable weight of lift for males
2. Maximum acceptable weight of lift for females
3. Maximum acceptable weight of lower for males
4. Maximum acceptable weight of lower for females
5. Maximum acceptable forces of push for males (initial and sustained)
6. Maximum acceptable forces of push for females (initial and sustained)
7. Maximum acceptable forces of pull for males (initial and sustained)
8. Maximum acceptable forces of pull for females (initial and sustained)
9. Maximum acceptable weight of carry (males and females)

8.6.2.2   Other Sources

Ayoub et al. (1978) and Mital (1984) have also published tables for maximum acceptable weights of lift. Even though their tables are similar in format and generally in agreement with those from Liberty Mutual, there are some differences. Possible sources for these differences may be differences in test protocol, differences in task variables, and differences in subject populations and their characteristics.

8.6.3   Experimental Procedures and Methods

For the sake of simplicity and convenience, the Liberty Mutual protocol for lifting or lowering and an excerpt from the lifting table will be used as examples for this section. The protocols used by Ayoub et al. (1978) and Mital (1984) were similar, but not exactly the same. The reader should refer to the original publications for details.

The Liberty Mutual experimental procedures and methods were succinctly reviewed in their most recent revision of the tables (Snook and Ciriello, 1991). The data reported in these revised tables reflect results from 119 second-shift workers from local industry (68 males, 51 females). All were prescreened to ensure good health prior to participation. These subjects were employed by Liberty Mutual for the duration of the project (usually 10 weeks). All received 4 to 5 days of conditioning and training prior to participation in actual test sessions.

Test subjects wore standardized clothing and shoes. The experiments were performed in an environmental chamber maintained at 21°C (dry bulb) and 45% relative humidity.

Forty-one anthropometric variables were recorded for each subject, including several isometric strengths and aerobic capacity.

A single test session lasted approximately 4 hours and consisted of 5 different tasks. Each task session lasted 40 minutes, followed by 10 minutes rest. Most subjects participated in at least two test sessions per week for 10 weeks. In general, a subject's heart rate and oxygen consumption were monitored during the sessions.

8.6.3.1 Lifting or Lowering Tasks

In a lifting or lowering task session, the subject was given control of one variable, usually the weight of the box. The other task variables would be specified by the experimental protocol. These variables include:

1. *Lifting zone* refers to whether the lift occurs between floor level to knuckle height (low), knuckle height to shoulder height (center), or shoulder height to arm reach (high).
2. *Vertical distance of lift* refers to the vertical height of the lift within one of these lifting zones. The specified values for distance of lift in the tables are 25 cm (10 in), 51 cm (20 in), and 76 cm (30 in). It is possible to use linear extrapolation for lift distances not exactly equal to one of these values.
3. *Box width* refers to the dimension of the box away from the body. The three values of box width are 34 cm (13.4 in), 49 cm (19.3 in), and 75 cm (29.5 in). It is possible to use linear extrapolation between these values.
4. The final task variable is *frequency of lift*. The frequencies are expressed as one lift per time interval and include intervals of 5 seconds, 9 seconds, 14 seconds, 1 minute, 2 minutes, 5 minutes, and 8 hours.

These same definitions apply to a lowering task, except the word "lower" is substituted for "lift." The test protocol for lowering was essentially identical to that for lifting, and the results are reported in a similar format. It should be noted, however, that the test protocols for lifting and lowering involved using a special apparatus that returned the box to its original specified location, so that the subject *only* lifted or lowered, not both.

Per the instructions, the subject was to adjust the weight of the box, according to his or her own perceptions of effort or fatigue, by adding or removing steel shot or welding rods from a box. The box had handles and a false bottom to eliminate visual cues. Each task experiment was broken into two segments, so that the initial weight of the box could be randomly varied between high vs. low, so that the subject approached his or her maximum acceptable weight from above as well as below. If the results met a 15% test-retest criterion, the reported result was the average of these two values. If the results did not meet this criterion, they were discarded and the test repeated at a later time.

In reporting the results, it was assumed that the gender-specific maximum acceptable weights for a particular task were normally distributed. As a consequence, the results were reported as percentages of population, stratified by gender. The Liberty Mutual tables are organized around the following percentages: 90, 75, 50, 25, and 10% (Snook and Ciriello, 1991). The 90th percentile refers to a value of weight that 90% of individuals of that gender would consider a maximum acceptable weight (90% "acceptable"), while the 10th percentile refers to a value of weight that only 10% of individuals of that gender would find acceptable (10% "acceptable").

8.6.3.2 Important Caveats

Snook and Ciriello (1991) have identified several important caveats that should be remembered when using the Liberty Mutual tables.

1. The data for each experimental situation were assumed to be normally distributed when the maximum acceptable weights and forces acceptable to 10, 25, 50, 75, and 90% of the industrial population were determined.

2. Not all values in the tables are based on experimental data. Some values were derived by assuming that the variation noted for a particular variable for one type of task would be similar to that observed for another task (e.g., the effects on lowering would be similar to that on lifting).

3. The tables for lifting, lowering, and carrying are based on boxes with handles that were handled close to the body. They recommend that the values in the tables be reduced by approximately 15% when handling boxes without handles. When handling smaller boxes with extended reaches between knee and shoulder heights, they recommend reducing the values by approximately 50%.

4. Some of the reported weights and forces exceed recommended levels of energy expenditure if performed for 8 or more hours per day. These data are italicized in the tables.

5. The data in the tables give results for individual manual materials-handling tasks. When a job involves a combination of these tasks, each component should be analyzed separately, and the component with the lowest percent of capable population represents the maximum acceptable weight or force for the combined task. It should be recognized, however, that the energy expenditure for the combined task will be greater than that for the individual components.

Some recent data suggest that persons performing lifting tasks are relatively insensitive to the perception of high disk compression forces on the spine (Thompson and Chaffin, 1993). As a result, there may be some tasks in the tables that exceed recommended levels of disk compression.

8.6.4 Related Research

8.6.4.1 Task and Subject Variables

A variety of researchers have examined the effects of other task and subject variables using the psychophysical protocol. Most of these studies involve a small number (<10) of college students as test subjects. Some experiments used the Liberty Mutual protocol; others used the protocol described by Ayoub et al. (1978) and Mital (1984). These "refinements" are summarized in Table 8.3.

8.6.5 Recommended Applications

8.6.5.1 Job Evaluation

The Liberty Mutual tables were developed for the purpose of evaluating work, not workers (Snook, 1987). In particular, the tables are intended to help industry in the evaluation and design of manual materials-handling tasks that are consistent with worker limitations and abilities (Snook and Ciriello, 1991). The explicit goal is the control of low back pain through reductions in initial episodes, length of disability, and recurrences (Snook, 1987).

TABLE 8.3

Miscellaneous Task Variables Evaluated Using the Psychophysical
Methodology

Task Variable(s)	Reference(s)
Zone of lift	Snook, 1985a
Distance of lift	Snook and Ciriello, 1991
Frequency of lift	Ayoub et al., 1978
Box width	Mital, 1984b
	Ciriello and Snook, 1983
	Mital and Ayoub, 1981
	Asfour et al., 1984
	Garg et al., 1980
Extended work shifts	Mital, 1984a
Combinations of lift, carry, and lower	Ciriello et al., 1990
	Jiang et al., 1986b
Angle of twist	Asfour et al., 1984
Box length	Asfour et al., 1984
	Garg et al., 1980
Material density	Mital and Manivasagan, 1983
Location of center of gravity	
Center of gravity relative to preferred hand	
Sleep deprivation	Legg and Haslam, 1984
Bag versus box	
Fullness of bag (same weight)	
Bag ± handles	
Day 1-to-day 5 of work week	Legg and Myles, 1985
Asymmetrical loads	Mital and Fard, 1986
Asymmetrical lifting	Mital, 1987a
	Mital, 1992
	Drury et al., 1989
Emergency scenario	Legg and Pateman, 1985
Handle position	Drury and Deeb, 1986
Handle angle	
Duration of lifting	Mital, 1984a
	Fernandez et al., 1991
Overreach heights	Mital and Aghazadeh, 1987
Restricted vs. unrestricted shelf opening clearances	Mital and Wang, 1989
Experienced vs. inexperienced workers	Mital, 1987b
Non-standard or restricted postures	Gallagher et al., 1988
	Gallagher, 1991
	Gallagher and Hamrick, 1992
	Smith et al., 1992

To apply the tables in the context of job evaluation, it is first necessary to specify the task variables of the job. For a lifting task, this would include the lift zone, distance of lift, box width, frequency of lift, and the presence or absence of box handles. In addition, it would be necessary to measure the weight of the object to be handled, perhaps using a scale or dynamometer. Once these variables are specified, the measured weight can be compared to the data in the table to determine the percent of capable population for males and females. The procedure is similar for pulling or pushing. The required force can be measured with a dynamometer.

Consider the following example. The task is to lift a 49-cm wide box that weighs 20 kg, once every minute, from floor level to knuckle height for a distance of 51 cm. Referencing Table 8.4, an excerpt from the Liberty Mutual tables, it is seen that the weight of the box, 20 kg, is exactly equal to the maximum acceptable weight of lift for 75% of males (i.e.,

152

Muscle Strength

TABLE 8.4

Excerpt from the Liberty Mutual Tables for Maximum Acceptable Weight of Lift (kg) for Males and Females

Gender	Box Width (cm)	Distance of Lift (cm)	Percent Capable	Floor Level to Knuckle Height One Lift Every							
				5 sec	9 sec	14 sec	1 min	2 min	5 min	30 min	8 h
Males	49	51	90	7	9	10	14	16	17	18	20
			75	10	13	15	20	23	25	25	30
			50	14	17	20	27	30	33	34	40
			25	18	21	25	34	38	42	43	50
			10	21	25	29	40	45	49	50	59
Females	49	51	90	6	7	8	9	10	10	11	15
			75	7	9	9	11	12	12	14	18
			50	9	10	11	13	15	15	16	22
			25	10	12	13	16	17	17	19	26
			10	11	14	15	18	19	20	22	30

Note: Italicized values exceed 8-hour physiological criteria (energy expenditure).

75% of males would consider this task "acceptable"). By contrast, the highest maximum acceptable weight of lift reported for females is 18 kg. As a result, this task is "not acceptable" to over 90% of females.

8.6.5.2 Job Design

To apply the tables in the context of job design, the process is essentially identical. All task-specific parameters must be identified, except the required weight or force (that is what you are determining). You select a desired percent of capable of population, noting gender effects, and then identify the maximum acceptable weight or force that corresponds to that desired percent. This is the value recommended for job design.

As an example, suppose you wish to design a lifting task that requires a 49-cm wide box that must be lifted 51 cm once per minute within the floor-to-knuckle zone. You desire to design this job to accommodate 75% of females. According to the data in Table 8.4, you would recommend that the box weigh no more than 11 kg. This weight would be acceptable to 75% of females and over 90% of males.

Multiple task analysis, consisting of a lift, carry, and lower, has also been investigated for the Liberty Mutual data (Ciriello et al, 1990). In this circumstance, it was observed that the maximum acceptable weight for the multiple task was lower than that for only the carrying task when performed separately, but not significantly different from the lifting or lowering maximum acceptable weights when performed separately. For this type of a multiple task, the maximum acceptable weight for the task should be the lowest maximum acceptable weight of the lift or lower as if it were performed separately. One should be careful, however, because the energy expenditure for the multiple task is probably underestimated when compared to performing the tasks separately. Similar results were reported by Jiang et al. (1986).

8.6.6 Evaluation According to Criteria for Physical Assessment

8.6.6.1 Is It Safe to Administer?

According to Snook (1992), there has been one compensable injury among the 119 industrial worker test subjects. This single episode involved a chest wall strain associated with a high lift. It was also associated with four days restricted activity, but no permanent disability.

8.6.6.2   Does the Protocol Give Reliable Quantitative Values?

The Liberty Mutual protocol incorporates a criterion for test-retest reliability (maximum difference of 15%). Legg and Myles (1985) reported that 34% of their data did not meet this criterion. In contrast, Gallagher (1991) reported that only 3% of tests in their study had to be repeated because of violating the 15% test-retest criterion. Clearly, the maximum acceptable weights and forces are quantitative.

8.6.6.3   Is It Practical?

There are two major sources of impracticality associated with this type of strength assessment: (1) it is conducted in a laboratory, and (2) the duration of testing is somewhat prolonged compared to other strength assessment methods. It is possible, however, to have the subjects use objects that are actually handled in the workplace. Equipment is not very costly.

8.6.6.4   Is It Related to Specific Job Requirements (Content Validity)?

The content validity of this method of strength assessment is one of its greatest assets. One potential weakness, however, is its insensitivity to bending and twisting.

8.6.6.5   Does It Predict Risk of Future Injury or Illness (Predictive Validity)?

The results of two epidemiological studies suggest that selected indices derived from the psychophysical data are predictive of risk for contact injury, musculoskeletal disorders (excluding the back), and back disorders (Ayoub et al., 1983; Snook et al., 1978). These indices are correlated to the severity of these injuries. A third study demonstrated predictive value (Herrin et al., 1986). It should be noted, however, that at high frequencies, test subjects selected weights and forces that often exceeded consensus criteria for acceptable levels of energy expenditure. In addition, test subjects may also select weights and forces that exceed consensus levels of acceptable disk compression.

8.7   Summary

In spite of advances in measurement techniques and an explosive increase in the volume of research, our understanding of human strength remains in its introductory stages. It is clear that muscle strength is a highly complex and variable function dependent on a large number of factors. It is not surprising, therefore, that there are large differences in strength, not only between individuals, but even within the same individual tested repeatedly on a given piece of equipment. The issue is compounded by the fact that correlations of strength among different muscle groups in the same individual are generally low, and that tests of isometric strength do not necessarily reflect the strength an individual might exhibit in a dynamic test. As a result of these and other influences, it is evident that great care needs to be exercised in the design, evaluation, reporting, and interpretation of muscular strength assessments.

Traditionally, tests of muscular strength were in the domain of the orthopedist, physical therapist, and exercise physiologist. However, such tests are also an important tool for the ergonomist, due to the high-strength demands required of workers in manual materials-handling tasks. In some cases, it has been shown that task demands may approach or even exceed the strength that an individual is voluntarily willing to exert in a test of strength. In such cases, there is evidence to suggest that the likelihood of injury is significantly greater than when the task demands lie well within an individual's strength capacity. Because the relationship between strength capabilities, job demands, and musculoskeletal injury has been established, it becomes apparent that tests of muscular strength may be of benefit to the ergonomist, both in the design of jobs and in ensuring that individuals have sufficient strength to safely perform physically demanding jobs. Several different strength assessment techniques have been employed for these purposes, each possessing unique characteristics and applicability to job design and/or worker selection procedures. The main purpose of this chapter has been to elucidate the strengths and weaknesses of some of these procedures, so that tests of strength may be properly applied in the design of jobs and the selection of workers.

References

Asfour, S.S., Ayoub, M.M., and Genaidy, A.M. 1984. A psychophysical study of the effect of task variables on lifting and lowering tasks, *J. Human Ergol. (Tokyo)*, 13, 3–14.

Ayoub, M.M. and Mital, A. 1989. *Manual Materials Handling* (London: Taylor and Francis), pp. 241–242.

Ayoub, M.M., Bethea, N.J., Devanayagam, S., Asfour, S.S., Bakken, G.M., Liles, D., Mital, A., and Sherif, M. 1978. Determination and modeling of lifting capacity, final report. HEW (NIOSH) Grant No. 5-R01-OH-00545-02.

Ayoub, M.M., Selan, J.L., and Liles, D.H. 1983. An ergonomics approach for the design of manual materials-handling tasks, *Hum. Factors*, 25(5), 507–515.

Caldwell, L.S., Chaffin, D.B., Dukes-Dobos, F.N., Kroemer, K.H.E., Laubach, L.L., Snook, S.H., et al. 1974. A proposed standard procedure for static muscle strength testing, *Am. Ind. Hyg. Assoc. J.*, 35, 201–206.

Chaffin, D.B. 1975. Ergonomics guide for the assessment of human static strength, *Am. Ind. Hyg. Assoc. J.*, 36, 505–511.

Chaffin, D.B. and Andersson, G.B.J. 1991. *Occupational Biomechanics*, 2nd ed. (New York: John Wiley & Sons), pp. 105–106.

Chaffin, D.B., Herrin, G.D., and Keyserling, W.M. 1978. Preemployment strength testing: an updated position, *J. Occup. Med.*, 20(6), 403–408.

Ciriello, V.M. and Snook, S.H. 1983. A study of size, distance, height, and frequency effects on manual handling tasks, *Hum. Factors*, 25(5), 473–483.

Ciriello, V.M., Snook, S.H., Blick, A.C., and Wilkinson, P.L. 1990. The effects of task duration on psychophysically determined maximum acceptable weights and forces, *Ergonomics*, 33, 187–200.

Drury, C.G. and Deeb, J.M. 1986. Handle positions and angles in a dynamic lifting task. Part 2. Psychophysical measures and heart rate, *Ergonomics*, 29(6), 769–777.

Drury, C.G., Deeb, J.M., Hartman, B., Wooley, S., Drury, C.E., and Gallagher, S. 1989. Symmetric and asymmetric manual materials handling. Part 1. Physiology and psychophysics, *Ergonomics*, 32(5), 467–489.

Fernandez, J.E., Ayoub, M.M., and Smith, J.L. 1991. Psychophysical lifting capacity over extended periods, *Ergonomics*, 34(1), 23–32.

Gallagher, S. 1991. Acceptable weights and psychophysical costs of performing combined manual handling tasks in restricted postures, *Ergonomics*, 34(7), 939–952.

Gallagher, S. and Hamrick, C.A. 1992. Acceptable workloads for three common mining materials, *Ergonomics*, 35(9), 1013–1031.

Gallagher, S., Marras, W.S., and Bobick, T.G. 1988. Lifting in stooped and kneeling postures: effects on lifting capacity, metabolic costs, and electromyography of eight trunk muscles, *Int. J. Ind. Erg.*, 3, 65–76.

Garg, A., Mital, A., and Asfour, S.S. 1980. A comparison of isometric and dynamic lifting capability, *Ergonomics*, 23(1), 13–27.

Herrin, G.D., Jaraiedi, M., and Anderson, C.K. 1986. Prediction of overexertion injuries using biomechanical and psychophysical models, *Am. Ind. Hyg. Assoc. J.*, 47(6), 322–330.

Jiang, B.C., Smith, J.L., and Ayoub, M.M. 1986a. Psychophysical modeling of manual materials-handling capacities using isoinertial strength variables, *Hum. Factors*, 28(6), 691–702.

Jiang, B.C., Smith, J.L., and Ayoub, M.M. 1986b. Psychophysical modeling for combined manual materials-handling activities, *Ergonomics*, 29(10), 1173–1190.

Jones, L.A. 1986. Perception of force and weight: theory and research, *Psychol. Bull.*, 100(1), 29–42.

Keyserling, W.M., Herrin, G.D., and Chaffin, D.B. 1980. Isometric strength testing as a means of controlling medical incidents on strenuous jobs, *J. Occup. Med.*, 22(5), 332–336.

Kroemer, K.H.E. 1982. Development of LIFTEST: a dynamic technique to assess the individual capability to lift material, Final Report, NIOSH Contract 210-79-0041. (Blacksburg, VA: Ergonomics Laboratory, IEOR Department, Virginia Polytechnic Institute and State University).

Kroemer, K.H.E. 1983. An isoinertial technique to assess individual lifting capability, *Hum. Factors*, 25(5), 493–506.

Kroemer, K.H.E. 1985. Testing individual capability to lift material: repeatability of a dynamic test compared with static testing, *J. Safety Res.*, 16(1), 1–7.

Kroemer, K.H.E., Marras, W.S., McGlothlin, J.D., McIntyre, D.R., and Nordin, M. 1990. On the measurement of human strength, *Int. J. Ind. Ergon.*, 6, 199–210.

Legg, S.J. and Haslam, D.R. 1984. Effect of sleep deprivation on self-selected workload, *Ergonomics*, 27(4), 389–396.

Legg, S.J. and Myles, W.S. 1985. Metabolic and cardiovascular cost, and perceived effort over an 8 hour day when lifting loads selected by the psychophysical method, *Ergonomics*, 28(1), 337–343.

Legg, S.J. and Pateman, C.M. 1985. Human capabilities in repetitive lifting, *Ergonomics*, 28(1), 309–321.

McDaniel, J.W. 1994. Personal communication.

McDaniel, J.W., Shandis, R.J., and Madole, S.W. 1983. Weight lifting capabilities of Air Force basic trainees, AFAMRL-TR-83-0001. Dayton, OH: Wright-Patterson AFBDH, Air Force Aerospace Medical Research Laboratory.

Mital, A. 1984a. Maximum acceptable weights of lift acceptable to male and female industrial workers for extended work shifts, *Ergonomics*, 27(11), 1115–1126.

Mital, A. 1984b. Comprehensive maximum acceptable weight of lift database for regular 8 h shifts, *Ergonomics*, 27, 1127–1138.

Mital, A. 1987a. Maximum weights of asymmetrical loads acceptable to industrial workers for symmetrical lifting, *Am. Ind. Hyg. Assoc. J.*, 48(6), 539–544.

Mital, A. 1987b. Patterns of differences between the maximum weights of lift acceptable to experienced and inexperienced materials handlers, *Ergonomics*, 30(8), 1137–1147.

Mital, A. 1992. Psychophysical capacity of industrial workers for lifting symmetrical loads and asymmetrical loads symmetrically and asymmetrically for 8 hour work shifts, *Ergonomics*, 35(7/8), 745–754.

Mital, A. and Aghazadeh, F. 1987. Psychophysical lifting capabilities for overreach heights, *Ergonomics*, 30(6), 901–909.

Mital, A. and Ayoub, M.M. 1981. Effect of task variables and their interactions in lifting and lowering loads, *Am. Ind. Hyg. Assoc. J.*, 42, 134–142.

Mital, A. and Fard, H.F. 1986. Psychophysical and physiological responses to lifting symmetrical and asymmetrical loads symmetrically and asymmetrically, *Ergonomics*, 29(10), 1263–1272.

Mital, A. and Manivasagan, I. 1983. Maximum acceptable weight of lift as a function of material density, center of gravity location, hand preference, and frequency, *Hum. Factors*, 25(1), 33–42.

Mital, A. and Wang, L.-W. 1989. Effects on load handling of restricted and unrestricted shelf opening clearances, *Ergonomics*, 32(1), 39–49.

Myers, D.O., Gebhardt, D.L., Crump, C.E., and Fleishman, E.A. 1984. Validation of the Military Entrance Physical Strength Capacity Test (MEPSCAT), U.S. Army Research Institute Technical Report 610, NTIS No. AD-A142 169.

Ostrom, L.T., Smith, J.L., and Ayoub, M.M. 1990. The effects of training on the results of the isoinertial 6-foot incremental lift strength test, *Int. J. Ind. Ergon.*, 6, 225–229.

Parnianpour, M., Nordin, M., Kahanovitz, N., and Frankel, V. 1988. The triaxial coupling of torque generation of trunk muscles during isometric exertions and the effect of fatiguing isoinertial movements on the motor output and movement patterns, *Spine*, 13(9), 982–992.

Schanne, F.T. 1972. Three Dimensional Hand Force Capability Model for a Seated Person, unpublished Ph.D. dissertation, University of Michigan, Ann Arbor.

Smith, J.L., Ayoub, M.M., and McDaniel, J.W. 1992. Manual materials handling capabilities in non-standard postures, *Ergonomics*, 35(7/8), 807–831.

Snook, S.H. 1978. The design of manual handling tasks, *Ergonomics*, 21, 963–985.

Snook, S.H. 1985a. Psychophysical acceptability as a constraint in manual working capacity, *Ergonomics*, 28(1), 331–335.

Snook, S.H. 1985b. Psychophysical considerations in permissible loads, *Ergonomics*, 28(1), 327–330.

Snook, S.H. 1987. Approaches to the control of back pain in industry: job design, job placement, and education/training. *Spine: State Art Rev.*, 2, 45–59.

Snook, S.H. 1992. Assessment of human strength: psychophysical methods, Roundtable presentation at the American Industrial Hygiene Conference and Exposition, Boston.

Snook, S.H., Campanelli, R.A., and Hart, J.W. 1978. A study of three preventive approaches to low back injury, *J. Occup. Med.*, 20(7), 478–481.

Snook, S.H. and Ciriello, V.M. 1991. The design of manual handling tasks: revised tables of maximum acceptable weights and forces, *Ergonomics*, 34(9), 1197–1213.

Stevens, S.S. 1957. On the psychophysical law, *Psychol. Rev.*, 64, 153–181.

Stevenson, J.M., Bryant, J.T., French, S.L., Greenhorn, D.R., Andrew, G.M., and Thomson, J.M. 1990. Dynamic analysis of isoinertial lifting technique, *Ergonomics*, 33(2), 161–172.

Stobbe, T.J. 1982. The Development of a Practical Strength Testing Program in Industry, unpublished Ph.D. dissertation, University of Michigan, Ann Arbor.

Stobbe, T.J. and Plummer, R.W. 1984. A test-retest criterion for isometric strength testing, in *Proceedings of the Human Factors Society 28th Annual Meeting*: Oct 22–26, 1984, San Antonio, TX, pp. 455–459.

Thompson, D.D. and Chaffin, D.B. 1993. Can biomechanically determined stress be perceived? in *Human Factors and Ergonomics Society, Proceedings of the 37th Annual Meeting*, Seattle, WA, pp. 789–792.

9

Isokinetic Strength Testing: Devices and Protocols

Zeevi Dvir

CONTENTS

ABSTRACT   Isokinetic dynamometry (ISD) is widely regarded as the standard method for assessing muscle performance. Patented during the late 1960s, this technology has evolved from the use of passive hydraulic resistance systems to computer-controlled, servomotor-based actuators, enabling the recording and measurement of concentric and eccentric contractions using the accommodating resistance principle. In spite of its wide application, as reflected by well over 10,000 units installed worldwide, the complexity of muscle performance analysis, which is further compounded by behavioral aspects that complicate the conduction of tests and even more so the interpretation of findings, makes it a very involved subject. This chapter surveys the basic principles of ISD, while dwelling on particular issues like reproducibility and effort optimality.

9.1 Ergonomic Relevance

Human activity invariably involves the exertion/absorption of forces on and from the surrounding environment. These forces require the production of muscular output, which may vary along the full spectrum of available tension levels and involve concentric as well as eccentric contractions. Isokinetic dynamometry is widely recognized as the most advanced method of assessing muscle output, with respect to either single or repetitive bouts. Because body–force interactions are so vital to ergonomic assessment and design, knowledge regarding the measurement of such tension levels and their magnitude is of direct relevance.

9.2 Introduction

Among the different existing methods for measuring muscle strength, isokinetic dynamometry is probably the only technique that allows measurement of muscle *performance*, a concept that is inclusive of various parameters such as muscle moment (strength), muscle work, muscle power, and muscle impulse. Measurement of repeated muscular contractions also enables the endurance (fatigue) profiling of muscles. Moreover, by suitably setting the dynamometer's control parameters, isometric measurements may be conducted. Not surprisingly, therefore, isokinetic dynamometry is considered the standard tool in muscle performance testing.

Isokinetic dynamometers (ISD) are specifically dedicated to measuring dynamic contractions. Their particular location within the various muscle testing methods is illustrated in Figure 9.1.

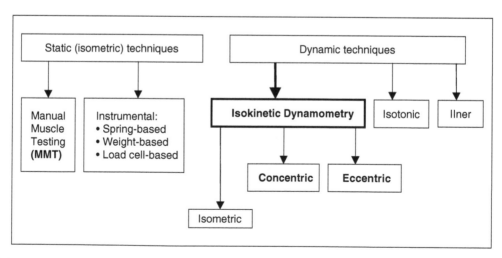

FIGURE 9.1
General classification of muscle strength testing methods. IIner = isoinertial.

9.3 Design Principles and the Moment–Angular Position Curve

The most characteristic of all muscle performance parameters is strength. By definition, muscle strength is the N·m value of the maximum point on the strength curve that describes the variation of the moment generated by the muscle group under examination, when the muscles actively contract at their *maximal* capacity to move the joint along an angular sector, the so-called range of motion (RoM). With respect to dynamic action, the term "maximal capacity" means that, at any point along the RoM, the contracting muscle group encounters the maximal resistance it can withstand while either shortening (concentric contraction) or lengthening (eccentric contraction). The resulting relationship, which is termed "moment-angular position (MAP) curve," is presented in Figure 9.2.

As has been noted in previous chapters, the moment generated by a contracting muscle group depends on two major mechanical parameters: the tension at a particular length and the lever arm of the muscle, as well as on its level of neurophysiological excitation. As one segment of the joint moves relative to the other(s), both of these parameters vary. However, whereas the muscle is capable of developing the highest tension (force) when it is in its longest configuration, the lever arm changes in a less predictable manner but normally attains its largest length at a point within the range of motion. Thus, the muscle moment curve typically assumes the shape of an inverted U (see Figure 9.2).

To continuously measure the *maximal* potential of the muscle along the total RoM requires a mechanism that provides a dynamically varying, continuously accommodating resistance, from one end of the RoM to the other. To that end, no "isotonic" device is an

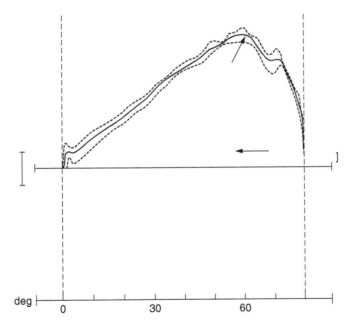

FIGURE 9.2
Consecutive isokinetic concentric MAPs of the quadriceps. The bar on the left is equal to 50 N·m. Arrow pointing to the maximal value of the ensemble, 156 N·m.

appropriate tool, because the resistance provided is constant. Equally, spring-based resistance cannot address the above requirement. Eccentric wheel + weights–based assembly may be suitable under very specific conditions, as long as the movement does not involve marked acceleration or deceleration components. However, such an assembly requires meticulous adjustment for each muscle group, a situation that denies any practical application.

The design of the original isokinetic dynamometer, which dates back to the mid-1960s, was based on hydraulic resistance. A piston, which was driven by the distal end of a body segment, forced the fluid (oil), which filled the cylinder, to move through an outlet nozzle. Thus, when a concentrically acting muscle pushed a lever arm that was connected to the piston, the resistance created by the fluid was equal in magnitude and opposite in direction to the muscular force. The rate at which the fluid was pushed through the nozzle was dictated by the nozzle cross section. Therefore, once the rate of fluid discharge reached its steady state (due to the incompressibility of the fluid), the counter-resistance provided by the fluid matched the force at any point along the piston's path. By using a mechanical selector, the nozzle cross-sectional area could be changed and, consequently, the rate of fluid discharge (i.e., the piston's velocity). This in turn was translated into the angular velocity of the relevant joint. Incorporation of a load cell between the distal segment and the piston enabled recording and measurement of the muscle output. Therefore, ISD was from its very outset a velocity-based muscle strength measurement method.

Due to technological limitations, hydraulic technology was used for another 15 years. However, as a passive device, the main limitation of the hydraulic dynamometer was that it was capable of measuring *concentric* exertions only — namely when the piston was forced to move by a larger force, which was supplied by the contracting muscle. The lack of an independent force (torque) generator as an integral part of the dynamometer meant that *eccentric* contractions, which occur when muscles are overcome by an external resistance in spite of their active contraction, could not be measured.

This limitation has been solved using computer-controlled servomotors, whose moment and power specifications are sufficient to overcome any human muscular capacity, whether the latter is acting concentrically or eccentrically. All modern ISDs incorporate an active device of this sort, where the angular velocity is accurately controlled by a PC, which also doubles as the data processing and display unit. The incorporation of the eccentric module is now regarded as critical to the understanding of muscle function and the assessment of muscle insufficiency.

9.4 The Moment–Angular Velocity Curve

If the *in vitro* relationship between force and length, as represented by the MAP curve, is the most basic mechanical parameter of skeletal muscle, the other basic parameter is the relationship between the force, or tension, developed within the muscle and its velocity or speed of contraction. This relationship has been described by Hill (1953):

$$(T + a)(v + b) = (T_o + a)/b$$

where T is the tension, T_0 is the isometric tension, v is the speed of shortening, and a and b are constants, such that $v_{max} = bT_0/a$, where v_{max} is the speed of shortening with no load.

Dynamic in vivo muscle performance may be measured by either:

1. Controlling the external load and measuring and/or calculating the resulting velocities and accelerations, or
2. Controlling the velocity and measuring the force (moment) output.

ISDs are particularly suitable using the second approach. However, in studies based on live subjects, the terms "force" and [linear] "velocity" are no longer appropriate. The former is replaced by the moment (also widely known by the term torque), and the latter is replaced by the angular velocity (ω) of the lever arm, which represents (but is not equal to) the angular velocity of the joint. It should also be emphasized that this statement is valid for angular motion testing, which is also known by the erroneous term open kinetic chain, but it should not be applied to so-called closed kinetic chain testing. In addition, the terms shortening and (active) lengthening are replaced by "concentric" and "eccentric contractions."

The isokinetically-derived moment-angular velocity (M–ω) curve (Figure 9.3) has been described in numerous publications. It is now widely recognized that this curve has two branches, relating to the concentric and eccentric strengths, respectively. Figure 9.3 depicts a typical M–ω curve from which the following relationships will be evident when the imposed test velocity is increased:

1. For concentric contractions, there is a decrease in the maximal moment developed by the muscle group. This inverse relationship is not confined to the case of a single joint. It is valid for two-joint muscles like the hamstring and, in a more global form, to multijoint systems like the extensors of the trunk.
2. For eccentric contractions, the maximal moment may rise initially, but at higher velocities, it remains stable or even decreases.

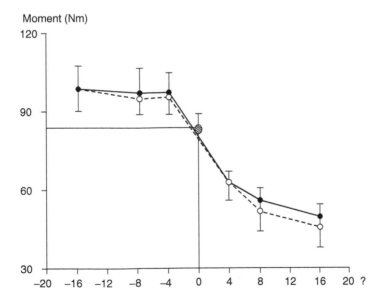

FIGURE 9.3
The *in vivo* M–curve of isokinetic grip. On the x-axis 4, 8, and 16 refer to the angular velocity of the lever arm when the fingers are concentrically flexing (dynamic grip), whereas the parallel negative values correspond to the forced extension of the fingers, namely, the eccentric component (adapted from Dvir, Z., *Clin. Biomech.*, 12, 472–481, 1997).

Additionally,

1. For the same velocity, the eccentric strength is greater than the concentric strength.
2. According to the principle proposed by Elftman (1966), the order of strength, dependent on contraction mode, is: eccentric > isometric > concentric. This order is graphically depicted in Figure 9.3 with respect to grip strength.

9.5   The Basic Components of an Isokinetic Dynamometer

A typical commercially available ISD consists of the following components (Figure 9.4):

a. *The force acceptance attachment* is the interface between the subject and the system. It consists of a metallic attachment on the lever arm, with or without foam padding, which connects to the lever arm via the load cell. The location of the unit along the lever arm is individually adjusted.
b. *The load cell* converts the force signal into an electric signal. The load cell may be part of the above attachment or may be located directly on the axis.
c. *The lever arm* provides the base for the force acceptance attachment and moves radially about a fixed axis.

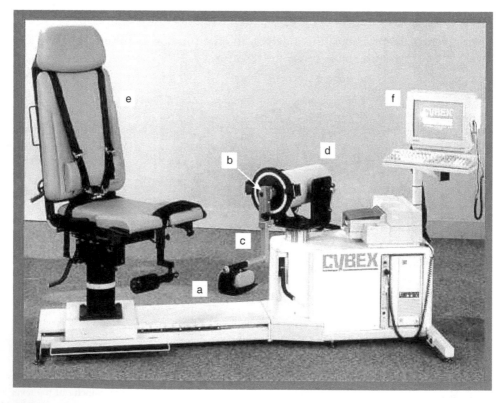

FIGURE 9.4
A modern ISD (Cybex): (a) force acceptance attachment, (b) load cell (axial), (c) lever arm, (d) head assembly + motor, (e) seat, and (f) computer.

d. *The head assembly* houses the motor responsible for the motion of the lever arm. The head may be moved up or down using an electric motor, for the purpose of alignment.

e. *The seat, or plinth*, serves to position the subject. The seat must have a stable frame and independent vertical and horizontal (forward/backward) alignment options.

f. *The control unit* consists of a personal computer and its associated peripheral equipment. The mode of operation and various other parameters (see below) are fed into the computer using the keyboard. The same computer is also responsible for the real-time data processing.

g. *Specific attachments* are required for the various applications of the isokinetic dynamometer.

9.6 Factors Affecting Isokinetic Findings

A large number of factors are involved in isokinetic testing. As a human performance test, isokinetic dynamometry brings in elements derived from the specific technology used (type of dynamometer), the protocol, the test environment, subjective factors such as motivation and apprehension, the specific pathology in case of patients, and the examiner's command of the technology and his experience. These factors are outlined in Table 9.1.

Each of the above-mentioned factors has received specific attention, some even considerable. It is clearly beyond the scope of this chapter to analyze the individual effects of these factors. Moreover, the complex interactions among them have a direct influence on the reproducibility of isokinetic test findings, an issue that is at the core of this technology. However, a glimpse of the problems involved in measurement of muscle performance using ISD will be presented with respect to two of the major test factors: range of motion and test velocity, both of which are entirely under the control of the examiner and common to all ISDs.

TABLE 9.1

Factors Affecting Isokinetic Test Findings

Factors Group	Subfactors
ISD	Type of dynamometer, calibration, acceleration ramping characteristics
Test procedure	Warm-up, stabilization, axis alignment, positioning, gravitational correction, instructions to subject, use and type of feedback
Test protocol	Range of motion, velocities, isometric preactivation, type of contraction (concentric/eccentric), number of criterion contractions, intercontraction pause, intervelocity pause, interside (R/L) pause
Environment	Ambient noise and temperature, testing time in the day
Subject/patient	Sex, age, level of activity, past injuries or disorders, presence of pain, side dominance, familiarity with isokinetic testing ("learning effect"), level of motivation, comprehension of instructions, apprehension, secondary gain
Examiner	Level of command of the technology, experience with testing patients

9.6.1   Range of Motion

The range of motion (RoM) is the most basic parameter relating to angular isokinetic testing. It describes the allowable angular displacement of the *lever arm*, which should not be confused with the motion taking place in the *biological joint*.

In specifying the RoM, the examiner determines the points of motion initiation and termination. However, this arc is not equal to the angular sector in which the lever arm moves at a constant velocity (the preset angular velocity, PAV) — the isokinetic sector (IS). The IS is always smaller than the total test RoM (as mentioned above), since at the beginning and the end of the test RoM the angular velocity of both the lever arm and the moving segment is, by definition, $0°/s$. In order for the lever arm to reach the PAV, it has to be accelerated. This means that a certain angular distance and a period of time elapse until the lever arm's velocity is equal to the PAV. The same rule applies during deceleration. Research with different ISDs and protocols has clearly indicated that the PAV occupies a limited sector, further confirming that an increase in the velocity seriously compromises the length of the PAV (Iossifidou and Baltzopoulos, 1996).

There is evidence that the size of the RoM has a direct effect on isokinetic performance. In a study by Narici et al. (1991) the effect on the peak moment (PM) of the RoM and isometric preactivation (IPA, see below) was explored. Maximal contractions of the quadriceps were performed at 180, 240, and $300°/s$. The experimental conditions consisted of a RoM of $90°$ without IPA (condition A), $90°$ with IPA equal to 25% of maximal voluntary contraction (condition B), and RoM of $120°$ with same IPA as in B (condition C). It was revealed that condition C resulted in significantly higher PM (up to 22% at the highest velocity) compared to condition A. Condition B resulted also in higher PM compared to condition A, but the differences were not equally striking. The improvement in PM was attributed to the longer *time* available for tension development as well as a greater *neural activation*. Thus, although a larger RoM seems to be responsible for higher moment output, the mediating factor — time — may be responsible for this enlargement.

9.6.2   Test Velocity

The second input parameter is the test angular velocity (ω), measured in degrees per second ($°/s$). In the case of linear motion, the unit of measurement is centimeter per second (cm/s) or inch per second (in/s). As indicated by Iossifidou and Baltzopoulos (1996), the PAV is not absolutely constant even within the IS. Indeed, their study indicated that differences between the actual velocity and the PAV were significant. Such differences are likely to characterize other commercially available dynamometers, reflecting technological limits.

9.6.3   The RoM–Angular Velocity–Contraction Time Paradigm

Regardless of the muscle–joint system, the commonly employed isokinetic test velocities typically consist of multiples of $30°/s$: 60, 90, 120, $180°/s$, etc. This practice has no biological, let alone, functional justification. Its roots may historically be traced to the prototype isokinetic dynamometer, the hydraulic Cybex, which was capable of measuring isometric and concentric contractions. This dynamometer was designed principally for testing knee musculature, and hence the use of $90°$ as a "standard" RoM was a natural choice. Bearing in mind that, at the time of its development, the idea of incorporating a computer as a controller and processor of this dynamometer was still primordial, the solution for varying the lever arm velocity had to be mechanical and limited to a discrete number of velocities.

It can therefore be assumed that, by allowing some six optional equispaced velocities (30, 60, 90, 120, 150, and 180°/s), the designers succeeded in both using existing valve technology (velocity controller) and accommodating contraction speeds that could be described from slow to fast. However, as the typical RoM was 90°, the hidden factor was, in fact, lever arm movement time. Hence, assuming ideal isokinetic conditions, and converting velocity into its precursors, time and distance, a test performed at 30°/s over a RoM of 90° should take 3 s using the formula:

$$RoM/\text{"constant"} \text{ velocity} = \text{"time"} = 90/3 = 3s$$

This can be considered a "slow" test. On the other hand, covering the same RoM at the highest velocity of 180°/s would require 0.5 s — a "fast" test.

Since the hydraulic technology was in use well into the mid-1980s, the practice of multiples of 30°/s became so well entrenched and reported in scientific publications, that little thought was given to its actual meaning. This situation is even more perplexing, given the digital capacity to achieve test velocities that may be more relevant in nature. However, time, as the truly other independent factor in the equation, almost lost any specific value in spite of the above-mentioned study by Narici et al. (1991), which underlined its singular significance. Moreover, time is rarely displayed on any dynamometer's "Results" screen. Whether this is an outcome of the reluctance of manufacturers to reveal that the IS is really limited or due to tacit understanding that the time factor is not what users are after cannot be answered with any degree of certainty.

An attempt to find out whether time is indeed the common denominator entails the use of different RoM–velocity combinations that would result in exactly the same lever-arm movement time. Moreover, such a protocol should be applied to different RoM sectors within the allowable joint RoM. For instance, if a RoM–velocity of 90°–90°/s is to be compared with a 30°–30°/s (which should ostensibly result in same movement time, see below), then the location of the short RoM sector (30°) within the long one (90°) is definitely material. Obviously, achieving such combinations while ensuring exactly the same movement time is quite impossible. It should also be borne in mind that *lever arm movement time* is just an estimate of the *contractile period of the muscle*, so that the introduction of an IPA will affect any functional relationship.

An alternative approach, admittedly less accurate to this problem, has recently been taken by Dvir et al. (2002). Shoulder concentric and eccentric flexion strength was tested in a group of normal subjects using two distinct RoMs: 80° (long, LR), which started at about 50°, and 16° (short, SR), which started at 82° and was located in the middle of the previous (Figure 9.5). To each RoM, a set of three angular velocities was adjusted as follows. For the LRoM the velocities were 40, 80, and 160°/s, whereas for the SRoM they were 8, 16, and 32°/s. The adjustment was based on the equivalence of the so-called nominal lever-arm movement times (NMT), defined as RoM/angular velocity. Thus, the experimental conditions consisted of the following NMTs: 80/40 = 16/8 (NMT = 2 s), 80/80 = 16/16 (NMT = 1 s), and 80/160 = 16/32 (NMT = 0.5 s). No attempt was made to actually compare the NMTs, since the particular dynamometer (a KinCom125E+) was not equipped with an auxiliary system for measuring the real NMT. Rather, the purpose of this experiment was to render a crude first approximation for the role of time in the outcome measure: peak moment. The findings of this study are depicted in Figure 9.6.

First, it should be realized that locating different velocities and RoMs on the same coordinate system (Moment–NMT as illustrated in Figure 9.6) was possible due to the use of the NMT as the independent variable and common denominator. Second, it is recognized that the NMT units are approximations rather than exact expressions for the actual movement times of the lever arm. With these qualifications in mind, the findings reveal

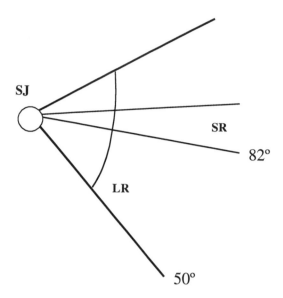

FIGURE 9.5
Short (16°) RoM (SR) vs. long (80°) RoM (LR) shoulder flexion testing (after Dvir et al. 2002).

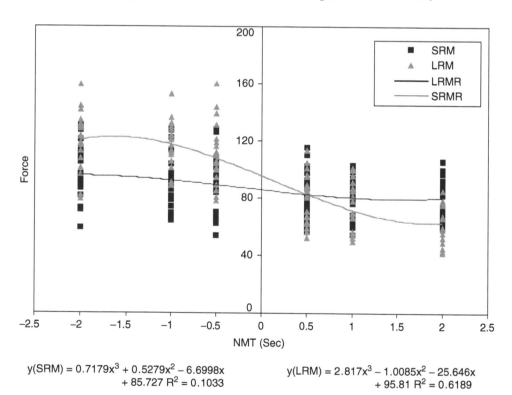

$$y(SRM) = 0.7179x^3 + 0.5279x^2 - 6.6998x + 85.727 \ R^2 = 0.1033$$

$$y(LRM) = 2.817x^3 - 1.0085x^2 - 25.646x + 95.81 \ R^2 = 0.6189$$

FIGURE 9.6
The approximate moment-angular velocity (M–ω) curves of the shoulder flexors obtained according to the protocol by Dvir et al. (2002). The numbers on the x-axis refer to the NMT. Positive and negative x-axis relates to the concentric and eccentric branches of the M–ω curve, whereas the y-axis depicts the moment of the flexors. The units on this axis are in fact expressed in N and should be multiplied by 0.4 for the moment value. The equations at the bottom of the graphs represent polynomial approximations of the Short RoM (left) and Long RoM (right) findings.

three prominent features. First, the M-NMT curves relating to both LRoM and SRoM resemble the well-known sigmoidal shape of the M–ω (moment-angular velocity) curve and are approximated by third-order polynomials. Interestingly, this third-order polynomial was also adjusted to M–ω curves in a previous study. (Tis et al., 1993). Second, for both RoMs, the eccentric strength values were higher than their concentric counterparts, although this was more conspicuous for the LR protocol. Third, if the isometric strength is calculated from the polynomial equations (by setting "X" to zero), then its values are 85 N·m and 95 N·m according to SR and LR protocols, respectively. It is worth noting that the SR did not coincide with a particularly optimal sector of the total flexion strength curve. Thus, it could be argued if the SR was located nearer to the initial part of the motion, an even better agreement between the derived SR and LR isometric strengths could be achieved. At any rate, when combined, these features add significantly to the validity of using NMT as a common basis for analyzing isokinetic strength values. Moreover, using the NMT concept may allow the scaling of previously collected data, which consisted of concentric and eccentric tests in order to establish a unified model.

Equally, if indeed testing using a short RoM provides comparable data to long or full RoM testing, it opens a serious new venue in the practice of isokinetics. Short-RoM testing has at least three distinct advantages over common techniques. First, by limiting the RoM, muscle strength may be tested effectively outside painful arcs or away from joint movement zones that may jeopardize various anatomical structures due to excess stress or strain. Second, limited RoM reduces appreciably the effect of machine–joint misalignment, since the larger the arc of movement, the less is the coincidence between the instantaneous axis of rotation and the motor axis. Third, in certain instances like trunk testing, long RoMs introduce significant gravitational effect such as may be found in trunk flexion and extension that span a RoM of, for example, 70°. By limiting trunk motion to a SR that is centered around the upright position, the values obtained may effectively reflect the net muscle strength without the need for gravitational correction. Studies along these lines indicated indeed that trunk extension strength scores were very well within previously reported scores derived from much longer RoMs (Dvir and Keating, 2001, 2003).

9.7 Measurement Units

Several outcome parameters are commonly used in isokinetic dynamometry.

9.7.1 Peak Moment

The maximal value of the MAP curve is termed the peak moment (PM), which is also synonymous with the term "strength." The peak moment does not involve specifying its location. The identification of the peak moment is not always straightforward. This is particularly apparent in two instances:

1. Eccentric contractions sometimes involve oscillations of varying amplitude in the moment–angular position curve. In pain-free subjects, repeated trials and averaging of the superimposed ensemble of curves normally serve to smooth out the curve, yet in certain instances the difficulty persists.
2. If the sampling rate of the data processing unit is fixed at 100 Hz (100 readings per second, the isokinetic industry standard) curves based on very high test

velocities appear as dotted rather than solid lines. In this case the accuracy of determining the peak moment may be compromised. Higher sampling rates solve this problem.

9.7.2 Average Moment

Strength is often expressed in terms of the average moment (AM), rather than the peak moment. The average moment is obtained from summing the moment values at each sampling point in the relevant IS and dividing the result by the number of points. The use of average moment as an alternative strength parameter demonstrates the controversy over the definition of the theoretical construct "strength": should strength be represented by a single moment value (i.e., the peak moment), or by the totality of all moment values that produce a single contraction? Since at present there is no unequivocal definition of strength (Rothstein et al., 1987), particularly one that mandates the use of the PM, the AM is a perfectly legitimate descriptor of strength.

9.7.3 Contractional Work

A parameter that is closely associated with average moment is the contractional work (CW), whose unit of measurement is the joule (J). It is a measure of the work done, or energy expended, by the muscle(s) under test. It is equal to the area under the MAP curve or, alternatively, to the average moment times the angular displacement

9.7.4 Contractional Power

Contractional power (CP), which is measured in watts (W), is an important performance parameter, which relates to the average time rate of work (P=W/t). The importance of this parameter derives from the fact that it reflects aspects other than strength, although it bears a close relationship to the latter. For instance, although the concentric peak moment is inversely related to the test velocities (moment–angular velocity relationship), contractional power may be positively related to the latter, as has been indicated with respect to the ankle plantar flexors (Gerdle and Fugl-Meyer, 1985). This phenomenon results from the nonlinear (decelerated) decay in the moment–velocity curve. Although it is theoretically valid to relate to an instantaneous P (the power based on a small angular sector), it probably has no value for the purpose of interpreting test findings based on normal or patient populations.

9.7.5 Contractional Impulse

Contractional impulse (CI) is the product of the moment multiplied by the time for which it acts, where I is the value of the impulse, namely:

$$I = M_{\text{average}} \times T$$

Impulse is measured in N·m·s. Studies of athletes and patients have shown that impulse has a special significance. Sale (1991) analyzed the performance of sprinters vs. cross-country skiers, using knee extension performance. The contractional impulse at 180°/s was the best discriminator between the two groups, while the peak moment at 30°/s revealed no differences. In another study in patients suffering from patellofemoral pain syndrome, the contractional impulse was highly correlated with the subjective pain ratings, whereas the average moment was not (Dvir et al., 1991). Both studies have suggested the wider application of this parameter.

9.8   Outcome Parameters and Reproducibility of Isokinetic Test Findings

The main operational concepts in isokinetics are *difference* and *change*. The first relates to either a bilateral strength (or other mechanical output parameter) difference, such as may exist as a normal variant in predominantly "unilateral" individuals (e.g., tennis players, with respect to shoulder function) or the difference between symmetrical muscle groups when one is injured. Alternatively, differences in dynamic muscle performance may be found upon comparison of an individual subject and a reference group (e.g., in trunk testing due to the absence of a contralateral segment) or for screening purposes. Change may be the objective of the test when one is expected — namely as a result of training in either normal subjects or patients. It is particularly with respect to the latter group that isokinetics has become such a powerful clinical tool.

The indices used for determination of either a difference or a change are generally based on bilateral comparison of the strength of relevant muscle. For instance, in a case of tear (partial or complete) of the shoulder rotator cuff, measurement of the internal and external rotators of the humerus as well as that of the humeral elevators is indicated. Using identical protocol — e.g., subject position (seated), arm elevated to the same degree and in the scapular plane, range of motion, velocities, concentric as well as eccentric efforts, etc. — the strength deficiency of the involved side should be calculated according to the following formula:

$$\% \text{ strength loss} = [(SUS - SIS)/SUS] \times 100$$

where SUS and SIS stand for strength of the unimpaired side and strength of the impaired side, respectively. The term strength is inclusive of the PM and AM and refers equally to concentric and eccentric strengths. Figure 9.7 depicts the results of an concentric and eccentric isokinetic test of a patient with a compromised quadriceps due to motor accident and fracture of the left femoral shaft.

Based on these the information in Figure 9.7(A) and 9.7(B), the deficiency of the L quadriceps relative to its R counterpart is:

$$\text{Concentric at } 120°/s = [(347 - 202)/347] \times 100 = 42\%$$

$$\text{Concentric at } 30°/s = [(402 - 225)/402] \times 100 = 44\%$$

$$\text{Eccentric at } 120°/s = [(417 - 272)/417] \times 100 = 35\%$$

$$\text{Eccentric at } 30°/s = [(376 - 242)/376] \times 100 = 36\%$$

Based on all four tests the average strength deficiency is therefore 39%.

As evident from this example, the strength deficiency based on the concentric test is not equal to that derived from its eccentric counterpart. Since eccentric efforts are less voluntarily controllable than their concentric counterparts, it could be argued that they reflect in a more accurate way the true capacity (weakness) of the relevant muscles. Hence the

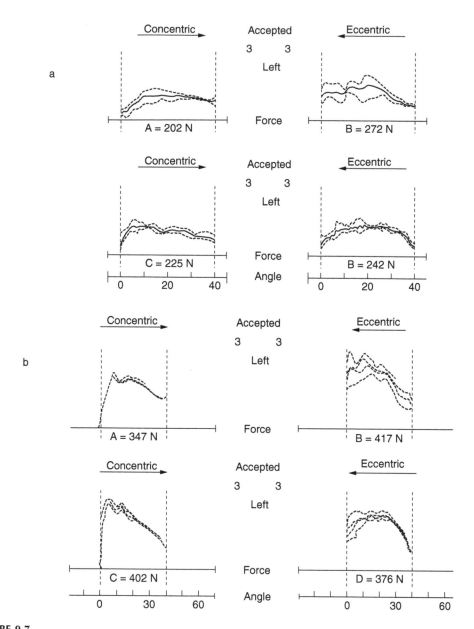

FIGURE 9.7
(a) MAP curves of the involved (L side) quadriceps. Upper graphs refer to the concentric (on the left) and eccentric (on the right) tests at 120°/s. Lower graphs refer similarly to tests performed at 30°/s. The numbers at the bottom of the graphs refer to the AM (in N). (b) MAP curves of the uninvolved (R side) quadriceps. Upper graphs refer to the concentric (on the left) and eccentric (on the right) tests at 120°/s. Lower graphs refer similarly to tests performed at 30°/s. The numbers at the bottom of the graphs refer to the AM (in N).

eccentric rather than the concentric capacity should serve as the criterion. On the other hand, the lingering practice among clinicians is to view concentric contractions as the true representative of muscle potential. Moreover, if tests are carried out using more than one velocity, the likelihood of obtaining identical deficiency figures is extremely low. As a result, averaging of the concentric and eccentric deficiencies is a possible solution that should be looked into very carefully. Such an averaging process should include all test conditions (i.e., two distinct velocities or NMTs).

In the case where a contralateral segment does not exist, as is the case with trunk sagittal movements (flexion and extension), some practitioners have argued in favor of comparing the individual strength figures with reference scores. It should, however, be borne in mind that such a comparison necessitates databases that are compartmentalized at least along the sex, age, and activity level of the subject and, where patients are concerned, also in terms of the specific pathology. Such a database has never been established, and therefore the justification for applying it is at best low.

Beside bilateral comparison of same parameter, be it PM, AM, CW, CP, or CI, an alternative index utilizes antagonistic muscle activity. The rationale behind this approach is the existence of a strength balance between antagonistic muscles. It was argued that imbalance, as indicated by a relative weakness, could precipitate joint movement disorders. One of the first such ratios to be tested isokinetically related to the hamstring–quadriceps relationship, the so-called H/Q ratio. When it was initially used (Osternig et al., 1983) the ratio was based on the concentric performance of these muscles. This ratio was, as expected, velocity dependent and provided little insight in terms of rehabilitation of knee muscles, particularly in patients presenting with anterior cruciate ligament (ACL) problems, either chronic insufficiency or the postoperative status. Specifically, it was apparent that this ratio lacked differentiating power. The same principle was later applied to shoulder internal vs. external rotators and ankle invertors vs. evertors but leads to equally disappointing results.

An alternative approach has been proposed by Dvir et al. (1989), which was based on the actual way antagonistic muscles operate. Previous work by Solomonow et al. (1987) has indicated that coactivation of the quadriceps and hamstring took place through opposite contraction modes, the quadriceps contracting concentrically and the hamstring eccentrically. Therefore, in order to assess the balancing nature of the hamstring, particularly in the ACL-deficient knee the hamstring–quadriceps ratio should correctly be H_e/Q_c (i.e., the eccentric strength of the hamstring divided by the concentric strength of the quadriceps), the so-called dynamic control ratio (DCR). In their study of chronic ACL patients, the values of the dynamic control ratio, H_e/Q_c, were compared with same-contraction-mode HQ ratios, H_{con}/Q_{con} and H_{ecc}/Q_{ecc}, at the test velocity of 30°/s. The findings indicated that, whereas H_{con}/Q_{con} and H_{ecc}/Q_{ecc} differed by no more than 3% between the involved and sound knees, there was a significant difference in the dynamic control ratio. A reduction of 11% in the concentric strength of the deficient-side quadriceps accounted for this finding. The DCR proved to be a useful index for differentiating compromised shoulders (Bak and Magnusson, 1997) but failed to reveal differences between unstable and stable ankles (Kaminski and Hartsell, 2002; Kaminski et al., 2001).

When bilateral deficiencies are identified, how large should they be to indicate a true difference? A similar question applies when variations in performance are recorded with respect to the same muscle group. This is one of the most critical issues in isokinetics, which has received astonishingly little attention. One of the first attempts to answer this question was by Sapega (1990). In a leading article, it was suggested that in expected weakness (e.g., following injury), a bilateral difference of 10–20% is possibly abnormal, whereas a difference of 20% or more should be considered as "almost certainly pathological." Strength differences of less than 10% could be largely attributed to factors like measurement error and/or side dominance. However, although widely quoted and used, these benchmarks were not subjected to a comprehensive and systematic probing.

Obviously, if the measurement error associated with isokinetic dynamometry had been zero, any change or variation would have been considered significant. Thus, telling whether a change has indeed occurred is intimately related to the problem of reproducibility. The importance of quantitatively defining reproducibility has gathered increasing recognition in recent years, due to the awareness of the limitations associated with existing correlational techniques [Pearson's r, ICC (intraclass correlation coefficient)] as well as to the emergence of alternative parameters and methods such as the standard error of measurement (SEM), the limits of agreement (LOA), and the coefficient of variation CVp (Bland and Altman, 1986, 1990; Neville and Atkinson, 1997). The principle governing these methods is the search for a number or numbers that will enable the user of isokinetic dynamometry to relate absolute moment values to the expected fluctuations that occur due the factors enumerated in the previous section. In spite of their widespread use, neither the ICC nor Pearson's r provide such insight. Moreover, choice of the samples reported in the scientific literature has often been complicated by a large heterogeneity, which produced inflated scores. Thus, the mixing of sexes and activity levels resulted in extreme strength scores which invariably led to artificially high correlations. This in turn was interpreted as endorsing the use of a particular protocol or ISD.

A different approach looked into the absolute error and interpreted the test-retest results in terms of expected error. For instance, using the SEM, Keating and Matyas (1998) were able to indicate that the error incurred in measuring strength by an ISD increased proportionally with the absolute strength. As a result, stronger subjects need to achieve greater absolute strength difference in order to demonstrate a significant change. This measurement property is termed heteroscedasticity, whereas no association between the absolute value and the error is termed homoscedasticity. If SEM is the parameter of choice, the magnitude of change necessary to indicate significance is commonly assumed to be 1.96 SEM (i.e., at 95% level of confidence). For instance, if for a specific patient (subject) group, muscle group, ISDn and protocol, the SEM stands at 15 N·m, this means that no less than a difference of 30 N·m must be recorded to indicate change. A similar approach is exemplified by the CVp, which sets the limits of change. Using this statistical tool, Madsen (1996) was able to calculate the necessary strength variations required to indicate significant change in trunk flexion and extension. The figures proposed by Madsen were exceedingly high, ranging into 30%.

Another approach is based on the use of the LOA, a method that subtracts the results of the retest with those of the test and computes the SD of the differences. Figure 9.8 is a presentation of a LOA procedure performed on a set of isokinetic data that underlines the heteroscedastic nature of the findings. From these pictorial representations, the error may be estimated and limits set appropriately.

Analyzing reproducibility of ISD-based findings leads to two reflections. First, there is a pressing need to decrease the measurement error in all sorts of isokinetic measurements. This entails adherence to very strict protocols and reduction of the effect of the various factors that were hitherto discussed. It is possible, for example, that limiting the range of motion and avoiding the use of high velocities (low NMTs) could be crucial in this respect. Second, there is an even more basic question related to the difference between statistical and clinical significance. Reducing the isokinetic measurement error to the minimal possible could theoretically result in zero error, as pointed out before. Does this mean that any change is of clinical/functional significance?

No affirmative answer is yet available, and one cannot be stated unless judged against another set of variables. For instance, if the strength of the trunk extensors of an individual who is required to lift heavy loads varies from 110 N·m before rehabilitation to a significant (statistically) 115 N·m after the intervention, does it confirm reemployment at the same task? One way to tackle this question is to look for general guidelines for this work station and decide that, if the job demand calls for 112 N·m as a benchmark, then this particular worker is probably ready to resume a normal work schedule. However, if the current results prevail, an increase in strength of about 20% (Dvir, 2004), which seems to be a practical guideline, does not permit such a step. Using the more stringent guidelines suggested by Madsen (1996) puts this employee even further from return to work. The

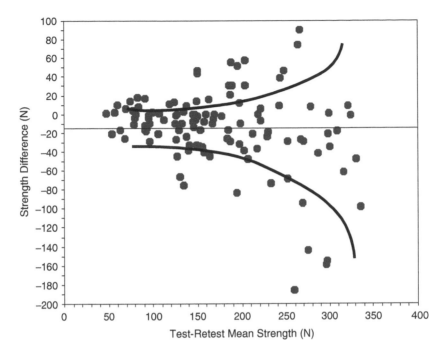

FIGURE 9.8
Heteroscedastic distribution of the differences of isokinetic strength scores. The mean individual strength appears on the x-axis, whereas the corresponding differences appear on the y-axis. The black curves are qualitative approximations.

final decision may therefore be "functionally" based by incorporating the isokinetic test results with other tests and factoring in elements such as worker motivation and enhanced safety features.

9.9 Special Topics

Absent from this chapter is an examination of the application of isokinetic dynamometry to specific joint–muscle systems. In addition the clinical ramifications of isokinetic findings have been mentioned only in passing. However, one of the persistent questions asked very often by professionals, as well as by people outside the profession, is how one knows whether a subject or a patient performed optimally. This question has serious implications, and it touches the very foundations of isokinetic dynamometry as a system for measuring human performance. As stated by Polatin and Mayer (1994), if *full effort* (in the very broad sense of the word) cannot be confirmed, no inference or judgment can be made based on the measured trait. This statement holds true irrespective of the modality that is being measured, such as muscle strength and fatigability, hearing, or vision.

There are a number of reasons for not producing a maximal muscular effort. First, if the subject perceives the test to be threatening or painful, the level of tension would be lowered. Furthermore, it may well be the case that, in the first exposure to the test, the subject may indeed produce a maximal effort but the experience of pain may deter him from repeating the same level of exertion in the next series of contractions. However, in spite of pain, subjects may still produce what may be interpreted a maximal level of effort

(Mayer et al., 1985). Second, the subject may have difficulty in comprehending the instructions — a problem with people of limited intellectual ability or perception of body parts. Third, since muscle weakness is a recognized impairment, there may be a motive to underperform in order to gain. It is the latter reason that rendered the medicolegal applications of isokinetics of such importance.

In terms of isokinetic measurements, the evolution of research relating to sincerity of muscular effort has followed two streams. One focused on the consistency of performance using the coefficient of variation (CV = [SD/mean] ×100) as the criterion. The other and more recent approach relates to the use of physiologically derived parameters, mainly the difference between the eccentric-to-concentric strength ratios obtained at relatively high and relatively low test velocity (Dvir, 2004). As for the first approach, the principle of consistency of performance as a criterion for optimality (sincerity) of effort has now been all but abandoned. Almost without exception, all studies so far have indicated that although the peak- or average-moment-based CV of an ensemble of MAP curves was significantly greater in submaximal compared with maximal effort, no distinct cutoff CV score could be established that distinguished effectively between the two levels of performance (Hamilton Fairfax et al., 1997). Moreover, it was demonstrated that normal subjects could reproduce submaximal effort very well.

A compact alternative approach is based on the incorporation of concentric and eccentric contractions. This method has been tested in a variety of muscles groups and in both genders and applied to knee extensors, flexors of the elbow, extensors of the trunk, grip muscles, and flexors of the shoulder. This method was also applied to clinical populations comprising shoulder, knee, and hand (grip) disorders and chronic low back-pain patients. For a comprehensive description see Dvir (2004).

The physiological rationale for the method derives from the different mechanical output of eccentric and concentric contractions, especially when performed at a submaximal level. When maximal effort is produced, muscle strength is governed by the M–ω function. As will be remembered, in concentric effort, strength is generally reduced with increase in the angular velocity of the relevant joint/s. On the other hand, the strength produced in eccentric contractions tends to change only slightly, with the majority of studies reporting an increase in strength when the angular velocity increases. Consequently in healthy subjects, the eccentric to concentric (Ecc/Con) strength ratio, which is defined with respect to the same velocity, is in principle greater than 1 under all isokinetic test conditions. If the test velocity is very low, namely approximating isometric conditions, the Ecc/Con ratio approaches unity and, in some instances, invariably indicating the presence of some disorder, it may be even less than 1.

These relationships are much accentuated when subjects are instructed to produce a submaximal effort (SE), irrespective of the nature of instruction (although not necessarily to the same extent). It seems that the ability to control muscle tension under SE differs significantly between the two contraction modes, reflecting, most probably, the effect of distinct neurophysiological demands. Whereas concentric contractions bring on the required motion, eccentric contractions are inhibitory and defensive in nature, operating in order to restrain motion. This specific activity means, by necessity, a significantly reduced precision in tension production compared to the level of control exercised during concentric contractions. Moreover, a higher joint angular velocity is probably interpreted as a more threatening situation, resulting in a larger motor unit pool recruitment and greater muscular tension. Indeed, neurophysiological studies indicate that eccentric activity may even be modulated by different brain structures (Enoka, 1996).

The variations in response pattern to increased joint velocity are well reflected upon comparison of SE and maximal effort (ME). As far as concentric activity is concerned, when subjects are asked to perform submaximally the difference between their strength output at low and high velocities is quite considerable. In contrast, the drop in eccentric output is rather moderate.

Based on these variations it was suggested that, by comparing Ecc/Con ratios related to velocities that are set a sufficiently large gradient, one could potentially differentiate between ME and SE. As a result, a parameter termed DEC (Difference between Ecc/Con ratios) has been defined by the following formula:

$$DEC = (Ecc/Con)_{high} - (Ecc/Con)_{low}$$

where the "high" and "low" refer to the test angular velocities. The gradient employed, 1:4, was not arbitrary but, in most of the cases, corresponded nominally (in terms of °/s) to half and twice the joint range of motion (RoM) for the "low" and "high" velocities, respectively. For example, for a RoM of 50° the "low" and the "high" velocities are 25°/s and 100°/s, respectively. Likewise a RoM of 12° leads to "low" and "high" velocities of 6°/s and 24°/s. It has also been observed that a reduction in the RoM necessitated a greater gradient in order to retain a comparable efficiency of differentiation between ME and SE (Dvir et al., 2002).

In a series of studies exploring the differentiation capacity of the DEC, subjects were asked to exert ME and then to perform submaximally according to different instructions. In the initial studies the instructions related to a percent of effort or to what the subjects perceived as a level of effort they could best reproduce (Dvir, 1997; Dvir and David, 1996). In later studies (Dvir, 1999; Dvir and Keating, 2001; Dvir et al., 2002) subjects were presented with a vignette describing a fictitious injury. They were instructed to attempt and convince the examiner that this injury was still affecting them in the form of muscle weakness. In all studies there was a highly significant difference between the strength levels recorded in ME and SE. However more important, the DEC scores differed in an equally significant way. Using tolerance intervals, a statistical tool which allows establishment of cutoff values, it was possible to state at a given level of confidence if the effort was maximal or submaximal. This statement is based on either staying below the cutoff (ME) or exceeding it (SE).

Consider, for instance, the strength curves depicted in Figure 9.7(B). The DEC score is calculated for the involved and uninvolved sides as follows:

$$DEC\ (L) = (Ecc/Con)_{120} - (Ecc/Con)_{30} = 272/202 - 242/225 = 1.35 - 1.08 = 0.27$$

$$DEC\ (L) = (Ecc/Con)_{120} - (Ecc/Con)_{30} = 417/347 - (376/402) = 1.20 - 0.94 = 0.26$$

Both DECs are well below the cutoff at 95% (1.02), and the effort in this case was therefore judged to be maximal.

References

Bak, K. and Magnusson, S.P. 1997. Shoulder strength and range of motion in symptomatic and pain-free elite swimmers, *Am. J. Sports Med.*, 24, 454–459.

Bland, J.M. and Altman, D.G. 1986. Statistical methods for assessing agreement between two methods of clinical measurement, *Lancet*, 1, 307–310.

Bland, J.M. and Altman, D.G. 1990. A note on the use of the intraclass correlation coefficient in the evaluation of agreement between two methods of measurement, *Comput. Biol. Med.*, 20, 337–340.

Dvir, Z. 1997. The measurement of isokinetic fingers flexion strength, *Clin. Biomech.*, 12, 472–481.

Dvir, Z. 2004. *Isokinetics: Muscle Testing, Interpretation and Clinical Applications*, 2nd ed. (Edinburgh, UK: Churchill Livingstone).

Dvir, Z., Eger, G., Halperin, N., and Shklar, A. 1989. Thigh muscle activity and anterior cruciate ligament insufficiency, *Clin. Biomech.*, 4, 87–91.

Dvir, Z., Halperin, N., Shklar, A., and Robinson, D. 1991. Quadriceps function and patellofemoral pain syndrome. Part I: pain provocation during concentric and eccentric isokinetic activity, *Isokinet. Exerc. Sci.*, 1, 26–30.

Dvir, Z. and Keating, J. 2001. Identification of feigned isokinetic trunk extension effort: an efficiency study of the DEC, *Spine*, 26, 1046–1051.

Dvir, Z. and Keating, J. 2003. Trunk extension effort in chronic low back dysfunction patients, *Spine*, in press.

Elftman, H. 1966. Biomechanics of muscle, *J. Bone Joint Surg.*, 48A, 363–373.

Gerdle, B. and Fugl-Meyer, A.R. 1985. Mechanical output and iEMG of isokinetic plantarflexion in 40–64 years old subjects, *Acta Physiol. Scand.*, 124, 210–211.

Hamilton Fairfax, A., Balnave, R., and Adams, R. 1997. Review of sincerity of effort testing, *Saf. Sci.*, 25, 237–245.

Hill, A.V. 1953. The mechanics of active muscle, *Proc. R. Soc. Lond. Biol Sci.*, 141B, 104–117.

Iossifidou, A.N. and Baltzopoulos, V. 1996. Angular velocity in eccentric isokinetic dynamometer, *Isokinet. Exerc. Sci.*, 6, 65–70.

Kaminski, T.W., Buckley, B.D., Powers, M.E., Hubbard, T.J., Hatzel, B.M., and Ortiz, C. 2001. Eversion and inversion strength ratios in subjects with unilateral ankle instability, *Med. Sci. Sports Exerc.*, 33S, 135.

Kaminski, T.W. and Hartsell, H.D. 2002. Factors contributing to chronic ankle instability: a strength perspective, *J. Athl. Train.*, 37, 394–405.

Keating, J.L. and Matyas, T.A. 1998. Unpredictable error in dynamometry measurements: a quantitative analysis of the literature, *Isokinet. Exerc. Sci.*, 7, 107–121.

Madsen, O.R. 1996. Trunk extensor and flexor strength measured by the Cybex 6000 dynamometer. Assessment of short-term and long-term reproducibility of several strength variables, *Spine*, 21, 2770–2776.

Mayer, T.G., Smith, S.S., Keeley, J., and Mooney, V. 1985. Quantification of lumbar function. Part 2: sagittal plane trunk strength in chronic low-back pain patients, *Spine*, 10, 765–772.

Narici, M.V., Sirtori, M.D., Mastore, P., and Mognoni, P. 1991. The effect of range of motion and isometric preactivation on isokinetic torques, *Eur. J. Appl. Physiol.*, 62, 216–220.

Neville, A.M. and Atkinson, G. 1997. Assessing agreement between measurements recorded on a ratio scale in sports medicine and sports science, *Br. J. Sports Med.*, 31, 314–318.

Osternig, L.R., Hamill, J., Sawhill, J., and Bates, B.T. 1983. Influence of torque and limb speed on power production in isokinetic exercise, *Am. J. Phys. Med.*, 62, 163–171.

Polatin, P.B. and Mayer, T.G. 1992. Quantification of function in low back pain, in *Handbook of Pain Assessment*, D.C. Turk and R. Melzack, Eds. (New York: Guilford Press).

Rothstein, J.M., Lamb, R.L., and Mayhew, T.P. 1987. Clinical uses of isokinetic measurements: critical issues, *Phys. Ther.*, 67, 1840–1844.

Sale, D.G. 1991. Testing strength and power, in *Physiological Testing of the High Performance Athlete*, 2nd ed., J.D. MacDougall, H.A. Wenger, and H.J. Green, Eds. (Champaign, IL: Human Kinetics).

Sapega, A.A. 1990. Current concepts review: muscle performance evaluation in orthopaedic practice, *J. Bone Joint Surg.*, 72A, 1562–1574.

Solomonow, M., Baratta, R., Zhou, B.H., Shoji, H., Bose, W., Beck, C., and D'Ambrosia, R. 1987. The synergistic action of the anterior cruciate ligament and thigh muscles in maintaining joint stability, *Am. J. Sports Med.*, 15, 207–213.

10

Hand Strength

Mircea Fagarasanu and Shrawan Kumar

CONTENTS

ABSTRACT Despite a large number of published materials that deal with upper extremity disorders, hand muscle strength has not been extensively addressed from an ergonomic point of view. This chapter represents a review of the state of knowledge in this area. Injury risk factors, grip and pinch force production, as well as the effect of upper extremity joints deviation on hand strength are presented. Older worker concerns and industrial ergonomic solutions are also addressed.

10.1 Ergonomic Relevance

Having a complete synopsis of hand injuries risk factors, as well as tabulated data for grip and pinch force values, would provide valuable information for ergonomic programs. Also, clarifying the still unknown relationship between shoulder/elbow/wrist position and hand strength, the interconnection between fingers' force, and the differences between young and older workers would allow for targeted ergonomic interventions.

10.2 Introduction

The introduction of power tools replaced some of the physical work needed from workers, yet the majority of work tasks that involve the upper extremities experience an elevated stress at the hand–tool interface. With 27 bones, 39 muscles, including 15 muscles which serve the thumb and index finger, and more than 25 degrees of freedom, the hand is the most complex musculoskeletal part of the human body. Due to its complex anatomical structure and intense usage (Muralidhar et al., 1999) and the fact that all occupational musculoskeletal injuries are biomechanical in nature (Kumar, 2001), ergonomic interventions are vital. Also, due to the fact that the hand is the main vehicle of motor activity and the most important sensory and tactile organ (Napier, 1956), perfect matching between tools/devices and hand characteristics is essential in all workplaces (Imrhan and Loo, 1989).

Most of the injuries that occur to the hand, forearm, and arm are due to poor design (Fransson and Winkel, 1991). The rapid transformation in workplace and task, in both office and industrial environments, causes high stress levels on the hand musculoskeletal system. These include inappropriate force exertions, highly repetitive tasks, and awkward postures, sometimes for a prolonged time with little rest (Kumar, 2001). Work-related hand musculoskeletal disorders develop from such exposures. The continuous increase in the incidence of upper extremity disorders constitutes strong evidence supporting the idea that workers are either anatomically ill-adapted to sustain such demanding tasks physically, or mentally incapable of coping with elevated psychosocial stresses (Kumar, 2001).

With scarcity of information regarding submaximal grip and pinch exertions (Radwin et al., 1992), the interaction between different joint deviations and upper extremity muscle's loads, and fatigue onset and progression, it is difficult to design safe tasks and tools. The need for reporting force application (Kroemer, 1970) and hand force production data is evident. Job and device redesign that take into account ergonomic data should be implemented in order to have a reduction in both work-related musculoskeletal disorders and ensuing claims. Although the initial costs may increase the financial burden for a period of time, the ergonomic modifications will enhance both productivity and workers' safety in the long term, thus affecting the economy positively.

10.3 Magnitude of the Problem

Although the majority of studies (Gerber, 1998; Kumar, 2001; Kumar and Simmonds, 1994; Muralidhar and Bishu, 2000; Muralidhar et al., 1999; Sukthankar and Reddy, 1995)

TABLE 10.1

Wrist, Hand, and Fingers Nonfatal Injuries as Percentages from Total Number of Work-Related Upper Extremity Disorders (WRUED) Involving Days Away from Work for 1998–2000 Interval

Year	WRUEDs per 10,000 Workers	Wrist Injuries	Hand Injuries (Except Fingers)	Fingers + Fingernails Injuries	Total (Wrist, Hand, and Fingers)
1998	45.2	21.01%	18.36%	36.72%	76.09%
1999	43.9	21.18%	17.76%	37.58%	76.52%
2000	41.7	22.30%	18.22%	36.21%	76.73%

Adapted from Bureau of Labor Statistics, 2002.

supports the relationship between office/industrial work and hand/wrist musculoskeletal disorders, there is still a scarcity of specific data regarding the recommended posture and pace for any task. Many industrial and office jobs involve extensive forearm and hand/fingers repetitive movements with an important force component. Increased prevalence of musculoskeletal disorders is present in almost all these activities (Hansson et al., 1996). In industries where there is an extensive use of tools, such as meat processing and packaging, the poultry industry, and automobile upholstering, the work-related upper extremity disorder prevalence is even greater (McGorry, 2001).

In 1984, 9% of all compensable work-related injuries were due to tool use, with over 50% related to the upper extremities (Mital, 1991). From 1987 to 1989 there was a 100% increase in the number of cases of cumulative trauma disorders (Wiker et al., 1989), reaching over 50% of all occupational illnesses in the United States (Sommerich et al., 1993). Cumulative trauma disorders were one of the biggest health problems in the 1990s (Halpern and Fernandez, 1996). Cumulative trauma disorders were the cause for 62% of all work-related illnesses reported in the United States in 1992 (Bureau of Labor Statistics, 1992). This constitutes a 38% increase compared to figures from 1981 (Halpern and Fernandez, 1996). According to the National Institute for Occupational Safety and Health (NIOSH) (1997), in the United States, 32% (705,800) of all work-related injuries arose from repetitive motion or overexertion. Of those, 92,576 cases were due to repetition (typing, repetitive grasping, repetitive hand tool use, etc.), with 55% affecting the wrist.

During the last few years, a slight decrease in the absolute number of work-related upper extremity disorders (WRUEDs) requiring days away from work was observed (BLS, 2002), the percentage of injuries that affect wrist, hand, and fingers is on rise (Table 10.1).

The impact of work-related upper extremity disorders on industrial performance has been noted by Hashemi et al. (1998), who assessed that, in a large workers' compensation carrier, WRUEDs accounted for 3.6% of all claims and 6.4% of all costs. Hand injuries caused the longest absence from work (Bureau of Labor Statistics, 2002) and hence were associated with higher costs (cost per claim 13 times greater for the WRUED when compared to the overall average).

10.4 Risk Factors for Hand and Wrist Injuries

A wide variety of hand/wrist injuries risk factors were described in several studies (Chaparro et al., 2000; Drury et al., 1985; Muralidhar and Bishu, 2000; Muralidhar et al., 1999; Silverstein et al., 1986; Stal et al., 1999). In almost all industrial and office work activities the hand–tool interface has the most elevated hazard for the hands (Muralidhar and Bishu, 2000). Fransson-Hall and Kilbom (1993) noted that the thenar area, the area around the pisiform bone, and the region between thumb and index finger are the hand regions most sensitive to externally applied surface pressure (EASP). This information is supported by Muralidhar and Bishu (2000), who determined that the lowest tolerance to pressure is present at the top of finger V, between fingers I and II, and at the base of finger II. Localized pressure on hand and wrist has been proven to be associated with musculoskeletal disorders, with a direct relationship between continuous pressure exerted by tools on the palm base and carpal tunnel syndrome (CTS) (Ketola et al., 2001). Since pain in the hand caused by extensive external pressure has been stated as a limiting factor in the performance of work that requires handheld tools use (Fraser, 1980), one should act promptly to remedy it. An adaptation to the tasks could allow one to work beyond a certain discomfort level that may elevate risks. Since the majority of occupational musculoskeletal disorders are caused by overexertion (Kumar and Simmonds, 1994), continuous forceful hand exertions (Bernard, 1997), nonneutral wrist positions, and awkward hand and finger postures (Ketola et al., 2001; Richards, 1997) constitute important risk factors for hand injuries. Due to their viscoelastic properties, prolonged loading causes permanent tissue deformation (Kumar, 2001). The capacity of producing greater force, in order to compensate for the decrease due to awkward postures, might be absent in pathologic conditions (Richards, 1997), raising the risk of injuries in such cases. Asymmetric activities lead to overload of several muscles, causing disproportionate stress concentration on tissues (Kumar, 2001).

In a study that assessed the role of repetitiveness and force application in work-related injuries, Silverstein et al. (1986) observed that the high forcefulness–high repetitiveness combination was the most hazardous, followed by high repetitiveness and low forcefulness, low repetitiveness and high forcefulness, and low repetitiveness and low forcefulness. Repetitive movements of the hand and wrist in both office and industrial tasks have also been cited as important risk factors (Ketola et al., 2001; Stal et al., 1999). Close causal relationship between age and hand performance has been assessed by previous studies (Imrhan and Loo, 1989; Mathiowetz et al., 1985; Ranganathan et al., 2001). There is a significant reduction in wrist ulnar deviation, flexion, and grip strength for men between age 60 and 90. For the same wrist deviations and force exertions, the elderly are the end of wrist range of motion (ROM), leading to increased muscle load (Chaparro et al., 2000). In long term, joint kinematics and muscle load sharing patterns different than the physiologic ones occur, leading to increased risk of injury (Kumar, 2001). The most important risk factors for hand and wrist injuries are presented in Table 10.2.

TABLE 10.2

Personal and Occupational Risk Factors for Hand and Wrist Injuries

Hand and Wrist Injuries Risk Factors		
	Occupational	
Personal	**Industrial**	**Office**
Age over 50 years	Localized pressure	Repetition
Female gender	Repetitive movements	Prolonged deviated postures
Previous injuries	Awkward postures	Percentage of time typing
Hand preference	Excessive hand force production	Lack of rest
Menopausal women	Working with cold hands	Preference for certain fingers
Obesity	Vibration	
	Fatigue	

10.5 Hand Force Production

10.5.1 Factors that Affect Hand Strength

Almost all working tasks involve a certain degree of hand force application. Both productivity and the risk of injury are affected by workers' hand performance. There is a complex relationship between industrial design and applied force and/or hand dexterity. On one hand, poor ergonomic design determines a decrease in hand force and precision; on the other hand, continuous feedback from experienced workers is useful to ensure proper interaction between man and machine.

Hand force production is generated by a combination of intrinsic and extrinsic muscles. The median and ulnar nerves control the hand intrinsic muscles, which are: 7 interossei (4 dorsal and 3 palmar), hypothenar muscles (flexor digiti quinti, opponens digiti quinti, and abductor digiti quinti), adductor pollicis muscle, the thenar muscles (oponens pollicis, abductor pollicis brevis, and flexor pollicis brevis), and the lumbrical muscles (Kozin et al., 1999). Hand extrinsic muscles are represented by: flexor pollicis longus (FPL), flexor digitorum superficialis (FDS), flexor digitorum profundus (FDF), abductor pollicis longus (APL), extensor pollicis longus (EPL), extensor pollicis brevis (EPB), extensor digitorum (ED), extensor digiti minimi (EDM), and extensor indicis (EI). The principal actions of extrinsic muscles are flexion and extension of the fingers and wrist deviation in sagittal (flexion and extension) and frontal plane (ulnar and radial deviation). Their names reflect the actions on the fingers. The extrinsic hand muscles roles in wrist movement are presented in Table 10.3.

Both active and passive components affect total muscle force production (Keir et al., 1996). This is also affected by muscle length (Keir et al., 1996). Muscles operating the hand have the mechanical properties of skeletal muscles: length–tension and force–velocity relationships (Dvir, 1997). During manual task completion, the position of upper extremity joints should cause optimal length for the muscle(s) being used. L_0 (optimal muscle length) is the length at which maximal isometric tension is exerted (Close, 1972; Gordon et al., 1966). Since tendon tissue has stiffness that is two orders of magnitude higher than that of muscle tissue (Keir et al., 1996), the greatest part of the segment excursion is due to muscular tissue elongation. Because the angles of pennation in all forearm muscles, except flexor carpi ulnaris (FCU), are less than 10°, the differences in the passive force present in these muscles are not due to pennation variability (Keir et al., 1996). Precise force application is caused by balanced forearm muscles contractions. The maximum applicable

TABLE 10.3

Extrinsic Hand Muscles Role in Wrist Movements

Muscle	Flexion	Extension	Radial Dev.	Ulnar Dev.
Thumb's Extrinsic Muscles				
Flexor pollicis longus (FPL)	X		X	
Abductor pollicis longus (APL)	X		X	
Extensor pollicis brevis (EPB)	X		X	
Extensor pollicis longus (EPL)		X	X	
2nd–5th Fingers' Extrinsic Muscles				
Flexor digitorum superficialis (FDS)	X			
Flexor digitorum profundus (FDP)	X			
Extensor digitorum (ED)		X		
Extensor indicis (EI)		X		
Extensor digiti minimi (EDM)		X		X

The table has a spanning header "Action At Wrist Level" over Flexion, Extension, Radial Dev., Ulnar Dev.

muscle force is directly proportional to the physiologic cross-sectional area (Kozin et al., 1999). Chau et al. (1997) stated that hand strength was more correlated with arm muscle cross sectional area than the gender and body mass index.

Hand performance is affected by burden level. Pfitzer et al. (1972) noted that different loading affects the hand performance variably. A 40% increase in the strength corresponds to 10% increase on an efficiency impairment scale, while a 60 and 80% increase corresponds to a 17 and 18% increase, respectively. Ergonomic interventions should consider required force exertion. The maximum strength application is limited by the weakest segment or joint implicated in that particular activity (Wells and Greig, 2001). Moreover, the optimal solutions are not universal. Due to different daily use, there is an alteration in the mechanical and physiological properties of skeletal muscle. In the dominant first dorsal interosseous (FDI) muscle, there were lower values for recruitment threshold, initial firing rate, discharge variability, and target force firing rate when compared to nondominant FDI muscle. This may be due to an increased percentage of slow twitch fibers in the preferentially used (dominant) muscle (Adam et al., 1998).

Although the difference in force between dominant and nondominant hand was assessed to be between 10 and 13% in the dominant hand's favor (Lunde et al., 1972), the nonsignificant effect of hand laterality on hand strength is sustained by the majority of studies (Imrhan and Loo, 1989; Mathiowetz et al., 1986; Reikeras, 1983). Rice et al. (1998) found no significant differences between dominant and nondominant hand for grip and pinch strength. Since, according to Schmidt and Toews (1970), 28% of the subjects had higher grip strength values in nondominant hand, the use of "10% rule" (the dominant hand is 10% stronger than the nondominant hand) is not well established.

Hand use capacity is also influenced by: anatomic integrity, strength, coordination, mobility, age (Chau et al., 1997; Mathiowetz et al., 1986), gender (Mathiowetz et al., 1986; Swanson et al., 1978), associated diseases (McPhee, 1987; Rice et al., 1998), and shoulder, elbow, and wrist condition and postures (McPhee, 1987). During grasping, muscles acting on shoulder and elbow joints (upper extremity spatial positioning) and distal (hand) muscles (fingers shaping according to object weight and height) are used (Kuhtz-Buschbeck et al., 1998). Table 10.4 presents the most important factors that affect hand performance. The peak hand strength is reached in the mid-twenties (Fisher and Birren, 1947; Schmidt and Toews, 1970) with a force decrease of 16.5% after 60 years age (Chau et al., 1997). In children, development (Ager et al., 1984) achievement in physical education, and breathing capacity (Weiss and Flatt, 1971) are correlated with hand performance. Christ et al. (1992) noted an important decrease in wrist and finger flexor muscles maximal voluntary contraction (MVC) for women in the 45–49 age group. For all age groups,

TABLE 10.4

The Correlation between Personal Factors and Hand Performance

Factors Positively Correlated	Factors Negatively Correlated
Muscle strength	Age over 50
Movement coordination	Female gender
Body height	Associated disease
Muscle optimal length	Small hand size
Body weight	Nondominant hand
Mobility	Triceps skinfold thickness
Overall development (children)	Young age (children)
Breathing capacity	
Back extensor strength	

women exert less force than men (Conti, 1998; Mathiowetz et al., 1986), with 35% of the gender difference being explained by hand size (Fransson and Winkel, 1991).

Associated pathology causes an important decrease in hand force. Rheumatoid arthritis causes a 90% decrease in the grip strength compared to a healthy person. For pinch strength, there is a 75% decrease (Rice et al., 1998). Although severe stages are incompatible with work, the acute drop in hand abilities leads to lessened hand force and precision, increasing the risk of injuries. Median nerve paralysis present in workers with CTS determines a loss in thumb motion and coordination, leading to perturbed opposition during pinch and grasp applications (Kozin et al., 1999). Also, hand imbalance with metacarpophalangeal hyperextension and interphalangeal flexion are present in cases with ulnar nerve paralysis. Since intrinsic muscles extend the interphalangeal joints and are the prime flexors of the metacarpophalangeal joint, a loss in their performance causes extrinsic muscle overload with asynchronous grasp exertions. In order to design ergonomic jobs and devices, one should bear in mind the effect of each factor on hand performance. Given the important variability among workers, adjustable tools are essential.

10.5.2   Hand Muscles Fatigue

Although the introduction of mechanized tools reduced the necessary force, there are still many tasks that require excessive physical exertion. In these cases fatigue is a common phenomenon among workers. In order to understand its development, intimate fatigue mechanisms and difference among workers should be studied.

Endurance is defined as the ability to sustain continuous dynamic contraction or isometric contraction for a prolonged period of time. When analyzing endurance one should take into account duration, intensity, and frequency. Without these, comparison between different outcomes is impossible (Wallstrom and Nordenskiold, 2001). Endurance can be studied by either maintaining a certain percentage of MVC for a period of time or measuring the strength magnitude that follows repetitive or sustained contractions for a predetermined time. Previous studies assessed endurance levels determining repetitive submaximal (75 and 50% of the MVC) contractions (Wolf et al., 1996), sustained submaximal (40 and 50% of the MVC) contractions (Aniansson et al., 1983; Chatterjee and Chowdhuri, 1991), or sustained maximal contractions (Nwuga, 1975). During the first repetitive contractions there is an important decrease in muscle strength, named by Ohtsuki (1981) as the fatigue phase. Once a certain level is reached, muscle strength decreases at a lower rate (the endurance level) (Wallstrom and Nordenskiold, 2001). Rohmert (1960) proposed the equation

$$T_s = -90 + 126/P - 36/P^2 + 6/P^3$$

as the relationship between human static strength and endurance time, where T_s represents the endurance time expressed in seconds, −90 is a constant, and P is the percentage of maximum strength. Kroemer (1970) questioned this equation, noting that its utility depends upon its usefulness in isometric exertions tasks.

During muscle fatigue, decline in the maximal contractile force, inability to maintain targeted force, and increased effort during muscle contraction occur (Blackwell et al., 1999). The maintenance of constant force is accomplished either by recruiting more motor units or increasing the discharge frequency in the active ones (Carpentier et al., 2001). During muscle fatigue, higher activation rates are required for constant force, leading to increased risk of injury due to lack of rest (Fuglevand et al., 1999). In contradiction with these findings, Zijdewind and Kernell (2001) noted important decrease in the activation rate due to fatigue. Although there is a consensus in the literature regarding the recruitment of new motor units, contradictory results are reported about the change in discharge frequency in active motor units during fatiguing contractions. Frequency has been reported to increase (Dorfman et al., 1990), remain constant (Maton and Gamet, 1989), or decrease (Gantchev et al., 1986). Fatigue is not greatest in the motor units that exerted the largest forces. Fast-contracting motor units are not more exposed to fatigue than slowly contracting units (Fuglevand et al., 1999). In addition to the local feedback and regulatory mechanisms, the central control plays an important role in adaptation by the motor units (Carpentier et al., 2001).

Muscles not directly involved in the force production undergo fatigue, too (Aymard et al., 1995; Zijdewind et al., 1998). Activation in both ipsilateral and contralateral muscles occurs during prolonged muscular contractions (Gandevia et al., 1993). Zijdewind and Kernell (2001) reported an increase in force and electromyographic activity in the contralateral muscle during both submaximal and MVC fatiguing contractions. This coactivation increases the risk of injury by two mechanisms: (a) due to accumulated fatigue in muscles that are not primary effectors during a specific task, the change in position and/ or pattern would find the new primary muscle already fatigued and would cause an increased stress resulting in muscle overload and overexertion; (b) due to fatigue in muscles other than the target one, unintended contractions of fatigued muscles could induce loss of precision, increasing the risk of errors and accidents. Proportional relationship between contralateral activation and targeted muscle activity was also demonstrated (Zijdewind and Kernell, 2001). Differences in moment arm determine higher forces at the proximal site when compared to forces at the distal site (Danion et al., 2001). Since, at the upper extremity level, the distal regions are more vulnerable, even forces lower than those exerted at proximal levels could induce musculoskeletal injuries. This causal relationship is even more evident if one works in awkward postures and highly repetitive tasks in which muscle overload and coactivation are ubiquitous.

Due to its particular characteristics, adductor pollicis muscle was extensively studied (Carpentier et al., 2001; Fulco et al., 1999, 2001). Its unique properties are: high proportion of slow-twitch high oxidative fibers and complete motor unit recruitment (Fulco et al., 2001). Merton (1954) noted that in adductor pollicis muscle, voluntary activation accounts for all force produced in both rested and fatigued muscle. During fatigue, for low-threshold (< 25% MVC) motor units, the first dorsal interosseus presents an increase in both mean twitch force and recruitment threshold. For high-threshold (> 25% MVC) motor units, both twitch force and activation decreased (Carpentier et al., 2001). Afferent feedback differences in muscle implicated in sustained contraction may explain the different behavior of low- and high-threshold motor units. Although Herbert and Gandevia (1996) showed that 90% of adductor pollicis force was explained by voluntary activation and there are no differences between genders, Fulco et al. (2001) was the first to assess a gender difference in muscle performance under hypoxic conditions. If in normoxia and hypoxia, men had higher MVC force for rested muscle when compared to women (Fulco et al., 2001), during sustained muscle contractions, women present a slower decrease in force. The fatigue rate in men was approximately two-fold faster in normoxia (-8 ± 2 vs. -4 ± 1 N/min, respectively, $p < .01$) and approximately 2.5 fold faster in hypoxia (-13 ± 2 vs. -5 ± 1 N/min, respectively, $p < .01$) than for women. Furthermore, the decrease in adductor pollicis force after one minute of exercise for women was less ($93 \pm 1\%$) compared to men ($80 \pm 3\%$). Also, the endurance time to exhaustion was double in women compared to men (14.7 ± 1.6 min vs. 7.9 ± 0.7 min, $p < .05$). Wallstrom and Nordenskiold (2001) noted that during the first 90 seconds there was a decrease of 33 and 30% for women and men respectively, whereas the decrease between seconds 90 and 180 was 12% for women and 13% for men.

Since the slow-twitch high-oxidative fibers proportion in adductor pollicis muscle is equal in both men and women, the women's superior muscle performance in tasks requiring total motor unit recruitment might be due to the fast-twitch fibers lowered capacity for oxidative phosphorylation in men (Fulco et al., 2001). Another point of view is that in women the adductor pollicis muscle contains a higher percentage of slowly fatigable fast-twitch oxidative fibers than in men, with differences in adductor pollicis muscle properties determined by muscle generating-capacity variance (Fulco et al., 1999). Gender differences demonstrated in adductor pollicis determine a higher oxidative capacity in women and a less impaired muscle capacity under hypoxic conditions. The assessment of differences between women and men should lead to ergonomic modifications for demanding activities where males are predominant. All these findings dictate the need for important differences between devices and workplaces for men and women. Also, differences in task completion pattern should be taken into account.

Due to the anatomy of the hand, changing the force application point along the finger axis might provide an important variation in the participation of muscles for force exertion, protecting them from overload and overexertion (Danion et al., 2001). Fatigue can be avoided by implementing training programs that would increase awareness among workers. Rest pauses and alternative postures could also avoid muscle overload. The result is a decreased injury rate due to better hand force production and precision.

10.6 Grip Force

10.6.1 Classification

In the past, different criteria have been used in order to classify grip force application. Significant diversity led to difficult-to-compare results and testing procedures. McBride (1942), considering the parts of the hand used, proposed grasping with the hand as a whole, grasping with both the thumb and fingers, and a combination of the palm and finger grasping as the most important subtypes of gripping applications. Griffiths (1943), based on the object shape, classified hand prehension into cylindrical grip, ball grip, ring grip, pincer grip, and pliers grip. Cutkosky and Wright (1986) divided gripping exertion into circular, when the thumb and fingers are placed radially around the object, and spherical, in which the fingers oppose the thumb. Napier (1956) introduced for the first time the terms hook grip, power grip, precision grip, and combination grip. During power grip the thumb is adducted at the carpometacarpal and metacarpophalangeal joints, and fingers are ulnarly deviated, laterally rotated, and flexed (Pryce, 1980). In precision grip the thumb is abducted and rotated, and the fingers are flexed and abducted at the metacarpophalangeal joints (Napier, 1956). There is not a distinct separation between power and precision grip while working. Often they are combined during job task completion. Landsmeer (1962) proposed the substitution of the term precision grip with precision handling.

Kamakura et al. (1980), in a study involving healthy volunteers, noted 14 patterns: 5 for power grip (involving areas of the palm, hand, and volar surfaces of the digits with the fingers IV and V flexed more than the radial fingers), 4 intermediate grip patterns (the contact area with the object is represented by the radial faces of the index and middle fingers), 4 prehension grip patterns (with the object between the fingers and the pulp of the thumb), and one prehension without the thumb. Kapandji (1970), in terms of digital segments involved in the force exertion, proposed the introduction of the following terms:

palmar prehension, prehension by digito-palmar opposition, prehension by subtermino-lateral opposition, prehension by subtermino opposition, prehension by termino opposition, and prehension between two sides of the finger.

Finally, Sollerman and Sperling (1978) proposed the Hand-Grip Classification, in which four prehension patterns were described: transverse grip (the object is held between the thumb and fingers at 90° to the hand margins), diagonal grip (the object is held between thumb and all four fingers with a diagonal object–palm contact interface), extension grip (the object is held with interphalangeal extension), and spherical volar grip (the object is surrounded by the thumb and fingers with palm contact). None of the above grip classifications are better than the others. They are suitable for describing grip applications regarding the tool being used, hand position, required force and/or precision, and hand regions involved in force exertion. In order to ensure unbiased data, grip classification should be chosen in concordance with particularities of the task being analyzed.

10.6.2 Force Exertion

Grip strength is widely used in many industrial tasks. During grip exertion the most exposed areas are the metacarpal regions (Muralidhar and Bishu, 2000). Also, elevated stress on the common extensor tendon is present, due to the increased passive forces in the digital extensors (Keir et al., 1996). Grip force is produced by the thumb flexors exerting force in opposition to the total force produced by other fingers' flexors. Imrhan and Loo (1989) noted that since the force is applied at metacarpophalangeal joints level, during gripping the finger flexors are more advantaged than the thumb flexors. There is an important variation among different reporting regarding the most exposed hand and fingers areas while gripping. Table 10.5 presents the zones of the hand with maximum risk of being injured during gripping applications. Given that some regions (distal pha-

TABLE 10.5

Hand and Finger Areas Most Exposed to Injury Due to Grip Force Applications

Author(s)	Exposed Zones	Comments
Yun et al., 1992	Thenar area Metacarpophalangeal joints Distal phalanges fingers I-V Proximal phalange digit II	Zones exposed to risk of injury while executing gripping tasks involving power tools, a knife, and hammer.
Chao et al., 1989	Distal phalanges for digits II-V Proximal phalange for finger II	Outcome is based on calculations regarding force applied while using different phalangeal distribution.
Cochran and Riley, 1986	Distal phalanges of the II, III, and IV fingers	They used force-sensing resistors and adjustable handles.
Fellows and Freivalds, 1989	Index and thumb metacarpophalangeal joints Thumb proximal phalange Distal phalanges digits I-IV	EMG, force-sensing resistors and subjective measurements were performed.
Iberall, 1987	Distal phalanges digits I, II, III II, III, IV metacarpophalangeal joints Proximal and middle phalanges for finger II (lateral side) Proximal and middle phalanges for finger III	The degree of stress on a certain hand region is influenced by the nature of grip being used.
Fransson-Hall and Kilbom, 1993	Thenar area Pisiform bone Area between digits I and II	These areas are the most likely to present pain during localized high pressure.

langes for fingers II–V, thenar area) are cited by the majority of authors, grip applications could be significantly limited by localized pressure in these regions.

The most important factors that influence grip force are: age (Carmelli and Reed, 2000; Mathiowetz et al., 1985), gender (Desrosiers et al., 1995; Richards, 1997), handedness (Crosby et al., 1994; Richards, 1997), tool handle surface (McGorry, 2001; Westling and Johansson, 1984), object shape, intended use (Pryce, 1980), body position (Martin et al., 1984; Teraoka, 1979), object weight and size (Frederick and Armstrong, 1995; Kinoshita et al., 1996), dynamometer setting, time between tasks (Netscher et al., 1998), upper extremity posture (Dawson et al., 1998), total number of muscle fibers, percentage of fibers activated, muscle section area, fiber tension (Carmelli and Reed, 2000), and hobby demand (Crosby et al., 1994). In all studies men were consistently stronger than females (Desrosiers et al., 1995; Richards, 1997; Wallstrom and Nordenskiold, 2001). Dawson et al. (1998) found lower values in females for all wrist positions. Su et al. (1994) noted that for males, the 20 to 39 years age group had the highest grip strength. For women, the peak was recorded in the 40 to 49 years age interval, with an ulterior decrease due to age. After 60 years of age there is a 20% decrease in grip force for both genders (Carmelli and Reed, 2000). Grip strength values obtained in different studies are presented in Table 10.6. Important variations among reportings are due to sample characteristics, experimental setup, and recording measurements being used.

Grip force is also subject to variation due to body and upper extremity position. Previous studies showed that grip forces while supine are weaker than grips measured in standing posture (Martin et al., 1984; Teraoka, 1979). Although Martin et al. (1984) and Richards (1997) determined no difference between grip force measured in supine and sitting subjects, Teraoka (1979) recorded higher values for the later posture. Decrease in gripping force has been reported for supination greater than 70 degrees. No effect of supination on force exertion has been noted. All these influences are explained by the muscles length–tension relationship (LaStayo et al., 1995). Also, the force produced on gripping is directed in order to stabilize the upper extremity (Richards, 1997). An important condition for grip force exertion is the presence of wrist stiffness. During finger flexion, the flexor tendons, which cross the wrist, provide an increase in wrist stabilization (Dawson et al., 1998). If the upper limb needs to be stabilized, less force may become available for producing grip force. Thus the safe limits may be crossed leading to musculoskeletal disorders.

The grip force necessary to work with a certain tool is equal to the grip force component normal to the handle surface multiplied by the coefficient of static friction between the hand and the tool. Cutkosky and Wright (1986) noted a significant decrease in the applied force using a screwdriver when a high-friction handle was used compared with an aluminum (low-friction) handle. In order to avoid acute accidents, the workers exert more grip force than required. Westling and Johansson (1984) saw the difference between necessary and applied force as a buffer. At high loads, workers exert no more than required force because of fatigue considerations. At low-loads, the available wide area of variation between the required force and the MVC value determines an important increase in the applied force (Frederick and Armstrong, 1995), keeping the risk elevated. The risk of injury is even more increased for subjects with CTS. In their case, due to decreased sensibility, coordination is almost absent, leading to significantly greater grip-moment ratio (Kozin et al., 1999; McGorry, 2001).

LaStayo et al. (1995) found a drop in grip strength due to fatigue. There was a significant drop of 17.2 pounds in grip force after 5 seconds of force exertion. The decrease was not linear, with a 4.8-pound decrease in the first second. Due to the high force required during industrial work, the rate of fatigue that occurs during prolonged/repetitive gripping activities is more important than the maximum grip force exerted. During repetitive grip exertions, muscle contraction is highly influenced by both the anaerobic metabolism and

TABLE 10.6

Grip Strength Values for Different Age Intervals

Study	Gender								Force								
		5	10	15	20	25	30	35	40	45	50	55	60	65	70	75	80+ years
Chaparro et al., 2000	Male													41.29		32.55	16.09
	Female													25.52		17.83	10.58
Mathiowetz et al., 1985	Male		—		54.0	53.9	54.3	53.4	52.1	49.0	50.7	45.1	40.0	40.6	33.6		29.3
	Female		—		31.4	33.2	35.1	33.0	31.4	27.7	29.3	25.5	24.5	22.1	22.1		19.0
Imrhan and Loo, 1989	Male	15.2		—			49.7				—				30.0		
	Female	12.9		—			31.4				—				21.5		
Voorbij and Steenbekkers, 2001	Male		—			54.08		53.57		52.04		48.97		43.36		38.26	31.63
	Female		—			33.67		31.16		31.63		29.08		26.53		21.42	18.36
Kumar and Simmonds, 1994	Female		—				25.0										
Desrosiers et al., 1995	Male						—							45.6		42.4	34.5
	Female						—							25.3		23.7	20.0
Mathiowetz et al., 1986	Male	18.7		34.5	48.2						—						
	Female	15.7		25.9	31.9						—						
Shiffman, 1992	Male						41.2		—			39.5		27.9			25.5
	Female																
Ager et al., 1984	Male	8.4	16.1	23.5							—						
	Female	7.3	14.5	21.4							—						
Boatright et al., 1997	Male		—		57.1			53.57	52.23			53.1		35.71	37.05		31.25
	Female				33.48			33.03	32.14			32.58		23.21	22.32		19.19

the proportion of type II (fast twitch) muscle fibers (Capodaglio et al., 1997). Grip endurance time depends on the fiber type composition, muscle blood flow, maximum force for the muscle being used, and individual range of motion. Given that all these factors are improved by training, different tasks should be assigned to experienced workers when compared to new employees. Mitigated strength capacity may lead to injuries that could possibly be prevented by using training programs and introducing rest pauses.

Information regarding muscle activity during grip force application could be used in order to ergonomically design new devices and/or working techniques. Berguer et al. (1999) noted that the muscle electrical activity amplitude while using the palm grip was decreased in the flexor digitorum superficialis (FDS), thenar compartment (TH), and extensor digitorum communis (EDC), unchanged in the extensor carpi ulnaris (ECU) and flexor carpi ulnaris (FCU), and elevated in flexor digitorum profundus (FDP) compared with the finger grip during laparoscopic instruments use. More visible differences between EMG aspects were seen during high force conditions. Furthermore, for the same object the use of above radial grip requires less force than lateral grasping from the side (Kinoshita et al., 1996). During above radial grip, in addition to the distal phalanx, the middle phalanx pulp was used, decreasing the localized mechanical pressure on the hand surface (Kinoshita et al., 1996). One should use 4- or 5-finger grips in order to be protected by muscle finger overload. Tool diameter is very important in grip strength application (Imrhan and Loo, 1988). Since the muscle cross-bridge attachments are at their maximum level when the muscle is near resting length, moderate diameters determine highest grip forces. When the muscle is very short or very long, the number of attachments decreases and the resulting force is lessened (Blackwell et al., 1999). Kinoshita et al. (1996) noted that there was an increase in the grip force with the increase in object weight and variations of diameter above and less than 7.5 cm. Also, smallest grip forces were assessed when extreme diameters were used (Blackwell et al., 1999). Based on available data, moderate diameters of handles with high friction coefficient should be used at the workplace. In this way, through inexpensive ergonomic modifications, important reductions in muscle load as well as safer working techniques are promoted.

10.7 Pinch Force

10.7.1 Classification

Sollerman and Sperling (1978) classified pinch applications in four finger grip (pinch) types: pulp pinch (involve thumb and index or middle finger), lateral pinch (thumb and radial side of the index finger), tripod pinch (thumb, index, and middle fingers) and five-finger pinch, which occurs when the thumb and all the fingers are used. Brorson et al. (1989) divided three-point pinch into tip pinch and palmar pinch. In tip pinch the device/tool is grasped between the tips of the thumb, index, and middle finger; whereas in palmar pinch the pinch meter is grasped between the pads of the thumb, index, and the middle finger. Two-point pinch includes tip, palmar, and lateral pinch. In lateral pinch (key pinch) the force is exerted between the pad of the thumb and the lateral side of the middle phalanx of the index finger. The interaction between intrinsic and extrinsic muscles is evident during lateral pinch applications. Both thumb and intrinsic muscles act for thumb positioning and force exertion against the flexor pollicis during pinch exertion (Kozin et al., 1999).

10.7.2 Force Exertion

The use of pinch force is needed in the majority of industrial tasks. During pinch, the most exposed hand regions are the top of fingers I, II, and III (Muralidhar and Bishu, 2000). Imrhan and Loo (1989) noted that during pinching, the force is applied at the tips or pads of the fingers, increasing the risk of injury at these levels. Localized reduction in sensibility may develop, leading to lack of feedback and inappropriate force exertions. Previous studies assessed the ratio between pinch and grip force of being 1:4 (Imrhan and Loo, 1989) to 1:5 (Kumar and Simmonds, 1994). Crosby et al. (1994) noted that pulp pinch was 16% and key pinch was 22% of grip maximum values.

Pinch strength is influenced by: hand dominance (pinch grip force is consistently less in the nondominant hand compared to dominant hand), occupation, range of motion, pain sensation, and self-perception of function (Fowler and Nicol, 2001). Also, pinch strength could be highly influenced by experimental conditions (Imrhan and Loo, 1989), with learning effects affecting both MVC and submaximal contractions. The ratio between dominant hand and nondominant hand was 1.12 ± 0.13 (mean $\pm$ SD) for both males and females, with no effect of age on its value (Brorson et al., 1989). Chong et al. (1994) found pinch (tip, palmar, and key) strength positively correlated with gender, body height and weight, mid-arm and mid-forearm circumference, and negatively correlated with age and triceps skinfold thickness. Positive correlation between finger length and pinch strength is also reported (Brorson et al., 1989). Armstrong and Chaffin (1979) proposed the $F_t = kF_L$ equation for the finger flexor tendon force estimation, where F_t = finger flexor tendon force, F_L = pinch force, and k = 2.8–4.3, being influenced by the object and person hand sizes. Data could be used for the estimation of stress level at wrist level. According to Chau et al. (1997), for pinch strength the highest correlation was obtained with gender and muscle area. These anthropometric values are easy and not costly to assess and should be included in the hand strength assessment techniques. In all studies, males were stronger compared to females in terms of pinch strength application (Brorson et al., 1989; Chau et al., 1997; Imrhan and Loo, 1989). The difference between pinch strength in males and females is smaller in children (female–male ratio = 0.89) than in adults (0.69), with force values increasing in this order: female children, male children, female elderly, male elderly, female adults, male adults (Imrhan and Loo, 1989). Pinch mean values assessed in previous studies for different age intervals are presented in Table 10.7.

During key handling, the lateral pinch forces are in a constantly maintained balance (Wells and Greig, 2001). Due to their important role in stabilizing thumb-tip force during unstable pinch, there is an important increase in abductor pollicis brevis and extensor pollicis longus. Their action is independent of force magnitude (Johanson et al., 2001). If prolonged precision tasks are performed, there is an elevated risk for abductor pollicis brevis and extensor pollicis longus overload with consecutive musculoskeletal injury. One should alternate between high-force and precision tasks in order to avoid the risk for localized fatigue/discomfort/injury.

Among all pinch types the strength values were, from the highest to the lowest: key pinch, palmar pinch, and tip pinch (Chong et al., 1994). Imrhan and Loo (1989) found the same magnitude order and noted that the relationships between forces exerted in different pinch types are constant regardless experimental conditions. The finger used in opposition to the thumb influences the force exerted during pulp pinch. The force increases in the following order: digit V (little finger), digit IV (the ring finger), digit II (the index finger), and digit III (the middle finger) (Swanson et al., 1970). Similar finger strength proportion was found during the fixed total pinch force task. The average contribution of each finger was 33, 33, 17, and 15% for index, middle, ring, and small finger, respectively (Radwin et al., 1992). During pinch exertion with index finger opposing the thumb, the joint position

TABLE 10.7

Pinch Types Values (kg) in Previous Studies (M = male, F = female)

Study	Pinch Type	Force														
		5	10	15	20	25	30	35	40	45	50	55	60	65	70	75+ Years
Mathiowetz et al., 1985	Tip (Pulp)	—	—	—	8.0 M / 4.9 F	8.1 M / 5.3 F	7.8 M / 5.6 F	8.0 M / 5.1 F	7.9 M / 5.1 F	8.3 M / 5.8 F	8.1 M / 5.5 F	7.4 M / 5.2 F	7.0 M / 4.5 F	7.5 M / 4.7 F	6.1 M / 4.5 F	6.2 M / 4.2 F
	Key (lateral)	—	—	—	11.6 M / 7.8 F	11.9 M / 7.9 F	11.7 M / 8.3 F	11.6 M / 7.4 F	11.4 M / 7.4 F	11.5 M / 7.8 F	11.9 M / 7.4 F	10.8 M / 7.0 F	10.3 M / 6.9 F	10.4 M / 6.6 F	8.6 M / 6.4 F	9.1 M / 5.6 F
	Palmar (chuck)	—	—	—	11.8 M / 7.6 F	11.6 M / 7.9 F	11.0 M / 8.6 F	11.6 M / 7.8 F	10.9 M / 7.5 F	10.7 M / 7.9 F	10.6 M / 7.7 F	10.5 M / 7.1 F	9.7 M / 6.6 F	9.5 M / 6.3 F	8.0 M / 6.4 F	8.3 M / 5.3 F
Imrhan and Loo, 1989	Tip (pulp)	2.7 M / 2.4 F				7.3 M / 4.7 F				—				4.3 M / 3.0 F		
	Key (lateral)	4.2 M / 3.6 F				9.4 M / 7.0 F				—				6.7 M / 4.9 F		
	Palmar (chuck)	4.0 M / 3.6 F				9.4 M / 7.0 F				—				5.9 M / 4.6 F		
Kumar and Simmonds, 1994	Key (lateral)		—					6.3 F					—			
Mathiowetz et al., 1986	Tip (pulp)	3.2 M / 2.9 F	4.4 M / 4.3 F	6.6 M / 5.3 F						—	—					
	Key (lateral)	5.0 M / 4.2 F	6.8 M / 6.3 F	10.4 M / 7.7 F						—	—					
	Palmar (chuck)	4.4 M / 4.0 F	6.2 M / 6.0 F	9.9 M / 7.9 F						—	—					
Shiffman, 1992	Tip (pulp)						6.0		—	—	—	5.4		—	4.7	4.2
	Key (lateral)						9.3		—	—	—	9.1		—	7.6	6.7
	Palmar (chuck)						8.3		—	—	—	8.3		—	6.7	6.0

(continued)

TABLE 10.7 (CONTINUED)

Pinch Types Values (kg) in Previous Studies (M = male, F = female)

Study	Pinch Type							Force								
		5	10	15	20	25	30	35	40	45	50	55	60	65	70	75+ Years
Ager et al., 1984	Tip (pulp)	1.6 M / 1.3 F	4.1 M / 4.1 F							—						
	Key (lateral)	2.9 M / 2.4 F	6.4 M / 5.5 F							—						
	Palmar (chuck)	2.7 M / 2.2 F	7.4 M / 7.0 F							—						
Boatright et al., 1997	Tip (pulp)								—							
	Key (lateral)			—		8.4 M / 6.6 F		8.0 M / 6.2 F		8.4 M / 5.8 F		10.2 M / 7.5 F		6.6 M / 4.9 F		4.9 M / 3.1 F
	Palmar (chuck)								—							
Chong et al., 1994	Tip (pulp)			—				6.7 M / 4.3 F		6.4 M / 4.2 F		5.8 M / 4.1 F			5.3 M / 3.5 F	
	Key (lateral)			—				9.8 M / 6.5 F		9.6 M / 6.6 F		9.0 M / 6.4 F			8.8 M / 5.6 F	
	Palmar (chuck)			—				9.0 M / 6.5 F		8.4 M / 6.2 F		8.1 M / 5.9 F			7.9 M / 5.3 F	
Halpern and Fernandez, 1996	Tip (pulp)			—			6.2 M					—				
	Key (lateral)			—			8.8 M					—				
	Palmar (chuck)			—			8.3 M					—				

is balanced in order to optimize the posture in which slipping is almost impossible (Radwin et al., 1992). This reduces MVC and increases safety. Imrhan and Loo (1989) noted the need for proper size handles if safe lateral and chuck pinches are desired. Armstrong and Chaffin (1979) showed that the index finger pinch strength was 42 to 93% greater when the digits 3, 4, and 5 were flexed and extended respectively. Also, increasing the force exertion level from 10 to 30% MVC causes an elevation of middle finger contribution from 25 to 38% from total finger force exertion (Radwin et al., 1992). The uneven load distribution among fingers leads to increased risk for stronger fingers, while little and ring fingers remain exposed due to anatomical characteristics.

In order to implement valid ergonomic interventions, one should be aware not only of hand musculoskeletal structures exposed to elevated stresses during repetitive and forceful applications but also of pinch variability among workers. The wide variety of job factors that influence force exertion should also be considered. For example, Frederick and Armstrong (1995) noted that increasing tool handle friction reduces required pinch force for tasks requiring more than 50% of pinch strength MVC. Pinch strength assessment provides an accurate determination of hand function (Fowler and Nicol, 2001). Information could be used for targeted tool/task design as well as for choosing the most appropriate muscle–tendon load transfer technique.

10.8 Differences Due to Shoulder, Elbow, and Wrist Position

The majority of work and daily living activities require positions different than the neutral one (Richards et al., 1996). The influence of upper extremity joint position was extensively noted in ergonomic literature (Berguer et al., 1999; Drury et al., 1985; Keir et al., 1996; LaStayo et al., 1995; Marley and Wehrman, 1992; Mathiowetz et al., 1985). The further the joint is, compared to the hand, the less well documented is its relation to hand performance. Furthermore, body posture has been shown to influence grip strength (Kuzala and Vargo, 1992). McPhee (1987) noted that the hand functional capacity is closely correlated with the upper extremity proximal portion capacity to position the hand in an ergonomic posture. Also, there is a strong relationship between awkward posture leading to indirect vision of the tool/working place and decreased performance (Berguer et al., 1999). Since long flexors and extensor muscles of the fingers act at the same time for intermediate joints stabilization and for maximum force exertion, any variation in their total length leads to significant decrease in the ability to contract with maximum performance (Richards et al., 1996). The influence of extrinsic finger and wrist musculature on hand movement and posture during wrist and finger flexion was also studied by Keir et al. (1996). Hand muscles are multiarticular, and fully deviated joints determine muscle overstretch.

Due to the dynamic aspect of almost all the tasks required during work, the relationship between grip force exertion and wrist/forearm position is very important (LaStayo et al., 1995). Previous studies addressed the impact of wrist position on grip strength (Drury et al., 1985; Fong and Ng, 2001; Lamoreaux and Hoffer, 1995; Melvin, 1977; O'Driscoll et al., 1992; Pryce, 1980). Outcomes are not consistent. Wrist extension was shown to either increase (Mathiowetz et al., 1985) or decrease (O'Driscoll et al., 1992) grip strength. Kraft and Detels (1972) demonstrated that the grip strengths recorded at 0°, 15°, and 30° wrist extension were not significantly different. Also, Pryce (1980) noted no differences in grip strength for the 0° and 15° wrist ulnar and/or extension deviation. For the 15° wrist flexion and 30° ulnar deviation, the values were significantly lower when compared to the neutral position. Both Pryce (1980) and Kraft and Detels (1972) noted significantly lower values at 15° wrist flexion when compared to the neutral position. Contrary to these findings is the study in which no differences in grip strength were found between neutral, 15°, and 30° wrist extension (Kraft and Detels, 1972). Hazelton et al. (1975) noted that 21° ulnar deviation and 14° radial deviation determine an increase in grip strength, and the 30° ulnar deviation allows for the highest grip strength. In contrast, Terrell and Purswell (1976) found a decrease in grip strength of 15 and 18% for 20° ulnar deviation and 20° radial deviation, respectively.

Because larger moment arms characterize wrist flexors compared to extensors, larger forces would require active extensors to maintain the wrist posture (Keir et al., 1996), leading to increased risk of injury for this muscle group while working with flexed wrist. Passive muscle forces, always present in antagonist muscles, elevate the risk even more. The tensions recorded in wrist extensors were between 5 and 10 N. These values represent between 5 and 36% of the maximal force. Berguer et al. (1999) noted ineffective finger grip while the wrist is flexed at 90°. When the wrist is fully extended or flexed, there is a loss of flexor tendon force due to friction and contact with the wrist structure. This causes a significant decrease in pinch strength (Halpern and Fernandez, 1996). Furthermore, wrist deviation in the coronal plane decreases grip strength due to the change in angles between the tendons and their insertions. Compression of tendons against the carpal tunnel structures is present as well (Fong and Ng, 2001). The risk of injury is raised, especially when repetition and/or high forces are present. Extensor muscles overload is likely to appear during grips involving large wrist flexion angles, such as tip pinch, briefcase grip, and key pinch. Alternating between these hand/finger positions and working postures that require wrist extension could reduce muscle fatigue, alleviating the risk of injury. In order to maintain a balance between wrist extensors and finger flexors during large objects grasp, there is a need for wrist flexion, whereas during grasping smaller objects, the wrist is extended (O'Driscoll et al., 1992). When designing jobs and devices, one should allow for the role of tool shape and size on hand function. Deviations from the wrist neutral position cause compression of carpal tunnel elements against the surrounding structures. Muscle length variations followed by hand/finger mechanical disadvantage are also present (Pryce, 1980). Adjustable and/or customized utensils should be promoted at the workplace for worker safety.

According to Pryce (1980), the wrist positions that led to the highest grip strength values were: 0° ulnar deviation and 15° extension, 15° ulnar deviation and 15° extension, 15° ulnar deviation and 0° extension, and 0° ulnar deviation without wrist extension. The differences between them were not significant. On the contrary, Fong and Ng (2001) reported that the grip strength recorded at 15° or 30° wrist extension and 0° ulnar deviations were significantly higher than the grip strength at 0° ulnar deviation and 0° wrist extension or 15° ulnar deviation with or without wrist extension. Maximum grip strength was recorded in the self-selected posture (35 ± 2° extension and 7 ± 2° ulnar deviation) without any effect of gender on the subjectively selected wrist posture (O'Driscoll et al., 1992). The beneficial effect of moderate wrist ulnar deviation on gripping force is also supported by Lamoreaux and Hoffer (1995), who noted that there is a decrease in grip strength when the wrist is radially deviated. No effect on pinch strength was recorded. There is a tied relationship between the wrist deviations in extension–flexion and ulnar–radial deviation planes. Pryce (1980) reported a significant interaction between ulnar deviation and wrist flexion–extension. Although the wrist might be positioned in the proper position in one plane, in order to obtain maximum force, it is necessary to keep it within the appropriate deviation range in the other plane, too. Differences in strength exertions among studies may be due to different elbow and/or shoulder position, which represent an important factor in hand performance (Kuzala and Vargo, 1992; Su et al., 1994).

In order to maintain gripping stability and strength, the wrist muscles contract in a balanced manner, positioning the wrist in the optimal posture for a given task (Dawson et al., 1998). During wrist stabilization, an important role is played by the wrist musculature, carpal bones, and ligaments (LaStayo et al., 1995). The finger flexors muscles EMG was approximately the same for wrist deviation within the 5° radial deviation – 10° ulnar deviation range with significant increase in myoelectrical activity for extremely deviated postures (Drury et al., 1985). Furthermore, the EMG activity in the left hand was 27% higher than for the right hand, with wider variations as a function of wrist angle. The increased variation in the nondominant hand could be explained by the effect of "occupational training" on the dominant hand in a world designed for right-handed workers.

Hand performance is also highly affected by forearm position (degree of supination or pronation). Grip and pinch strengths are increased or not changed by supination (Agresti and Finlay, 1986) and decreased by forearm pronation (Marley and Wehrman, 1992). Richards et al. (1996) assessed grip force exertion in pronation as being the weakest, followed by neutral position and forearm supination. The drop in gripping force during forearm pronation is explained by the loss in force generation of the long finger flexors (LaStayo et al., 1995). In this position the muscles are stretched, leading to mitigated strength. Fraser (1980) noted that the maximum pinch strength during supination is due to biceps brachii's role of forearm stabilization. This provides support for forearm digital flexors to contract at their maximum capacity. High risk of musculoskeletal injury is present during work that involves repetitive changes from supination to pronation concomitant with important force demand. During the shift between supination and pronation, the direction of pulls of the flexor muscles that originate from the radius and rotate around the ulna changes (Richards et al., 1996), making it even more difficult to maintain the muscular balance. Almost all studies test hand force in setups that lead to maximum strength. Due to variability of forearm positions used during work, forearm supination should not be the only position tested in grip strength tests.

Both grip and pinch forces are significantly affected by elbow and shoulder posture (Capodaglio et al., 1997; Halpern and Fernandez, 1996; Kuzala and Vargo, 1992; Marley and Wehrman, 1992; Su et al., 1994). No consensus has been reached regarding the upper extremity position that provides the highest hand force. Because the flexor digitorum superficialis crosses the elbow joint, elbow position influences its strength performance. Kuzala and Vargo (1992) and Marley and Wehrman (1992) reported significantly stronger grip strength with extended elbow (0° flexion) when compared to elbow flexion (90° flexion has been shown to allow for the highest force values by other studies). Mathiowetz et al. (1985) found higher grip values when elbow was 90° deviated compared to 0° position. It has been demonstrated that grip strength was higher at 90° elbow flexion than at 130° elbow flexion or no flexion. Maximum hand force recorded at 0° elbow flexion could be explained by the relation between joint deviation and muscle length. The more flexed the elbow is, the shorter is the flexor digitorum superficialis, leading to a decrease in force exertion (Kuzala and Vargo, 1992). Shoulder and/or body stabilization could account for elevated hand force exertion with flexed elbow. Higher torque mean values were recorded during grip with the elbow adducted (no shoulder flexion) and flexed at an angle of 90° than in the tests performed with arm abducted (shoulder flexion) and extra-rotated and the elbow flexed at an angle of 90°. This difference might be due to a better hand/forearm stabilization and wrist maintenance within the neutral zone (Capodaglio et al., 1997). In a study that assessed hand strength in four different positions (elbow fully extended with 0°, 90°, and 180° shoulder flexion and elbow 90° flexed with 0° shoulder flexion), Su et al. (1994) showed that 180° shoulder flexion with elbow fully extended was the position which provided the highest grip force, whereas the weakest strength was recorded during 90°

elbow flexion with 0° shoulder flexion. The most used positions while performing working tasks, 90° elbow flexion with 45° and 90° shoulder flexion were not studied.

Extensive studies in this area are urgently needed in order to assess the most appropriate upper extremity position while exerting hand force. Joints should not be viewed as individual entities. Their interrelation is what allows for the significant upper extremity mobility and, more important, for posture compensation when working in awkward postures. The majority of studies are static, with subjects adjusting their upper extremities in order to exert maximum grip strength (LaStayo et al., 1995). In order to obtain applicable data, dynamic studies in which industry-like postures and frequency are present should be carried out. Once the relationship between hand force and upper extremity musculoskeletal complex is established, job/workstation redesign could be performed based on scientific data. Lessened hazard levels and increased productivity may follow.

10.9 Individual Finger Strength

Individual finger contribution to the total hand force has been studied by different authors (Danion et al., 2001; Fransson and Winkel, 1991; Li et al., 1998a,b; Radwin et al., 1992), yielding inconsistent results. There is a consensus regarding the index and middle finger being stronger than the ring and little finger (Ejeskar and Ortengren, 1981; Swanson et al., 1970), with the middle one being the strongest (Ejeskar and Ortengren, 1981). Ring finger contribution greater that index finger was assessed only by Fransson and Winkel (1991), who described the distribution of forces as being 21.2, 33.6, 26.5, and 18.1% for digits II, III, IV, and V, respectively. Radwin et al. (1992) showed that for object weights below 1 kg, the finger force magnitude from the highest to the smallest was: index, middle, ring, and little fingers. For weights above 2 kg, the order was middle, index, ring, and little fingers, with thumb force equal to the others four fingers' force sum. Also, an increase of 1.5 kg force demand, from 0.5 kg to 1.0 kg, determined an increase in the thumb, middle, and ring fingers' contribution and a decrease for index and little fingers (Kinoshita et al., 1996). Although the load reduction on the little finger is a useful protective tool against overexertion, the redistribution of elevated force on the other fingers, including the ring finger, could lead to increased risk of injury.

The sum of each finger's maximum force is bigger than the force of fingers II, III, IV, and V acting in parallel (Danion et al., 2001). The sum of the maximum force from every finger yields 183 N, which is 83% more than the average pinch strength using all five fingers simultaneously (Radwin et al., 1992). The fingers act as a veritable complex (tied communication between its components) when hand force demand variations and/or change in hand and fingers posture take place. There is a consistent force sharing among fingers regardless of the total force production (Danion et al., 2001). When a finger is removed from the grasping application, the biggest variation in applied force is seen in the fingers adjacent to the removed finger (Kinoshita et al., 1996). Injury due to sudden change in loading may develop. During maximal voluntary contractions, the activation of one finger inhibits the activity of adjacent fingers (force deficit). This sharing pattern could be explained by the reduction of load per digit leading to decreased muscular activation. The sharing pattern among fingers may be explained by a minimization of secondary moments about the longitudinal functional axis of the hand (Li et al., 1998a,b). Central neuromuscular control could also play a role in individual finger force exertion.

Due to their highly repetitive and intensive force component, work-related hand activities determine localized muscular fatigue, with important changes in the strength production pattern of muscles. Danion et al. (2000) noted an enslaving process in which, during finger contraction, the other fingers produced force too. Enslaving remained unchanged during fatiguing exercises, when force was measured at the site involved in fatigue, and increased when another site was the zone for force production. An increased risk of injury is present due to a lack of rest and muscle overload. The central contribution to force exertion control is supported by Danion et al. (2001), who found large transfer of fatigue across fingers, culminating with the removal of the fatigued finger from the force-application complex. Excluding the fatigued finger from the force production allows it to recover and to enter later into the synergy.

Both enslaving and force deficit phenomena might be due to the presence of multifinger forearm muscles and intertendinous connection (Danion et al., 2000). When designing tools and working techniques, one should consider that, due to their interaction, fingers constitute a musculoskeletal complex. The flexor digitorum profundus and flexor digitorum superficialis muscles, which contribute to several fingers' flexion, and juncturae tendinum, which links together the digits (Fransson and Winkel, 1991), allow fingers to act in a simultaneous and complementary manner. Taking into account the significant drop (25%) in finger strength for all fingers due to fatigue (Danion et al., 2000) and the fact that individual finger strength was decreased by the participation of more fingers (Radwin et al., 1992), it is indicated to design tasks that involve the simultaneous use of fingers. This protective technique should be applied even when the job could be completed using only one or two fingers. In this way the force exerted will be split between all fingers, reducing the muscle load and keeping the work within safer limits.

10.10 Older Workers

The proportion of elderly workers in the working population is increasing, stressing the importance of preventive interventions for this specific group. The baby boom generation trend will continue in the twenty-first century (Rahman et al., 2002). In the United States, in 2030 the number of elderly (65 and over) will reach 70 million, twice the number in 1996 (Resources Services Group, 1997). In order to work at its best and in a safe environment, this segment of the working population requires customized workstations. Targeted design modifications based on scientific data are the only valid solution that could address this issue. When designing jobs and workstations, it is currently assumed that the same movement and force patterns are used by elderly and young populations alike (Shiffman, 1992).

Although Crosby et al. (1994) did not find a significant effect of age on hand force exertion, many studies (Brorson et al., 1989; Chaparro et al., 2000; Desrosiers et al., 1995; Mathiowetz et al., 1985) have shown differences between old and young workers in both force/endurance and precision. From the last years of the first decade of life, which is the period when the hand prehensile development ends (Kuhtz-Buschbeck et al., 1998), to death the hand force capacity is in a continuous transformation, with periods of both development and involution. A curvilinear relationship between grip strength and age, with a peak between 25 and 59 years and a decline thereafter, was noted (Desrosiers et al., 1995; Shiffman, 1992). For tip, key, and palmar pinch, the values were constant from 20 to 59 years, with a decline from 60 to 79 years (Mathiowetz et al., 1985). Ranganathan et al. (2001) noted a reduction of 30% for gripping force in elderly (65–79 years) compared to young subjects (20–35 years). The decrease in grip and pinch strength occurs in both genders (Voorbij and Steenbekkers, 2001).

Females exert less grip force than males, with the difference between forces increasing with age. Age does not affect the greater grip strength values in men (Crosby et al., 1994). Female grip strength was 61.8% of the male value for the 60–69 years age group and decreased to 46.7% for age 90+ (Chaparro et al., 2000). Furthermore, Ranganathan et al. (2001) found a 43% grip strength decrease in older women compared to older men, vs. 34% less grip strength in young women when compared to young men. The relationship between age and force exertion control was stressed by Ranganathan et al. (2001), who showed that aging not only reduces the MVC but also mitigates the ability to maintain steady submaximal force. The impact of magnitude is not similar on pinch and grip strengths. The effect of age on hand strength was more pronounced in grip compared to pinch applications (Chong et al., 1994; Mathiowetz et al., 1985; Ranganathan et al., 2001). Due to its ubiquitous usage, pinch strength does not vary as much as grip strength as a function of age. This could be explained by the training effect of daily activities on pinch force (Chong et al., 1994). This idiosyncrasy could be viewed as an advantage for older workers and should be used to replace, when possible, the grip force demand. The degenerative effect of age on hand performance might be due to changes in both peripheral (muscle, nerves) and central (central nervous system, circulator system) regulation mechanisms. A complete list of changes that determine the important drop in hand performance is presented in Table 10.8. Given that the body weight is a good indicator for hand strength (Desrosiers et al., 1995), the assessment of the relationship between age and grip/pinch maximum force should allow for the possible increase in weight that counterbalances the decreasing effect of age on strength (Boatright et al., 1997). In this case, although the muscle suffered degenerative modification due to age, the values are inflated due to increased body weight.

The decline in hand strength interferes with both office/industry responsibilities (hand tool handling, typing, etc.) and daily task activities, such as opening a medicine bottle, drinking, eating, etc. The impact is even more important if one takes into account the important reduction in joint mobility at this level. Aging could account for up to 40% reduction in range of motion (ROM) compared to a younger worker (Chaparro et al., 2000), elevating the risk of injury, especially while working in awkward postures for a prolonged period of time. All these modifications affect pinch and grip precision and

TABLE 10.8

Central and Peripheral Causes for Reduced Hand Muscle Performance in Elderly

Peripheral	Central
Central nervous system degradation (loss in muscle coordination capacity)	Reduction in hand tactile sensation (lack of feedback)
Endocrine changes (decreased endocrine communication among apocrine and epicrine systems)	Muscle fibers reduction (especially fast twitch fibers – type II) (selective atrophy)
Protein metabolism perturbation (decrease in protein quality)	Muscle mass atrophy (changes in muscle size)
Perturbation in circulatory system (intramuscular flux reduction, mitigated effort capacity)	Local vascular degenerative changes (arteriosclerosis)
	Incomplete muscle innervation
	Muscle–nerve plate junction degenerative changes
	Contractile proteins degradation
	Drop in functional muscle fibers proportion

determine an increase in task completion time (Rahman et al., 2002), leading to a drop in performance if appropriate ergonomic modifications are not implemented.

When designing jobs for elderly workers, one should consider that, during fine motor movements, the muscle activation is increased even more at this age (Chaparro et al., 2000). The force applied by elderly individuals is bigger than the one applied by younger group in the same task, especially in activities that require high-precision movements (Rahman et al., 2002). These differences could be explained by changes in muscle activation pattern, skin properties, and central nervous system, which lead to lack of feedback and confidence during precise tasks. The risk for localized muscle fatigue and overload that follows high physical and mental stress is even more pronounced than in the general working population. The decrease in hand sensibility in the elderly also causes a drop in their capacity to assess the objects' slipperiness, increasing the risk of errors and accidents (Ranganathan et al., 2001). The introduction of exercises/training methods for older workers would lead to a reduction in the risk involved in different tasks. Increase in performance and productivity due to a mitigated completion time and lack of unsuccessfully repetitive movements will follow.

10.11 Hand Performance Measurement Techniques

In order to implement ergonomic changes based on valid data, the need for hand performance measurement devices is evident. Their usefulness is proven by their wide usage. In order to objectively measure the hand function, the Jebsen Test of Hand Function (JTHF) was proposed in 1969 in the United States (Jebsen, 1973). The test includes hand movements that are present in activities of daily living (ADL). Grip and pinch strength assessment tools were used in the past in order to study the neuropsychological status of brain-damaged patients and the effectiveness of surgical treatment. Return-to-work capacity used these tests, too (Chong et al., 1994). Isokinetic dynamometry has been proved to be efficient in identifying feigned efforts (Dvir, 1999), playing an important role in legal issues. Giampaoli et al. (1999) showed that handgrip assessment tools are valid for incident disability prediction in men 77 years or older.

Among all devices for grip strength measurement, the Jamar dynamometer and Martin vigorimeter are the most known. The Jamar dynamometer has a sealed hydraulic system, with a gauge calibrated in pounds and kilograms, and five different settings. It was shown to give the most accurate measure of grip strength by the majority of studies (Ashford et al., 1996; Chong et al., 1994; Desrosiers et al., 1995; Mathiowetz et al., 1985; Shechtman et al., 2001). Moreover, the California Medical Association Committee recommended the Jamar Dynamometer as the best measuring device for grip strength (Kuzala and Vargo, 1992). Ashford et al. (1996) noted inaccuracy less than ±3% for the Jamar dynamometer, which is even lower than the one indicated by the manufacturer (±5%). These results stress its accuracy. The other grip strength assessment device is the Martin vigorimeter. It is not as well known as the Jamar dynamometer, but several studies used this tool. It has a rubber bulb connected by a tube to a manometer calibrated in kilopascals. It is very suitable for grip force measurement in people with arthritis, since it eliminates any stress on joints (Melvin, 1977). Because the subjects have to compress the rubber bulb, muscle isometric activity is involved during strength measurement (Desrosiers et al., 1995). Desrosiers et al. (1995) noted that, although the Martin vigorimeter measures the grip pressure, not the force applied, a high correlation between the Jamar dynamometer and the Martin vigorimeter was found. For pinch strength measurement, the B&L pinch gauge presents the highest accuracy. Due to the wide variety of devices that was used, it is very difficult to compare results from different studies. For example, the Osco pinch meter is no longer commercially available (Mathiowetz et al., 1985). Although it may be more convenient to use a certain type of measurement device, researchers should take into account that using only compatible tools outcomes could increase usefulness and applicability. Also, Chadwick and Nicol (2001) noted that of all types of grip measuring devices (pneumatic, hydraulic, mechanical, and strain gauge), the ones that are designed to assess only the maximal force and have only one degree of freedom are not valid.

The American Society of Hand Therapists (ASHT) concluded that upper extremity position influences hand strength tests. They recommended that during testing the subject should be seated with the shoulder adducted and neutrally rotated, 90° elbow flexion, and the forearm and wrist in neutral position (Mathiowetz et al., 1985). Given that there is no difference in grip force between supine and sitting positions when the upper extremity is maintained in the position recommended by the ASHT (Richards, 1997), the two positions could be interchanged when one is not available. Although ASHT recommended the posture for grip assessment, grip strength assessment in different positions is needed in order to determine which are the safest and the most hazardous postures. In order to be able to compare data from different studies, standardized alternative postures should be used (Mathiowetz et al., 1985). Obtaining high grip values is not everything. The upper limb posture during force exertion is even more important. Introducing design modifications based only on maximal hand force values, without correlating the outcome with the posture in which it was recorded, would lead to long-term musculoskeletal problems.

During maximum-strength assessment, subjects should gradually increase the exertion until the maximum is reached and maintain this level for three seconds (Caldwell et al., 1974). Also, considering that, as any index of human performance, there is an important variation in strength applications, repeated measurements are essential (Young et al., 1989). The majority of studies recommend the use of three recordings (Desrosiers et al., 1995; Mathiowetz et al., 1985). Chaparro et al. (2000) proposed repeating the exertions until two maximum values vary within 10%. The greater value is used. The use of means from three trials provides a higher reliability (0.89 and 0.93 for the right and left hand, respectively) compared with only one trial (0.79 for right and 0.86 for left). This procedure is even more important if one considers that no learning or fatigue effects are present during the use of three consecutive trials (Mathiowetz et al., 1985). Crosby et al. (1994) noted that the repeated testing procedure is not necessary because over 50% of the subjects had decreased values when the test was repeated. The consistent decline in force might be due to short resting breaks between trials. When the study is carried out over a prolonged period of time, serial measurements are even more needed. Young et al. (1989) assessed a variation between 5.1 and 8.4 pounds (19.2–23.7%) for grip strength for 6 measurements performed in 3 weeks. For lateral pinch strength, the fluctuation was between 2.6 and 3.8 pounds (13.8 and 17.6%).

There is an important interindividual variation in terms of device setting. Although the Jamar dynamometer has five settings, in order to save time the majority of studies used only one setting (II). This choice was made based on previous data and did not take into account the subjects' characteristics. The proportion of research participants that exerted maximum force when setting II was used varies from 60% (Crosby et al., 1994) to 89% (Firrell et al., 1996). It has been shown that individuals who had maximal values at setting I had lower body weight and height (Firrell et al., 1996). Preliminary hand and body measurements should be made in order to designate the right setting for a certain worker. If only setting II is used, biased (decreased) values will be obtained for subjects that would have exerted higher forces if the proper setting had been available to them (Firrell et al.,

1996). O'Driscoll et al. (1992) found a linear and inverse correlation between the Jamar dynamometer setting and wrist extension. This relationship was not true for ulnar and radial deviations. The resting length position for fingers flexors coincides with a moderate flexion in MCP–IP joints (Dvir, 1997). All the positions that require excessive joints deviations, such as Jamar dynamometer positions I and V, determine a decrease in the number of filaments overlapping, with a consequential drop in strength. While different settings should be used in order to match various hand sizes, due to variability in force direction and hand-device interaction surface, only data obtained from the same setting should be compared. Even when the same settings are used, differences between manufacturers determine various grip dimensions, leading to incompatible data. For example, the dynamometer used in Bechtol's study measures 1.50 in. at setting II while the Jamar dynamometer has 1.75 in. at setting II (Firrell et al., 1996).

In order to assess hand/wrist position while exerting force, joint deviation measurements are also essential. The goniometer outcome for wrist deviations differs significantly from data obtained manually (Marshall et al., 1999). Observers underestimate wrist nonneutral postures (Ketola et al., 2001). Therefore, the use of electrogoniometers is indicated. Obtaining research-based force limits for the most-used wrist deviations would provide vital data for ergonomic design programs. In addition to the above-mentioned devices, electromyography and subjective magnitude estimation are used for grasping exertion level assessment. Due to its inability to measure muscle activity during complex manual work and considering that it is not specific for individual finger activity, electromyography is suitable only for static exertions and fixed postures (Radwin et al., 1992). McGorry (2001) noted that the wide variation in EMG–grip ratio determined by wrist/upper extremity posture and grip type makes the use of electromyography in grip force estimation unreliable. Self-rating introduces an important bias in hand performance assessment. Subjective magnitude estimation is very inaccurate and depends on the participant objectivity (Radwin et al., 1992). Porac and Coren (1981) showed a 74% concordance between the responses given in a questionnaire regarding hand preference and the actual skill performance. This outcome reveals that there is a bias in self-reporting.

Hand performance and hand proficiency vary considerably from one type of task to another (Borod et al., 1984). Therefore, the use of several hand performance assessment tests is better, in order to have a complete hand capability assessment. For example, the difference between dominant and nondominant hands is very well seen in handwriting tests but presents an important overlap in gripping strength test (Provins and Magliaro, 1993). This outcome contradicts the results of Reikeras (1983), who noted that under pathological conditions, when it is impossible to determine both-hand performance, the assessment of the other hand with consequent use of data is a useful procedure. Although both dynamic and static phases play a role in dexterity hand capacity, the majority of prehension patterns assess only the static components. Including tasks present in work and daily life activities would increase the test validity. In order to obtain a comprehensive overview of the subject hand function, the grip/pinch strength and range of motion assessment should be accompanied by a questionnaire regarding other aspects of work/daily living tasks (Fowler and Nicol, 2001).

One should be aware that whenever volunteers are involved in a study, there are high chances to have subjects that thought they might do well. A biased outcome with higher hand strength force values is possible. Rigorous sample size formation increases the external validity of the study, assuring a superior power and generalizability. Another concern when using hand strength measurement tools is the lack of attention given to the quality of movements that are performed (Conti, 1998). Triangulating with different parallel measurement techniques (hard tools, observations, etc.) would ensure an objective assessment (Fagarasanu and Kumar, 2002). Although standard testing positions are required in order to have comparable data, alternative postures with different wrist/ elbow/shoulder deviations should be performed in order to have normative data regarding the grip strength during deviated working postures (Fong and Ng, 2001). Considering higher correlation between hand strength/range of motion and biomechanical trial data, the force assessment represents a cost- and time-efficient method of hand-function assessment. Normative data for grip and pinch force exertions could be used in engineering design, rehabilitation programs parameters, performance assessment, and training programs development (Chaparro et al., 2000; Giampaoli et al., 1999).

10.12 Industry Relatedness — Ergonomic Solutions

The work–men interface is influenced by both task/workstation design and workers' individual characteristics (adaptation capacity, endurance, maximum strength, skills, dexterity). Targeted ergonomic interventions based on valid data, as well as training programs that increase awareness among workers, represent legitimate solutions for work-related primary and secondary prevention.

Both in industry and office activities the limits are set arbitrarily, and no connection between applied force and awkward posture is made (Ketola et al., 2001). During industrial tasks, poor ergonomic design produces elevated localized pressure, leading to increased risk of injury. For example, the use of laparoscopic instruments for a prolonged time leads to thenar nerve palsies (Horgan et al., 1997; Kano et al., 1993; Majeed et al., 1993), arms muscle fatigue, and increased forearm muscle overload compared to laboratory experiments (Berguer et al., 1997). The effect of design on performance is highlighted also by the difference between the laparoscopic instruments (tip force transmission ratio of 1:3) (Sukthankar and Reddy, 1995) and the standard surgical instruments where the transmission ratio is 3:1 (Gerber, 1998). Perceived hand pain is a limiting factor in work with handheld tools. The most sensitive regions are the most likely to be injured if one exceeds the safer limit during repetitive and/or forceful tasks (Muralidhar et al., 1999). There are wide variations in the force applied on a tool's handle: for cylindrical handles, a radial force is present, while, for an elliptical or rectangular handle cross section, the maximum grip force is exerted along the major axis with unequal force along its length. Finally, a shearing force component is present during the use of tools that produce a moment about the long axis (screwdriver) (McGorry, 2001). Although it is very difficult to assess the amount of forces applied with or by hand tools, because of its importance, the quantification of force exerted at hand–tool interface should be included in the ergonomic evaluations. Kumar and Simmonds (1994) noted that, with the exception of 40% MVC level, there was a consistent bias in perception of force exerted at all graded contractions. Perceptions were lower at 60 and 80% of MVC, but were higher at 20% compared to their objective values. As a consequence, repetitive tasks requiring forces below 40% MVC will lead to overestimation of applied force and to hazardous levels of exposure, promoting musculoskeletal injuries. Tasks that require force application beyond the 40% level will be performed with force exertions lower than the strength necessary to handle the tool under safe conditions. Accidents due to drops and inappropriate grip are likely to appear.

Force applied is highly influenced by tool handle surface and shape (Berguer et al., 1999; Chadwick and Nicol, 2001; Kinoshita, 1999; Muralidhar et al., 1999). For tool slips to be avoided, forces greater than the tangential loads should be applied. Safer limits could be easily crossed (Jenmalm et al., 2000), especially when using tools with inappropriate handles. Due to the hand glabrous skin properties (high density of specialized mechano-receptors) (Salimi et al., 1999), tactile sensors are very important in maintaining the grip force above slip force level (Kinoshita, 1999). The gripping force is adjusted for both the weight and the object texture, with elevated grip forces being recorded for lower a coefficient of friction (Salimi et al., 1999). The important role of hand sensibility is demonstrated by the fact that anesthesia of a digit increases force production in the other fingers. This may be due to lack of sensitive feedback and/or to shifting to nonanesthetized digits as a compensation for the lack of sensitive information from that finger. This is very important for workers suffering from CTS who continue to work with partial/total anesthesia of one or more fingers. The still unaffected fingers will exert compensatory force and will be overloaded and at high risk for musculoskeletal injury (Kozin, 1999). The equation $F_f = \mu F_n$ (Amonton's Law of Friction), in which friction force is equal to normal contact force multiplied by coefficient of friction, could be used for applied force prediction in tasks that involve frictional coupling between object handle and hand. The modified equation would be: $F_p = W/2\mu$, where friction force equals weight divided by the coefficient of friction multiplied by 2. This equation is valid only in cases in which the frictional force is equally distributed on both sides of the handle (Frederick and Armstrong, 1995). Using this equation, one could predict the required force, making it possible to take the necessary actions in order to reduce the stress level on hand musculoskeletal system.

A tool's handle shape causes important variations in working patterns and posture. Because the middle finger is the strongest and the little finger is the weakest, during cross-action tool usage the small finger has the longest lever arm, and the index finger has the shortest lever arm. Reversed grip, although it may not increase the grip force exertion, constitutes a safer working technique, reducing the risk of injury (Fransson and Winkel, 1991). Kadefors et al. (1989) noted spontaneous use of reverse grip among workers. A certain size diameter cannot be used for all tools. Consideration of applied forces, required postures, moment, and force applications should be taken into account. Also, adjustable handles should be implemented in the workplace. In this way, small fingers will be at their proper position. If not, high load requirements are present on a finger that is not capable of maximal contraction due to poor design (Blackwell et al., 1999). The lower grip strength values for females are due to both muscle force and hand size. Therefore, ergonomic redesign interventions should promote not only a reduction of the amount of force required to complete the task but also a tool resizing (Muralidhar et al., 1999). Jenmalm et al. (2000) noted lessened grip force with increased handle tool curvature. This may be due to deviated working postures during which the wrist stabilization process is extremely complicated, especially if dynamic movements are involved (LaStayo et al., 1995). Data regarding applied grip force and moments during hand tool use would bring important information about the individual adaptation, individual responses to exposures, and elevated-risk office and industry activities.

Although gloves have been used in many industrial tasks as protective devices, their extensive exploit also has negative features. Gloves affect hand performance influencing: task time (Muralidhar and Bishu, 1994), dexterity (Bradley, 1969; McGinnis et al., 1973; Banks and Goehring, 1979), grip strength (Hertzberg, 1955; Cochran et al., 1986), and range of motion (Griffin, 1944). Uniform thick gloves introduce more hazards such as insecure grasp, loss of sensory feedback, reduction of range of motion and mitigated hand dexterity (Muralidhar et al., 1999). These modifications produce changes in working patterns leading to elevated musculoskeletal and mental stress and awkward postures. Although thick gloves provide better protection against vibration and toxic agents, due to the cutaneous sensation mitigation, increased applied force was recorded (Kinoshita, 1999). Ergonomic (selective thickness) gloves provide an elevated protection, especially for exposed areas, without increasing bulk; increase grip strength; and do not mitigate productivity compared to conventional gloves. They represent the solution that permits the work at higher pressure for a longer period of time before discomfort appears (Muralidhar and Bishu, 2000). Moreover, considering the wide variation in pressure-discomfort threshold over the palm, it is suggested to have proper protection in the critical areas rather than have several complete layers of material. Even with selective thickness gloves there are several exposed hand areas. Further work is needed in order to eliminate the low pressure-discomfort threshold assessed for the top of finger IV and V and the base of finger IV (Muralidhar and Bishu, 2000).

In addition to workstation/tool redesign, job rotation programs should also be used in order to reduce the prevalence of occupational musculoskeletal injuries. A relocation of workers suffering from work-related disorders is desirable. In its absence, employees that continue to work in the same job position as the one that caused the injury, will suffer continued tissue degradation, leading to decreased productivity and an increase in work claims and lost days (Sande et al., 2001). Also, training programs that promote minimum required force applications should be implemented in order to educate workers to work within the safe limits. Finally, the cumulative effect of prolonged awkward postures and extensive force application must be emphasized.

10.13 Summary and Conclusions — Future Research

Hand strength has not been thoroughly addressed from an ergonomic point of view. The majority of studies support the relationship between work and hand musculoskeletal disorders. Wrist, hand, and finger musculoskeletal disorders due to work are still on the rise, with all industrial and office risk factors still acting at elevated levels. Hand performance is affected by muscle strength, hand size, gender, body weight and height, age, associated diseases, and hand dominancy. Fatigue is a common phenomenon among workers, causing declines in the maximal contractile force, increased effort during muscle contraction, and inability to maintain targeted force. Owing to the dynamic aspect of almost all the tasks performed during work, there is a tied relationship between wrist/elbow/shoulder position and hand strength. There is not any consensus regarding the optimal upper extremity posture. There is a consensus regarding the index and middle fingers being stronger than the ring and little fingers. The thumb force equals the other four fingers' force sum. Older workers represent an important and growing segment of the actual working force. In order to avoid an increase in musculoskeletal pathology, their special needs should be addressed from an ergonomic point of view. Evidence-based ergonomics intervention should stay at the forefront of all device and/or job (re)design.

Almost all studies use "healthy university students." Different study samples should be used in order to ensure an increased external validity. The use of real workers could reveal aspects that are not obvious in university students. Both on-site workers' musculoskeletal adaptation and changes in posture while performing specific tasks due to prolonged work are important factors that modify the risk factors exposure level.

Although previous studies determined grip and pinch strength in several elbow and shoulder positions, more research is needed in this area in order to assess the force application during positions that are used in real work. An increment of 5° should be used for wrist/elbow/shoulder deviations, with different combinations between them. Recording data only while the upper extremity is in the standard posture recommended by ASHA will not provide data that can be applied for further ergonomic job and workstation design.

Furthermore, almost all studies focused on static measurements. While this setting is easier to use, the utilization of dynamic recordings would provide the difference between static and dynamic force exertions. While the hand and fingers areas are exposed to high risk for musculoskeletal injuries, extensive work is required in order to reduce the elevated hazard from localized pressure on the area between thumb and digit II, the distal end of digit IV, and between the metacarpal bone and tip of finger V. Hand protection should be accomplished using combinations of different glove materials in order to ensure an important reduction in localized pressure at the hand–device interface, without resulting in loss of precision. To facilitate both the perfect glove fit and the adjustability between workers, stretchy materials seem a suitable solution and should be tested.

The well-documented differences between right and left hand should not be viewed only in terms of applied force. In order to ensure an appropriate grip or pinch, the fingers/wrist/elbow postures present important variations between right and left sides. Living in a right-hand-designed world, the use of the same devices and workstations impose a greater risk for left-handed workers. Further research in this area is needed in order to ensure targeted ergonomic interventions. The data difference in hand muscle fatigue and recovery pattern between men and women should be used to facilitate gender-customized devices. A closer collaboration between data generators (researchers) and data users (designers) would allow a reduction in work-related musculoskeletal injury with consequent cost saving.

Finally, followup studies addressing the capacity of returning workers to cope with the new/modified jobs are of extreme importance in order to reduce companies' costs and to ensure successful return to work.

References

Adam, A., De Luca, C., and Erim, Z. 1998. Hand dominance and motor unit firing behaviour, *J. Neurophysiol.*, 80, 1373–1382.

Ager, C.L., Olivett, B.L., and Johnson, C.L. 1984. Grasp and pinch strength in children 5 to 12 years old, *Am. J. Occup. Ther.*, 38, 107–113.

Agresti, A. and Finlay, B. 1986. *Statistical methods in social research*, 2nd ed. (San Francisco: Dellon Publishing), pp. 315–356.

Aniansson, A., Sperling, L., Rundgren, A., and Lehnberg, E. 1983. Muscle function in 75-year-old men and women: a longitudinal study, *Scand. J. Rehabil. Med. Suppl.*, 9, 92–102.

Armstrong, T.J. and Chaffin, D.B. 1979. Carpal tunnel syndrome and selected personal attributes, *J. Occup. Med.*, 21, 481–486.

Ashford, R.F., Nagelburg, S., and Adkins, R. 1996. Sensivity of the Jamar dynamometer in detecting submaximal grip effort, *J. Hand Surg.*, 21A, 402–405.

Aymard, C., Katz, R., Laffite, C., Le Bozec, S., and Penicaud, A. 1995. Changes in reciprocal and transjoint inhibition induced by muscle fatigue in man, *Exp. Brain Res.*, 106, 418–424.

Banks, W.W. and Goehring, G.S. 1979. The effects of degraded visual and tactile information on diver work performance, *Hum. Factors*, 21(4), 409–415.

Berguer, R., Gerber, S., Kilpatrick, G., Remler, M., and Beckley, D. 1999. A comparison of forearm and thumb muscle electromyographic responses to the use of laparoscopic instruments with either a finger grasp or a palm grasp, *Ergonomics*, 42, 1634–1645.

Bernard, B.P. 1997. *Musculoskeletal Disorders and Workplace Factors: A Critical Review of Epidemiologic Evidence for Work-Related Musculoskeletal Disorders of the Neck, Upper Extremity, and Low Back* (Cincinnati, OH: National Institute for Occupational Safety and Health (NIOSH)).

Blackwell, J.R., Kornatz, K.W., and Health, E.M. 1999. Effect of grip span on maximal grip force and fatigue of flexor digitorum superficialis, *Appl. Ergon.*, 30, 401–405.

Boatright, J.R., Kiebzak, G.M., O'Neil, D.M., and Peindl, R.D. 1997. Measurement of thumb abduction strength: normative data and a comparison with grip and pinch strength, *J. Hand Surg.*, 22A, 843–848.

Borod, J.C., Caron, H.S., and Koff, E. 1984. Left-handers and right-handers compared on performance and preference measures of lateral dominance, *Brit. J. Psychol.*, 75, 177–186.

Bradley, J.V. 1969. Effect of gloves on control operation time, *Hum. Factors*, 11(1), 13-20.

Brorson, H., Werner, C.O., and Thorngren, K.G. 1989. Normal pinch strength. *Acta Ortop. Scand.*, 60, 66–68.

Bureau of Labor Statistics. 1992. Occupational injuries and illnesses in the United States by industry, 1990. Washington, DC: U.S. Department of Labor, Bureau of Labor Statistics, Bulletin 2399, April 1992.

Bureau of Labor Statistics. 2002. www.bls.gov.

Caldwell, L., Chaffin, D.B., Dukes-Dobos, F.N., Kroemer, K.H.E., Laubach, L.L., Snook, S.H., and Wasserman, D.E. 1974. A proposed standard procedure for static muscle strength testing, *AIHA*, 35, 201–206.

Capodaglio, P., Maestri, R., and Bazzini, G. 1997. Reliability of a hand gripping endurance test, *Ergonomics*, 40, 428–434.

Carmelli, D. and Reed, T. 2000. Stability and change in genetic and environmental influences on hand-grip strength in older male twins, *J. Appl. Physiol.*, 89, 1879–1883.

Carpentier, A., Duchateau, J., and Hainaut, K. 2001. Motor unit behaviour and contractile changes during fatigue in the human first dorsal interosseus, *J. Physiol.*, 534, 903–912.

Chadwick, E.K.J. and Nicol, A.C. 2001. A novel force transducer for the measurement of grip force, *J. Biomech.*, 34, 125–128.

Chao, E.Y.S., An, K.N., Cooney, W.P., and Linscheid, R.L. 1989. Biomechanics of the Hand — A Basic Research Study, Singapore World Scientific.

Chaparro, A., Rogers, M., Fernandez, J., Bohan, M., Choi, S.D., and Stumpfhauser, L. 2000. Range of motion of the wrist: implications for designing computer input devices for the elderly, *Disabil. Rehabil.*, 22, 633–637.

Chatterjee, S. and Chowdhuri, B.J. 1991. Comparison of grip strength and isometric endurance between the right and left hands of men and their relationship with age and other physical parameters, *J. Hum. Ergol.*, 20, 41–51.

Chau, N., Petry, D., Huguenin, P., Remy, E., and Andre, J.M. 1997. Comparison between estimates of hand volume and hand strengths with sex and age with and without anthropometric data in healthy working people, *Eur. J. Epidemiol.*, 13, 309–316.

Chong, C.K., Tseng, C.H., Wong, M.K., and Tai, T.Y. 1994. Grip and pinch strength in Chinese adults and their relationship with anthropometric factors, *J. Formos Med. Assoc.*, 93, 616–621.

Christ, C.B., Boileau, R.A., and Slaughter, M.H. 1992. Maximal voluntary isometric force production characteristics of six muscle groups in women aged 25 to 74 years, *Am. J. Hum. Biol.*, 4, 537–545.

Close, R.I. 1972. Dynamic properties of mammalian skeletal muscles, *Physiol. Rev.*, 52, 129–197.

Cochran, D.J. and Riley, M. 1986. The effects of handle shape and size on exerted forces, *Hum. Factors*, 28, 253–265.

Conti, G.E. 1998. Clinical interpretation of "Grip strengths and required forces in accessing everyday containers in a normal population," *Am. J. Occupat. Ther.*, 52, 627–628.

Crosby, C.A., Wehbe, M.A., and Mawr, B. 1994. Hand strength: normative values, *J. Hand Surg.*, 19A, 665–670.

Cutkosky, M.R. and Wright, P.K. 1986. Modeling and manufacturing grips and correlations with the design of robotic hands, *Proceedings 1986 IEEE International Conference on Robotics and Automation*, San Francisco, CA, pp. 1533–1539.

Danion, F., Latash, M.L., Li, Z.M., and Zatsiorsky, V.M. 2001. The effect of a fatiguing exercise by the index finger on single- and multi-finger force production tasks, *Exp. Brain Res.*, 138, 322–329.

Dawson, N.M., Felle, P., and O'Donovan. 1998. A new manual power grip, *Acta Anat.*, 163, 224–228.

Desrosiers, J., Bravo, G., Hebert, R., and Dutil, E. 1995. Normative data for grip strength of elderly men and women, *Am. J. Occup. Ther.*, 49, 637–644.

Desrosiers, J., Hebert, R., Bravo, G., and Dutil, E. 1995. Comparison of the Jamar dynamometer and the Martin vigorimeter for grip strength measurement in a healthy elderly population, *Scand. J. Rehabil. Med.*, 27, 137–143.

Dorfman, L.J., Howard, J.E., and McGill, K.C. 1990. Triphasic behavioural response of motor units to submaximal fatiguing exercise, *Muscle Nerve*, 13, 621–628.

Drury, C.G., Begbie, K., Ulate, C., and Deeb, J.M. 1985. Experiments on wrist deviation in manual materials handling, *Ergonomics*, 28, 577–589.

Dvir, Z. 1997. The measurement of isokinetic fingers flexion strength, *Clin. Biomech.*, 12, 473–481.

Ejeskar, A. and Ortengren, R. 1981. Isolated finger flexion force: a methodological study, *Hand*, 13, 223–230.

Fagarasanu, M. and Kumar, S. 2002. Measurement Instrument and Data Collection: A Consideration of Constructs and Biases in Ergonomics Research, *Int. J. Ind. Ergon.*, 30, 251–265.

Fellows, G.L. and Freivalds, A. 1989. The use of force sensing resistors in ergonomic tool design. *Proc. 33rd Annual Meeting of the Human Factors Society* (Santa Monica, CA: Human Factors Society), pp. 713–717.

Firrell, J.C. and Crain, G.M. 1996. Which setting of the dynamometer provides maximal grip strength? *J. Hand Surg.*, 21A.

Fisher, M.B. and Birren, J.E. 1947. Age and strength, *J. Appl. Psychol.*, 31, 490–497.

Fong, P.W.K. and Ng, G.Y.F. 2001. Effect of wrist positioning on the repeatability and strength of power grip, *Am. J. Occup. Ther.*, 55, 212–216.

Fowler, N.K. and Nicol, A.C. 1999. Measurement of external three-dimensional interphalangeal loads applied during activities of daily living, *Clin. Biomech.*, 14, 646–652.

Fowler, N.K. and Nicol, A.C. 2001. Functional and biomechanical assessment of the normal and rheumatoid hand, *Clin. Biomech.*, 16, 660–666.

Fransson, C. and Winkel, J. 1991. Hand strength: the influence of grip span and grip type, *Ergonomics*, 34, 881–892.

Fransson-Hall, C. and Kilbom, A. 1993. Sensitivity of the hand to surface pressure, *Appl. Ergon.*, 24, 181–189.

Fraser, T.M. 1980. *Ergonomic Principles in the Design of Hand Tools, Occupational Safety and Health Series No. 44.* (Geneva: International Labor Office).

Frederick, L.J. and Armstrong, T.J. 1995. Effect of friction and load on pinch force in a hand transfer tasks, *Ergonomics*, 38, 2447–2454.

Fuglevand, A.J., Macefield, V.G., and Bigland-Ritchie, B. 1999. Force-frequency and fatigue properties of motor units in muscles that control digits of the human hand, *J. Neurophysiol.*, 81, 1718–1729.

Fulco, C.S., Rock, P.B., Muza, S.R., Lammi, E., Braun, B., Cymerman, A., Moore, L.G., and Lewis, S.F. 2001. Genders alters impact of hypobaric hypoxia on adductor pollicis muscle performance, *J. Appl. Physiol.*, 91, 100–108.

Fulco, C.S., Rock, P.B., Muza, S.R., Lammi, E., Cymerman, A., Butterfield, G., Moore, L.G., Braun, B., and Lewis, S.F. 1999. Slower fatigue and faster recovery of the adductor pollicis muscle in women matched for strength with men, *Acta Physiol. Scand.*, 167, 233–239.

Gandevia, S.C., Macefield, V.G., Bigland-Ritchie, B., Gorman, R.B., and Burke D. 1993. Motoneuronal output and gradation of effort in attempts to contract accurately paralyzed leg muscle in man, *J. Physiol. (London)*, 471, 411–427.

Gantchev, G.N., Gatev, P., Ivanova, T., and Tankov, N. 1986. Motor unit activity during fatigue, *Biomed. Biochem. Acta*, 45, 69–75.

Gerber, S. 1998. A comparative study of forces involved with manipulation of standard and laparoscopic surgical instruments (in press).

Giampaoli, S., Ferrucci, L., Cecchi, F., Noce, C.L., Poce, A., Dima, F., Santaquilani, A., Vescio, M.F., and Menotti, A. 1999. Hand-grip strength predicts incident disability in nondisabled older men, *Ageing*, 28, 283–288.

Gordon, A.M., Huxley, A.F., and Julian, F.J. 1966. The variation in isometric tension with sarcomere length in vertebrae muscle fibers, *J. Physiol.*, 184, 170–192.

Griffin, D.R. 1944. Manual dexterity of men wearing gloves and mittens. Fatigue Lab, Harvard University, Report No. 22.

Griffiths, H.E. 1943. Treatment of the injured workman, *Lancet*, 1, 729–731.

Halpern, C.A. and Fernandez, J.E. 1996. The effect of wrist and arm postures on peak pinch strength, *J. Hum. Ergol.*, 25, 115–130.

Hansson, G.A., Balogh, I., Ohlsson, K., Rylander, L., and Skerfving, S. 1996. Goniometer measurement and computer analysis of wrist angles and movements applied to occupational repetitive work, *J. Electromyogr. Kinesiol.*, 6, 23–35.

Hashemi, L., Webster, B.S., Clancy, E.A., and Courtney, T.C. 1998. Length of disability and cost of work-related musculoskeletal disorders of the upper extremity, *J. Occup. Environmental Med.*, 40, 261–269.

Hazelton, F.T., Smidy, G.L., Flatt, A.E., and Stephens, P.L. 1975. The influence of wrist position on the force produced by the finger flexors, *J. Biomech.*, 8, 301–306.

Herbert, R.D. and Gandevia, S.C. 1996. Muscle activation in unilateral and bilateral efforts assessed by motor nerve and cortical stimulation, *J. Appl. Physiol.*, 80, 1351–1356.

Hertzberg, T. 1955. Some contributions of applied physical anthropometry to human engineering, *Ann. N.Y. Acad. Sci.*, 63, 621–623.

Horgan, L.F., O'Riordan, D.C., and Doctor, N. 1997. Neuropraxia following laparoscopic procedures: an occupational injury, *Minimally Invasive Ther. Allied Technol.*, 6, 33–35.

Iberall, T. 1987. The nature of human prehension: three dexterous hands in one, *IEEE Proceedings International Conference on Robotics and Automation*, Vol. 2.

Imrhan, S.N. and Loo, C.H. 1988. Modelling wrist-twisting strength of the elderly, *Ergonomics*, 31, 1807–1819.

Imrhan, S.N. and Loo, C.H. 1989. Trends in finger pinch strength in children, adults, and the elderly, *Hum. Factors*, 31, 689–701.

Jebsen, R. 1973. An objective and standardized test of hand function, *Arch. Phys. Med. Rehabil.*, 54, 129–135.

Jenmalm, P., Dahlstedt, S., and Johansson, R.S. 2000. Visual and tactile information about object-curvature control fingertip forces and grasp kinematics in human dexterous manipulation, *J. Neurophysiol.*, 84, 2984–2997.

Johanson, M.E., Valero-Cuevas, F.J., and Hentz, V.R. 2001. Activation patterns of the thumb muscles during stable and unstable pinch tasks, *J. Hand Surg.*, 26A, 698–705.

Kadefors, R., Areskoug, A., Dahlman, S., and Wikstron, L. 1989. Verktyg for hander-Beteendestudier, *Svenska Lakaresallskapets Handlingar, Hygia*, 98, 115.

Kamakura, N., Matsuo, M., Ishii, H., Mitsuboshi, F., and Miura, Y. 1980. Patterns of static prehension in normal hands, *Am. J. Occup. Ther.*, 34, 437–445.

Kano, N., Yamakawa, T., and Kasugai, H. 1993. Laparoscopic surgeon's thumb (letter comment), *Arch. Surg.*, 128, 1172.

Kapandji, H. 1970. *The physiology of the joints*, 2nd ed. (Edinburgh, UK: Livingstone).

Keir, P.J., Wells, R.P., and Ranney, D.A. 1996. Passive properties of the forearm musculature with reference to hand and finger postures, *Clin. Biomech.*, 11, 401–409.

Ketola, R., Toivonen, R., and Viikari-Juntura, E. 2001. Interobserver repeatability and validity of an observation method to assess physical loads imposed on the upper extremities, *Ergonomics*, 44, 119–131.

Kinoshita, H., Murase, T., and Bandou, T. 1996. Grip posture and forces during holding cylindrical objects with circular grips, *Ergonomics*, 39, 1163–1176.

Kinoshita, H. 1999. Effect of gloves on prehensile forces during lifting and holding tasks, *Ergonomics*, 42(10), 1372–1385.

Kozin, S.H., Porter, S., Clark, P., and Thoder, J.J. 1999. The contribution of the intrinsic muscles to grip and pinch strength, *J. Hand Surg.*, 24A, 64–72.

Kraft, G.H. and Detels, P.E. 1972. Position of function of wrist, *Arch. Phys. Med. Rehabil.*, 53, 272–275.

Kroemer, K.H.E. 1970. Human strength: Terminology, measurement, and interpretation, *Hum. Factors*, 12, 297–313.

Kuhtz-Buschbeck, J.P., Stolze, H., Johnk, K., Boczek-Funcke, A., and Illert, M. 1998. Development of prehension movements in children: a kinematic study, *Exp. Brain Res.*, 122, 424–432.

Kumar, S. 2001. Theories of musculoskeletal injury causation, *Ergonomics*, 44, 17–47.

Kumar, S. and Simmonds, M. 1994. The accuracy of magnitude production of submaximal precision and power grips and gross motor efforts, *Ergonomics*, 37, 1345–1353.

Kuzala, E.A. and Vargo, M.C. 1992. The relationship between elbow position and grip strength, *Am. J. Occup. Ther.*, 46, 509–512.

Lamoreaux, L. and Hoffer, M.M. 1995. The effect of wrist deviation on grip and pinch strength, *Clin. Orthop.*, 314, 152–155.

Landsmeer, J.M.F. 1962. Power grip and precision handling, *Ann. Rheum. Ther.*, 21, 164–170.

LaStayo, P., Chidgey, L., and Miller, G. 1995. Quantification of the relationship between dynamic grip strength and forearm rotation: a preliminary study, *Ann. Plast. Surg.*, 35, 191–196.

Li, Z.M., Latash, M.L., Newell, K.M., and Zatsiorsky, V.M. 1998a. Motor redundancy during maximal voluntary contraction in four-finger tasks, *Exp. Brain Res.*, 122, 71–78.

Li, Z.M., Latash, M.L., and Zatsiorsky, V.M. 1998b. Force sharing among fingers as a model of the redundancy problem, *Exp. Brain Res.*, 119, 276–286.

Lunde, B.K., Brewer, W.D., and Garcia, P.A. 1972. Grip strength of college women, *Arch. Phys. Med. Rehabil.*, 491–493.

Majeed, A.W., Jacob, G., Reed, M.W., and Johnson, A.G. 1993. Laparoscopist's thumb: an occupational hazard (letter comments), *Arch. Surg.*, 128, 357.

Marley, R. and Wehrman, R. 1992. Grip strength as a function of forearm rotation and elbow posture, in *Proceedings of the 36th Annual Meeting of the Human Factors Society*, 791–795.

Marshall, M.M., Mozrall, J.R., and Shealy, J.E. 1999. The effects of complex wrist and forearm posture on wrist range of motion, *Hum. Factors*, 41, 205–213.

Martin, S., Neale, G., and Elia, M. 1984. Factors affecting maximum momentary grip strength, *Hum. Nutr. Clin. Nutr.*, 39C, 137–147.

Mathiowetz, V., Kashman, N., Volland, G., Weber, K., Dowe, M., and Rogers, S. 1985. Grip and pinch strength: normative data for adults, *Arch. Phys. Med. Rehabil.*, 66, 69–74.

Mathiowetz, V., Rennells, C., and Donahoe, L. 1985. Effect of elbow position on grip and key pinch strength, *J. Hand Surg.*, 10A, 694–697.

Mathiowetz, V., Wiemer, D.M., and Federman, S.M. 1986. Grip and pinch strength: norms for 6 to 19 year olds, *Am. J. Occup. Ther.*, 40, 705–711.

Maton, B. and Gamet, D. 1989. The fatigability of two agonist muscles in human isometric voluntary submaximal contraction: an EMG study, *Eur. J. Appl. Physiol.*, 58, 369–374.

McBride, E.D. 1942. *Disability Evaluation* (Philadelphia, PA: J.B. Lippincott).

McGinnis, J.S., Bensel, C.K., and Lockhar, J.M. 1973. Dexterity Afforded by CB Protective Gloves, Report No. 73-35-PR, U.S. Army Natick Laboratories, Natick, MA.

McGorry, R. 2001. A system for the measurement of grip forces and applied moments during hand tool use, *Appl. Ergon.*, 32, 271–279.

McPhee, S.D. 1987. Functional hand evaluations: a review, *Am. J. Occup. Ther.*, 41, 158–163.

Melvin, J.L. 1977. *Rheumatic Disease: Occupational Therapy and Rehabilitation* (Philadelphia, PA: F.A. Davis).

Merton, P.A. 1954. Voluntary strength and fatigue, *J. Physiol. (Lond)*, 123, 553–564.

Mital, A. 1991. Hand tools: injuries, illness, design and usage, in: *Workspace, Equipment and Tool Design*, Mital, A. and Karwowski, W., Eds. (New York: Elsevier), pp. 219–256.

Muralidhar, A. and Bishu, R.R. 1994. Glove evaluation: a lesson from impaired hand testing, in: *Advances in Industrial Ergonomics and Safety VI*, Aghazadeh, F., Ed. (London: Taylor and Francis).

Muralidhar, A. and Bishu, R.R. 2000. Safety performance of gloves using the pressure tolerance of the hand, *Ergonomics*, 43, 561–572.

Muralidhar, A., Bishu, R.R., and Hallbeck, M.S. 1999. The development and evaluation of an ergonomic glove, *Appl. Ergon.*, 30, 555–563.

Napier, J.R. 1956. The prehensile movements of the human hand, *J. Bone Joint Surg. Br.*, 38-B, 902–913.

National Institute for Occupational Safety and Health (NIOSH). 1997. Musculoskeletal Disorders and Workplace Factors: A Critical Review of Epidemiologic Evidence for Work-Related Musculoskeletal Disorders of the Neck, Upper Extremity, and Low Back.

Netscher, D., Steadman, A.K., Thornby, J., and Cohen, V. 1998. Temporal changes in grip and pinch strength after open carpal tunnel release and the effect of ligament reconstruction, *J. Hand Surg.*, 23A, 48–54.

Nwuga, V.C. 1975. Grip strength and grip endurance in physical therapy students, *Arc. Phys. Med. Rehabil.*, 56, 296–300.

O'Driscoll, S.W., Horii, E., Ness, R., Cahalan, T.D., Richards, R.R., and An, K.N. 1992. The relationship between wrist position, grasp size, and grip strength. *J. Hand Surg.*, 17A, 169–177.

Ohtsuki, T. 1981. Inhibition of individual fingers during grip strength exertion, *Ergonomics*, 24, 21–36.

Pfitzer, J.T., Ellis, N.C., and Johnston, W.L. 1972. Grip strength, level of endurance, and manual task performance, *J. Appl. Psychol.*, 56, 278–280.

Porac, C. and Coren, S. 1981. *Lateral Preferences and Human Behaviour* (New York: Springer-Verlag).

Provins, K.A. and Magliaro, J. 1993. The measurement of handedness by preference and performance tests, *Brain Cogn.*, 22, 171–181.

Pryce, J.C. 1980. The wrist positioning between neutral and ulnar deviation that facilitates the maximum power grip strength, *J. Biomech.*, 13, 505–511.

Radwin, R.G., Oh, S., Jenson, T.R., and Webster, J.G. 1992. External finger forces in submaximal five-finger static pinch prehension, *Ergonomics*, 35, 275–288.

Rahman, N., Thomas, J.J., and Rice, M.S. 2002. The relationship between hand strength and the forces used to access containers by well elderly persons, *Am. J. Occup. Ther.*, 56, 78–85.

Ranganathan, V.K., Siemionow, V., Sahgal, V., and Yue, G.H. 2001. Effects of aging on hand function, *J. A. Geriatr. Soc.*, 49, 1478–1484.

Reikeras, O. 1983. Bilateral differences of normal hand strength, *Arch. Orthop. Trauma Surg.*, 101, 223–224.

Resources Services Group, American Association of Retired Persons and Administration on Aging, U.S. Department of Health and Human Services. 1997. A profile of older Americans (Washington, DC: AARP).

Rice, M.S., Leonard, C., and Carter, M. 1998. Grip strengths and required forces in accessing everyday containers in a normal population, *Am. J. Occup. Ther.*, 52, 621–626.

Richards, L.G. 1997. Posture effects on grip strength, *Arch. Phys. Med. Rehabil.*, 78, 1154–1156.

Richards, L.G., Olson, B., and Palmiter-Thomas, P. 1996. How forearm position affects grip strength, *Am. J. Occup. Ther.*, 50, 133–138.

Rohmert, W. 1960. Ermittlung von Erholungspausen fuer statische Arbeit des Menschen, *Int. Z. Augewandte Physiol.*, 18, 123–164.

Salimi, I., Brochier, T., and Smith, A.M. 1999. Neuronal activity in somatosensory cortex of monkeys using a precision grip. I. Receptive fields and discharge patterns, *J. Neurophysiol.*, 81, 825–834.

Sande, L.P., Coury, H.J.C.G., Oishi, J., and Kumar, S. 2001. Effect of musculoskeletal disorders on prehension strength, *Appl. Ergonomics*, 32, 609–616.

Schmidt, R.T. and Toews, J.V. 1970. Grip strength as measured by the Jamar dynamometer, *Arch. Phys. Med. Rehabil.*, 51, 321–327.

Shechtman, O., MacKinnon, L., and Locklear, C. 2001. Using the BTE Primus to measure grip and wrist flexion strength in physically active wheelchair users: an exploratory study, *Am. J. Occup. Ther.*, 55, 393–400.

Shiffman, L.M. 1992. Effects of aging on adult hand function, *Am. J. Occup. Ther.*, 46, 785–792.

Silverstein, B.A., Fine, L.J., and Armstrong, T.J. 1986. Hand, wrist cumulative trauma disorders in industry, *Br. J. Ind. Med.*, 43, 779–784.

Sollerman, C. and Sperling, L. 1978. Evaluation of activities of daily living function — especially hand function, *Scand. J. Rehabil. Med.*, 10, 139–143.

Sommerich, C., McGlothlin, J.D., and Marras, W.S. 1993. Occupational risk factors associated with soft tissue disorders of the shoulder: a review of recent investigations in the literature, *Ergonomics*, 36, 697–717.

Stal, M., Hansson, G.A., and Moritz, U. 1999. Wrist positions and movements as possible risk factors during machine milking, *Appl. Ergon.*, 30, 527–533.

Su, C.Y., Lin, J.H., Chien, T.H., Cheng, K.F., and Sung, Y.T. 1994. Grip strength in different positions of elbow and shoulder, *Arch. Phys. Med. Rehabil.*, 75, 812–815.

Sukthankar, S.M. and Reddy, N.P. 1995. Force feedback issues in minimally invasive surgery, *Interactive Technology and the New Paradigm for Healthcare*. (San Diego, CA: IOS Press).

Swanson, A.B., Goran-Hagert, C., and DeGroot-Swanson, G. 1978. Evaluation of impairment of hand function (Chapter 4), in *Rehabilitation of the Hand*, J.M. Hunter, L.H. Schneider, E.J. Mackin, and J.A. Bell, Eds. (St. Louis, MO: CV Mosby).

Swanson, A.B., Matev, I.B. and De Groot, G. 1970. The strength of the hand, *Bull. Prosthet. Res.*, 9, 387–396.

Teraoka, T. 1979. Studies on the peculiarity of grip strength in relation to body positions and aging, *Kobe J. Med. Sci.*, 25, 1–17.

Terrell, R. and Purswell, J. 1976. The influence of forearm and wrist orientation on static grip strength as a design criterion for hand tools, *Proceedings of the International Ergonomics Association* (College Park, MD: IEA), pp. 28–32.

Voorbij, A.I.M. and Steenbekkers, L.P.A. 2001. The composition of a graph on the decline of total body strength with age based on pushing, pulling, twisting and gripping force, *Appl. Ergon.*, 32, 287–292.

Wallstrom, A. and Nordenskiold, U. 2001. Assessing hand grip endurance with repetitive maximal isometric contractions, *J. Hand Ther.*, 14, 279–285.

Weiss, M.W. and Flatt, A.E. 1971. A pilot study of 198 normal children pinch strength and hand size in the growing hand, *Am. J. Occup. Ther.*, 25, 10–12.

Wells, R. and Greig, M. 2001. Characterizing human hand prehensile strength by force and moment wrench, *Ergonomics*, 44, 1392–1402.

Westling, G. and Johansson, R.S. 1984. Factors influencing the force control during precision grip, *Exp. Brain Res.*, 53, 277–284.

Wiker, S., Langalf, G., and Chaffin, D. 1989. Arm Posture and Human Movement Capability, *Hum. Factor*, 31, 421–441.

Wolf, L.D., Matheson, L.N., Ford, D.D., and Kwak, A.L. 1996. Relationships among grip strength, work capacity, and recovery, *J. Occupat. Rehabil.*, 6, 57–70.

Young, V.L., Pin, P., Kraemer, B.A., Gould, R.B., Nemergut, L., and Pellowski, M. 1989. Fluctuation in grip and pinch strength among normal subjects, *J. Hand Surg.*, 14A, 125–129.

Yun, M.H., Kotani, K., and Ellis, D. 1992. Using force sensing resistors to evaluate hand tool grip design, *Proceedings of Human Factors Society 36th Annual Meeting*, 806–810.

Zijdewind, I. and Kernell, D. 2001. Bilateral Interactions During Contractions of Intrinsic Hand Muscles, *J. Neurophysiol.*, 85, 1907–1913.

Zijdewind, I., Zwarts, M.J., and Kernell, D. 1998. Influence of a voluntary fatigue test on the contralateral homologous muscle in humans? *Neurosci. Lett.*, 253, 41–44.

Additional Reading

Anderson, P.A., Chanoski, C.E., Devan, D.L., McMahon, B.L., and Whelan, E.P. 1990. Normative study of grip and wrist flexion strength employing a BTE Work Simulator, *J. Hand Surg.*, 15A, 420–425.

Brunnstrom, S. 1980. *Clinical Kinesiology* (Chicago: F.A. Davis).

Buchholz, B. and Armstrong, T.J. 1991. An ellipsoidal representation of human hand anthropometry, *Hum. Factors*, 33, 429–441.

Castro, M.C.F. and Clinquet, A. 2000. Artificial grasping system for the paralyzed hand, *Artif. Organs*, 24, 185–188.

Chao, E.Y.S., An, K.N., Cooney, W.P., and Linscheid, R.L. 1989. *Biomechanics of the Hand — A Basic Research Study* (Singapore: World Scientific).

Kinoshita, H., Murase, T., and Bandou, T. 1996. Grip posture and forces during holding cylindrical objects with circular grips, *Ergonomics*, 39, 1163–1176.

Kumar, S., Narayan, Y., and Chouinard, K. 1997. Effort reproduction accuracy in pinching, gripping, and lifting among industrial males, *Int. J. Ind. Ergon.*, 20, 109–119.

Mital, A. 1991. Hand tools: injuries, illness, design and usage, in *Workspace, Equipment and Tool Design*, A. Mital and W. Karwowski, Eds. (New York: Elsevier), pp. 219–256.

Sande, L.P., Coury, H.J. C.G., Oishi, J., and Kumar, S. 2001. Effect of musculoskeletal disorders on prehension strength, *Appl. Ergon.*, 32, 609–616.

11

Grasp at Submaximal Strength

B.J. Kim and R.R. Bishu

CONTENTS

ABSTRACT Human hands, as one of the most complex musculoskeletal structures, have the most delicate dexterity, tactility, and ability of manipulation. In everyday life people use hands for grasping or carrying things. The exerted force for these activities does not require maximal musculoskeletal strength, but submaximal strength, because the amount of effort exerted is usually determined by levels of loads and psychomotor coordination. When grasping, people usually overexert at the start and gradually reach a stable exertion level in order to minimize muscle fatigue. This can be regarded as a natural biological effort to minimize the energy cost function. Grasp strengths tend to change with such factors as weight of load lifted, types of gloves, surface friction of gloves, handle types, handle orientation, and individual safety perception. How these factors affect the submaximal strength grasp will be discussed.

11.1 Importance of the Hand

Along with the brain, the human hand is probably one of the most complex and sophisticated organs that can uniquely perform exploration, prehension, perception, and manipulation. The importance of the hand to human culture can be recognized by its depiction in art and sculpture, its reference frequency in vocabulary and phraseology, and its importance in communication and expression. The human hand is distinguished from that of the primates by the presence of a strong opposable thumb that enables humans to accomplish tasks requiring precision and fine control. The hand provides humans with both mechanical and sensory capabilities.

11.2 Hand Strength

11.2.1 Prehensile Capabilities of the Hand

Hand movements are categorized into prehensile and nonprehensile movements by Napier (1956). In prehensile hand movements, an object is seized and held partly or wholly within the compass of the hand. Nonprehensile movements do not involve object grasping and seizing, but pushing or lifting motions of the hand as a whole or of the digits individually. Later, Landsmeer (1962) classifies hand grasping further into power grips and precision handling. In power grips, a dynamic initial phase is distinguished from a static terminal phase. The dynamic phase includes opening of the hand, positioning of the fingers, and grasping of the object. Meanwhile, there is no static terminal phase in precision handling. Westling and Johansson (1984) find that the force control during precision grip is affected by factors of friction, weight, and individual safety margin. The study shows that the grip force is related to the frictional conditions of the previous trials when running multiple trials. It is found that the grip applied when holding small objects stationary in the space is critically balanced such that accidental slipping between the skin and the object does not occur and the grip force does not reach exceedingly high values. This sense of critical balance, from the perspective of the amount of force applied for grasping, is important because too much firm grasping can result in a destruction of fragile objects and also serve as a possible cause of injury to the hand, lead to muscle fatigue, and interfere with further manipulative activity imposed upon the hand. Sensory perception in the hand is based on the presence of mechanoreceptors distributed all over the palmar area, especially at the tips of the fingers. Thus, feedback from the hand plays a critical role for grasping tasks and enables the hand to control the amount of force exerted. Anything that blocks the transmission of impulses from the hand interferes with the feedback cycle and affects grip force control. Gloves do interfere with the feedback cycle of the hand.

11.2.2 Maximal Grip Strength

Human strength has been a subject for a considerable amount of research because, as Kroemer et al. (2000) state, it is dependent on a number of parameters. Maximum grip strength or maximum voluntary contraction (MVC) has always been used as a design measure for tasks involving hand exertions, even though little research has been performed to confirm the validity of the measure. Astrand and Rodahl (1986) report a variation of maximal grip strength by 10% in day-to-day muscular strength without any evidence or studies to validate the report. Bishu and Klute (1995) also find that maximal exertion is unstable and deviates as much as 30%. Hand exertion capabilities for an extended period of time are studied using force endurance (Bishu et al., 1994; Monod and Scherrer, 1965; Rohmert, 1960). According to these studies, maximum exertion can be sustained only for a few seconds. Meanwhile, the sustenance of exertions can be improved by reducing maximal exertion 10 to 15%. Maximal grip strength is also dependent upon postures (Terrel and Purswell, 1976) and gloves (Bishu and Klute, 1995: Bishu et al., 1993, 1994; Cochran et al., 1986; Lyman and Groth, 1958).

Hand maximal strength that is measured by grip strength or grasp strength has been used as a metric to assess hand capabilities at work places. Although submaximal grasps are required in most work environment more than are maximal grasps, little research has focused on submaximal types of grasps (see Grant, 1994; Lyman and Groth, 1958; Radwin and Oh, 1992; Westling and Johansson, 1984).

11.2.3 Submaximal Grasp

A series of experiments on submaximal grasps are performed by one of the authors of this chapter. A generic hand dynamometer is fabricated as shown in Figure 11.1. Two load cells are fixed to the dynamometer to record the grasp force. This dynamometer is used in a number of experiments to understand all aspects of submaximal grasps.

When grasping at submaximal levels, it is typical that people overexert initially to a peak level and then slowly reduce the grasp to a stable level (Bishu et al., 1994; Bronkema et al., 1994; Kim and Bishu, 1997). Bronkema et al. (1994) report the peak force, stable force, and ratio of stable force to load grasped for several loads as shown in Table 11.1. The study shows that the difference between the peak force and the stable force for the loads investigated is almost the same regardless of the loads lifted, which may imply a similar pattern of muscle strength transition from overexertion to stable. Grant reports a similar study result of overexertion grasping (Grant, 1994). For the study, 30 male participants are asked to lift with bare hands a cylindrical aluminum handle attached with various levels of loads. The grasp force is compared with the theoretical grasp force that is calculated on the basis of the friction factor between hand and handle, the mass of the

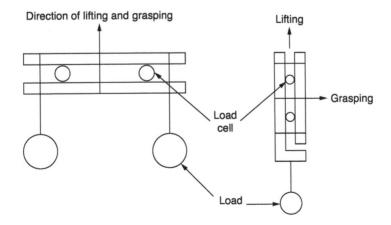

FIGURE 11.1
Horizontal grasp vs. vertical grasp. (From Kim, B. and Bishu, R. 1997, in *Proceedings of 13th Triennial Congress of the International Ergonomics Association*, Tampere, Finland, pp. 546–548. With permission.)

TABLE 11.1

Load Effect on Peak and Stable Forces

Load	Peak Force	Stable Force	Ratio of Stable/Load
0.5	15	9.6	19.2
5.5	22	17	3.1
10.5	29.7	22	2.1
15.5	35	29	1.9
20.5	40.8	34	1.7

Source: Bronkema, L., Bishu, R.R., Garcia, D., Klute, G., and Rajulu, S. 1994. In *Proceedings of the 38th Annual Conference of the Human Factors and Ergonomics Society,* (Nashville, TN), pp. 597–601. With permission.

object, and the vertical movement acceleration. The result shows that the actual grasp force tends to be higher than the theoretical grasp force, especially at the low loads. It can be inferred from the result that the initial grasp force is conservatively estimated from the perception of loads lifted, and that when the estimated grasp force exceeds a threshold for keeping a balance between slipping force and friction force, the overexertion is gradually released up to the threshold level and is stabilized.

11.2.4   Factors for Submaximal Grasp

The extent of overgrasp and the subsequent reduction to the stable level appears to be dependent on the following factors.

11.2.4.1   *Load Lifted*

Several levels of loads are examined to find an effect of load lifted on submaximal grasp force. Bronkema et al. (1994) and Bishu et al. (1994) report a high ratio of stable force to load lifted at the low levels of loads compared to the high levels of loads as shown in Figure 11.2. Although Figure 11.2 is scaled appropriately, it is apparent that the ratio of stable force to load lifted becomes smaller as the loads lifted increase. One issue of interest is the amount of force overexerted per pound lifted on average. Bishu et al. report that the average grasp force exerted per pound of load lifted is 18.4 pounds and 1.7 pounds for loads of 0.5 pounds and 20.5 pounds, respectively. One of the possible reasons of the higher overexertion at the lower load is that there will be no cost for people to overexert because they consider the low load to be within their physical strength capabilities (Kumar, 2003). Another possible reason is that people may exhibit a grasp control based on an estimate of the maximum voluntary contraction employed for the load, and that their estimates tend to be less stable at the lower load than the higher load. It means that the perception and employment of the small fraction of the respective maximum voluntary contraction will be more difficult than will that of the large fraction of the maximum. More research is needed on this topic.

11.2.4.2   *Gender Effect*

Bronkema et al. (1994) report that females show lower peak force, stable force, and ratio of stable force to load lifted than do males, as shown in Table 11.2. One would expect a gender difference in maximal exertion force. However, it is worth noting a gender difference in submaximal exertion, because the difference would not be expected unless a grasp

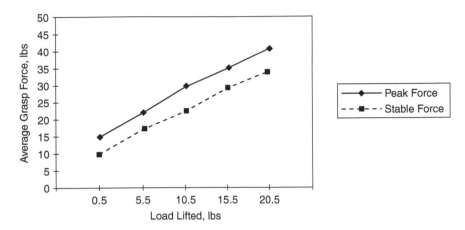

FIGURE 11.2
Effects of load on grasp force. (From Bronkema, L., Bishu, R.R., Garcia, D., Klute, G., and Rajulu, S. 1994, in *Proceedings of the 38th Annual Conference of the Human Factors and Ergonomics Society*, Nashville, TN, pp. 597–601. With permission.)

TABLE 11.2

Gender Effect on Peak and Stable Force

	Peak	Stable	Stable/Load
Male	16.24	12.61	2.37
Female	12.45	10.05	1.85

Source: Bronkema, L., Bishu, R.R., Garcia, D., Klute, G., and Rajulu, S. 1994. In *Proceedings of the 38th Annual Conference of the Human Factors and Ergonomics Society*, (Nashville, TN), pp. 597–601. With permission.

control behavior in terms of overexertion and stable grasp is employed as a proportion of maximal strength. The peak force, stable force, and ratio of stable to load lifted for females at submaximal grasp are as 76, 78, and 73%, respectively, of those of males.

11.2.4.3 Glove Effect

Glove types, as shown in Figure 11.3, have a significant effect on holding force. Bronkema et al. (1994) experiment with two different types of gloves for finding effects of donning gloves. Significantly different results are found from the cotton glove, but not the leather glove, compared to the bare hand. Bishu et al. (1994) also report that thick shuttle gloves that have beads on the palmar surface yield the same grasp force as bare hands. The similar grasp force between the leather gloves and the bare hands is probably due to a similar friction between them. The material characteristic of leather may make the subjects sense a considerable amount of friction available on the surface of the glove regardless of the thickness of the material. Glove thickness, however, affects fatigue rates or muscular exertions of the flexors and extensors (Batra et al., 1994; Bensel, 1993; Hallbeck and McMullin, 1991; Plummer et al., 1985). Sudhakar et al. (1988) state that "thicker gloves impact the biomechanics of the hand by causing an apparent increase in the length and the thickness of the digits, thereby causing the forearm flexors and extensors to exert more force."

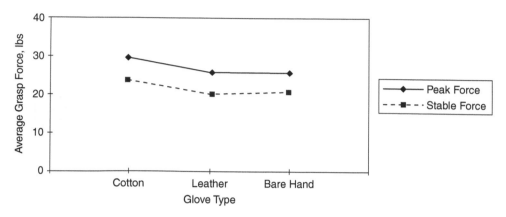

FIGURE 11.3
Effect of glove type on grasp force. (From Bronkema, L., Bishu, R.R., Garcia, D., Klute, G., and Rajulu, S. 1994, in *Proceedings of the 38th Annual Conference of the Human Factors and Ergonomics Society*, Nashville, TN, pp. 597–601. With permission.)

11.2.4.4   Handle Orientation

Buhman et al. (2000) examine the effect of handle length and handle orientation on submaximal grasp force by using two sizes of handles as shown in Figure 11.4. In their experiment, the subjects are instructed to grasp the handle in a lateral orientation and a transverse orientation. The handles are oval-shaped in the grip section. Grasping in the direction of the major axis is named lateral, and grasping in the direction of the minor axis transverse. The following results are observed from the study. First, the pattern of peak and stable forces are similar to other conditions discussed above. Second, the large-sized handle tends to need more grasp force than does the small handle. Finally, the grasp force using the transverse orientation needs less force than does that using the lateral orientation.

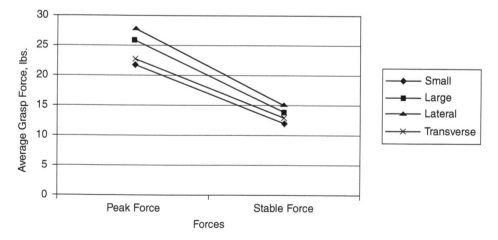

FIGURE 11.4
Effect of handle on grasp. (From Buhman, D.C., Cherry, J.A., Bronkema, L., and Bishu, R.R. 2000. *Int. J. Ind. Ergon.* 25, 247–256. With permission.)

11.2.4.5 Grasp Direction

Bronkema and Bishu (1996) show that when the lifting is executed with a perpendicular posture between the directions of arm muscle action and lifting, the exerted grasp force is higher than the weight of load lifted, and the ratio of the exerted grasp force to the weight of load is decreased by increasing the weight of load lifted. However, some earlier studies report that the capacity to generate force can vary as the angles that muscles act during maximal voluntary exertion differ (Brand, 1985; Hazelton et al., 1975). The directions of arm muscle action and lifting can be considered as another factor that influences the exerted grasp force during submaximal or "just holding" type grasp. Kim and Bishu (1997) examine the effect of the direction, as shown in Figure 11.1 as a horizontal grasp between muscle action and lifting on submaximal voluntary exertion. In addition to the factors considered in the previous studies, they investigate the change of grasp force in the sequence of lifting. The sequence of lifting is considered by such four stages as: stage 1 for initial grasp, stage 2 for transition grasp between initial grasp and stable grasp, stage 3 for stable grasp, and stage 4 for grasp during release. A high ratio of stable force-to-load-lifted is also found at the low loads, but no effect of the friction of glove surfaces is observed over the loads considered. The lack of friction effect is mainly due to having the same direction between muscle action and lifting. One thing of interest on the high ratio is that the transition of grasp force occurs more quickly at the low loads than at the high loads, as shown in Figure 11.5. For low loads, people tend to adjust overexertion and reach a threshold of grasp force quickly.

11.2.4.6 Surface Friction

Bronkema and Bishu (1996) investigate the effect of friction on grasp force by applying two different sizes of silicone pads to the glove surface. As indicated in Table 11.3, the application of silicon to the surface of the glove significantly affects the peak and stable holding force. Just one pad of silicon covering 4.8 square inches of the glove surface area

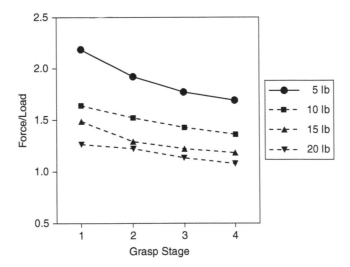

FIGURE 11.5
Ratio of stable force to load lifted vs. grasp stage. (From Kim, B. and Bishu, R. 1997, in *Proceedings of 13th Triennial Congress of the International Ergonomics Association*, Tampere, Finland, pp. 546–548. With permission.)

TABLE 11.3

Friction Effect on Ratio of Stable Force to Load Lifted

Friction Levels	Ratio of Stable Force to Load Lifted
Friction 1 (no silicon)	6.54
Friction 2 (4.8 square inches silicon)	4.85
Friction 3 (7.2 square inches silicon)	4.78

Source: Bronkema, L. and Bishu, R.R. 1996. In *Proceedings of the 40th Annual Meeting of the Human Factors and Ergonomic Society,* (Philadelphia, PA), pp. 702–706. With permission.

(friction level 2) reduces the holding force involved to an average of 82% of the holding force using the glove condition with no silicon (friction level 1). The same effect is shown in Figure 11.6. It appears that subjects *choose* to exert less force on the object being held when wearing the glove with more friction, and that people have a good capability of sensing the friction of the glove and of adjusting their grip strength accordingly. It can be an important finding for glove designers to consider friction effects to reduce the overexertion force, because a frequent application of overexerted force is regarded as one of risk factors for cumulative trauma disorders (Putz-Anderson, 1991). However, it is necessary to investigate the results with some cautions before any applications. The grasp force measured in this experiment is not the minimum force required for holding the object, but the actual force observed. It implies that the subjects may not exert the exact amount of force that is necessary to hold the object. In fact, they may tend to overexert to prevent the object from any accidental slips. This tendency can be defined as a "safety factor" when investigating grasps at submaximal exertions. The minimum required amount of force is not considered in that study. The relationship between the minimum required amount of force and the grasp force exerted is unknown. It can be linear or nonlinear. Thus, it is still not clear how much force people overexert even if people exert less force using gloves with surface frictions than without surface frictions. As long as the level of overexertion when using gloves with frictions is not a significant issue, people show more relaxed grasps with gloves that have high friction coefficient surfaces.

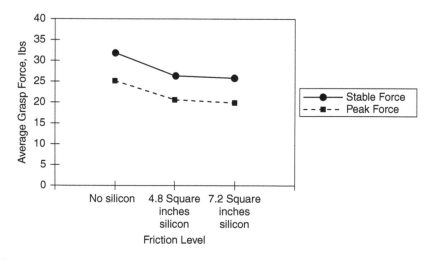

FIGURE 11.6

Effect of friction on grasp force. (From Bronkema, L. and Bishu, R.R. 1996, in *Proceedings of the 40th Annual Meeting of the Human Factors and Ergonomic Society,* Philadelphia, PA, pp. 702–706. With permission.)

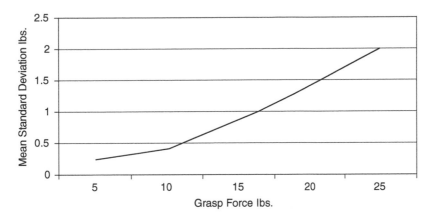

FIGURE 11.7
Grasp force effect on grasp control. (From Wilhem, G.A. and Bishu, R.R. 1997, in *Advances in Occupational Ergonomics and Safety*, B. Das and W. Karwowski, Eds., Amsterdam, IOS Press, pp. 297–300. With permission.)

11.2.4.7 Grasp Control

Wilhem and Bishu (1997) examine the stability of grasp force at various levels of exertion and hand condition and define the stability as grasp control. The grasp control is measured by the amount of variance of grasp force from average. The larger the variance is, the less stability the grasp force has. The study reports that people can keep the grasp strength more stable at the low loads than at the high loads, as shown in Figure 11.7. It implies that keeping a stable force is more difficult as the strength needed approaches to the maximal exertion because the muscles get fatigued more easily at the high loads. The study also reports that males are able to control the grasp better than are females.

11.3 A Theoretical Perspective

Given that people tend to overexert, and that the overexertion increases as the loads lifted decreases, is it possible to predict how much force people will overexert for a given load?

It is possible to calculate the amount of force theoretically necessary to lift an object using a handle interface as shown in Figure 11.1. From the friction force equation:

$$F = \frac{W}{\mu}$$

where:

F = force exerted by the hand perpendicular to the handle
μ = coefficient of friction between the hand and the handle interface
W = weight of the object lifted

The coefficient of friction between two surfaces, for instance a handle and a glove, is a constant. When lifting a handle with a glove, two possible cases can be considered for the

TABLE 11.4

Coefficient of Friction

	Friction-Cotton	Friction-Leather
Friction 1 (no silicon)	0.37	0.53
Friction 2 (4.8 square inches silicon)	0.47	0.65
Friction 3 (7.2 square inches silicon)	0.61	0.81

Source: Cherry et al. 2000. With permission.

friction, that is, one friction between the glove and the hand and the other between the glove and the handle. Cherry et al. (2000) investigate the coefficients of friction for the glove conditions that are used in the study performed by Bronkema et al. (1994). The coefficients of friction are shown in Table 11.4. Based on the coefficients of friction obtained, they calculate the theoretical grasp force for the levels of loads. For the purpose of simplicity, the friction between the glove and the hand is ignored in that study.

Figure 11.8 shows the comparison between the theoretical force calculated and the actual force observed at different load conditions. It is worthwhile to note two things from Figure 11.8. One thing is that the actual force, using the cotton gloves, is higher than that used with the leather glove for all levels of loads. The other one is that the difference between the two forces under the cotton-glove condition starts deviating around the middle level of the load, and that the deviation becomes wider. Meanwhile, the difference under the leather-glove condition becomes narrower and finally converges.

Cherry et al. (2000) try to adjust the theoretical force to the actual force by introducing a *correction* factor to the theoretical force as defined below:

$$F_{actual} = C \times F_{theoretical}$$

where:

C = correction factor or rate of exertion above theoretical force.

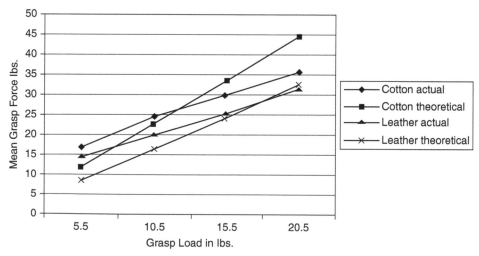

FIGURE 11.8
Theoretical vs. actual grasp. (From Cherry et al., 2000. With permission.)

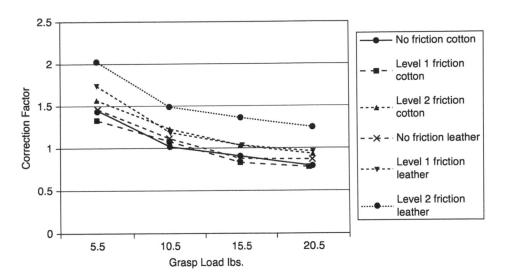

FIGURE 11.9
Correction factor. (From Cherry et al., 2000. With permission.)

A plot of the correction factor can provide some information about the nature of the relationship between the theoretical force and the actual grasp force. Figure 11.9 shows the plots of the correction factors for the two gloves. The correction factor for the leather glove shows more consistent relation to the load lifted than does that for the cotton glove. As the load lifted increases, the actual grasp force is quite close to the theoretical force when using the leather glove with friction level 1 and level 2. Thus, it may be possible to estimate the actual grasp force from the friction force equation if a leather glove having a high friction coefficient is used to lift a high load. In the case of the cotton glove, the correction factors for the three friction conditions become less than 1 as the load lifted increases. It implies that people do not need as much force as the theoretically predicted to grasp the load.

The theoretical force required is based on the weight of the object and the contact of one flat surface. When lifting the handle attached with a load in the transverse orientation, however, the contact is not single surface but multiple. Also, the surface is not flat but shaped. Based on these circumstances, it may be possible to explain why the use of the cotton gloves exerts less than the theoretically needed grasp force as the load lifted increases. One plausible reason is due to the fact that it is much easier for fingers to flex the cotton glove to make a firm grip (or a barehanded grip shape) than for a leather glove. Consequently, the grip with the cotton glove becomes less flat, resulting in more friction effects than the calculated coefficient. Since it becomes more difficult to have a firm grip with a leather glove attached with a piece of silicon when increasing the load, the friction surface becomes flatter. Thus, the actual grasp force with the leather gloves approaches to the theoretically calculated force. Another reason is the difference of the level of feedback of grasping perception through the gloves. It will be much easier to receive a feedback of grasp force when using the cotton gloves than the leather gloves, especially at the high loads. The leather gloves attached with pieces of silicones will block the feedback of grasp force.

11.4 Conclusions

The characteristics of grasps at the submaximal levels can be summarized as follows:

1. At the beginning of grasp, the grasp force at the low loads tends to be more overexerted than does that at the high loads.
2. The ratio of the overexertion for the low loads declines faster than does that for the high loads.
3. The ratio of the stable force to the load lifted is higher at the low loads than at the high loads. It approaches unity as the load increases.
4. The difference between the peak force level and the stable force level is almost constant over the all levels of loads considered.
5. Gender differences are observed for the submaximal exertion as readily as for the maximal exertion.
6. Wearing cotton gloves tends to require more grasp force.
7. Grasp force differs according to the handle orientation.
8. The grasp control (i.e., a stability of grasp) can be obtained more easily at the low loads.
9. The additional friction on gloves can make the actual grasp force the same as the theoretically needed force. However, it is dependent upon type of glove.
10. The correction factor that is a coefficient relating actual grasp force to theoretical grasp force approaches unity or lower values with increasing levels of loads. More research is needed before any generalizations.

References

Astrand, P. and Rodahl, K. 1986. *Textbook of Work Physiology*, 3rd ed. (New York: McGraw-Hill).

Batra, S., Bronkema, L.A., Wang, M.J., and Bishu, R.R. 1994. Glove attributes: can they predict performance? *Int. J. Ind. Ergon.*, 14, 201–209.

Bensel, C.K. 1993. The effects of various thicknesses of chemical protective gloves on manual dexterity, *Hum. Factors*, 36, 687–696.

Bishu, R.R., Bronkema, L.A., Garcia, D., Klute, G., and Rajulu, S. 1994. Tactility as a function of grasp force: effects of glove, orientation, pressure, load, and handle, NASA Technical Paper 3474, NASA (Houston, TX; Johnson Space Center).

Bishu, R.R. and Klute, G. 1993. Investigation of the effects of extravehicular activity (EVA) gloves on performance, NASA Technical Paper 3401, NASA (Houston, TX: Johnson Space Center).

Bishu, R.R. and Klute, G. 1995. The effects of extra vehicular activity (EVA) gloves on human performance, *Int. J. Ind. Ergon.*, 16, 165–174.

Bishu, R.R., Klute, G., and Kim, B. 1993. The effects of extra vehicular activity (EVA) gloves on dexterity and tactility, in *Proceedings of the 37th Annual Meeting of the Human Factors and Ergonomic Society* (Seattle, WA), pp. 826–830.

Bishu, R.R., Klute, G., and Kim, B. 1994. Force endurance capabilities of extra vehicular activity gloves at different pressure levels, in *Proceedings of IEA 94, Volume 2: Occupational Health and Safety* (Toronto), pp. 80–82, Aug. 1994.

Brand, P.W. 1985. *Clinical Mechanics of the Hand* (St. Louis: C.V. Mosby).

Bronkema, L. and Bishu, R.R. 1996. The effects of glove frictional characteristics and load on grasp force and grasp control, in *Proceedings of the 40th Annual Meeting of the Human Factors and Ergonomic Society* (Philadelphia, PA), pp. 702–706.

Bronkema, L., Bishu, R.R., Garcia, D., Klute, G., and Rajulu, S. 1994. Tactility as a function of grasp force: effects of glove, orientation, load and handle, in *Proceedings of the 38th Annual Conference of the Human Factors and Ergonomics Society* (Nashville, TN), pp. 597–601.

Buhman, D.C., Cherry, J.A., Bronkema, L., and Bishu, R.R. 2000. Effects of glove, orientation, pressure, load and handle on submaximal grasp force, *Int. J. Ind. Ergon.* 25, 247–256.

Cherry, J.A., Christensen, A.D., and Bishu, R.R. 2000. Glove comfort vs. discomfort: Are they part of a continuum or not? A multidimensional scaling analysis, in *Proceedings of the 44th Annual Conference of the Human Factors and Ergonomics Society* (San Diego, CA), pp. 95–98.

Cochran, D.J., Albin, T.J., Bishu, R.R., and Riley, M.W. 1986. An analysis of grasp force degradation with commercially available gloves, in *Proceedings of the 30th Annual Meeting of the Human Factors Society* (Dayton, OH), pp. 852–855.

Grant, K.A. 1994. Evaluation of grip force exertion in dynamic manual work, in *Proceedings of the Human Factors Society 38th Annual Meeting* (Santa Monica, CA), pp. 549–553.

Hallbeck, M.S. and McMullin, D.L. 1991. The effect of gloves, wrist position, and age on peak three-jaw chuck pinch force: a pilot study, in *Proceedings of the 35th Annual Meeting of the Human Factors Society* (Santa Monica, CA), pp. 753–757.

Hazelton, F.T., Smidt, G.L., Flatt, A.E., and Stephens, R.I. 1975. The influence of wrist position on the force produced by the finger flexors, *J. Biomech.*, 8, 301–306.

Kim, B. and Bishu, R. 1997. Grasp force: how dependent is it on the load grasped? in *Proceedings of 13th Triennial Congress of the International Ergonomics Association* (Tampere, Finland), pp. 546–548.

Kroemer, K., Kroemer, H., and Kroemer, E. 2000. *Ergonomics: How to Design for Ease and Efficiency* (Englewood Cliffs, NJ: Prentice Hall).

Kumar, S. 2003. Personal communication.

Landsmeer, J.M.F. 1962. Power grip and precision handling, *Ann. Rheum. Dis.*, 21, 164–169.

Lyman, J. and Groth, H. 1958. Prehension force as a measure of psychomotor skill for bare and gloved hands, *J. Appl. Psychol.*, 42, 18–21.

Monod, H. and Scherrer, J. 1965. The work capacity of a synergic muscular group, *Ergonomics*, 8, 329–338.

Naper, J.R. 1956. The prehensile movements of the human hand, *J. Bone Joint Surg.*, 38, 902–913.

Plummer, R., Stobbe, T., Ronk, R., Myers, W., Kim, H., and Jaraiedi, M. 1985. Manual dexterity evaluation of gloves used in handling hazardous materials, in *Proceedings of the 29th Annual Meeting of the Human Factors Society*, pp. 819–823.

Putz-Anderson, V. 1991. *Cumulative Trauma Disorders, A Manual for Musculoskeletal Diseases of the Upper Limb*, (London: Taylor and Francis).

Radwin, R.G. and Oh, S. 1992. External finger forces in submaximal five-finger static pinch prehension, *Ergonomics*, 35, 275–288.

Rohmert, W. 1960. Ermittlung von Erholungspausen fur Statische Arbeit des Menschen. *Int. Z. Angew. Physiol. Einsehl. Arbeitsphysiol.*, 18, 123–164.

Sudhakar, L.R., Schoenmarklin, R.W., Lavender, S.A., and Marras, W.S. 1988. The effects of gloves on muscle activity, in *Proceedings of the 32nd Annual Meeting of the Human Factors Society* (Santa Monica, CA), pp. 647–650.

Terrel, J. and Purswell, 1976. The influence of forearm and wrist orientation on static grip strength as a design criteria for hand tools, in *Proceedings of the 20th Annual Meeting of the Human Factors Society* (Santa Monica, CA), pp. 28–32.

Westling, G. and Johansson, R.S. 1984. Factors influencing the force control during precision grip, *Exp. Brain Res.*, 53, 277–284.

Wilhem, G.A. and Bishu, R.R. 1997. Grasp control: effect of grasp force, in *Advances in Occupational Ergonomics and Safety*, B. Das and W. Karwowski, Eds. (Amsterdam, IOS Press), pp. 297–300.

12

Shoulder, Elbow, and Forearm Strength

Tyler Amell

CONTENTS

ABSTRACT This chapter elucidates the complex structural and functional characteristics of the human shoulder, elbow, and forearm in terms of muscular strength. Strength is discussed in the context of work through focusing upon and synthesizing the current state of clinical and experimental research concerning the muscles and joint systems in these parts of the upper limb. Such information is important when designing work, because all manual labor involves muscular activity about these joints in order to facilitate movements involving the hands. In addition, the upper limb is commonly injured in the process of work, and in some cases this may be due to job tasks that are designed beyond the safe working capacity of the upper limbs.

12.1 Chapter Outline

The first part of this chapter introduces and discusses the importance of possessing information concerning the motions and muscular strength capability about the shoulder, elbow, and forearm. The next section consists of a brief review of the relevant functional anatomy of those aspects of the upper extremity crucial to understanding muscular strength. This is followed by a discussion of the kinematics and kinetics of the shoulder, elbow, and forearm. The summary and conclusions revisit the primary topics in the chapter and summarize the information presented.

12.2 Ergonomic Relevance

The muscular strength capability about the shoulder, elbow, and forearm is a very important consideration in the design of any work task involving manual activity. Tasks involving manual handling, control operation, etc. require some degree of muscular strength, and depending upon the upper extremity posture necessitated by the task design, some tasks may impose significantly more load upon the worker than need be. In such circumstances, when the design variables exceed the safe working capacity of the worker, musculoskeletal injury may ensue. Hence as a means of primary prevention of such conditions, information on the muscular strength about these joints is of great importance.

12.3 Introduction

All manual work is, by definition, performed through action of the hands. All acts of handling loads (lifting, lowering, pushing, pulling, etc.), using tools (hammers, jackhammers, screwdrivers, etc.), as well as operating controls (wheels, levers, buttons, etc.) involve the hands as the point of contact between the worker and the equipment. The point of contact results in an external force, moment, or load being imparted or countered through internal muscular action. The hands, in turn, are positioned in three-dimensional space by motions about several other musculoskeletal structures. The body is often referred to as a system of linked musculoskeletal segments, with each segment corresponding to the major articulation(s) involved in producing muscular strength in the area (Chaffin and Andersson, 1999; Kroemer, 1999). The ankle, knee, hip, spine, shoulder, elbow, and wrist all act in concert to facilitate work. For example, the lower limbs (ankle, knee, hip) are responsible for ambulation and moving the body to a position that is suitable for performing a job task. The spine (thoracic and lumbar) can flex and rotate in order to position the upper limbs in a more advantageous working posture. The last link in this system is the upper limbs, which have the important function of positioning the hands at the final position and fine-tuning, in a sense, the coarser movements of the lower limbs and trunk. The hands are the most mobile and dexterous of the musculoskeletal structures, and as a result, the final opportunity to fine-tune the movement resides in their domain. Once the positioning of the hands is complete, which includes the muscular activity to support the upper limb mass, the reserve strength capacity of the involved musculature is now available for work; hence upper limb positioning is a limiting factor in strength production.

The versatility of the workers' ability to perform manual labor is heavily reliant upon motions about the shoulder and elbow joints, two of the three major joint systems in the upper limb. Without the muscular strength and coordination of movements about these two joints as well as the forearm, the hands cannot be placed in those meaningful positions required by work. Therefore, information pertaining to the muscular strengths workers are capable of generating about the shoulder, elbow, and forearm is of paramount importance for the design of manual work. The important roles of hand and wrist strength, as well as grip strength, are discussed elsewhere in Chapters 10 and 11, respectively.

Muscular strength varies as a function of upper limb position, due to the length–tension relationship, as well as speed of contraction, due to the force–velocity relationship of skeletal muscle (Gulch, 1994; Lieber, 1992; Lieber and Bodine-Fowler, 1993). Generally speaking, for the purpose of work tasks and work design, muscular strength is at a maximum near the midrange of a given joint's range of motion (ROM) in a given plane. Also, muscular strength is greatest at zero velocity and tends to decrease with increasing velocity of limb movement. Therefore, the motions about the shoulder, elbow, and forearm are highly relevant to any discussion of the muscular strength capability about these joints. The range and patterns of motion about these joints is best described with reference to the functional anatomy of the area.

12.4 Functional Anatomy of the Shoulder, Elbow, and Forearm

The upper limb is composed of the *shoulder* (articulation of the arm and trunk), *arm* (brachium) area between the shoulder and elbow, *forearm* (antebrachium) between the elbow and wrist, *wrist* (carpus) between the forearm and the hand, and the *hand* (manus). Principal bones in the area are the *clavicle* (collar bone) and *scapula* (shoulder blade) in the shoulder, which form the pectoral girdle; the *humerus* in the arm; the *radius* and *ulna* in the forearm; and the *carpal*, *metacarpal*, and *phalanges* in the wrist and hand (Moore, 1992). This review is limited to discussion of the shoulder, elbow, and forearm regions, as the wrist and hand are discussed elsewhere (Chapters 10 and 11). Only those major joints with an integral function in work and the production of strength and mobility, as measured externally, are discussed and hence are limited to the glenohumeral, humeroulnar, humeroradial, proximal, and distal radioulnar joints. The sternoclavicular and acromioclavicular joints are not discussed.

The shoulder (glenohumeral) joint is a multiaxial ball and socket type of synovial joint formed by the articulation of the proximal head of the humerus and the glenoid cavity of the scapula. The elbow joint is a hinge type of synovial joint formed by the articulation of the distal humerus and the proximal ends of the radius and ulna. The distal radioulnar joints are formed from the distal aspects of the radius and ulna. The shoulder is capable of motion in a variety of axes and planes, and as a result, multiple muscle systems are required in order to produce such complex motion patterns. Table 12.1 through Table 12.3 summarize the motions and musculature of the scapula, shoulder, elbow, and forearm, respectively. Figure 12.1 through Figure 12.3 depict the motions of the shoulder, elbow, and forearm. The elbow is only capable of flexion and extension, while the distal radioulnar joint only permits supination and pronation at the distal forearm/wrist. A detailed anatomical review is beyond the scope of this chapter, and hence only pertinent information for strength production is presented.

The articulations between the bones of the upper limb permit significantly more movement than the corresponding articulations of the lower limb due to fact that the former articulations are not weight bearing, whereas the latter are. As a result, in the upper limb, stability is sacrificed for mobility, while the opposite is true for the lower limb. For instance, the shoulder and hip joints are both multiaxial ball-and-socket synovial joints; however

TABLE 12.1

Motions and Musculature Acting about the Scapula

Motion	Contributing Muscles
Elevation of scapula	Levator scapulae[a]
	Trapezius[a] (superior fibers)
	Serratus anterior (superior fibers)
Depression of scapula	Latissimus dorsi[a]
	Pectoralis major[a]
	Pectoralis minor
Protraction of scapula	Pectoralis minor[a]
	Serratus anterior[a]
	Levator scapulae
Retraction of scapula	Rhomboids[a]
	Trapezius[a]
	Latissimus dorsi
Superior rotation of scapula	Serratus anterior[a] (inferior fibers)
	Trapezius (superior and inferior fibers)
Inferior rotation of scapula	Levator scapulae[a]
	Pectoralis minor[a]
	Rhomboids[a]
	Latissimus dorsi
	Pectoralis major

[a] Muscles that are principle contributors to motion (Moore, 1992).

one-third of the head of the humerus is encapsulated within the glenoid cavity of the scapula and held in place via relatively lax ligaments and surrounding musculature. This allows for a wide range of movement about the shoulder complex; however, the system is inherently unstable and significantly reliant upon musculature for support. Conversely, in the hip joint, the head of the femur is fully encapsulated in the comparatively deep acetabulum of the ilium and held in place via very strong ligaments and surrounded by powerful muscles. This system allows for less overall movement but gains a high degree of stability. Similarly, the only bony articulation between the entire upper limb and the axial skeleton exists at the sternoclavicular joint, another example of the innate mobility of the upper limb gained as a result of a loss in stability. The resulting high degree of mobility and less stability of the upper limb are issues that have serious implications for work ability and strength, as well as impaired functioning when this body part is injured.

12.5 Kinematics and Kinetics of the Shoulder, Elbow, and Forearm

Kinematics describes the motions of a body, while kinetics describes the forces producing those motions. In this chapter, kinematics describes the motion about the upper extremity joints in question, and kinetics describes the forces about the joints in question producing the observed motions. A succinct review of the pertinent kinematics of the relevant anatomy is provided in Section 12.5.1, while a more thorough review of the pertinent kinetics is provided in Sections 12.5.2 through 12.5.4. The information of interest, in terms of work design, is the total external joint moment strengths with respect to posture and speed of movement of the shoulder, elbow, and forearm. When manual work is performed, such as when a load is lifted/lowered, the hands grasp the object to form a rigid link, which requires grip strength (see Chapter 11 for more

TABLE 12.2

Motions and Musculature Acting about the Shoulder

Motion	Contributing Muscles
Adduction of humerus, vertical plane; ROM about 50° from midline	Latissimus dorsi[a] Pectoralis major (both heads)[a] Coracobrachialis Infraspinatus[b] Teres major Triceps brachii (long head)
Abduction of humerus, vertical plane; ROM about 180°	Deltoid (middle fibers)[a] Supraspinatus[a,b]
Medial rotation of humerus, vertical and horizontal plane; ROM about 100°	Deltoid (anterior fibers)[a] Latissimus dorsi[a] Pectoralis major (both heads)[a] Subscapularis[b] Teres major
Lateral rotation of humerus, vertical and horizontal plane; ROM about 30° past neutral	Deltoid (posterior fibers)[a] Infraspinatus[b] Teres minor[b]
Flexion of humerus, vertical and horizontal plane; ROM about 180°	Coracobrachialis[a] Deltoid (anterior fibers) Pectoralis major (clavicular head) Biceps brachii (short head)
Extension of humerus, vertical and horizontal plane; ROM about 60° past neutral	Deltoid (posterior fibers)[a] Latissimus dorsi[a] Pectoralis major (sternocostal head) Teres major Triceps brachii (long head)

Note: Also included is the aggregate maximum range of motion (ROM) in degrees (Nordin and Frankel, 2001; Chaffin and Andersson, 1999; Whiting and Zernicke, 1998; Moore, 1992; Eastman Kodak Company, 1986).

[a] Muscles that are principle contributors to motion.
[b] Rotator cuff muscles (supraspinatus, infraspinatus, teres minor, subscapularis).

TABLE 12.3

Motions and Musculature Acting about the Elbow and Forearm

Motion	Contributing Muscles
Flexion of forearm; ROM between 140° and 150°	Biceps brachii[a] Brachialis[a] Brachioradialis[a] Pronator teres Extensor carpi radialis longus Flexor carpi radialis
Extension of forearm; ROM between 140° and 150° (10° hyperextension)	Triceps brachii[a] Anconeus Brachioradialis
Supination of forearm; ROM between 80° and 100°	Supinator[a] Biceps brachii[a]
Pronation of forearm; ROM about 70°	Pronator quadratus[a] Pronator teres[a] Flexor carpi radialis Anconeus

Note: Also included is the aggregate maximum range of motion (ROM) values in degrees (Nordin and Frankel, 2001; Chaffin and Andersson, 1999; Whiting and Zernicke, 1998; Moore, 1992; Eastman Kodak Company 1986).

[a] Muscles that are principle contributors to motion.

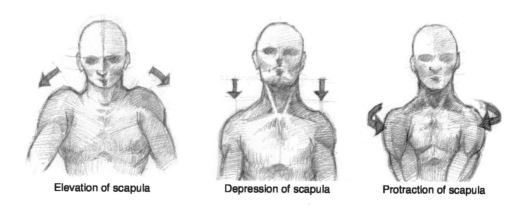

| Elevation of scapula | Depression of scapula | Protraction of scapula |

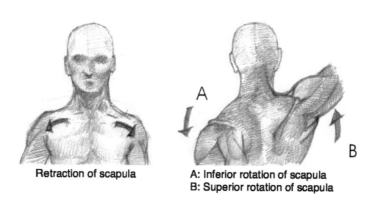

Retraction of scapula   A: Inferior rotation of scapula
B: Superior rotation of scapula

FIGURE 12.1
Motions of the scapula. (Figures courtesy of Adrien Cho.)

information), and then the other links in the musculoskeletal system take over. Any number of motions may occur at the shoulders, along with flexion or extension of the elbow or pronation or supination of the forearm. Therefore, the total external moments about each joint are considered the "strength." It is these values that must be considered in the design of manual tasks.

12.5.1   Kinematics of the Shoulder, Elbow, and Forearm

The approximate ranges of motion (ROM) for the shoulder, elbow, and distal forearm joints are described in Table 12.2 and Table 12.3, with reference to Figure 12.1 through Figure 12.3. A variety of sources were combined in order to produce these values, due to the fact that numerous methodologies have been used for measuring these ROMs, and these values are highly dependant upon testing procedures and starting anatomical positions (Chaffin and Andersson, 1999; Eastman Kodak Company, 1986; Hagberg et al., 1995; Nordin and Frankel, 2001). They are only provided as an interpretative aid for the kinetics of the shoulder, elbow, and forearm. These values are the aggregate means and are conservative; 5th and 95th percentile values can be found in Chaffin and Andersson (1999). Please refer to Nordin and Frankel (2001) for more detailed information on this subject. In addition, the nomenclature used to describe the movement about the shoulder in

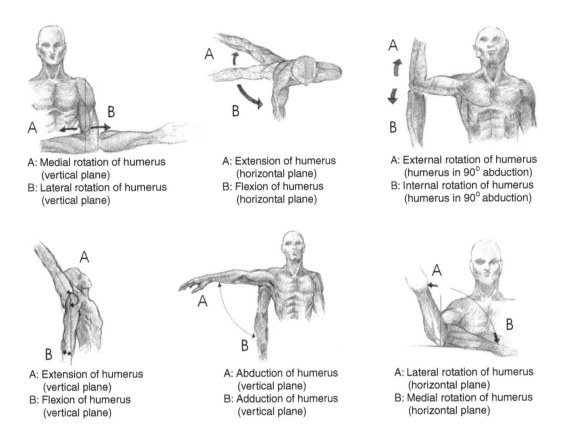

A: Medial rotation of humerus (vertical plane)
B: Lateral rotation of humerus (vertical plane)

A: Extension of humerus (horizontal plane)
B: Flexion of humerus (horizontal plane)

A: External rotation of humerus (humerus in 90° abduction)
B: Internal rotation of humerus (humerus in 90° abduction)

A: Extension of humerus (vertical plane)
B: Flexion of humerus (vertical plane)

A: Abduction of humerus (vertical plane)
B: Adduction of humerus (vertical plane)

A: Lateral rotation of humerus (horizontal plane)
B: Medial rotation of humerus (horizontal plane)

FIGURE 12.2
Motions of the shoulder. Medial and lateral rotation of the humerus are also sometimes referred to as internal and external rotation of the humerus (as shown with the humerus in 90° of abduction). (Figures courtesy of Adrien Cho.)

particular is inconsistent, as evidenced in Figure 12.2. Wherever possible, common alternative nomenclature is presented for the purpose of clarity.

12.5.2 Kinetics of the Shoulder

Due to the role shoulder strength plays in work, particularly manual labor, as well as sport, the kinetics about this joint has been the focus of several biomechanical studies (Engin, 1980; Giroux and Lamontagne, 1992; Kuhlman et al., 1992; Mayer et al., 1994; Walmsley, 1993; Walmsley and Szybbo, 1987). As noted above, the kinetics of the shoulder (external joint strength) is highly dependant upon the motions and planes of observation, as well as the velocity of movement. Information describing the strength capability about the shoulder for the pertinent motions described in Figure 12.2 is discussed in Sections 12.5.2.1, through 12.5.2.3. This information is drawn from clinical, experimental, and functional research sources and may not be entirely applicable to work due to the artificial nature of some of the studies, such as those performed under isokinetic conditions. Furthermore, each testing protocol is quite different, which leads to systematic differences in the strength measurement. Since functional work strength is the aggregate of motion about each joint system, which may or may not be in accordance with past research studies or clinical observations, some interpretation and estimation on behalf of the end users of such information may be necessary.

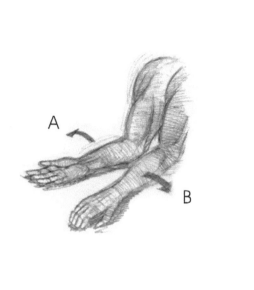

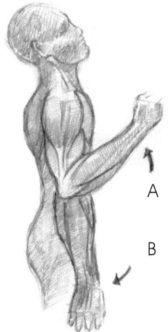

A: Supination of forearm
B: Pronation of forearm

A: Flexion of forearm
B: Extension of forearm

FIGURE 12.3
Motions of the forearm and elbow. (Figures courtesy of Adrien Cho.)

12.5.2.1 Abduction and Adduction Strength of the Shoulder

Shoulder abduction strength may be important for overhead work, as well as when reaching to handle a load or operate a control. Shoulder adduction strength may be important when handling a load, turning a wheel, or operating a lever. Information on shoulder abduction and adduction strength has been reported by several authors (Amell et al., 2000; Gil Coury et al., 1998 ; Kuhlman et al., 1992; Stobbe, 1982; Takala and Viikari-Juntura, 1991) and can be found in several compendiums (Chaffin and Andersson, 1999; Hagberg et al., 1995). Table 12.4 and Table 12.5 list the maximum abduction and adduc-tion strengths, respectively, from a wide variety of published reports. It is evident from this information that shoulder and elbow position, as well as the functionality of the measurement, impact profoundly upon the strength outcome. Some strength studies involve the joint moment strengths directly, with all other joints isolated, for example the work of Stobbe (1982) and Kuhlman et al. (1992). As with all strength measures, these authors report results that when compared exhibit a high degree of variability between them. This observation will remain true for all studies discussed in this chapter. For instance, the joint moments provided by Kuhlman et al. (1992) seem to equate with the 5% percentile data provided by Stobbe (1982), in both male and female populations. This result is probably due to different testing positions and populations. From these data it is evident that the maximum static (isometric) abduction moment about the shoulder ranges from approximately 40 to 100 N•m in males, and from 15 to 60 N•m in females. Static abduction strength about the shoulder can also be measured indepen-dent of axis of rotation and can be expressed in terms of the force, as provided by Takala and Viikari-Juntura (1991). Also included in Table 12.4 are isokinetic data. Isokinetics is

TABLE 12.4

Shoulder Abduction Strengths

Source	Variable	Unit	Strength ± (SD) ♂	♀	Description
Stobbe, 1982; Chaffin and Andersson, 1999; Hagberg et al., 1995	5%ile 50%ile 95%ile	(N•m)	43 71 101	15 37 57	25 ♂, 22 ♀; Measured isometrically in 90° vertical shoulder abduction, elbow at 90°, hand supine relative to head, cuff placed medial to medial surface of forearm
Takala and Viikari-Juntura, 1991; Hagberg et al., 1995	Right Left	(N)		96 (28) 88 (29)	Measured isometrically with a sling placed on lateral epicondyle of the elbow
Essendrop et al., 2001	Right Left	(N)	195 (87) 208 (97)	195 (87) 208 (97)	6 ♂, 13 ♀; Data from each sex are pooled; measured isometrically in the seated position with the elbow flexed to 90° and the dynamometers placed bilaterally over the epicondyle of the humeri
Mayer et al., 1994	Isometric 60°/sec 180°/sec 240°/sec 300°/sec	(N•m)	47 (12) 38 (7) 33 (8) 31 (7) 30 (8)	28 (6) 22 (4) 18 (4) 19 (4) 18 (4)	32 ♂, 19 ♀; Data were recorded under concentric conditions on the dominant side; also included in this study are nondominant side and eccentric data; measured using standard testing procedures for the Lido Active system
Kuhlman et al., 1992	20° 45° 90° 90°/sec 30° 60° 90° 120° 90°/sec 210°/sec	(N•m)	56 (13) 59 (13) 52 (13) 47 (10) 73 (10) 74 (9) 65 (10) 47 (10) 70 (10) 58 (11)	32 (6) 32 (8) 27 (6) 27 (3)	9 ♂ Age 51–65; 9 ♀ Age 51–65; Measured isometrically in 30° of horizontal flexion anterior to the coronal plane; deg/sec values measured isokinetically 21 ♂ Age 19–30; Measured isometrically in 30° of horizontal flexion anterior to the coronal plane; deg/sec values measured isokinetically

Note: N•m = Newton meter; N = Newton; ♂ = male; ♀ = female.

an artificial motion whereby limb velocity is held constant and the residual force not required for limb movement is converted into a moment of force. The attempted acceleration beyond the preset velocity is the measured moment. These strength values are less than the values recorded under isometric conditions, since some of the joint strength capacity must be devoted to movement rather than producing force. This is evident in Table 12.4. Such conditions are not typically found in manual labor, yet do provide insight into human strength capability.

Contrary to the information on abduction strength, the adduction strength data found in Table 12.5 are functional in nature. The information provided by Amell et al. (2000) and Gil Coury et al. (1998) involves the static bilateral functional adduction strength about the shoulder and elbow with respect to squeezing a metal box (with a load cell inside). Figure 12.4 depicts the postures tested by Amell et al., (2000), while Figure 12.5 depicts the

TABLE 12.5

Shoulder Adduction Strengths

Source	Variable	Unit	Strength ± (SD) ♂	♀	Description
Stobbe, 1982; Chaffin and Andersson, 1999; Hagberg et al., 1995	5%ile 50%ile 95%ile	(N•m)	35 67 115	13 30 54	25 ♂, 22 ♀; Measured isometrically in 90° vertical shoulder abduction, 0° horizontal shoulder, elbow at 90°, cuff placed on inferior surface of upper arm 2 cm medial to elbow joint
Gil Coury et al., 1998	S-30/ E90° S0/E60° S0/E90° S0/E120° S30/E30° S30/E60° S30/E90° S60/E0° S60/E60° S90/E0° S90/E30° S90/E60°	(N)	72 (21) 85 (21) 89 (19) 89 (24) 88 (22) 85 (24) 87 (22) 86 (23) 85 (23) 60 (16) 75 (21) 91 (28)	33 (8) 37 (9) 41 (11) 44 (13) 40 (9) 42 (9) 39 (11) 36 (11) 35 (10) 25 (9) 32 (9) 35 (11)	15 ♂, 15 ♀; Measured isometrically; mean bilateral strength measured while squeezing a 40 ×20 × 20 cm metal box. The box was located directly in front of the participants. Variable angles refer to shoulder angle (0° is neutral) and elbow angle, respectively
Amell at el., 2000	S-30/ E90° S0/E60° S0/E90° S0/E120° S30/E30° S60/E0° S90/E0°	(N)	79 (22) 86 (24) 72 (17) 77 (23) 81 (26) 69 (17) 66 (17)		10 ♂; Measured isometrically; mean bilateral strength measured while squeezing a 40 × 20 × 20 cm metal box. The box was located directly in front of the participants. Variable angles refer to shoulder angle (0° is neutral) and elbow angle, respectively
Mayer et al., 1994	Isometric 60°/sec 180°/sec 240°/sec 300°/sec	(N•m)	72 (17) 53 (11) 48 (13) 48 (17) 45 (18)	34 (9) 29 (8) 22 (6) 24 (7) 23 (7)	32 ♂, 19 ♀; Data were recorded under concentric conditions on the dominant side; also included in this study are non-dominant side and eccentric data; measured using standard testing procedures for the Lido Active system
Pentland et al., 1993	Average 60°/sec 120°/sec Peak 60°/sec 120°/sec	(N•m) (N•m)	 44 (18) 41 (18) 53 (23) 48 (22)	 44 (18) 41 (18) 53 (23) 48 (22)	18 ♂, 12 ♀; Data from each sex are pooled; data were recorded under isokinetic conditions on the dominant side; also included in this study are eccentric data and data from a serial testing session (only day-1 data are listed); measured using standard testing procedures for the Kin-Com system

Note: N•m = Newton meter; N = Newton; ♂ = male; ♀ = female.

adduction force as a function of limb position, showing the variation in strength as the angles change. This information is integral to manual handling tasks when the coupling between the load and hand is poor, such as when lifting a box without handles. In such cases, the upper extremity must produce enough adduction force to maintain the static friction coefficient imparted by the surface, as well as the load when it is lifted. These data again show that upper extremity position has a significant impact upon strength.

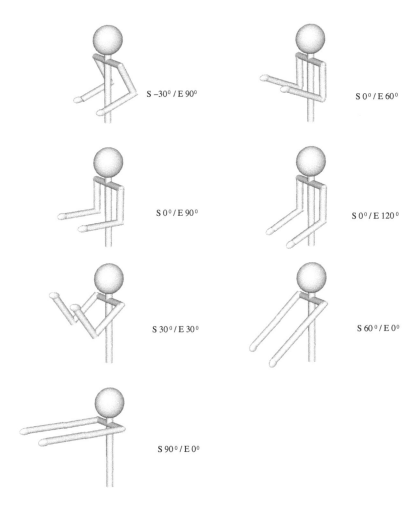

FIGURE 12.4
Upper extremity positions. (Studied by Amell, T.K., Kumar, S., Narayan, Y., and Gil Coury, H.C. 2000, *Ergonomics*, 43(4), 512–527.)

12.5.2.2 Medial and Lateral Rotation Strength of the Shoulder

Medial and lateral rotation about the shoulder may be necessary when manually handling loads, such as carrying a box, or when actuating a lever or a wheel. It may also be necessary when reaching for controls, or when the manual activities such as cleaning, painting, or scrubbing are requirements of the job task. Table 12.6 and Table 12.7 list some of the past work on medial (internal) and lateral (external) rotary strengths in a variety of shoulder postures (Chaffin and Andersson, 1999; Hagberg et al., 1995; Kuhlman et al., 1992; Mayer et al., 1994; Stobbe, 1982; Walmsley and Dias, 1995; Walmsley and Szybbo, 1987). Fiftieth-percentile isometric medial rotation strengths for males and females are reported to be about 50 and 20 N•m, respectively, while isokinetic data at a variety of velocities do not exceed these values and therefore may reflect that the slight decrease in strength with added movement is indicative of an "average" population in these studies (Table 12.6).

Participants in those studies listed in Table 12.7 could not elicit external rotation strengths that were as great as internal rotation strengths, indicating that this motion pattern is

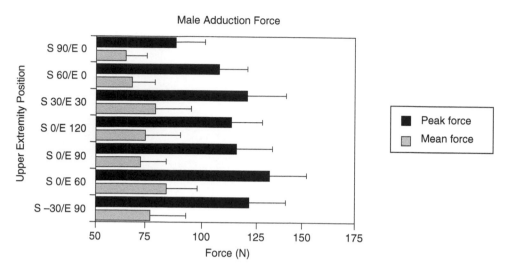

FIGURE 12.5

Male adduction force as a function of upper extremity position. S = Shoulder angle; E = Elbow angle. (From Amell, T.K., Kumar, S., Narayan, Y., and Gil Coury, H.C. 2000, *Ergonomics*, 43(4), 512–527.)

weaker from a biomechanical perspective. Reasons for this can be traced to the functional anatomy of the area. Medial rotation of the humerus is produced by the powerful deltoid (anterior fibers), latissimus dorsi, and pectoralis muscles, to name a few, while lateral rotation is produced only the posterior fibers of the deltoid and muscles of the rotator cuff. This results in isometric strength values that are approximately 70% of those measured under conditions of internal rotation. Therefore a design recommendation minimizing the level of external rotation about the shoulder would be considered optimal, based upon these observations and the available data.

12.5.2.3   Flexion and Extension Strength of the Shoulder

Work tasks may require flexion and extension of the shoulder in order to be performed adequately. Tasks in which the work is carried out at high heights or at a distance from the chest may require forward flexion (in the vertical plane), while those at low heights or at close distance to the chest may require extension about the shoulder in the vertical plane. Similarly in the horizontal plane, tasks requiring sweeping motions such as pulling cable or operating controls may also require horizontal flexion or extension. Several studies have investigated these strengths (Chaffin and Andersson, 1999; Hagberg et al., 1995; Mayer et al., 1994; Pentland et al., 1993; Stobbe, 1982; Walmsley, 1993). See Table 12.8 and 12.9 for more information. In studies conducted by Mayer et al. (1994), isometric flexion strength about the shoulder in males and females was approximately 70 and 45 N•m, respectively. Hagberg et al. reported values in accordance with these.

Mayer et al. (1994) reported isometric values that were 30% greater in males and 20% greater in females for shoulder extension than for flexion, which again indicates a preference for one pattern of motion over another in terms of strength capability. As was observed in medial and lateral rotation shoulder strength, where medial rotary strength was greater, shoulder extension strength is greater. In the medial rotary pattern, more load can be drawn inward toward the body than can be pushed outwards away from the body (externally). Similarly, more load can be drawn toward the body through extension of the shoulder than through rotating away from the body in the vertical plane, or through flexion.

TABLE 12.6

Shoulder Medial (Internal) Rotation Strengths

Source	Variable	Unit	♂	♀	Description
			Strength ± (SD)		
Stobbe, 1982; Chaffin and Andersson, 1999; Hagberg et al., 1995	5%ile 50%ile 95%ile	(N•m)	28 52 83	9 21 33	25 ♂, 22 ♀; Measured isometrically in 90° vertical shoulder abduction, 0° horizontal shoulder, elbow at 90°, cuff placed on medial to the wrist
Mayer et al., 1994	Isometric 60°/sec 180°/sec 240°/sec 300°/sec	(N•m)	43 (12) 42 (8) 37 (8) 35 (8) 34 (8)	21 (7) 23 (5) 20 (4) 18 (5) 19 (3)	32 ♂, 19 ♀; Data were recorded under concentric conditions on the dominant side; also included in this study are nondominant side and eccentric data; measured using standard testing procedures for the Lido Active system
Walmsley and Szybbo, 1987	Neutral 60°/sec 120°/sec 180°/sec	(N•m)		30 (7) 28 (7) 26 (7)	12 ♀; Data were recorded isokinetically under concentric conditions on the dominant side; measured using standard testing procedures for the Cybex II system, neutral = 15° of abduction of the shoulder; 90A = 90° shoulder abduction; 90F = 90° shoulder flexion
	90A 60°/sec 120°/sec 180°/sec 90F 60°/sec 120°/sec 180°/sec			25 (5) 23 (5) 23 (5) 27 (6) 26 (5) 24 (6)	
Walmsley and Dias, 1995	Kin-Com 60°/sec 120°/sec 180°/sec Lido 60°/sec 120°/sec 180°/sec Cybex 60°/sec 120°/sec 180°/sec	(N•m)	38 (16) 34 (15) 33 (13) 41 (20) 44 (19) 40 (17) 43 (15) 42 (15) 39 (14)	38 (16) 34 (15) 33 (13) 41 (20) 44 (19) 40 (17) 43 (15) 42 (15) 39 (14)	9 ♂, 6 ♀; Measured isokinetically in 90° humeral abduction on three separate isokinetic dynamometers; data from each sex are pooled

Note: N•m = Newton meter; N = Newton; ♂ = male; ♀ = female.

Differing values were observed with respect to both flexion and extension strengths of the shoulder in the horizontal plane. No motion pattern produced clearly superior strength values. There was no definitive pattern, with the exception of 50th percentile isometric values for males (Table 12.8 and Table 12.9).

12.5.3 Kinetics of the Elbow

Elbow strength has been one of the most studied of all human strengths, due to the relative ease of measuring this parameter. Since the degrees of freedom about the elbow joint are limited to motion in one plane, about one axis, it is relatively easy to isolate, and its strength measurement is relatively straightforward and not nearly as complex as that of

TABLE 12.7

Shoulder Lateral (External) Rotation Strengths

Source	Variable	Unit	Strength ± (SD) ♂	♀	Description
Stobbe, 1982; Chaffin and Andersson, 1999; Hagberg et al., 1995	5%ile 50%ile 95%ile	(N•m)	23 33 51	13 19 28	25 ♂, 22 ♂; Measured isometrically in 90° vertical shoulder abduction, 0° horizontal shoulder, elbow at 90°, cuff placed on medial to the wrist
Walmsley and Szybbo, 1987	Neutral 60°/sec 120°/sec 180°/sec	(N•m)		19 (4) 17 (4) 16 (3)	12 ♀; Data were recorded isokinetically under concentric conditions on the dominant side; measured using standard testing procedures for the Cybex II system, neutral = 15° of abduction of the shoulder; 90A = 90° shoulder abduction; 90F = 90° shoulder flexion
	90A 60°/sec 120°/sec 180°/sec 90F 60°/sec 120°/sec 180°/sec			19 (3) 18 (3) 17 (4) 24 (5) 20 (5) 18 (5)	
Kuhlman et al., 1992	-60°	(N•m)	33 (7)	15 (3)	9 ♂ Age 51–65; 9 ♀ age 51–65; Measured isometrically in 30° of horizontal flexion anterior to the coronal plane; deg/sec values measured isokinetically
	-45°		33 (7)	16 (6)	
	-30° 0° 90°/sec 210°/sec		32 (9) 30 (6) 28 (6) 24 (5)	17 (5) 18 (4) 16 (4) 15 (3)	
	-60°		45 (8)		21 ♂ Age 19–30; Measured isometrically in 30° of horizontal flexion anterior to the coronal plane; deg/sec values measured isokinetically
	-30° 0° 30° 60° 90°/sec 210°/sec		44 (7) 39 (6) 37 (5) 32 (6) 39 (8) 34 (6)		
Walmsley and Dias, 1995	Kin-Com 60°/sec 120°/sec 180°/sec Lido 60°/sec 120°/sec 180°/sec Cybex 60°/sec 120°/sec 180°/sec	(N•m)	37 (11) 36 (13) 33 (11) 28 (14) 29 (12) 27 (10) 39 (10) 37 (9) 35 (8)	37 (11) 36 (13) 33 (11) 28 (14) 29 (12) 27 (10) 39 (10) 37 (9) 35 (8)	9 ♂, 6 ♀; Measured isokinetically in 90° humeral abduction on three separate isokinetic dynamometers; data from each sex are pooled

Note: N•m = Newton meter; N = Newton; ♂ = male; ♀ = female.

the shoulder. Elbow strength nevertheless must conform to the principles of limb posture and positioning, as well as velocity of movement, as the strength about this joint will be affected by these two variables.

TABLE 12.8

Shoulder Flexion Strengths

Source	Variable	Unit	Strength ± (SD) ♂	Strength ± (SD) ♀	Description
Flexion					
Hagberg et al., 1995	45°	(N•m)	69 (14)	32 (6)	Average of concentric and eccentric strengths
	0°		90 (20)	43 (10)	
Mayer et al., 1994	Isometric	(N•m)	68 (13)	45 (13)	32 ♂, 19 ♀; Data were recorded under
	60°/sec		41 (8)	29 (5)	concentric conditions on the dominant
	180°/sec		41 (8)	25 (5)	side; also included in this study are
	240°/sec		42 (9)	24 (5)	nondominant side and eccentric data;
	300°/sec		39 (9)	24 (5)	measured using standard testing procedures for the Lido Active system
Pentland et al., 1993	Average	(N•m)			18 ♂, 12 ♀; Data from each sex are pooled;
	60°/sec		41 (17)	41 (17)	data were recorded under isokinetic
	120°/sec		37 (16)	37 (16)	conditions on the dominant side; also included in this study are eccentric data and data from a serial testing session
	Peak	(N•m)			(only day-1 data are listed); measured
	60°/sec		55 (24)	55 (24)	using standard testing procedures for the
	120°/sec		48 (23)	48 (23)	Kin-Com system
Horizontal Flexion					
Stobbe, 1982;	5%ile	(N•m)	44	12	25 ♂, 22 ♀; Measured isometrically in 90°
Chaffin and	50%ile		92	40	vertical shoulder abduction, 0° horizontal
Andersson, 1999;	95%ile		119	60	shoulder, elbow at 90°, hand prone
Hagberg et al.,					relative to floor; cuff placed on medial to
1995					the nearest surface of the forearm

Note: N•m = Newton meter; N = Newton; ♂ = male; ♀ = female.

12.5.3.1 *Flexion and Extension Strength of the Elbow*

The role the elbow plays in work is pivotal, particularly in manual tasks when the humerus (shoulder joint) is capable of resting vertically. Such tasks include working at a correctly positioned bench, whereby the motions may predominantly be flexion and extension about the elbow, coupled with internal (medial) or external (lateral) rotation about the shoulder. Other examples include manual handling tasks such as lifting and carrying a load. Typically the shoulder is held in neutral position (vertically), and the elbow is flexed to near 90°, which is its strongest position, while the load is carried. Where permitted by shape and coupling, the load could be lowered so that the weight of the load is transferred straight to the shoulders with the elbow near 180°, because this is more efficient. However, not all loads can be handled in such a manner.

Table 12.10 lists some of the relevant studies of flexion and extension strength about the elbow (Chaffin and Andersson, 1999; Hagberg et al., 1995; Kroemer and Marras, 1981; Pentland et al., 1993; Stobbe, 1982). Isometric elbow strength is greater in flexion than extension. For example, 50th-percentile males and females are capable of generating approximately 75 and 40 N•m, respectively, in flexion, while only 45 and 30 N•m, respectively, in extension. As expected, dynamic isokinetic values are lower. The participant population in other studies reviewed corresponds to approximately the 5th to 50th percentile values of those reported by Stobbe, 1982. Only studies investigating concentric types of muscular contraction were included in this review, because these types of

TABLE 12.9

Shoulder Extension Strengths

Source	Variable	Unit	Strength ± (SD) ♂	Strength ± (SD) ♀	Description
Extension					
Mayer et al., 1994	Isometric	(N•m)	93 (24)	54 (12)	32 ♂, 19 ♂; Data were recorded under concentric
	60°/sec		67 (16)	38 (10)	conditions on the dominant side; also included
	180°/sec		57 (12)	32 (8)	in this study are nondominant side and eccentric
	240°/sec		56 (14)	31 (9)	data; measured using standard testing
	300°/sec		55 (15)	32 (7)	procedures for the Lido Active system
Pentland et al., 1993	Average	(N•m)			18 ♂, 12 ♀; Data from each sex are pooled; data
	60°/sec		35 (17)	35 (17)	were recorded under isokinetic conditions on the
	120°/sec		31 (16)	31 (16)	dominant side; also included in this study are
					eccentric data and data from a serial testing
	Peak	(N•m)			session (only day-1 data are listed); measured
	60°/sec		46 (20)	46 (20)	using standard testing procedures for the Kin-
	120°/sec		42 (20)	42 (20)	Com system
Horizontal Extension					
Stobbe 1982; Chaffin and Andersson 1999; Hagberg et al. 1995	5%ile	(N•m)	43	19	25 ♂, 22 ♀; Measured isometrically in 60°
	50%ile		67	33	horizontal shoulder angle, elbow at 90°, hand
	95%ile		103	57	prone relative to floor; cuff placed on medial to
					the nearest surface of the forearm

Note: N•m = Newton meter; N = Newton; ♂ = male; ♀ = female.

contractions are far more common in work tasks, so it can be safely stated that the flexion pattern of motion about the elbow is capable of generating up to 40 N•m moments in 50% of the general population. Since the objective of design is to accommodate the greatest range possible and to reduce the load, in terms of strength, so as to maximize this potential where practical, this would mean that a moment value for elbow flexion of about 15 N•m would be optimum.

12.5.4   Kinetics of the Forearm

This discussion of the kinetics of the forearm is limited to the motions of pronation and supination about the distal radioulnar joint. At this time, relatively few studies have been conducted on the strength of pronation and supination. Considering the prevalence of poorly designed everyday tasks, including some work tasks that require this movement, this observation is surprising. One probable cause for this lack of information is that the strength capability about the distal radioulnar joint is low, often leading to quick localized neuromuscular fatigue when work tasks require this movement. Furthermore, the motion is somewhat awkward, particularly if the movement must be repeated in a reciprocating manner with insufficient rest (e.g., manual screw driving). As a result, it is probably best to avoid work designs that place relatively significant loads, or require repetitive loading and extensive pronation, supination, or reciprocating movement.

TABLE 12.10

Elbow Flexion and Extension Strengths

Source	Variable	Unit	Strength ± (SD) ♂	Strength ± (SD) ♀	Description
Flexion					
Stobbe, 1982; Chaffin and Andersson, 1999; Hagberg et al., 1995	5%ile	(N•m)	42	16	25 ♂, 22 ♀; Measured isometrically in 90° of elbow flexion, forearm horizontal and hand semi-prone relative to floor; cuff placed medial to the wrist
	50%ile		77	41	
	95%ile		111	55	
Kroemer and Marras, 1981	Level	(N)			20 ♂, 20 ♀; Measured isometrically while seated in 90° of elbow flexion, forearm horizontal; cuff placed medial to the wrist. MVC = maximum voluntary contraction
	100% MVC		335 (80)	141 (45)	
	75% MVC		167 (70)	75 (32)	
	50% MVC		122 (52)	60 (30)	
	25% MVC		80 (40)	40 (18)	
	Peak	(N)			
	100% MVC		363 (83)	162 (45)	
	75% MVC		211 (81)	97 (40)	
	50% MVC		157 (62)	76 (36)	
	25% MVC		100 (45)	47 (22)	
Pentland et al., 1993	Average	(N•m)			18 ♂, 12 ♀; Data from each sex are pooled; data were recorded under isokinetic conditions on the dominant side; also included in this study are eccentric data and data from a serial testing session (only day-1 data are listed); measured using standard testing procedures for the Kin-Com system
	60°/sec		37 (16)	37 (16)	
	120°/sec		33 (15)	33 (15)	
	Peak	(N•m)			
	60°/sec		39 (17)	39 (17)	
	120°/sec		35 (15)	35 (15)	
Extension					
Stobbe, 1982; Chaffin and Andersson, 1999; Hagberg et al., 1995	5%ile	(N•m)	31	9	25 ♂, 22 ♀; Measured isometrically in 90° of elbow flexion, forearm horizontal and hand semi-prone relative to floor; cuff placed medial to the wrist
	50%ile		46	27	
	95%ile		67	39	
Pentland et al., 1993	Average	(N•m)			18 ♂, 12 ♀; Data from each sex are pooled; data were recorded under isokinetic conditions on the dominant side; also included in this study are eccentric data and data from a serial testing session (only day-1 data are listed); measured using standard testing procedures for the Kin-Com system
	60°/sec		30 (13)	30 (13)	
	120°/sec		27 (12)	27 (12)	
	Peak	(N•m)			
	60°/sec		31 (12)	31 (12)	
	120°/sec		29 (11)	29 (11)	

Note: N•m = Newton meter; N = Newton; ♂ = male; ♀ = female.

12.5.4.1 *Pronation and Supination Strength of the Forearm*

Perhaps the most obvious work-related example of pronation and supination in the forearm is found in the task of manual screw driving. This work task requires reciprocating movements about the long axis of the forearm in order to complete the task. The strength production ability of the user while using a screwdriver is highly dependant upon the tool used; hence any research related to this task was excluded from Table 12.11. The issue will, however, be discussed briefly below. Other job tasks may include turning doorknobs, which are inherently bad designs — door levers are far more desirable, because they

TABLE 12.11

Forearm Pronation and Supination Strengths

Source	Variable	Unit	Strength ± (SD) ♂	♀	Description
Pronation					
Armstrong, 1987; Hagberg et al., 1995	Key Handle	(N•m)	2.8 (0.9) 14.1 (3.4)	1.7 (0.4) 6.3 (2.1)	Key: pronating using a pinch grip Handle: supinating using hand grip on a t-bar
Supination					
Armstrong, 1987; Hagberg et al., 1995	Key Handle	(N•m)	2.6 (0.6) 12.9 (6.1)	1.5 (0.4) 5.9 (1.7)	Key: pronating using a pinch grip Handle: supinating using hand grip on a t-bar

Note: N•m = Newton meter; N = Newton; ♂ = male; ♀ = female.

decrease the torsional force requirements substantially and remove the necessity to pronate or supinate the forearm. Also, many new building codes in Canada require levers because they do not exclude the disabled or the elderly population from access. Other work tasks include the actuation of buttons or dials and the act of positioning the hands in the required position for manual handling of materials (e.g., supinating the forearm to place the hands in an advantageous position to grip a handle).

Table 12.11 lists some information on pronation and supination about the forearm relevant to design. It is evident that the maximal forces are quite low in comparison to the strength generation capabilities about the elbow and shoulder, and paltry compared to those observed in the lower extremity. Based upon the data in Table 12.11, males can exert up to 14 N•m on a handle in pronation and 13 N•m in supination, while females can exert approximately 50% of these values. The functional anatomy of the area is again quite important, as the various musculature may be in advantageous or disadvantageous states depending upon elbow angle. The biceps brachii, due to its biarticular nature, can significantly impact the strength of pronation and supination in the forearm (Nordin and Frankel, 2001).

In functional terms, the ability to use a screwdriver has been examined by several groups of individuals, including Mital and Channaveeraiah (1988) and Mital (1986). As noted above, since tools are involved, the torque production varies with the lever arm. Nevertheless, values range from 1 to 5 N•m for both males and females using several different screwdrivers with various handle diameters (Mital and Channaveeraiah, 1988).

Given the limited data presented in Table 12.11 as well as that presented above, it is evident that pronation and supination about the forearm is neither an efficient nor a strong motion pattern in humans, hence designs limiting the requirement for reciprocating movements should be used wherever practical.

12.6 Summary and Conclusions

Although the functional strengths of the shoulder, elbow, and forearm have not been described for every conceivable work task, the basic volitional strength capability in the common planes and motion patterns has, under both static and dynamic conditions. The shoulder, having sacrificed stability for mobility, exhibits varied amounts of external muscular strength, highly dependant upon the plane of observation and the velocity of contraction. It is strongest when the motion pattern is directed toward the body, such as in adduction, internal rotation, and extension in the vertical plane, or when acting close to the trunk. Designers should be cognizant of this observation and propose work tasks accordingly in the effort to design safer, more efficient job systems.

The elbow and forearm, although not possessing the same inherent multiple degrees of freedom from a mobility perspective that the shoulder possesses, also exhibit muscular strengths which vary according to limb position and velocity of movement. Design principles similar to those discussed above for the shoulder should be applied when designing work tasks for the elbow and forearm. The whole upper limb, because of biarticular muscular activity and total synergy of movement issues, must be considered when upper limb strength is a primary factor in the design of job tasks requiring manual activity.

References

Amell, T.K., Kumar, S., Narayan, Y., and Gil Coury, H.C. 2000. Effect of trunk rotation and arm position on gross upper extremity adduction strength and muscular activity, *Ergonomics*, 43(4), 512–527.

Chaffin, D.B. and Andersson, G.B.J. 1999. *Occupational Biomechanics*, 3rd ed. (New York: Wiley & Sons).

Eastman Kodak Company. 1986. *Ergonomic Design for People at Work*, Vol. 2 (New York: Van Nostrand Reinhold).

Engin, A.E. 1980. On the biomechanics of the shoulder complex, *J. Biomech.*, 13, 575–590.

Essendrop, M., Bente Schibye, B., and Hansen, K. 2001. Reliability of isometric muscle strength tests for the trunk, hands and shoulders, *Int. J. Ind. Ergon.*, 28, 379–387.

Gil Coury, H.J.C., Kumar, S., Rodgher, S., and Narayan, Y. Measurements of shoulder adduction strength in different postures, *Int. J. Ind. Ergon.*, 22, 195–206, 1998.

Giroux, B. and Lamontagne, M. 1992. Net shoulder joint moment and muscular activity during light weight-handling at different displacements and frequencies, *Ergonomics*, 35(4), 385–403.

Gulch, R.W. 1994. Force-velocity relations in human skeletal muscle, *Int. J. Sports Med.*, 15, s1–s10.

Hagberg, M., Silverstein, B., Wells, R., Smith, M.J., Hendrick, H.W., Carayon, P., and Pérusse, M. 1995. *Work Related Musculoskeletal Disorders (WMSDs): A Reference Book for Prevention*, I. Kuorinka and L. Forcier, Eds. (London: Taylor and Francis).

Kroemer, K.H.E. 1999. Assessment of human muscle strength for engineering purposes: a review of the basics, *Ergonomics*, 42(1), 74–93.

Kroemer, K.H.E. and Marras, W.S. 1981. Evaluation of maximal and submaximal static muscle exertions, *Hum. Factors*, 23(6), 643–653.

Kuhlman, J.R., Iannotti, J.P., Kelly, M.J., Riegler, F.X., Gevaert, M.L., and Ergin, T.M. 1992. Isokinetic and isometric measurement of strength of external rotation and abduction of the shoulder, *J. Bone Joint Surg. (Am)*, 74(9), 1320–1333.

Lieber, R.L. 1992. *Skeletal Muscle Structure and Function: Implications for Rehabilitation and Sports Medicine* (Baltimore: Williams & Wilkins).

Lieber, R.L. and Bodine-Fowler, S.C. 1993. Skeletal muscle mechanics: implications for rehabilitation, *Phys. Ther.*, 73(12), 844–855.

Mayer, F., Hostman, T., Rocker, K., Heitkamp, H.C., and Dicktuch, H.H. 1994. Normal values of isokinetic maximum strength, the strength velocity curve, and the angle at peak torque of all degrees of freedom in the shoulder, *Int. J. Sports Med.*, 15, S19–S25.

Mital, A. 1986. Effect of body posture and common hand tolls on peak torque exertion capabilities, *Appl. Ergon.*, 17, 87–96.

Mital, A. and Channaveeraiah, C. 1988. Peak volitional torques for wrenches and screwdrivers, *Int. J. Ind. Ergon.*, 3, 41–64.

Moore, K.L. 1992. *Clinically Oriented Anatomy,* 3rd ed. (Baltimore: Williams & Wilkins).

Nordin, M. and Frankel, V.H. 2001. *Basic Biomechanics of the Musculoskeletal System,* 3rd ed. (Philadelphia: Lippincott Williams & Wilkins).

Pentland, W.E., Lo, S.K., and Strauss, G.R. 1993. Reliability of upper extremity isokinetic torque measurements with the kin-com dynamometer, *Isokinet. Exerc. Sci.,* 3(2), 88–95.

Stobbe, T.J. 1982. The Development of a Practical Strength Testing Program in Industry. Ph.D. thesis, Ann Arbor: Industrial and Operation Engineering, Industrial Health Science, University of Michigan.

Takala, E., Viikari-Juntura, E., Loading of shoulder muscles in a simulated work cycle: comparison between sedentary workers with and without neck–shoulder symptoms, *Clin. Biomechan.,* 6, 145–52, 1991.

Whiting, W.C. and Zernicke, R.F. 1998. *Biomechanics of Musculoskeletal Injury* (Windsor: Human Kinetics).

Walmsley, R.P. 1993. Movement of the axis of rotation of the glenohumeral joint while working on the Cybex II dynamometer. Part II: abduction/adduction, *Isokinet. Exerc. Sci.,* 3(1), 21–26.

Walmsley, R.P. and Dias, J.M. 1995. Intermachine reliability of isokinetic concentric measurements of shoulder internal and external peak torque, *Isokinet. Exerc. Sci.,* 5, 75–80.

Walmsley, R.P., Szybbo, C. 1987. A comparative study of the torque generated by the shoulder internal and external rotator muscles in different positions and at varying speeds, *J. Orthop. Sports Phys. Ther.,* 9(6), 217–222.

13

Cervical Muscle Strength

Robert Ferrari

CONTENTS

ABSTRACT Cervical muscle strength has been investigated through various approaches. This chapter reviews how various researchers have addressed the technical requirements of quantifying cervical muscle strength, the quantitative results known to date, the relevance of this research to health and disease, and the unresolved areas of this research topic. The reader will learn that most of the early research methodologies have proven unreliable, and the main focus in recent research has been measuring cervical muscle strength via surface electromyography. A discussion of research priorities is included.

13.1 Ergonomic Relevance

The relevance of cervical muscle strength to ergonomics lies in the fact that an understanding of cervical muscle strength lends itself to possibilities of neck injury and neck pain prevention and treatment. Neck pain, most commonly thought to be "muscular pain" is second only to low back pain as the most common musculoskeletal symptom patients present to physicians. Occupations with maintained postures and repetitive work tasks are commonly associated with a high prevalence of neck and shoulder disorders. Although not well elucidated, because the cervical muscles support the weight of the head, weakness of the cervical muscles is thought to contribute to persistent pain in patients reporting chronic neck pain.

13.2 Introduction

In its role as the most proximate spinal region to the head, the neck is charged with the task of supporting the position and function of the head, allowing the special sensory organs therein to function most optimally. The muscles allow for free movement of the head in relation to the body and, with the special senses, assist in orientation, exploration, and reflex behaviors. The manner in which this is accomplished requires an understanding of the structure of the neck, including bones, nerves, joints, ligaments, and muscles. This chapter will deal with the strength of the cervical muscles as one of the more important elements in this constellation of organs and tissues.

The relevance of understanding and quantifying cervical muscle strength is manifold. Beyond the need to extend our understanding of anatomy and physiology as the underlying tenets of understanding health and disease, an understanding of cervical muscle strength lends itself to possibilities of neck injury and neck pain prevention and treatment. Neck pain, most commonly thought to be "muscular pain," is second to only low back pain as the most common musculoskeletal symptom patients present to physicians. Although not well elucidated, because the cervical muscles support the weight of the head, weakness of the cervical muscles is thought to contribute to persistent pain in patients reporting chronic neck pain, though the possibility that this is a result of neck pain and not a cause remains open (Erdelyi et al., 1988; Hagberg and Wegman, 1987; Lundervold, 1951; Veiersted and Westgaard, 1993).

Occupations with maintained postures and repetitive work tasks are commonly associated with a high prevalence of neck and shoulder disorders. Whiplash-associated disorders are a group of disorders associated with neck injury primarily in the setting of motor vehicle collisions. Engineering studies indicate that translational and rotational forces, combined with eccentric contraction of cervical muscles, likely lead to muscle injury. Whiplash injury has become an epidemic problem with significant health care and social costs. There remain many questions about the nature of the injury, the muscles involved, and the optimal therapies that may help to curb the chronic neck pain that too often arises. Cervical muscle strength research may help to shed light on some of these issues. Despite this promise, there is a relative dearth of information on cervical muscle strength. This may be due in part to the lack of gold standards to safely and accurately quantify cervical muscle strength. The cervical muscles have both static and dynamic functions, since they move the head in all directions and have a postural role as antagonist of gravity. Beyond muscle strength, these functions require feedback loops involving proprioceptive sensations originating in muscles, ligaments, tendons, labyrinths, and eyes. The posterior cervical muscles can be arbitrarily divided into four layers. The outermost layer is formed by the trapezius muscle, which extends from the superior aspect of the neck to the midthoracic spine region and is involved in neck, arm, and shoulder girdle motion. The second and third layers in depth, proceeding inward, are the splenius and semispinalis muscles, respectively. The multifidus and the small muscles located between the occiput and the first two cervical vertebrae form the innermost layer. The static function of the neck depends not only on neck muscle strength but also on endurance. The head is normally held pitched slightly forward and held there by tonic activity of the cervical muscles. The cervical muscles, however, also allow the head to move in almost any direction.

Although not an exclusive list, experimental experience indicates that accurate assessment of cervical muscle strength requires at least (a) stabilization of the torso, (b) measurement through a full range of motion, (c) correction for the influence of head weight during testing, and (d) standardization of testing procedures. The methodology of cervical muscle strength studies is a key concern. As will be seen, very few studies achieve these aforementioned criteria, and thus there remains a paucity of understanding and data concerning cervical muscle strength. This review will consider how various researchers have addressed the technical requirements of quantifying cervical muscle strength, the quantitative results known to date, the relevance of this research to health and disease, and the unresolved areas of this research topic.

13.3 Measuring Cervical Muscle Strength

In this review we are concerned with muscle activity. This is to be distinguished from muscle action and muscle function. Muscle action refers to the anatomical effect of the torques and forces a muscle exerts. The action is inferred from the geometry and anatomical relationships of body parts. Thus, it is known that all cervical muscles have the action of some degree of lateral flexion toward the side of the body on which they lie. Muscles that cross superior to the atlantoaxial joint have more prominent head-rotating actions than muscles confined to attachments below this level of the cervical spine. Table 13.1 lists the cervical muscles and their actions. Muscle function is the role the muscle plays in a specific behavior. Muscle function is concerned with why a muscle is recruited for action by the nervous system. Muscle activity, however, is the presence of observable physiological changes in the muscle as it is recruited by the nervous system. Palpation, electromyography, and other *in vivo* measurements deal with muscle activity. In the setting of measuring activity, one may measure muscle strength. Although closely tied to muscle strength, measuring muscle endurance is a different matter, since endurance reflects the amount of time a muscle is capable of performing a given task, and this depends on the rate of energy expenditure in relation to the amount of energy available and the rate at which it is replaced. This does not depend solely on the muscle, as it also reflects nutrient supply to and metabolic waste removal from active muscle.

Muscle activity data can be used to understand, perhaps, the basis for localized muscle fatigue at many occupations that are associated with relatively low activity levels. Muscle activation patterns may be used to test biomechanical models and, potentially, the diagnosis of muscle dysfunction syndromes. Various methods have been used to measure cervical muscle strength, each with its disadvantages and advantages. As one can gather, it is difficult to readily make comparisons between studies using different methodologies and reporting different calculated results from raw data. Vasavada et al. (2001) recently reviewed a number of cervical muscle strength studies with an attempt to standardize the result to moments. What can be confirmed from these heterogeneous studies is at least that the greatest cervical strength is in extension, followed by lateral flexion, flexion, and then axial rotation.

13.3.1 Manual Testing

As early as 1966, Krout and Anderson used manual muscle testing techniques to determine that patients with chronic neck pain syndromes were weaker in the cervical flexors than healthy controls. Patients responded to a strengthening program with increased range of

TABLE 13.1

The Cervical Muscles and Their Actions

Muscle	Origin	Insertion	Action	Readily Accessible to Surface EMG
Trapezius	From occiput to spinous process of T12	Lateral third of the clavicle, over the acromioclavicular joint onto the superficial process of the acromion, along the spine of the scapula, and onto the medial part of the inferior lip of the spine of the scapula	Retraction of scapula, elevation and depression of shoulder girdle, lateral rotation of the scapula, contralateral rotation, lateral flexion, and extension of the head and neck	Yes
Levator scapulae	Transverse processes of C1, C2, C3, C4	Vertebral border of the scapula	Elevation and protraction of shoulder girdle, internal rotation of shoulder girdle, ipsilateral rotation, lateral flexion and extension of the head and neck	Yes
Rhomboid minor and major	Nuchal ligament and spinous process of C7 and T1	Medial border of scapula	Scapular retraction	No
Splenius capitis	Nuchal ligament and spinous process of C7 and upper thoracic vertebrae	Lateral part of the superior nuchal line and the mastoid process	Ipsilateral rotation, lateral flexion, and extension of the head and neck	Yes
Splenius cervicis	Nuchal ligament and spinous process of C7 and upper thoracic vertebrae	Posterior tubercles of the cervical transverse processes		Yes
Semispinalis capitis	Transverse processes of C7–T6 and pars interarticularis of C4–C6	Deep to superior nuchal line	Extension of head can neck	Yes
Nuchal muscles (erector spinae, semispinalis capitis and transversospinalis, cervicis)	Varies with individual muscles	Varies with individual muscles	Extension of head and neck, resistance of flexion action of other muscles	No
Suboccipital muscles (rectus capitis posterior major and minor, obliquus capitis inferior and superior)			Ipsilateral rotation and extension	No

(continued)

Muscle			Function	
Interspinalis cervicis	Spinous processes C2–C7	Spinous processes C2–C7	Likely proprioceptive function	No
Sternocleidomastoid	Proximal clavicle and superior aspect of manubrium	Mastoid process and superior nuchal line	Flexion, lateral flexion, and contralateral rotation of head and neck	Yes
Scalenus anterior	1st rib	Transverse processes of C3–C6	Flexion, lateral flexion, and contralateral rotation of head and neck	Yes
Scalenus medius	1st rib	Transverse processes of C2–C6, and possibly atlas	Flexion, lateral flexion of head and neck	Yes
Scalenus posterior	2nd rib	Transverse processes of C3–C6	Lateral flexion of head and neck	Yes
Longus colli	Upper thoracic vertebral bodies Also transverse processes of C5–C7	Transverse processes of C5–C7 and also to atlas	Flexion, lateral flexion of lower cervical spine and cervicothoracic junction	No
Longus capitis	Various cervical transverse processes	Occiput	Flexion, lateral flexion of the upper cervical spine	No
Intertransversarii	Transverse processes from atlas to T1	Transverse processes from atlas to T1	Lateral flexion of the head and neck	No
Rectus capitii	Atlas	Occiput	Head flexion on atlas and lateral flexion	No

motion and strength. Manual muscle testing, however, simply allows for measurement of relative muscle strength and, furthermore, reflects the action of a number of muscles at once. Several studies (Agre and Rodriquez, 1989; Aitkens et al., 1989; Beasley, 1961) have shown that manual muscle testing is not reliable for strength assessment when compared to quantitative isometric techniques (see below). This methodology has thus generally been abandoned.

13.3.2   Dynamometers

Dynamometers have been used as a quantitative isometric methodology, but remain a manual technique, most often used for measuring limb muscle strength. A number of researchers have applied this methodology to measuring cervical strength (Alricsson et al., 2001; Blizzard et al., 2000; Conley et al., 1997; Mayer et al., 1994; Mayoux-Benhamou et al., 1995; Petrofsky and Phillips, 1982; Silverman et al., 1999; Staudte and Duhr, 1994; Vernon et al., 1992). In many cases, however, subjects wore a helmet attached to flexible steel bars. Subjects were seated, but the torso was not restrained and the system did not correct for the gravitational effect on torques due to head and helmet weight. In other studies, the subjects were standing, and the trunk was manually stabilized by the examiner. Finally, some researchers had subjects lie prone, and subjects applied extension forces against a manually applied force at the occiput.

Staudte and Duhr (1994) studied the largest group — 272 healthy persons between age 18 and 84. Their data shows a much higher isometric force for men than women, the women achieving about two-thirds the force for all ages, and due to comparable head weights between men and women at the various ages, this was not a factor affecting outcomes. Increasing age and female gender predicted lower force outputs. Of all the anthropometric data, the most important predictor of extension force was body weight rather than neck circumference, but height was not measured in this study. The authors caution against the fact that the measurements depend on the motivation of both the subject and the examiner.

Conley et al. (1997) used this same methodology to measure cervical muscle strength as part of a study examining the specificity of resistance training responses in neck muscle size and strength. Recruiting 22 healthy subjects and with the above methodology, they measured muscle strength before and then after a 12-week program of exercises done four times per week. The subjects were divided into three groups, a control group, one doing "conventional resistance exercises" for the trunk and limbs, and the last doing also head extension exercises. Only the group doing extension exercises increased their muscle strength in extension by 34%, and that group had an increase mostly in the cross-sectional area of the splenius capitis, semispinalis capitis, semispinalis cervicis, and multifidus muscle, this value increasing by about 25%. The levator scapulae, longissimus capitis, and longissimus cervicis, and scalenus muscles had lesser degrees of increase in size (on the order of 5 to 10%). Thus, general resistance exercises for the body did not increase cervical muscle mass, but specific neck exercises did. This suggests that if increased neck muscle strength is desired to prevent injuries in a given occupation or sport, general resistance exercises are not sufficient, and specific neck exercises need to be used to achieve a demonstrable increase in strength (see also Portero et al., 2001, Tsuyama et al., 2001).

Still, there are too many confounders to obtain reliable and reproducible cervical strength measurements via the dynamometer methodology.

13.3.3 Nonmanual Isometric Techniques

The disadvantage of the dynamometer-based studies is that, notwithstanding the apparent reliability in a group of coinvestigators, the strength values of the flexors and extensors may vary according to the strength of the investigator (Wikholm and Bohannon, 1991).

To increase the reliability and accuracy of muscle strength measurements, other researchers have used an approach that allows for torso restraint, seated upright posture, and control of the mass of the head and the influence of gravity on torque measurements (Barber, 1994; Jacobs et al., 1995; Jordan et al., 1999; Leggett et al., 1991; Peolsson and Oberg, 2001; Pollock et al., 1993).

Leggett et al. evaluated the reliability of a machine they designed to measure isometric cervical extension strength through a 126-degree range of motion in a sample of healthy men and women. They first established reliability measurements, then recruited 91 subjects. These subjects were participating in a study designed to develop strength in the cervical extensor muscles. Subjects were seated in a system that had a torso restraint applied to the chest and a resistance pad applied to the occiput, with shoulder harness and lap belt employed. The eight different positions ranged from zero degrees (full cervical flexion) to 126 degrees of extension, and maximal voluntary isometric cervical extension was measured at each position. Runs were conducted one day per week for 10 weeks with each subject. The subjects performed one set of variable resistance cervical extensions with a weight load that allowed about 8 to 12 repetitions to volitional fatigue, and when 12 repetitions could be exceeded, the weight was increased by 10%. Maximal voluntary torque was measured for extension. On average, at any given angle the isometric torque for men was about double that of women. The torque remained relatively unchanged until the starting position had moved 90 degrees in extension from the fully flexed position, and from that point to 126 degrees the torque values increased by approximately 25%. The technique was shown to have high reliability, and the researchers provided some normative data over the ranges of 18 to 50 for men and women. Training over the 10 weeks increased isometric extensor strength for individual subjects over a range of 6.3 to 14.3% increase. This was less than what the authors expected from a strength-training program, but the program was infrequent. The authors did show, however, that extensor isometric strength depends on the angle at which testing is done, and that in the greater angles of extension, lesser improvements in strength were noted. Jordan et al. (1999) and Peolsson and Oberg (2001) confirmed these observations and, specifically, that gender is an important predictor of cervical strength.

Like Conley et al. (1997), who used a dynamometer, Pollock et al. (1991) found that intensive training improves cervical flexor and extensor strength (by as much as 22%) and reduces neck pain and perceived disability in both male and female subjects, but they did so using the methodology of Legget et al., which is considered more accurate than the dynamometer method used by Conley et al.

13.3.4 Surface Electromyography

Because manual testing is not very reliable, and because the resistance testing methods indicated above do not address the strength of specific muscles, there has been a greater emphasis in recent years on electromyography analysis of muscle activity. Surface electrode placement has been preferred due to its noninvasive nature and absence of pain compared to needle electromyography, the latter having been used in a few studies of muscle action rather than strength. The problem with this area of research is that the goals of electromyography research may vary with the muscles of interest for study, as does the anatomical feasibility of muscle location. Some muscles are more sensitive to interventions, such as training programs, while others are more relevant to occupational tasks. There are, in theory, up to 20 pairs of muscles that can stabilize or move the head and neck. Only a few of those muscles are superficial enough to be accessed with surface electrodes. Most studies rely primarily on anatomy to predict the function of the neck muscles, but electromyography is a means by which to verify conclusions based on predictions from models. The present review focuses more specifically on experiments determining muscle strength, and the reader is referred to the review by Sommerich et al. (2000) concerning the use of EMG to identify muscle function. Falla et al. (2002) indicate that surface electrodes should be positioned over the lower portion of the muscles and not the midpoint, which has been commonly used in past studies.

An interesting prospective study of 55 female workers who had begun employment at a chocolate manufacturing company was conducted by Veiersted and Westgaard (1993). This prospective study allowed the researchers to determine the relationship between baseline EMG studies and future neck or shoulder pain in a cohort who did not recall pain in the previous year. Veiersted and Westgaard (1993) measured the electromyographic activity of the trapezius muscles during a specific work pattern. Bipolar surface electrodes were positioned bilaterally and parallel to muscle fibers, halfway between the acromion and the vertebra prominens. Subjects then recorded a pain diary, and strength was then measured every tenth week for a period of at least 6 months. Those who developed symptoms were labeled as patients and the rest as nonpatients. Conducting measurements in these two groups of subjects, Veiersted (1996) found that several EMG parameters were predictive of who developed neck pain in a population who had similar occupational exposures and in whom age, anthropometric data, and shoulder elevation force did not predict future onset of trapezius myalgia. All muscles need to rest after a period of activity. EMG gaps are interruptions in EMG activity during work, unconscious to the subject, and they are mostly of shorter duration than 0.5 seconds. Neck or shoulder pain patients have fewer EMG gaps than nonpatients, whether before contracting the trapezius, or during low-level repetitive work, but the predictive value was weak. The significance of this physiological phenomenon remains unknown. Veiersted (1996) examined the reliability of EMG parameters describing muscle activity of the trapezius muscles in repetitive light work in 12 female workers during 20-minute work sessions. Veiersted (1996) found that within-subject variability between sessions was substantial for gaps and also static muscle activity, but neither the time of day nor the day of the week influenced the measurements of static muscle activity, EMG gaps, or mean median frequency in this setting, and they found no increased frequency decline during the work session through the day or the week as a potential indication of local cumulative fatigue.

Queisser et al. (1994) investigated a group of 12 healthy men, sampling EMGs from the semispinalis, splenius capitis, levator scapulae, and trapezius muscles. With varying angles of flexion as the starting position, the supine and torso-stabilized subjects extended their necks with the head pressed against a force plate, and the researchers calculated the EMG–torque relationship of cervical extensors at low-force levels for various positions of the cervical spine. They found through study of innervation zones that it is necessary to consider endplates in neck muscles, but also that significant intersubject differences in the EMG torque relationship could not be eliminated by normalization of the EMG parameters and torques. The elimination of gravity, the continuous monitoring of positions, and the consideration of localization of motor endplate regions were essential prerequisites for the acquisition of reliable relationship between EMG of different neck muscles and external torques. Similar data has been obtained by others using this approach (Akesson et al., 1997; Barton and Hayes, 1996; Carlson et al., 1996). It has been shown that, when one muscle causes pain, it inhibits its pair and may result in bilateral weakness (Barton and Hayes, 1996; Choi and Vanderby, 1999).

Yet, it is not until the recently published work by Kumar et al. (2001) that we have had a more comprehensive database of normative results for cervical muscle strength (see Table 13.2). Kumar et al. measured the cervical isometric force generation capacity in men and women while seated in the upright neutral posture. They attempted to overcome the problems of unstandardized methodology and to include oblique plane movements. Sampling 20 men and 20 women, all healthy with no neck pain in the prior year, the subjects' torsos were stabilized in a chair that placed the head of the subject in direct contact with a resistance arm. The subjects then pushed against the resistance arm as hard as they could over a total of 5 seconds. Reliability was found to be high for this approach in a total of eight directions of effort, even when the sequence of positions was randomized. The strength-measuring device consisted of a telescopic 15-cm wide rectangular metal tube welded to the iron plate bolted to the floor. Any force exerted on the upholstered resistance arm was registered on a load cell. In both genders, peak and average strength measurements were highest in extension, followed by posterolateral extension, lateral flexion, anterolateral flexion, and flexion. When the strength values were normalized and compared between genders and the activities, a clear pattern emerged. The female strength values were 30 to 40% lower than the male strength values. As indicated above, there is a gradual increase in cervical muscle strength in both sexes as one moves from anterior to posterior direction on either side of the body. This agrees with recent work by Garces et al. (2002), which further shows that the ratio of strength in flexion to strength in extension remains at about 0.6 for both genders, through all ages, and at any angle tested. This suggests that a decrease in strength related to age evenly affects the cervical flexors and extensors.

Still, while Kumar et al. (2001) found a statistically significant correlation between cervical strength and age, body weight, and direction of force exertion, these correlations were not very high, the regressions explaining only 45% of the variance. The lack of high correlation between myoelectric activity and direction of force exertion suggests that individuals appear to use the cervical muscles in personally varying manners. In a more limited study, Moroney et al. (1988) had also found that the calculated muscle forces and the measured myoelectric activities were often relatively low for a number of cervical muscles, and this observation remains unexplained.

TABLE 13.2

Summary of Means of Cervical Strength Measurements (Newtons, N) in Multiple Directions

Direction of Measurements	Male	Female
Flexion	57	30
Anterolateral Flexion	54	32
Lateral Flexion	58	39
Extension	79	56
Posterolateral Extension	65	49

Note: Refer to Kumar et al. (2001) for statistical details of measurements and calculations. Since right and left-sided lateral, posterolateral and anterolateral measurements are statistically the same, the two are not distinguished here and right-sided values are shown.

Overall, examining the Kumar et al. (2001) data, the unassisted cervical muscle strength among young males ranged between 72 Newtons (16 pounds) for flexion to 100 Newtons (22.3 pounds) for extension; and 42 Newtons (9.3 pounds) for flexion and 72 Newtons (16 pounds) for extension among females. The values for cervical strength in flexion and extension are generally comparable to those of Staudte and Duhr (1994), Jordan et al. (1999), and Harms- Ringdahl and Schuldt (1988) but lower than those of Queisser et al. (1994). The differences can in part likely be attributed to methodological differences. Queisser et al. for example, had subjects lying supine, and the input from extrinsic muscles could not be controlled for. Kumar et al. (2001) used a gravity-neutral method and stabilized the torso by a four-point system, minimizing any contribution from extrinsic muscles and other segments of the body. This method and the very high reliability make this trial among the more useful for normative data on cervical muscle strength. This data have been further elaborated upon by Kumar et al. (2002) more recently and are summarized in Table 13.2.

13.4 Cervical Muscle Strength in Clinical Populations

13.4.1 Acute and Chronic Neck Pain

There has long been interest in understanding whether reduced cervical muscle strength is a risk factor for acute neck pain, acute injury, or transition from acute to chronic neck pain. There is also the possibility that maladaptive behaviors following acute neck pain result in reduced cervical muscle strength and potentiate chronic neck pain.

In assessing acute neck pain or stiffness prospectively, Blizzard et al. (2000) sampled 100 never-injured (neck injury) adults and examined both anthropometric characteristics and cervical muscle flexor strength. In these generally healthy subjects, a percentage of whom were found to have neck pain and stiffness with headaches over one-month follow-up, no difference was found in anthropometric and muscle strength measurements between those who had symptoms in the one month interval of tests and retests vs. those who did not have symptoms. Similarly, measuring cervical muscle extensor strength, Jacobs et al. (1995) examined recreational and experienced cyclists compared to noncyclists. Cyclists who reported problems with neck pain that they related to cycling had no difference in extensor strength than those with less frequent or no neck pain. This did not exclude the possibility of fatigue from sustained extension, but weakness per se was not evident. This suggests that, at least in the case of acute, and short-lived neck pain or headache symptoms, there is no overt role or result of decreased muscle strength. Instead, there may be other measurements that are more relevant to acute neck pain episodes than strength measurements. Harms-Ringdahl and Schuldt (1988) showed, for example, that the fraction of extensor strength used to counteract the load moment induced by the weight of the head and neck varies with posture. Studying ten healthy female subjects, they examined neck extensor strength at different joint angles in the sagittal plane, the subjects sitting upright, with the torso fixed. The resistive force during maximum neck extension was recorded with a strain gauge in four different positions of the lower cervical spine: extended, vertical, slightly flexed, and much flexed. For each of these four positions, the upper cervical spine was kept in three positions: flexed, neutral, and extended according to video analysis. From this approach, moments of force about the bilateral motion axes of the atlantooccipital joint and the C7–T1 motion segment were calculated. Their results suggested that the flexed cervical spine positions produce higher muscular load than vertical, even when taking muscular strength into account. They hypothesized that prolonged sitting postures with flexed cervical spine could contribute to neck pain via fatigue in the neck extensors.

In the prospective study of 55 female workers who had begun employment at a chocolate manufacturing company conducted by Veiersted and Westgaard (1993), results show that workers who developed symptoms (neck or shoulder pain) had fewer EMG gaps than nonpatients, whether before contracting the trapezius, or during low-level repetitive work. However, these results have not been reproduced by others, and the predictive value of this finding was weak.

Studies on chronic neck pain patients have, however, been focused more on attempting to ascertain the relevance of cervical muscle strength to chronic neck pain. As previously stated, Krout and Anderson used manual muscle testing techniques to show that patients with chronic neck pain syndromes were weaker in the cervical flexors that healthy controls, but the reliability of such measurements has only been improved in the last decade or so. Silverman et al. (1991) used a hand-held dynamometer to quantify anterior cervical flexor muscle strength in a population with chronic neck pain and in a healthy control group to determine whether the groups differed in the strength of these muscles. They too found that the chronic pain patients had a significantly weaker anterior cervical musculature than controls, even when controlling for age and gender. Vernon et al. (1992) added to these clinical studies by using their modified sphygmomanometer dynamometer for study on 40 healthy male subjects, and 24 patients (half of them male) with chronic neck pain. The subjects underwent paired trials that involved six ranges of movement including forward flexion, extension, right and left lateral flexions, and right and left lateral rotation. Barton and Hayes (1996) measured neck flexor muscle strength in a small sample of healthy controls and chronic neck pain patients. The results from both Vernon et al. (1992) and Barton and Hayes (1996) confirm that all force values were significantly lower in the neck pain subjects, with no significant difference in recordings from the right or left sterno-cleidomastoid muscles. There were no significant differences in muscular efficiency, EMG relaxation times, and force relaxation times between patients and healthy subjects (Barton and Hayes, 1996). But also, in pain subjects with unilateral symptoms, there were no differences in sternocleidomastoid EMG relaxation times between the affected and unaffected muscles. This suggests that when one muscle causes pain, it inhibits its pair and may result in bilateral weakness. Though these studies suggest a global weakness in chronic neck pain patients, the effects of apprehension and pain on performance remain unknown. This is particularly important, because the methodology allows for extrinsic muscles in other body segments to be recruited as an adaptation in the chronic neck pain patients. Akesson et al. (1997) showed, for example, in studying the work load in female dentists with pain disorders involving the neck, that surface EMG activity of the trapezius muscles in conducting specific work activities was about one-third lower in those with pain than those without pain. Thus, the load-saving behavior implied an adaptation to reduce muscle activity, not a structural weakness. It is also not known whether true muscle weakness exists as a cause of chronic neck pain or a result.

Highland et al. (1992) tested the effect of neck strength training specifically on neck pain populations. They measured the change in neck range of motion and muscle strength in 90 patients after an eight-week training program. For strength testing, patients were seated with the torso restrained, and they flexed and extended the neck through the 126 degree range, with strength calculations being similar in manner to that of Leggett et al. (1991) above. The patients performed isometric neck extension contractions from eight equidistant positions along the 126-degree span. After eight weeks (two sessions per week for four weeks then one session per week for four weeks), significant gains were seen in extensor strength as well as range of motion, and pain levels were significantly reduced. In a similar study, Berg et al. (1994) also found gains in extensor strength as well as pain levels being significantly reduced. They specifically examined 17 women in occupations that seemed to be associated with high rates of neck pain. In both the studies by Highland et al. (1992) and Berg et al. (1994), however, there are no control group comparisons, and thus it is unclear as to the true efficacy of these isometric strength training programs (see also Ylinen and Ruuska, 1994). It is also unclear which exercises are most useful if there is an effect, as Berg et al. (1994), used exercises in flexion, extension, and rotation, while Highland et al. (1992) used extension contractions only. Yet, both achieved similar results.

13.4.2   Cervical Spine Osteoarthritis

Gogia and Sabbahi (1994) have suggested that localized muscle fatigue can be assessed by monitoring the changes in frequency-domain properties of the myoelectric signals during sustained isometric contraction. They compared 25 patients with radiological evidence of osteoarthritis of the cervical spine and chronic neck pain to 25 radiologically normal controls, the normal subjects being obviously younger, since radiological changes are most prevalent in older age groups. The subjects were also asymptomatic. All subjects were seated, and had their torsos stabilized and limbs relaxed to try and reduce extrinsic confounders. The subjects exerted force in flexion and extension against inflated cuffs while surface electrodes monitored activity over the trapezius muscles and sternocleidomastoids. Osteoarthritis subjects exhibited more pronounced fatigue in both anterior and posterior cervical muscles, but there was no significant difference in the initial median frequency of the anterior and posterior muscles. They found, however, that higher slopes of median frequency values were seen in the osteoarthritis subjects. Higher slope values have been correlated with increased accumulation of metabolites, decreased intramuscular pH, reduced accumulation of calcium ions, and slowing of intramuscular conduction velocity (DeLuca, 1984). Age could be the factor explaining these measurements, and osteoarthritis and age correlate highly. The findings by Kumar et al. (2002), which included subjects in the fourth, fifth, and sixth decade, however, suggest that age is a weak predictor of at least cervical muscle strength. There have not been sufficient studies of cervical muscle fatigue to ascertain the effect of age to determine if exercises targeted at older age groups with neck pain should be different than those targeted for younger patients.

13.4.3   Chronic Obstructive Pulmonary Disease

Patients with chronic obstructive pulmonary disease (COPD) have high strains on the respiratory muscles because of increased airflow resistance, increased respiratory rate reducing "effective" lung compliance, and dynamic pulmonary hyperinflation causing changes in the chest wall geometry. The latter effect makes the inspiratory muscles operate at shorter-than-normal lengths and thereby reduces their ability to lower intrathoracic pressure. It is conventionally thought that some muscles of the neck (such as the sternocleidomastoid) adapt to this chronic work overload by developing hypertrophy. Using concentric needle electrodes, De Troyer et al. (1994) recorded EMG activity of the scalene, sternocleidomastoid, and trapezius muscles, aiming to measure muscle activity in 40 patients with COPD. In the seated posture, all subjects had inspiratory EMG activity in the scalenes, the activity starting with the beginning of inspiration and reaching its peak at the end of inspiration. This pattern was almost never seen in the trapezius and sternocleidomastoid muscles. EMG activity in these muscles was continuous and unrelated to respiratory phases, and could be related by changing head posture. Similar findings were obtained in the supine posture. Peche et al. (1996) then studied ten male patients with COPD and ten matched (age, weight, and height) healthy controls. They collected anthropometric data and assessed isometric flexion strength with a dynamometer. The subjects were supine, and the torso and pelvis were stabilized by straps. The researchers found that the flexion torques were statistically the same in both groups. The study design, however, does not isolate the sternocleidomastoid muscle itself, and thus it is simple cervical flexion that was tested. Whether some of the cervical flexors of the neck are stronger while others are weaker (due to general weight and fitness reduction in many COPD patients) remains to be determined.

13.5 Future Research Priorities

The relevance of muscle and strength to neck pain, especially in occupational settings, will likely only be fully explored if the loose concepts of postural constraints, muscle fatigue from lack of sufficient rest or poor fitness, and repetitive work natures are studied with objective methods. It would be ideal to have a larger data bank of normative measurements of cervical muscle strength and activity in activities of daily living, as well as various occupations, in prospective studies before neck pain onsets. Such prospective studies will provide the epidemiological data needed to detect causal relationships and thus preventative opportunities by developing guidelines for workloads and durations, pre-occupational fitness standards, and rehabilitative measures.

The accurate assessment of cervical muscle strength requires several problems to be resolved. First, the subjects must be studied in an environment and posture that allows for stabilization of the torso and limbs, either avoiding or at least accounting for extrinsic muscle contributions. Second, strength measurements should be obtained through a full range of motion in multiple directions, since occupations, sports, and activities of daily living engage a variety of head and neck movements and postures. Third, either the measurement setting should be gravity neutral, or there should be correction for the influence of head weight during testing. Finally, there must be standardization of testing procedures. The more recent work by Kumar et al. (2001, 2002) accomplishes much of this in the limited number of healthy subjects thus far studied. It may be helpful to monitor, via EMG, the torso and limb muscle activity as part of verifying the lack of or degree of extrinsic muscle activity in trials of cervical muscle strength. Studies that considered the effects of varying instructions on strength measurements are also worthwhile. There may be differences between strong encouragement and otherwise minimal input by the researcher.

Measurements of cervical strength are also of value to understanding the problem of injury threshold in acceleration-deceleration events, more commonly known as whiplash injury. Currently, most engineering analyses of experimental collisions assess injury threshold by calculating the change in velocity (delta V) of the target vehicle and correlating this to reported symptoms and physical examination findings. Since symptoms are subjective, it may be wiser to ask what loads are experienced by the neck muscles during a collision and how this compares to the findings of cervical strength measurements where symptoms are routinely absent.

13.6 Summary

There is much work to be done. We have learned that the accurate and useful measurement of cervical muscle strength will, at least for now, require more technical approaches, as manual testing and hand-held dynamometer measurements are less reliable and less standardized. They also do not allow one to determine the individual muscular contribution to actions such as flexion, extension, rotation, et cetera. Research thus far clearly indicates that EMG measurements are the superior approach. The sum total of this body of research indicates that cervical muscles generate more force in extension than they do in flexion or rotation, and women have a significantly lower muscle strength compared to men, while age and anthropometric data are not a strong predictor of cervical muscle strength. From a methodological point of view, it must be noted that isometric strength depends on the angle at which testing is done.

Clinical summations from limited research indicate that reduced cervical strength does not appear to be a risk factor for acute neck pain in the workplace, at least for lighter job tasks, but there is a clear association between muscle weakness and chronic neck pain. Whether this arises from pain inhibition or not is unclear, and whether the weakness bears a causal relationship to the transition from acute to chronic neck pain, or whether it is the result of maladaptive behaviors or compensation to reduce pain in some acute pain patients (who then go on to chronic pain) is unknown. Intensive training appears to improve cervical flexor and extensor strength and reduce neck pain, but further studies are needed to confirm this.

References

Agre, J.C. and Rodriguez, A.A. 1989. Validity of manual muscle testing in post-polio subjects with good muscle strength, *Arch. Phys. Med. Rehabil.*, 70, A17–A18.

Aitkens, S., Lord, J., Bernauer, E., Fowler, W.M., Jr., Lieberman, J.S., and Berck, P. 1989. Relationship of manual muscle testing to objective strength measurements, *Muscle Nerve*, 12, 173–177.

Akesson, I., Hansson, G.A., Balogh, I., Moritz, U., and Skerfving, S. 1997. Quantifying work load in neck, shoulders and wrists in female dentists, *Int. Arch. Occup. Environ. Health*, 69, 461–474.

Alricsson, M., Harms-Ringdahl, K., Schuldt, K., and Linder, J. 2001, Mobility, muscular strength and endurance in the cervical spine in Swedish air force pilots, *Aviat. Space Environ. Med.*, 72, 336–342.

Barber, A. 1994. Upper cervical spine flexor muscles: age related performance in asymptomatic women, *Aust. Physiother.*, 40, 167–172.

Barton, P.M. and Hayes, K.C. 1996. Neck flexor muscle strength, efficiency, and relaxation times in normal subjects and subjects with unilateral neck pain and headache, *Arch. Phys. Med. Rehabil.* 77, 680–687.

Beasley, W.C. 1961. Quantitative muscle testing: principles and applications in research and clinical services, *Arch. Phys. Med. Rehabil.*, 42, 398–420.

Berg, H.E., Berggren, G., and Tesch, P.A. 1994. Dynamic neck strength training effect on pain and function, *Arch. Phys. Med. Rehabil.*, 75, 661–665.

Blizzard, L., Grimmer, K.A., and Dwyer, T. 2000. Validity of a measure of the frequency of headaches with overt neck involvement, and reliability of measurement of cervical spine anthropometric and muscle performance factors, *Arch. Phys. Med. Rehabil.*, 81, 1204–1210.

Carlson, C., Wynn, K.T., Edwards, J, Okeson, J.P., Nitz, A.J., Workman, D., and Cassisi, J. 1996. Ambulatory electromyogram activity in the upper trapezius region. Patients with muscle pain versus pain-free control subjects, *Spine*, 5, 595–599.

Choi, H. and Vanderby, R., Jr. 2000, Muscle forces and spinal loads at C4/C5 level during isometric voluntary efforts, *Med. Sci. Sports Exerc.*, 32, 830–838.

Conley, M.S., Stone, M.H., Nimmons, M., and Dudley, G.A. 1997. Specificity of resistance training responses in neck muscle size and strength, *Eur. J. Appl. Physiol.*, 75, 443–448.

De Troyer, A., Peche, R., Yernault, J.C., and Estenne, M. 1994. Neck muscle activity in patients with severe chronic obstructive pulmonary disease, *Am. J. Respir. Crit. Care Med.*, 150, 41–47.

DeLuca, C.J. 1984. Myoelectrical manifestations of localized muscular fatigue in humans. *Crit. Rev. Biomed. Eng.*, 11, 251–279.

Erdelyi, A., Sihvonen, T., Helin, P., and Hanninen, O. 1988. Shoulder strain in keyboard workers and its alleviation by arm supports, *Int. Arch. Occup. Environ. Health*, 60, 119–214.

Falla, D., Dall Alba, P., Rainoldi, A., Merletti, R., and Jull, G. 2002, Location of innervation zones of sternocleidomastoid and scalene muscles — a basis for clinical and research electromyography applications, *Clin. Neurophysiol.*, 113, 57–63.

Garces, G.L., Medina, D., Milutinovic, L., Garavote, P., and Guerado, E. 2002, Normative database of isometric cervical strength in a healthy population. *Med. Sci. Sports Exerc.*, 34, 464–470.

Gogia, P.P. and Sabbahi, M.A. 1994. Electromyographic analysis of neck muscle fatigue in patients with osteoarthritis of the cervical spine, *Spine*, 19, 502–506.

Hagberg, M. and Wegman, D.H. 1987. Prevalence rates and odds ratios of shoulder-neck diseases in different occupational groups, *Br. J. Ind. Med.*, 44, 602–610.

Harms-Ringdahl, K., and Schuldt, K. 1988. maximum neck extension strength and relative neck muscular load in different cervical spine positions, *Clin. Biomech.*, 4, 17–24.

Highland, T.R., Dreisinger, T.E., Vie, L.L., and Russell, G.S. 1992. Changes in isometric strength and range of motion of the isolated cervical spine after eight weeks of clinical rehabilitation, *Spine*, 17 (Suppl), S77–S82.

Jacobs, K., Nichols, J., Holmes, B., and Buono, M. 1995. Isometric cervical extension of recreational and experienced cyclists, *Can. J. Appl. Physiol.*, 20, 230–239.

Jordan, A., Mehlsen, J., Bulow, P.M., Ostergaard, K., and Danneskiold-Samsoe, B. 1999. Maximal isometric strength of the cervical musculature in 100 healthy volunteers, *Spine*, 24, 1343–1348.

Krout, R.M., Anderson, T.P. 1966. Role of anterior cervical muscles in production of neck pain, *Arch. Phys. Med. Rehabil.*, 47, 603–611.

Kumar, S., Narayan, Y., and Amell, T. 2001, Cervical strength of young adults in sagittal, coronal, and intermediate planes, *Clin. Biomech.*, 16, 380–388.

Kumar, S., Narayan, Y., Amell, T., and Ferrari, R. 2002, Electromyography of superficial cervical muscles with exertion in the sagittal, coronal and oblique planes, *Eur. Spine J.*, 11, 27–37.

Leggett, S.H., Graves, J.E., Pollock, M.L., Shank, M., Carpenter, D.M., Homles, B., and Fulton, M. 1991. Quantitative assessment and training of isometric cervical extension strength, *Am. J. Sports Med.*, 19, 653–659.

Lundervold, A. 1951. Occupational myalgia: electromyographic investigations. *Acta Psychiatr. Neurol. Scand.*, 26, 359–369.

Mayer, T., Gathcel, R., Keeley, J., Mayer, H., and Richling, D. 1994. A male incumbent worker industrial database. Part II: cervical spinal physical capacity, *Spine*, 19, 762–764.

Mayer, T., Gathcel, R., Keeley, J., Mayer, H., and Richling, D. 1994. A male incumbent worker industrial database. Part III: lumbar/cervical fucntional testing, *Spine*, 19, 765–770.

Mayoux-Benhamou, M.A., Revel, M., and Vallee, C. 1995. Surface electrodes are not appropriate to record selective myoelectric activity of splenius capitis muscle in humans, *Exp. Brain Res.*, 105, 432–438.

Moroney, S.P., Schultz, A.B., and Miller, J.A.A. 1988. Analysis and measurement of neck loads, *J. Orthop. Res.*, 6, 713–720.

Peche, R., Estenne, M., Gevenois, P.A., Brassine, E., Yernault, J.C., and De Troyer, A. 1996. Sterno-cleidomastoid muscle size and strength in patients with severe chronic obstructive pulmonary disease, *Am. J. Respir. Crit. Care Med.*, 153, 422–425.

Peolsson, A. and Oberg, B. 2001, Intra- and inter-tester reliability and reference values for isometric neck strength, *Physiother. Res. Int.*, 6, 15–26.

Petrofsky, J.Y. and Phillips, C.A. 1982. The strength-endurance relationship in skeletal muscle: its application to helmet design, *Aviat. Space. Environ. Med.* 53, 365–369.

Pollock, M.L., Graves, J.E., Bamman, M.M., Leggett, S.H., Carpenter, D.M., Carr, C., Cirulli, J., Matkozich, J., and Fulton, M. 1993. Frequency and volume of resistance training: effect on cervical extension strength, *Arch. Phys. Med. Rehabil.*, 74, 1080–1086.

Portero, P., Bigard, A.X., Gamet, D., Flageat, J.R., and Guezennec, C.Y. 2001, Effects of resistance training in humans on neck muscle performance, and electromyogram power spectrum changes, *Eur. J. Appl. Physiol.*, 84, 540–546.

Queisser, F., Bluthner, R., Brauer, D, and Seidel, H. 1994. The relationship between the electromyogram-amplitude and isometric extension torques of neck muscles at different positions of the cervical spine, *Eur. J. Appl. Physiol.*, 68, 92–101.

Silverman, J.L., Rodriquez, A.A., and Agre, J.C. 1991. Quantitative cervical flexor strength in healthy subjects and in subjects with mechanical neck pain, *Arch. Phys. Med. Rehabil.*, 72, 679–681.

Sommerich, C.M., Joines, S.M.B., Hermans, V., and Moon, S.D. 2000, Use of surface electromyography to estimate neck muscle activity. *J. Electromyogr. Kinesiol.*, 10, 377–398.

Staudte, H.W., and Duhr, N. 1994. Age- and sex-dependent force-related function of the cervical spine, *Eur. Spine J.*, 3, 155–161.

Tsuyama, K., Yamamoto, Y., Fujimoto, H., Adachi, T., Nakazato, K., and Nakajima, H. 2001. Comparison of the isometric cervical extension strength and a cross-sectional area of neck extensor muscles in college wrestlers and judo athletes, *Eur. J. Appl. Physiol.*, 84, 487–491.

Vasavada, A.N., Li, S., and Delp, S.L. 2001. Three-dimensional isometric strength of neck muscles in humans, *Spine*, 26, 1904–1909.

Veiersted, K.B. 1996. Reliability of myoelectric trapezius muscle activity in repetitive light work, *Ergonomics*, 39, 797–807.

Veiersted, K.B. and Westgaard, R.H. 1993. Development of trapezius myalgia among female workers performing light manual work, *Scand. J. Work Environ. Health*, 19, 277–283.

Vernon, H.T., Aker, P., Aramenko, M., Battershill, D., Alepin, A., and Pennner, T. 1992. Evaluation of neck muscle strength with a modified sphygmomanometer dynamometer: reliability and validity, *J. Manip. Physiol. Ther.*, 15, 343–349.

Wikholm, J.B. and Bohannon, R.W. 1991. Hand-held dynamometer measurements: Tester strength makes a difference, *J. Orthop. Phys. Ther.*, 13, 191–198.

Ylinen, J. and Ruuska, J. 1994. Clinical use of neck isometric strength measurement in rehabilitation, *Arch. Phys. Med. Rehabil.*, 75, 465–469.

14

Trunk and Lifting Strength

Shrawan Kumar

CONTENTS

ABSTRACT Numerous parameters of "strength" have been described, explained, and discussed in the foregoing chapters. The current chapter will examine aspects of strength as they relate to the trunk strength in various directions and lifting strength. Within the confines of this chapter three other related topics will be addressed (need, rationale, and production and perception), which will lead up to the central theme of the chapter, trunk and lifting strength. The need for strength in a more general form and with a broader context has been dealt with in Chapter 1. Similarly, the determinants of strength in an overall context were discussed in Chapter 4. The reader is directed to review those chapters for a more comprehensive understanding.

14.1 Ergonomic Relevance

The relevance of strength to ergonomics is obvious because strength application is required in most industrial tasks. The magnitude may vary. The extensive epidemiological literature has linked the magnitude of force application with manifestation of musculoskeletal injuries of the relevant body part. Therefore, it becomes essential to understand the strength capability and its limitations for choosing and implementing strategies of injury control. Since the back is the most frequently injured part of the body, and it is also most costly, trunk and lifting strength gain an enhanced ergonomic relevance.

14.2 Introduction

14.2.1 Functional Need

The general need for strength was initially affected by the size principle (Shankland and Seaver, 2002) in order to keep the organism together and organized. Had the organism grown bigger without organization and supporting structures, all larger organisms would have been just heaps of cellular mass. This would be a considerable disadvantage to the organism. Having overcome the biological and environmental barriers when we reached the industrial age, muscle strength became the essential tool through which we could manipulate materials to develop products, guide processes, and achieve the necessary division of labor for economic growth and development. Industrial development entirely changed the life of *Homo sapiens*. To provide for man's insatiable appetite for greater numbers of more diverse products, industries have continually increased their production to meet the global need. The constant and repeated use of the trunk became an industrial need in many circumstances.

Hence, strength application became essential to carry out most of the industrial tasks, making it one of the primary needs of industry.

14.2.2 Rationale for Measurement of Trunk and Lifting Strength

The literature in the area of trunk and lifting is truly vast, and many books and reports have been written about the subject (Ayoub and Mital, 1989; Bernard and Fine, 1997; Frymoyer et al., 1991; National Academy of Science, 2001). In this context, the purpose of this chapter is not to do a thorough review of the area, rather to put forward a cogent, balanced, and scientifically supportable description. For a review of the area, the readers are directed to the above-cited references. Since the beginning of industrial revolution, it has been reported that heavy work, in particular involving lifting, is associated with low back pain. In fact, in earlier times the society in general was willing to accept back problems as the cost of earning a livelihood. In 1898, Taylor (cited by Copley, 1923) pushed the envelope of productivity in mining industry by changing the dimensions of shovels being used by workers and providing a financial incentive for higher production. With a combination of these measures, he was able to obtain the same work by 140 workers that took 400 to 600 workers to accomplish before. Such cost-saving and production-enhancing strategies became industry's darling. What is not widely reported is that the workers did not last for long in such an environment. However, industry was not affected, because there were other unemployed workers waiting to take the place of injured workers. Evidence of work-mediated injury is not specific only to North America. The World Health Organization (WHO) (1995) reported that 66% of all people above the age of 10 years spend 33% of their entire life at work. While there are many benefits of work, unfortunately it also causes 120 million occupational accidents and 200,000 fatalities worldwide. The Bureau of Labor Statistics (1999) reported that there were 1.8 million loss-of-time injuries in 1997 in the United States. Workers in the age bracket of 25 to 44 years constituted 57% of all cases. Sprains and strains were the most common causes of loss time accident/injury cases (43.6%), and most commonly these involved backs. Similarly, Statistics Canada (1995), in its latest report on work injuries, reported that 48% of all injuries involved sprains and strains. The largest group of cases in Canada involved the back (27%). Overexertion was determined to be the most common cause. Statistics of most developed countries indicate overexertion, sprain and strain, and involvement of human back as commonplace. At the same time, it is fair to say that establishment of a causal relationship with exposure that may meet all rigorous scientific criteria has been elusive. Since we cannot (and should not) subject a sample to mechanical loading with a view to cause injury and prove this relationship, some continue to advocate that back injuries are not mediated through mechanical stress (Bigos et al., 1991; Hadler, 2000). However, it is also fair to say that most authors who have jumped on the psychosocial bandwagon do not make a distinction between low back disability and low back injury as reported (Kumar, 2001; Pope, 1998). It is also disheartening to see poorly conducted studies gain recognition and acceptance. Videman et al. reported that low back injury is genetically regulated, and mechanical factors played little or no part (2%) in the affliction of this condition. The conclusion was based on 15-minute interviews with subjects who answered a set of simple (not known if suggestive) questions on history of lifetime loading. It is interesting to note Videman et al. (1990) claimed biomechanical association based on their dissection of 86 cadavers who had suffered back pain in life. In ensuing years, the imagination was captured by psychosocial factors (Bigos et al., 1991), only to give way to more recent genetic cause. Furthermore, Dempsey et al. (1997) made an argument that genetic factors do not deserve attention.

A causal association between biomechanical factors and low back pain makes intuitive, scientific, and engineering sense. Some of the early studies showed clear and convincing evidence to prove that the magnitude of the load and the frequency of lifting were among the most important risk factors for low back injury. Chaffin and Park (1973) reported that the ratio of the individual's lifting strength to the weight lifted at work was strongly correlated with low back pain. When the load lifted on job was 20% of the strength of the individual, there were less than 1.5 incidences per thousand person-week. When the load to be lifted increased to 80% of the strength of the worker the incidence jumped to 14 per thousand person-week (more than eight-fold increase). Chaffin et al. (1977) also found that, with increasing magnitude of the load, the severity of injury also increased in terms of lost time. Additionally, Chaffin et al. (1978) reported that, as the strength required on job (job strength rating) rose from 50% capacity to 100% or greater, the incidence of injuries linearly increased to three times that of the incidence in jobs with 50% strength requirement. Subsequently, Ayoub et al. (1983) proposed a Job Stress Index (JSI), which was based on a combination of load, lifting capacity, work duration, and lifting frequency.

Ayoub et al. (1983) reported that when JSI reached a value of 1.5 or more the injury incidences increased substantially.

The preponderance of evidence relating the mechanical loading of the back and back pain/injury/disorder is overwhelming. Kelsey (1984), on the basis of their epidemiological study, reported that the risk of acute lumbar disc prolapse was three times greater in jobs requiring lifting more that 25 lb in excess of 25 times per day as compared to lesser loads. Frymoyer and Cats-Baril (1986) reported that each year 5% of American adults experience a low back pain (LBP) episode, of which 400,000 are attributable to an occupational injury. Kumar (1990) investigated the association between cumulative load and back pain in a group of institutional aides with physically stressful jobs. Of a total sample of 173 institutional aides in the facility, 161 participated, giving a participation rate of 92.6%. The prevalence of low back pain in this sample was 62%. The mean number of years worked prior to the first episode of back pain was 14.3 years in men and 11.6 years in women. The individual job analysis and calculation of compression and shear loads at the thoracolumbar and lumbosacral intervertebral discs revealed that the cumulative compression and shear at both levels were significantly higher in institutional aides with pain compared to those without ($p < .05$–0.01). The pain and no-pain groups were demographically not significantly different. Frymoyer and Cats-Baril (1991) stated that, whereas the injury incidences to the low back may not have changed, the changes in cost due to the rate of disability have risen significantly.

An extensive epidemiological review of credible evidence was carried out by Bernard and Fine (1997) by reviewing over 600 epidemiological studies published by NIOSH. Prior to review, four criteria were established for inclusion of studies to provide for epidemiological rigor. The overwhelming conclusion that emerged from this exercise was that there is a consistent relationship between the musculoskeletal disorders (MSDs) and certain physical factors, specially at higher levels of exposure. In another epidemiological review of evidence linking physical factors to musculoskeletal injuries, Punnett (2000) states, "The epidemiological evidence linking physical ergonomics exposures at work with risk of MSDs is extensive, biologically plausible, and methodologically adequate to inform primary prevention." In another complementary review of recent experimental studies, Keyserling (2000) reported that the physical ergonomic factors (such as weight lifted, horizontal reach distance, posture, frequency, displacement, presence of handles, and shift duration) were significantly related to biomechanical and/or psychosocial measures of strain. These in turn have been related and have been demonstrated to be causally associated with low back pain/injury.

A reflective examination of the foregoing epidemiological and experimental evidence leaves a clear understanding that all physical ergonomic factors are, in the end, based on the strength capability of an individual. Based on the published evidence, Kumar and Mital (1992) argued that human organs and tissues do have a "margin of safety" within which they tend to operate. Different methods of indexing the margin of safety may give us different values, but a conservative selection criterion will tend to enhance system safety. Finally, Kumar (2001) proposed four theories of musculoskeletal injury causation. Of the four theories, three were based on the strength characteristics of the individual. These were: (a) Cumulative Load Theory, (b) Differential Fatigue Theory, and (c) Overexertion Theory. The Cumulative Load Theory argued that it was not just the instantaneous peak load that is causally related with injury. The viscoelastic nature of all biological tissues renders their mechanical properties time dependent. If the tissues make incomplete or inadequate recovery subsequent to a load exposure, that deficit will continue and will be compounded by the following exposures. Such a phenomenon will accentuate stress concentration, significantly increasing the potential of injury precipitation (Kumar, 2001).

In the context of this theory, it may be recognized that the material properties of the tissues modulated the strength of the tissues and strength conditioning due to previous exposure and acquisition of strength. Hence, the strength becomes a critical variable to endure exposure, or fail under the loading regime.

Under the "Differential Fatigue Theory" Kumar (2001) argued with experimental evidence that asymmetrical loading of the tissues results in different levels of fatigue in different muscles. Under appropriate circumstances this will lead to a force imbalance and also, due to motor unit synchronization, jerky contraction. Transmission of differential load through the system may exceed the threshold of tolerance of some tissues, due to the strain rate dependence of their mechanical properties, and potentiate precipitation of injury (Kumar, 2001). Finally the "Overexertion Theory," which was an extension of Kumar's (1994b) conceptual model, stated that the overexertion may occur due to force, or exposure, or motion, or a varying combination of all three. When the exposure exceeds the physiological tolerance limit, it results in overexertion with dire consequences. In any case, strength plays a significant role for all three theories. If we can measure the strength in any given condition, we will be in a more informed position to understand the margin of safety and, possibly, the manner in which injury could be potentiated. Such information gives us a useful tool to possibly control the injury risk.

14.3 Strength Production and Perception: Issues

When we consider strength production, we also need to consider what factor may be influencing the results we obtain. There are various determinants of muscle strength, which are described in Chapter 4. All those factors, as discussed before — gender, age, posture, speed, frequency, and mode of exertion — will modulate the data. These are individual related variables, which are of particular significance in lifting and, hence, important to industry. Lifting strength capability has been employed in numerous ways in industry, including employment preselection criterion (Chaffin et al., 1977). While it may have been a technically sound method, it was not acceptable in society, due to equal opportunity employment in large measure. However, there were numerous other ways strength testing could be and is being used in industry. It can be a valuable tool for job, product and process design. Its consideration may enable us to execute an efficient system or operation.

Kumar and Simmonds (1994) suggested that effort perception can be used as an ergonomic tool. They argued that overexertion injuries in contemporary industrial society are commonplace, as has been pointed out in this chapter in section 1.1.2. Therefore, a reduction in overexertion may result in a reduction in injuries, thus providing us a measure of control. Kumar and Simmonds (1994) argued that it was achievable with three steps of implementation process:

1. Determination of the critical level of exertion beyond which overexertion may occur.
2. Development of a valid and reliable method of determining of that level of effort. The ease with which the latter can be done will go a long way to ensure the success of implementation.
3. Incorporation of thus generated data in job design.

Psychophysical ratings have been extensively used in the design of manual materials handling (Ayoub et al., 1978; Mital, 1983; Snook, 1978). Such strategies were based on the observation that the psychophysical ratings were highly correlated with physiological stresses as measured by heart rate and oxygen uptake (Borg, 1962; Gamberale, 1972; Johanssen, 1986). Since these psychophysical observations and arguments were based on data obtained through the use of bicycle ergometers and measured heart rate and oxygen uptake, a lack of direct relevance was pointed out, because injuries always precipitate biomechanically. However, Ekblom and Goldbarg (1971) reported that the perception of exertion was not based only on the central cardiopulmonary factor, but also on a musculoskeletal local factor. Using autonomic blocking agents, they demonstrated that the heart rate was not correlated with perceived exertion. This finding was further supported by Pandolf et al. (1972).

So far, the psychophysical design of the manual materials-handling jobs have been based on the feedback of experimental subjects to the question of what they will consider maximum acceptable weight for lifting. Injury control, through this strategy, has not been entirely satisfactory. Could it be that maximum acceptable weight, in context of making as much money as workers can, may be interfering with the safety? Could another strategy, which determines the threshold level of exposure and keeps the workers' exposure below this level have a better chance? If yes, then a quick and simple method of measurement and extrapolation may assist the process. On this logic Kumar and Simmonds (1994) and Kumar et al. (1994, 1997) carried out a series of projects to have young adults (male and female) and industrial subjects provide their 20, 40, 60, 80, and 100% of contraction in pinch, grip, and stoop lift activities. The general pattern of perceived force generation was identical in male and female students (Figure 14.1). Furthermore, when the strengths produced were compared with the extrapolated values based on MVC, an interesting discovery was made. The entire sample consistently underestimated the value of 20% MVC for strength production and overestimated 60 and 80% MVC. As the level of force production increased, the difference from the perceived extrapolated value also increased. Interestingly the strength produced at 40% MVC was not significantly different from the 40% extrapolated value from MVC (Figure 14.2). Kumar et al. (1997) carried out the same protocol with industrial subjects. In this sample they reported mixed results. The pinch and power grips followed a similar pattern as the nonindustrial subjects for the average strengths. However, for lifting strength the pattern of phenomenon was entirely different (Figure 14.3). These subjects underestimated each of the submaximal values and produced significantly greater strength. The authors had repeated this experiment three times with different samples and obtained the same results. The authors surmised that the industrial subjects took pride in their lifting strength and continually masked their objective estimation ability to produce higher values of strength. While the psychophysical technique is scientifically valid, the question arises — is this technique appropriate for the industrial population on which it is continually applied? Perhaps objectively determined values may have greater potential of success.

14.4 Trunk Strength

Discussion of trunk strength in the context of lifting strength is essential. Lifting involves multiple organs: hands, wrists, elbows, shoulders, and the trunk. The lifting

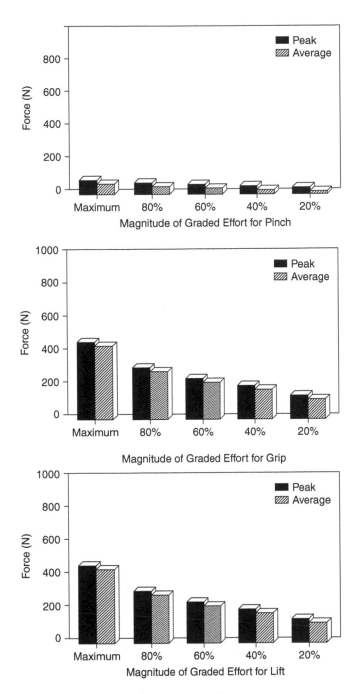

FIGURE 14.1
Magnitude of graded efforts in pinch grip, power grip, and stoop lifting.

strength will therefore be a composite of the strengths of all these organs. The strength of hand and wrist is discussed in Chapter 10, and the elbow and shoulder strength is presented in Chapter 12. This is an additional reason for discussing trunk strength, as it is the organ most frequently injured with significant social and economic consequences.

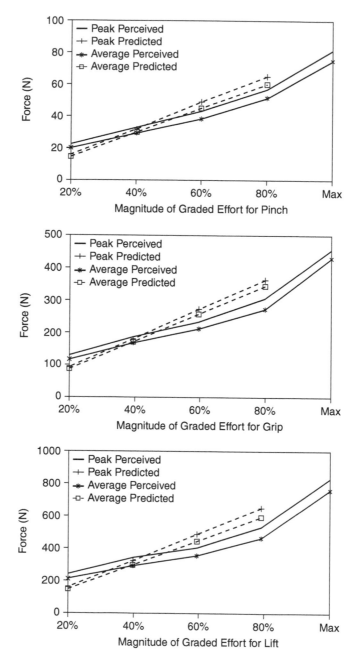

FIGURE 14.2

Comparison between the magnitudes of subjectively assessed and produced graded forces in pinch grip, power grip, and stoop lifting efforts in nonindustrial subjects with those predicted from respective maximal voluntary contractions (MVC).

14.4.1 Measurement

Human muscle strength has been studied in a variety of contexts by using different methodologies and equipment. The method of measurement has a significant influence on the results obtained (Kumar and Garand, 1992; Kumar et al., 1991). Commercially available devices and protocols for various kinds of strength testing have been presented

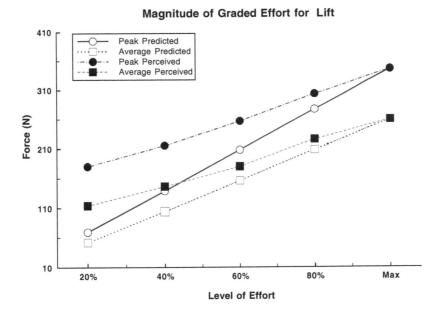

FIGURE 14.3
Comparison between the magnitudes of subjectively assessed and produced graded lifting strength in industrial male subjects with those predicted from their MVC.

in Chapter 8 and Chapter 9. A review of the devices and associated results was published by Beimborn and Morrissey (1988). This paper was largely restricted to load cells, tensiometers, dynamometers, and Cybex machines. Newton and Waddell (1993) focused on commercially available "iso machines" which consisted of Cybex II®, Cybex Back Testing System®, Kin-Com®, Biodex®, LIDO®, and B-200® Isostation. The authors had some misgivings regarding the use of these devices, which they listed. This list consisted of references to too many unpublished abstracts and internal reports, unstandardized or unclearly described protocols, lack of reliability studies, and a number of concerns regarding the studies published using these devices. However, Newton et al. (1993) conducted an experimental evaluation of these devices and concluded that the Cybex II Back Testing System was highly reliable. Based on their literature review, they suggested that their conclusion applied to other iso machines. However, it must be noted that most of the devices have some limitations. A common limitation is that none of the devices have the capacity to measure dynamic strength in the coronal plane. Stability of the hip and motion in the sagittal plane is difficult to control during measurement of extension and flexion with the Cybex® (Kumar, 1996a). This concern may be true for LIDO® as well. Although Kin-Com® and Biodex® have a better combination of posture and stabilization for sagittal plane activity, they have no provision for measurement in the transverse plane. The Cybex® device uses a stabilizing harness for the upper torso, which may interfere with measurement of trunk rotation (Kumar, 1996a). There is a sizable body of knowledge with respect to trunk strength that has not used any commercial device, but rather developed hardware to meet the design criteria of the investigator. Such devices designed, fabricated, and used in several studies by Kumar are presented below.

14.4.1.1   *Flexion, Extension, and Lateral Flexion Tester I (FELT-I)*

In consideration of LBP patients whose pain and functional levels could not be predicted in advance, the safety and comfort of these patients were given primary importance.

Therefore, a measurement system was conceptualized which would allow for a safe execution of the test. One of the concerns was the support of the trunk in test postures due to gravitational effect. Should a patient have difficulty in such tests, he/she should not have to support the weight of the trunk. Therefore Flexion/Extension and Lateral Flexion Tester I (FELT-I) was designed to counter the effect of gravity by supporting the entire body in supine. In this device patients could let go of their effort at any stage without need for support. In fact, from the beginning to the end, the entire weight of the body was supported in the lying position.

FELT-I consisted of two components, (a) an upper body receiver (UBR), and (b) a lower body receiver (LBR) (Figure 14.4). The LBR was bolted to the floor to prevent any movement. The UBR was placed on a polished stainless steel plate measuring 150 cm × 200 cm with a thickness of 0.5 cm. The bottom of the UBR was fitted with eight low-friction ball-bearing castors, which allowed easy and low resistance motion of the UBR on the steel plate. The UBR and LBR were hinged, incorporating a high-precision potentiometer such that the axis of the potentiometer was aligned with the center of rotation. The UBR and LBR had a center line to align them. The steel plate was marked in angles with reference to the center of the potentiometer. The shaft of the UBR joining with LBR was on a sleeve assembly that allowed sliding of the UBR with respect to LBR. A linear variable differential transducer (LVDT) was incorporated into the sleeve assembly to allow the measurement of the linear translation. The height of LBR and UBR was 65 cm to be a comfortable seating height. The tops of both components were fitted with high-density foam to make a comfortable surface to lie on. The UBR and LBR were 100 cm and 120 cm long, respectively, and both pieces were 80 cm wide. Halfway toward their joint, they tapered to meet in the middle at the hinge joint. Along their sides (on both sides of UBR and LBR) additional adjustable stabilizing devices (metal-based supports) 20 cm × 30 cm were affixed (Figure 14.4). They could be arranged to face and provide support in any direction. On the top surface of the UBR and LBR long slits were cut, through which wide nylon straps with Velcro were passed. The nylon straps could be slid along to appropriately position them to stabilize different-sized subjects. The slits for legs and thighs were 15 cm and 30 cm apart, respectively. On the UBR also the slits were 15 cm and 30 cm apart to allow accommodation of subjects in side lying and prone or supine lying. On one side of the UBR there was a bracket in which a rotating potentiometer was mounted to measure and record the changing angle between the UBR and load cell for correction. On the wider side of the UBR there was a lever which, when pushed down, could lock the UBR in any position. The frictional force between the UBR and the stainless steel plate was 40 N. This device could be connected to the Static Dynamic Strength Tester (SDST) through a cable and pulley system. By lying on FELT-I in prone or supine positions, lateral flexion could be tested, and by lying on it in side lying position, flexion/extension could be tested.

14.4.1.2   *Flexion, Extension, and Lateral Flexion Tester II (FELT-II)*

This device was specifically designed to isolate the spinal motion and eliminate any contamination from hip motion. In order to achieve this, a device to test flexion/extension in a seated position was designed and fabricated (Figure 14.5). FELT-II consisted of a height-adjustable chair with backrest removed and a footrest and leg separator mounted on a circular plate rotating on its center. This chair could rotate 360° and could be locked in increments of 5°. This circular rotating plate (CRP) was mounted on a rectangular sliding plate (RSP), which could be slid along the central longitudinal slit of a base plate measuring 125 cm by 50 cm and bolted to the floor. The RSP could be slid and locked in position in steps of 2.5 cm. There were four Velcro straps to stabilize the hip and lower extremities of the subject in seated position. The first ran across the hips between pubis and anterior superior iliac spine, the

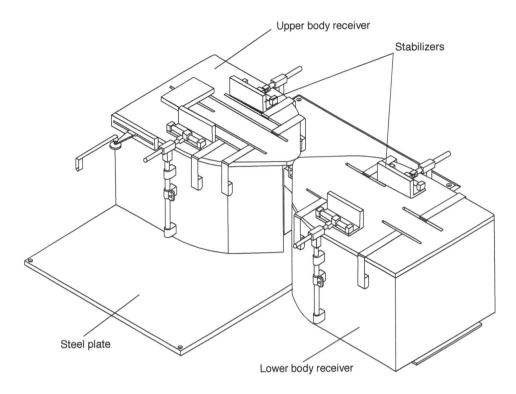

FIGURE 14.4
Schematic diagram of FELT-I. (From Kumar, 1996a.)

second ran across the distal end of the femur, the third ran across the proximal end of the tibia, and the fourth across the ankles. High-precision potentiometers were mounted on specially designed brackets to allow three-dimensional adjustment. The potentiometers were aligned with the centers of the hip and lumbosacral joints. The shafts of these potentiometers were fitted with tubular receptacles into which 50 cm-long lightweight hollow metal tubes could be easily inserted and aligned with the shoulders or spine (Figure 14.6). The shoulder tube was used to measure flexion/extension, and the spine tube was used to measure lateral flexion. These potentiometers and tubes acted like electrogoniometers.

14.4.1.3   Axial Rotation Tester (AROT)

The AROT was designed to measure bilateral axial rotation while eliminating the movement restriction created by the shoulder and chest harness of the Cybex device. The AROT (Figure 14.7) consisted of a rigid metal frame mounted on a metal base plate measuring 80 cm × 140 cm. The frame was 180 cm high and was cross-braced by a tension wire. Inside the frame mounted on the base plate was an adjustable chair that could be slid back and forth and adjusted vertically. The backrest of the chair was removed. Directly above the chair, supported by a long bar, was an adjustable shoulder harness mounted on a circular plate. This plate, in turn, was attached to a spring-loaded rod sliding within a sleeve with a locking screw to position it rigidly at any chosen position. The rod could rotate when the positioned subject underwent axial rotation. This rotation was measured by a high precision potentiometer. The potentiometer was mounted on a support plate beside the rod. The rod and the potentiometer were coupled through a set of gears. Mounted on the crossbar was a dial to read off the extent of rotation bilaterally.

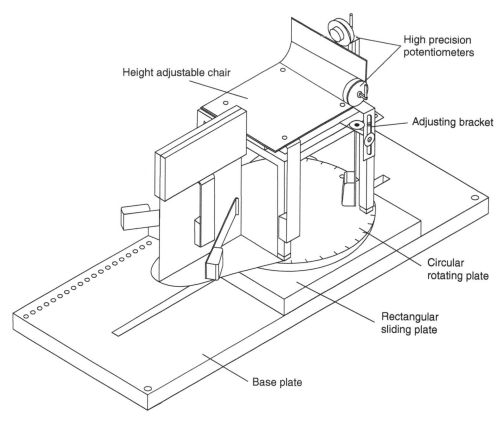

FIGURE 14.5
Schematic diagram of FELT-II. (From Kumar, 1996a.)

The chair seat of the AROT was mounted on a metal frame 26 cm high, which could be raised an additional 14 cm by a jack underneath the seat. The backrest of this plastic contoured seat was sawed off at a height of 10 cm. In front of the seat was a vertical plate to separate the legs. Four nylon and Velcro straps stabilized the hip, distal thigh, proximal tibia, and ankles. The shoulder harnesses could swing out and in to accommodate any width of shoulder. Each of the shoulder tongues could be opened or closed by a long screw bolt and nut assembly. The tongue's contact surfaces were contoured and upholstered to couple with the shoulders without causing localized discomfort and pressure. Once they were snugly fitted to the shoulders, the nut and bolt assembly was tightened to hold firmly. Further Velcro straps in front as well as back could be easily fastened to prevent any sliding. The tongue blades measured 33 cm long and were attached to the ends of the two arms of an inverted Y-shaped metal arm. These ends were 35 cm apart. At the junction of the two arms of the Y, there was a circular plate of 38 cm diameter with a groove in its rim. The base of the Y piece constituted a spring-loaded shaft measuring 65 cm that was capable of lowering and raising the entire shoulder harness assembly. This shaft was supported by a triangular metal support system (15 cm wide and 35 cm high) with three plates through which the sleeve and shaft passed for strong support and alignment. To this shaft, a precision potentiometer with one output to a computer and another to a dial was coupled through gear wheels.

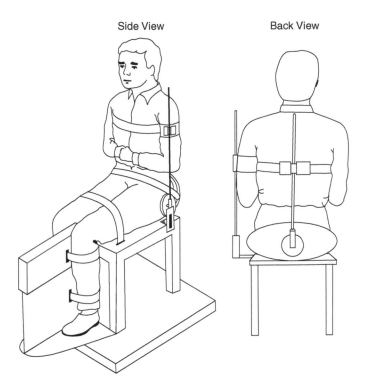

FIGURE 14.6
Alignment of a subject in FELT-II for testing flexion/extension (side view) and lateral flexion (back view). (From Kumar, 1996a.)

14.4.1.4 Static Dynamic Strength Tester (SDST)

The Static Dynamic Strength Tester (SDST) was developed by Kumar (1991a). The Static Dynamic Strength Tester was designed to do a variety of tasks in isometric and isokinetic modes at many variable preselected velocities. By coupling it with FELT-I, FELT-II, and AROT, the isometric and isokinetic strengths could be measured as required during extension, flexion, lateral flexion, and axial rotation. The SDST, in isokinetic mode, provided a constant velocity motion regardless of the force applied. The speed, however, could be adjusted by a velocity controller. The constant velocity was achieved by use of a one-way clutch to a take-up spool mounted on a shaft rotating at a fixed preset speed. When the speed threshold was reached, the clutch engaged the constant speed shaft and controlled the speed with a very high resistance. The SDST consisted of three components, (a) the framework, (b) the electromechanical drive, and (c) the measuring system. The framework (Figure 14.8) consisted of a 100 × 100 cm platform raised 35 cm. Two 2-m high vertical posts were affixed to the side of the platform. These were braced at the top by a crossbar. The vertical posts provided a mechanism to mount a sliding crossbar with two rollers in the middle through which the metal strap passed. On one side of the platform, a winch was mounted with an appropriate cable arrangement that allowed raising and lowering of the sliding crossbar. At the two ends of the sliding crossbar, bolts with wing nuts provided a mechanism to rigidly lock it in its place.

The electromechanical drive (Figure 14.9) consisted of an electronic speed control, a quarter-horsepower DC motor, and mechanical drive. The motor mounted on a worm gear was connected by a timing belt and a pulley system to a central rotating shaft. A cable drive drum was mounted on the constant-speed drive shaft coupled through a pair

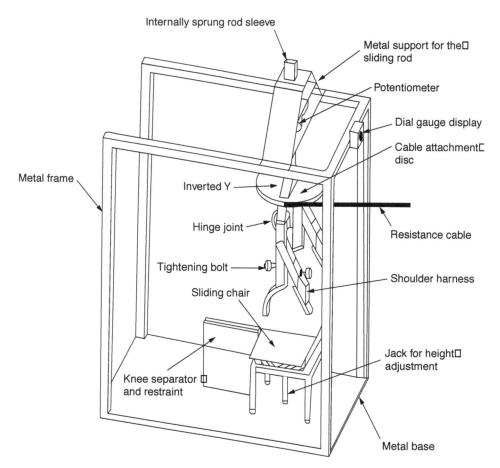

FIGURE 14.7
Schematic diagram of axial rotation tester (AROT). (From Kumar, 1996a.)

of internal one-way Sprague clutches. In the middle of this cable drum, a take-up spool for the flexible steel strap was mounted. One end of the strap was fixed to the spool and the other was attached to an SM 500 load cell after passing it through the double rollers of the sliding crossbar. Attached to the other end of the load cell was a 5-cm wide nonstretchable nylon loop to go around the shoulders of the prepositioned subjects for testing flexion/extension and lateral flexion strength using FELT-II. When using FELT-I or AROT, an airplane cable was used to couple those two devices. When force was applied, the steel strap tended to rotate the take-up spool and thereby the cable drum at a preset constant velocity. A velocity from 0 to 150 cm per second could be selected on the dial.

A displacement measuring potentiometer was mounted in parallel with the rotating drum. The velocity of the steel strap was measured by the tachometer coupled by a belt to the gearbox output. The tachometer signal was used as input for the SCR feedback control. A load cell (Interface Model SM 500–500 lb) with a natural frequency of 1.5 KHz was inserted between the metal strap and the attachment to FELTs or AROT. The output of the load cell was fed to a force monitor (ST-1, Prototype Design Fabrication Company).

14.4.1.5   Calibration of the Devices

To calibrate the testing devices, the output voltages from the load cell, the linear displacement potentiometer, and the angular displacement potentiometers were measured at

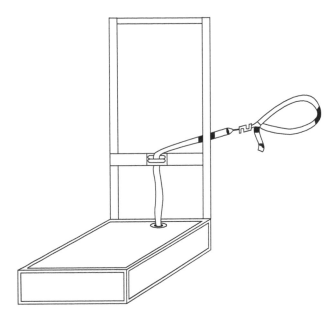

FIGURE 14.8
Framework of static dynamic strength tester. (From Kumar, 1991a, University of Alberta, Edmonton.)

regular intervals and plotted to verify accuracy. To perform the calibration, the output voltages were measured corresponding to several incremental accurately known inputs.

The SDST load cell was calibrated using input levels of 0, 10, 20, 30, 40, and 50 pounds. The SDST linear displacement potentiometer was calibrated using inputs of 0, 10, 20, 30, 40, 50, 60, 70, 80, 90, and 100 centimeters. The FELT-II angular displacement potentiometers were calibrated at input levels of –90, –60, –30, 0, 30, 60, and 90 degrees. The AROT angular displacement potentiometer was calibrated at input levels of –30, –20, –10, 0, 10, 20, and 30 degrees.

Linear regression techniques were applied to the data to find a linear equation relating output voltages to input levels. This equation was then utilized for conversion of the transducer's voltages to actual input units. The calibration plots for the load cell, linear displacement potentiomenter of the SDST, the potentiometer of the FELT-II, and potentiometer of the AROT are presented in Figure 14.10.

14.4.2 Flexion and Extension

All physical activities, whether activities of daily living (ADL) or occupational, are mediated through application of force. Low back pain in many instances is precipitated during such exertion and overexertion (NIOSH, 1981). The association of low back pain with physical activity and consequent suffering and time loss has considerably accentuated the interest in trunk strength measurement and interpretation. Initial studies presented somewhat conflicting results. Flint (1955) studied 19 female patients suffering from chronic LBP and compared them against 27 control subjects. She reported a significant reduction in back and abdominal muscle strength of the LBP patients. On the other hand, Nachamson and Lindh (1969) reported data from 63 patients who have been suffering LBP for less than a month. They found that male patients' strength values were not significantly lower than the controls, while females patients had significantly lower strength values compared to female controls. Berkson et al. (1977) supported the findings of Nachamson and Lindh (1969) through their isometric tests, which showed that male patients with acute LBP had

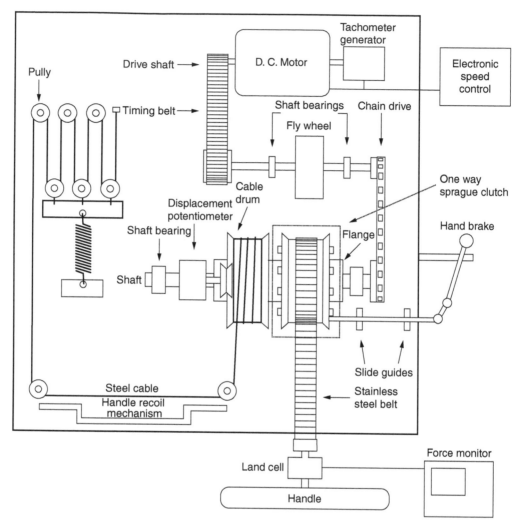

FIGURE 14.9

Plan view of static dynamic strength tester electromechanical drive. (From Kumar, 1991a, University of Alberta, Edmonton.)

close to normal strength in activities for which they could assume a posture, presumably not hindered by pain. Addison and Schultz (1980) however, through similar isometric tests, showed that patients seeking hospitalization had significantly reduced strength. In another study McNeill et al. (1980), again using the same isometric test protocol, reported that both male and female patients had their strength reduced to 60% of those of normal controls. They further found that the extensor capability had suffered the greatest loss. Suzuki and Endo (1983) reported no imbalance between trunk flexors and extensors in contrast to the previous findings. Their results supported the findings of McNeill et al. (1980) but contradicted those of Nachemson and Lindh (1969).

Using the isokinetic methodology (Cybex II). Mayer et al. (1985b) and Mayer et al. (1985c) studied the trunk strength in sagittal and transverse planes. They reported a significant reduction in strength capability of LBP patients in both planes, the sagittal plane reduction being more dramatic. In sagittal plane, Mayer et al. (1985b) reported the drop was significant for both flexion and extension, though the extensor strength was affected more severely. This finding was similar to that of McNeil et al. (1980). Pope et al. (1985) reported

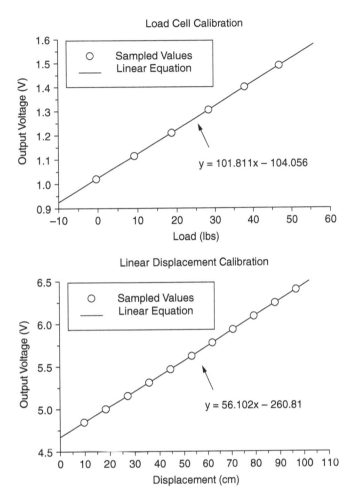

FIGURE 14.10A
Calibration of FELT-I, FELT-II, and AROT systems. (From Kumar, 1991a, University of Alberta, Edmonton.)

that LBP decreased both extensor and flexor strength but did not affect their ratio. Similar to Nachemson and Lindh (1969), the findings of Mayer et al. (1985b) found that females were affected more than males.

The foregoing review of literature makes it abundantly clear that isolated planar trunk strength tests have been largely motivated by clinical application. The measurements of clinical samples reported have varied greatly in sophistication. Schmidt et al. (1980) used manual stabilization of the ankles and hips of patients, side lying on a plinth. Patients and controls, in this position, were asked to flex or extend their trunks to eliminate the effect of gravity. Hasue et al. (1980) used Cybex dynamometer to measure trunk flexion and extension strengths in supine and prone postures, but they had to manually stabilize the feet and used a bar with roller attached to the lever arm of Cybex. Langarana et al. (1984) and Smith et al. (1985) also used Cybex for their measurements. Thus the literature presents a less than complete picture of isolated planar isometric and isokinetic strength capability. Lack of availability of appropriate "posture controlling and measuring" tools may have been a major hurdle in development of such a database.

With the noncommercial devices (designed for the purpose), Kumar (1991a, 1994a, 1996a,b,c) developed a large database. In a series of studies, Kumar (1991, 1994, 1995, 1996c) used 59 male and 43 female normal healthy subjects [male: mean age 31.8 (12.1),

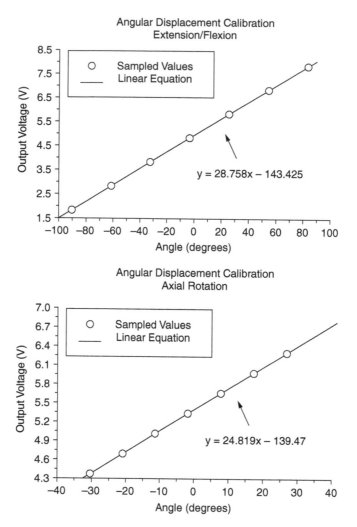

FIGURE 14.10B

Calibration of FELT-I, FELT-II, and AROT systems. (From Kumar, 1991a, University of Alberta, Edmonton.)

height 175.6 cm (6.1), and weight 75.9 kg (11.7); female: mean age 30.4 (10.1), height 165.8 cm (6.5), and weight 60.0 kg (9.2)] to develop the database. These subjects were tested for concentric isometric and isokinetic conditions of flexion and extension. The isometric flexion and extension were tested at 0°, 20°, 40°, and 60° of trunk flexion. For isokinetic extension, the subjects were placed at 60° flexion after appropriate stabilization. They were then required to extend their trunk with maximal continuous force (without jerk) to the neutral posture. The isokinetic flexion was initiated from the neutral posture and brought to 60° flexion. Throughout the range of motion in flexion and extension the subjects were required to maintain their maximal level of contraction. General observation of the results (Table 14.1 to Table 14.3) clearly indicates that male subjects were strongest in extension followed by flexion. The highest strength recorded was in isometric extension at 20° of flexion (359 Nm). All subjects were able to generate significantly higher torque in isometric as opposed to isokinetic mode. A multivariate analysis of variance revealed that there was a significant main effect due to the gender of the subjects, mode of activity (isometric vs. isokinetic), and type of activity (flexion or extension) ($p < .01$). Their various interactions were also significant ($p < .01$) A follow-up multiple comparison of isometric extension and

TABLE 14.1

Extensor and Flexor Torques of the Normal Males (Newton-Meters)

Effort	Parameter	0° M	0° SD	20° M	20° SD	40° M	40° SD	60° M	60° SD
Extension	Peak isometric	321	138	359	134	295	111	223	100
	Average isometric	259	107	288	109	232	85	171	73
	Isokinetic	210	115	210	100	116	78	41	55
Flexion	Peak isometric	194	63	185	56	151	48	98	37
	Average isometric	162	53	156	47	128	40	82	30
	Isokinetic	33	27	119	61	119	39	93	27

From Kumar, 1991a, University of Alberta, Edmonton.

TABLE 14.2

Extensor and Flexor Torques of the Normal Females (Newton-Meters)

Effort	Parameter	0° M	0° SD	20° M	20° SD	40° M	40° SD	60° M	60° SD
Extension	Peak isometric	185	85	230	89	181	74	131	50
	Average isometric	154	70	189	80	147	61	106	41
	Isokinetic	141	74	148	78	84	50	9	4
Flexion	Peak isometric	114	34	128	31	109	29	78	31
	Average isometric	98	31	109	27	93	25	66	24
	Isokinetic	25	23	100	42	85	32	69	2

From Kumar, 1991a, University of Alberta, Edmonton.

TABLE 14.3

Isokinetic Torque Produced by the Sample

Condition	Male Extension M	SD	Flexion M	SD	Female Extension M	SD	Flexion M	SD
Peak Isokinetic	247	103	156	47	162	77	114	40
Average Isokinetic	185	76	119	36	127	64	88	30

From Kumar, 1991a, University of Alberta, Edmonton.

flexion revealed that each of the factors (group, type, condition-postural angle) and their two-way interactions were significant (Table 14.4). Based on the above finding, a univariate analysis of variance was carried out for peak and average torques. For both these two variables the factors of group, type of activity, and postural angle (condition) were significant ($p < .01$). Their two-way interactions were also significant ($p < .01$). This implied that the strength data will be significantly different for flexion and extension between males and females for both isometric and isokinetic efforts. Among males the extensor torque produced at 20° flexed posture of the trunk was significantly different from all others. The extensor torque in this position was the maximum torque registered of all activities (Figure 14.11). The extensor torques at 0° and 60° trunk flexion, though not significantly different from each other, were significantly different from all other torques measured. The flexor torques at 60° and 40° of trunk flexion were significantly different from all other torques measured. The flexor torques at 60° and 40° of trunk flexion were significantly different

TABLE 14.4

Multivariate Comparison of Isometric Extension
and Flexion Torques

Hypothesis Term	F-Ratio	DF	PROB
Group	11.07	3	0.00
Type	58.42	3	0.00
Condition	52.82	9	0.00
Group × Type	6.06	3	0.00
Group × Condition	3.98	9	0.00
Type × Condition	8.66	9	0.00
Group × Type × Condition	.94	9	N.S.

From Kumar, 1991a, University of Alberta, Edmonton.

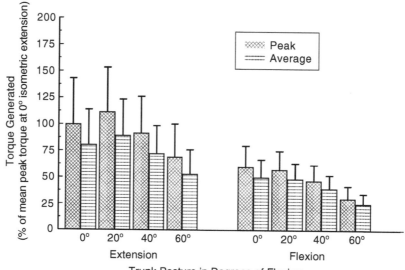

FIGURE 14.11
Isometric strength of normal males in extension and flexion. (From Kumar, 1991a, University of Alberta, Edmonton.)

from all other torques measured. The pattern of relationship between different conditions was the same for peak as well as the average torques.

In general terms, the extensor torques were invariably higher than those of the flexion at the corresponding trunk angles. In fact, the lowest extensor torque was higher than the highest flexor torque. Within the extension type of activities the male subjects were strongest at 20° of trunk flexion, and it declined on both sides of this position. When the values were normalized against trunk peak isometric extensor torque in neutral posture, the peak torque at 20° of trunk flexion was 112%, and at 40° it dropped by 20%. At 60° of trunk flexion it declined by 43%. The average isometric torque in extension had a similar pattern but a lower value ranging between 53 and 90%. The highest average torque of 90% was registered at 20° of trunk flexion, and the lowest value of 53% at 60° of trunk flexion. Many subjects felt that they reached their end of the range of motion and had difficulty in generating extensor torque.

The isometric flexor torques for normal males were between 25 and 60% of the isometric extensor torque in neutral posture. The maximal flexor torque was generated at 0° of trunk flexion for peak as well as average torque. The values at 20° flexion, though not significantly different from those at 0°, were lower than those recorded at 0°. Thus, the flexor torque was highest in neutral posture and declined with increasing trunk flexion, at 60° of flexion being only 50% of the value at 0° trunk flexion. The average flexor torque remained between 83 and 85% of the peak flexor torque at the corresponding trunk posture. Thus, the average torque is generally a fixed and high proportion of the peak strength. From the values recorded in extension and flexion for the male sample an extension to flexion ratio of 1:0.6, 1:0.52, 1:0.51, and 1:0.43 was obtained for 0°, 20°, 40°, and 60° of trunk flexion, respectively. The extension/flexion ratios were consistent for peak as well average torques at all trunk postures except at extreme flexed posture. In the latter posture, because of the difficulty of exerting strength due to the physical interference of the end of the range, the extension to flexion ratio may be less valid and reliable. Such a ratio has been of clinical use due to differential effect on extensor or flexor capability caused by a given injury. Recovery of this ratio may indicate full healing and functional restoration but may not represent the full potential of the subject/patient.

In normal females, like normal males, the extensor torque at 20° of trunk flexion was significantly different from all other conditions for peak as well as average torques. Extensor torques at 0° and 40° of trunk flexion were significantly different from all other conditions, except themselves. The flexor torque at 60° of trunk flexion was also significantly different from all other conditions. In addition, the flexor torques at 0°, 20°, and 40° of trunk flexion and the extensor torque at 60° of trunk flexion were significantly different from the extensor torque at 40° of trunk flexion. The peak isometric extensor torque among normal females also occurred at 20° of trunk flexion and was 71% of the reference normalizing value (Figure 14.11). Like normal males, the extensor torques among normal females also declined on both sides of this trunk position. At 0° trunk flexion it dropped by 13%, at 40° by 15%, and at 60° of trunk flexion it dropped by 30%. The peak extensor torque for normal females ranged between 41 and 71%, whereas the average torque ranged between 33 and 59%. The average torque also was highest at 20° of trunk flexion (59%) and declined on both sides of this posture. The average extensor torque was always between 80 and 83% of the corresponding peak extensor torque.

The pattern of variation of isometric flexor torque among normal females mirrored that of the extensor torque and the pattern described for normal males. The peak as well as the average flexor torques were highest at 20° of trunk flexion and declined on either side of this posture. The peak isometric flexor torque was 40% of the reference normalization value, and the least value was 24% at 60° of flexion. The average flexor torque measured in different trunk postures remained between 85 and 87% of the corresponding peak flexor torques. As in normal males, in normal females the average extensor and flexor torques also constitute a high and sustained proportion of the peak torques. Such a consistent relationship may allow the possibility of gleaning one from the other. At the four trunk postures (i.e., 0°, 20°, 40° and 60° of trunk flexion), the extension to flexion ratios were 1:0.62, 1:0.56, 1:0.60, and 1:0.58 for peak torque and 1:0.64, 1:0.57, 1:0.63, and 1:0.63 for average torque, respectively.

A multivariate ANOVA for the effect on the extension/flexion strength recorded in isokinetic mode revealed a significant effect of group, the type of activity (flexion or extension), and the condition (angle of trunk flexion at which the measurement was made) (Table 14.5). In addition, 2-way interactions (except Group × Type) and 3-way interactions were also significant. The factors of group, type of activity, and the trunk posture all affect the isokinetic torque in extension and flexion. Furthermore, the 2-way interactions (except Group × Condition) and 3-way interactions were all significant. Among males, the torque generated in neutral posture for extension and flexion were different from all others except at 60° for flexion. Among females, however, there were fewer significant differences.

TABLE 14.5

Multivariate Analysis of Variance of the
Isokinetic Extension/Flexion Strength in
Different Postures

Hypothesis Term	F-Ratio	DF	PROB
Group	3.48	3	0.02
Type	21.47	3	0.00
Condition	2480.28	9	0.00
Group × Type	1.27	3	N.S.
Group × Condition	2.83	9	0.00
Type × Condition	68.68	9	0.00
Group × Type × Condition	3.79	9	0.00

From Kumar, 1991a, University of Alberta, Edmonton.

The isokinetic torque registered among males for extension was highest in neutral posture, being 69% of the reference value (Figure 14.12). At increasingly higher trunk flexion the extensor torque declined, being 65, 36, and 13% at 20°, 40°, and 60° of trunk flexion, respectively. It is worth noting that the isokinetic trunk extension was initiated in 60° of trunk flexion and, hence, a considerably smaller value of the extensor torque. At the four trunk postures, the isokinetic extensor strength represented 69, 58, 39, and 18% of their corresponding isometric strength. The isokinetic flexor torques ranged between 10 and 37% of the reference value. The smallest value was recorded at 0°, where the isokinetic flexion would have been initiated. The highest values were found at 20° and 40° of trunk flexion, declining significantly at 60° of trunk flexion. The isokinetic flexor strength recorded at 0°, 20°, 40°, and 60° represented 16, 63, 78, and 96%, respectively, of their corresponding isometric strength. The extension/flexion ratios for isokinetic torques registered at 0°, 20°, 40° and 60° of trunk flexion were 1:0.14, 1:0.57; 1:1.02, and 1:2.23, respectively. However, it must be pointed out that flexion was initiated at 0°, and extension was initiated at 60°. In the initial phases of these activities, possibly due to lack of full

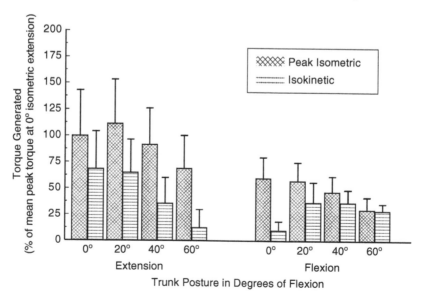

FIGURE 14.12

Isometric and isokinetic strengths of normal males at common trunk postures in extension and flexion. (From Kumar, 1991a, University of Alberta, Edmonton.)

force, the extension/flexion ratios may be exaggerated. Therefore, confidence cannot be placed in these. However, extension/flexion ratios for the peak and average isokinetic strengths were 2:0.64 and 1:0.63. The average isokinetic strength for extension as well as flexion were 75% of their respective peak values.

In female subjects, the isokinetic extensor torque ranged from 3 to 46%, the least being in full flexion and the most was recorded at 20° of trunk flexion. At 0° it was 44%, and at 40° only 26%. Thus, the highest extensor torque was generated by the normal female subjects in 20° of trunk flexion. The isokinetic extensor torques at 0°, 20°, 40°, and 60° were 75, 64, 46, and 7%, respectively, of their corresponding isometric peak extensor torques. The flexor torques were significantly lower than extensor torque ranging from 8 to 21% of the normalizing reference. The maximum isokinetic flexor torque was recorded at 20° of trunk flexion being 31% of the reference. This declined to 27 and 21% at 40° and 60° of the trunk flexion. The flexor torque at 0° was only 8%. The isokinetic flexor torques represented 22, 77, 79, and 87% of the corresponding isometric values at the four trunk angles studied. The extension/flexion ratios for the isokinetic values among females were similar to those of males: 1:0.2, 1:0.67, 1:1, and 1:7 for 0°, 20°, 40°, and 60° of trunk flexion, respectively. The first and the last ratios have the same inaccuracies among females as indicated for males. It is for this reason that the first and last ratios must not be accepted without question. However, extension/flexion ratios for the peak and average isokinetic strength were 1:0.68 and 1:0.69, respectively. The average isokinetic strengths for extension and flexion were 76 and 77%, respectively, of their corresponding peak strengths.

In the normal sample, a comparison of isometric peak torque (extension and flexion) in neutral posture and the isokinetic peak torque through a multivariate as well as univariate analyses of variance revealed that they were significantly affected by the group, mode, and type of activities and their 2-way interactions. This was done for peak as well as average torques. The multiple comparisons for males and females showed that the isometric extension was significantly different from all other conditions. Among males, the isokinetic flexion was also significantly different from other conditions. The isokinetic strengths for extension and flexion (measured at the four angles at which isometric strengths were measured) were also significantly different from the corresponding isometric strength and were significantly influenced by the factors of group, mode, type, and condition, and their various interactions.

The isometric extensor strength of females was between 33 and 71% of the reference normalizing value. However, in terms of proportion of the corresponding extensor strength of the male sample, it was between 58 and 65% (Table 14.6). The isokinetic extensor strength ranged from 3 to 46% of the reference value, but ranged from 23 to 72% of the corresponding extensor strength of the normal male sample. The extensor

TABLE 14.6

Extension and Flexion Strengths of Females Expressed as Percent of Corresponding Strength of Males

Activity	Variable	Trunk Posture in Degrees of Flexion			
		0°	20°	40°	60°
Extension	Peak isometric	58	63	60	59
	Average isometric	59	65	63	62
	Isokinetic	63	70	72	23
Flexion	Peak isometric	60	68	72	80
	Average isometric	62	69	72	84
	Isokinetic	80	83	72	72

From Kumar, 1991a, University of Alberta, Edmonton.

strength at 60° flexion in isokinetic mode, being very close to the initial test posture, did not register a value that can be taken as a true representation of the ability. The isometric flexor strength of females ranged from 21 to 40% of the reference value. As a percentage of corresponding value of the normal male sample, the female flexor torque represented 60 to 84%. The values recorded at 60° of trunk posture were 80% for peak and 84% for average flexor strength. In all other postures it was within 60 to 72%. Due to a larger drop in 60° trunk flexion posture (presumably due to lesser flexibility) among males, the female percentage rises significantly when considered in comparison. In isokinetic flexion, the females generated between 8 and 31% of the reference value, but between 72 and 83% of the corresponding values for normal males.

14.4.3 Lateral Flexion

In comparison to flexion and extension, little work has been done in lateral flexion. The earliest papers reporting some data on lateral flexion (McNeill et al., 1980; Thorstenson and Arvidson, 1982) also do not refer to any other paper having reported any lateral flexion data. McNeill et al. (1980) tested 27 healthy males and 30 healthy females. These subjects stood upright in the test apparatus and were stabilized at the hip and legs after a comfortable standing posture was achieved. A snug harness with cables was fitted on all subjects and connected to strain gauge load cells to measure the strength generated in isometric mode. The mean absolute lateral flexion strength was 400 N for males. They found that on average females developed about 60% of the force generated by men. The authors also noted that the force generated for left lateral flexion was generally greater than that for right lateral flexion. This finding was supported by Thorstenson and Arvidson (1982). For their study, Thorstenson and Arvidson (1982) had designed their own testing device, which neutralized the effect of gravity. They reported that the peak torque for both right and left sides was achieved when the trunk was laterally flexed to the contralateral side. There was no consistent difference between right and left sides. They found that the right/left torque ratios ranged between 0.88 and 1.14. However, the authors did not report if this difference was statistically significant. Thorstenson and Nilsson (1982) also reported results of their experiment on 14 normal healthy subjects for lateral flexion. The authors reported that the peak torque on either side was achieved when the active muscles were lengthened with respect to the neutral position. Authors did not find a consistent difference between torques to the left and right. In isokinetic mode, the authors recorded a reduction in torque. Parnianpour et al. (1989) tested isoinertial lateral flexion in nine normal and healthy volunteers and reported an absolute value of 126 (40) Nm.

Kumar (1991a, 1996c) and Kumar et al. (1995) tested 59 normal healthy males and 43 normal healthy females for right and left lateral flexion in isometric and isokinetic modes. The isometric tests were carried out in neutral posture and 10°, 20°, and 30° laterally flexed on both left and right sides. The isokinetic testing was done from the neutral posture to the end of the range of motion from which data for neutral, 10°, 20°, and 30° lateral flexion were extracted to compare with those recorded in isometric mode. The isometric and isokinetic lateral flexion torques for the left and right for male and female subjects are presented in Table 14.7 and Table 14.8, respectively. A multivariate analysis of variance conducted for the isometric lateral flexion strength revealed that it was significantly affected by the group and the condition of the test and their interaction ($p < .01$). To explore further, univariate ANOVA was carried out for peak as well as average strength. Both these variables were affected by group, condition, and their interaction ($p < .01$). However, there was no significant difference between the lateral flexion strength recorded for left and right sides. Therefore, the values of the left and right sides were pooled for further

TABLE 14.7

Torque Produced in Lateral Flexion by the Normal Males (Newton-Meters)

Direction of Effort	Parameter	Trunk Posture in Degrees of Lateral Flexion							
		0°		10°		20°		30°	
		M	SD	M	SD	M	SD	M	SD
Left	Peak isometric	162	40	154	40	140	36	118	35
	Average isometric	137	35	131	35	119	30	99	30
	Isokinetic	29	19	78	33	100	29	89	28
Right	Peak isometric	165	41	156	42	140	40	123	105
	Average isometric	141	36	134	36	120	35	38	33
	Isokinetic	20	6	71	31	105	34	86	30

From Kumar, 1991a, University of Alberta, Edmonton.

TABLE 14.8

Torque Produced in Lateral Flexion by the Normal Males (Newton-Meters)

Direction of Effort	Parameter	Trunk Posture in Degrees of Lateral Flexion							
		0°		10°		20°		30°	
		M	SD	M	SD	M	SD	M	SD
Left	Peak isometric	101	33	117	30	104	24	87	23
	Average isometric	88	29	100	26	89	21	73	19
	Isokinetic	23	17	56	32	73	33	64	22
Right	Peak isometric	117	32	118	31	106	29	85	22
	Average isometric	100	28	101	27	90	26	72	19
	Isokinetic	18	10	50	25	77	23	71	27

From Kumar, 1991a, University of Alberta, Edmonton.

analysis. The multiple comparison revealed that, among males, peak as well as average torque at each angle were different from all others with the exception of the 0° and 10°. Among females the 30° lateral flexion, both peak and average, was significantly different from the rest, and 20° was different from that of the 10°. The greatest isometric lateral flexion strength in males was recorded in the neutral posture. With the progression of postural, asymmetry the strength declined. The decline in strength was greatest at 30° of lateral flexion. The peak strength in neutral posture was 50% of the reference value and declined by 2 and 6% at 10° and 20° of lateral flexion, respectively. The decline at 30° of lateral flexion was 13%. A similar pattern of variation was found for average strength in lateral flexion. In neutral posture the average lateral strength was 43% and declining by 2, 6, and 12% for 10°, 20°, and 30° of lateral flexion among males. Among these subjects, the average strength at the four postural angles ranged from 83 to 86% of their corresponding peak strength. The extension and lateral flexion ratios for peak and average strength were 1:0.5 and 1:0.53, respectively. This ratio with flexion was considerably higher, 1:0.86 and 1:0.88 for peak and average lateral flexion strength, respectively.

Among females, the highest isometric lateral flexion strength was recorded in neutral posture and at 10° of lateral flexion. With further progression of the postural asymmetry, the strength declined. The isometric lateral flexion strength in males was recorded in the neutral posture. With the progression of postural asymmetry, the strength declined. The isometric lateral flexion strength at 20° of trunk lateral flexion declined only marginally, with additional decline at 30° of lateral flexion. The difference between the neutral and 30° laterally flexed posture was only about 10% for the peak as well as average isometric strengths. The average isometric strength ranged between 84 and 87% of their corresponding peak strengths. In terms of their ratio with the extensor torques, they were 1:0.62 and 1:0.56 for

TABLE 14.9

Female Lateral Flexion Strength as Percent of the
Corresponding Male Strength

Variable	Trunk Posture In Degrees of Lateral Flexion			
	0°	10°	20°	30°
Peak isometric	69	75	75	68
Average isometric	70	76	75	66
Isokinetic	100[a]	72	72	81[a]

[a] These values being at the end of the range of motion for the study
are likely to have errors.

From Kumar, 1991a, University of Alberta, Edmonton.

the peak and average isometric strengths, respectively. Their ratios with the flexor torque were 1:1 for peak as well as average isometric strength. The females lateral flexion strength in isometric mode ranged from 22 to 36% of the reference normalizing value, as compared to 31 to 52% for the males. However, the female isometric lateral flexion strength ranged between 66 and 75% of their corresponding males' values (Table 14.9).

The isokinetic lateral flexion strength measured at trunk angles at which isometric trials were conducted, when subjected to a multivariate ANOVA, revealed that they were significantly affected by the type of sample and the trunk posture in which the measurement was made. A follow-up univariate ANOVA for the isokinetic strength confirmed the findings of the multivariate ANOVA. The strength was significantly affected by the group and the condition of the trial (trunk posture). A multiple comparison of the postures revealed that 0° lateral flexion was different from all other angles for males and females both. Further, the 10° of lateral flexion was different from the 20° for both males and females, but was different from 30° among females only.

The isokinetic lateral flexion strength of the males ranged from 9 to 31% of the reference normalizing values, being lowest at 0° and highest at the 20° of trunk lateral flexion (Figure 14.13). The lower value at 0° is anticipated, being so close to the initiating posture of the test activity. This is further supported by the isokinetic strength at 0° making only 18% of the corresponding isometric strength, whereas the 10°, 20°, and 30° were 50, 70, and 75%, respectively, of their corresponding isometric strength. The female isokinetic lateral flexion strength ranged from 7 to 23% of the reference normalizing factor. The pattern of the variation of females strength was identical to that of the males, being lowest at 0° and highest at 20°. For the reason stated above, the 0° lateral flexion strength is not considered representative. It represented 21% of its corresponding isometric value, whereas the strengths at 10°, 20°, and 30° were 47, 71, and 75%, respectively, of their corresponding isometric strength. The females' isometric strength ranged from 72 to 100% of the normal females (Table 14.9). These ratios towards the end of the range of motion are considered faithful representation, due to the initial acceleration and the final deceleration. However, ratios between the peak and average isokinetic strengths for extension and lateral flexion were 1:0.49 and 1:0.46, respectively, for males. For females these were 1:0.55 and 1:0.48. These ratios with flexion strengths were 1:0.75 and 1:0.73 for males and 1:0.8 and 1:0.74 for females. The average isokinetic strength was 71% of peak isokinetic strength for males and 67% for females.

A comparison of isometric lateral flexion strength and isokinetic lateral flexion strength was first conducted for neutral posture and, subsequently, at trunk angles at which isometric measurements were made. In the neutral posture, a multivariate analysis of variance showed the group and the mode (isometric vs. isokinetic) to be highly significant factors. A follow-up univariate ANOVA also revealed that the group and mode were

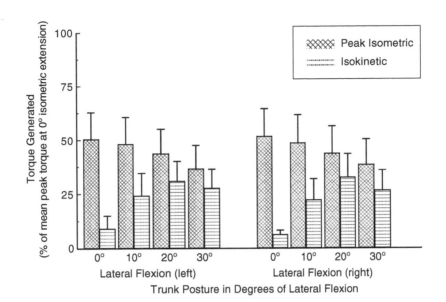

FIGURE 14.13
Isometric and isokinetic strength of normal males at common trunk postures in lateral flexion. (From Kumar, 1991a, University of Alberta, Edmonton.)

significant factors. They also had significant interactions. The actual values and the findings have been described above. The comparison of isokinetic strength with isometric strength in lateral flexion at 0°, 10°, 20°, and 30° of trunk lateral flexion showed a significant effect of group, mode of activity, and condition (trunk posture) and their two-way and three-way interactions, except interaction between group and condition. A multiple comparison showed that, among males, all isokinetic conditions were significantly different from all isometric conditions. Among females, however, the isokinetic lateral flexion strengths at 30° and 20° of lateral flexion were not significantly different from those of 30° isometric lateral flexion strength.

14.4.4 Axial Rotation

Rotation of the trunk is a common activity of daily living. Most occupational activities also require trunk rotation to varying degrees. Several studies reported in literature have linked trunk rotation to back injuries (Duncan and Ahmed, 1991; Frymoyer et al., 1980, 1983; Manning et al., 1984; Marras et al., 1993; McGill and Hoodless, 1990; Ralston et al., 1974; Schaffer, 1982; Snook, 1978; Snook, 1980). The U.S. Department of Labor has reported that of all injuries that occur during lifting, 33% are attributable to twisting as a single factor (Schaffer, 1982). Manning et al. (1984) through their epidemiological study demonstrated that twisting of the trunk was involved 11.4% of accidental back injuries and in 49% of nonaccidental injuries. Further they ranked rotation as the third most common body movement associated with low back pain. In more than 400 repetitive industrial lifting jobs in 48 different industries, trunk rotation was found to have the second most motion variables with significant odds ratio for low back disorders (Marras et al., 1993).

Although trunk torsion has been associated with back disorders for a long time, the measurement of the rotary strength of the human trunk has been studied only infrequently. When studied, the trunk rotation strength has been measured more frequently in isometric mode. Kumar (1994a) stated that the paucity of data for rotation and lateral bending was attributable directly to the lack of suitable, accurate, standardized, and affordable devices to permit such measurements. One of the first studies in the literature that reported rotary strength of the human trunk was that of Smith et al. (1985). They used a Cybex Back Tester for this. It is unclear from their study if the torso-stabilizing device allowed the necessary coupled lateral flexion during axial rotation, and if it allowed accommodation of appropriate torso shape change in movement and prevented any contribution from the hip. Smith et al. (1985) stated that, until that time data on axial rotation strength was "nonexistent." The authors reported a bilateral symmetry in axial rotation in both genders. This finding has since been confirmed by Kumar (1991a, 1994a), Kumar et al. (1995) and Newton et al. (1993). Several studies have reported that, in axial rotation strength, like many other strength parameters, men were stronger than women (Kumar, 1991a, 1994a, 1996c, 1997, 2001, 2002; McGill and Hoodless, 1990; Smith et al., 1988). A poor predictability of rotary strength from sagittal plane activities has been reported (Kumar et al., 1991; Kumar and Garand, 1992; Smith et al., 1985). Kumar et al. (1995) reported that, using anthropometric variables and some other strength parameters, the isometric axial rotation strength could be predicted, explaining up to 88% of variance ($p < .001$), but isokinetic strength was predicted poorly. Most published studies have reported a significantly higher torque production capacity in isometric mode compared to isokinetic (Kumar, 1991, 1994a; Kumar and Garand, 1992; Kumar et al., 1988, 1991, 1995; Smith et al., 1985). In a series of studies, Kumar (1991a, 1994a, 1996c) and Kumar et al. (1995) demonstrated that the axial rotation strength was the weakest of all movements that the trunk could produce in both genders. It represented 20 and 24% of exertion capability in isometric and isokinetic modes, respectively. With postural asymmetry in steps of 5° deviation, the axial rotation strength steadily declined in both genders. Pope et al. (1987) reported that the axial prerotation of the trunk by 30° in the direction of torque development decreased the developed torque marginally, whereas prerotation in the opposite direction marginally increased the developed torque. Kumar (1997) and Kumar et al. (2002) presented results from a study where 50 subjects performed axial rotation from neutral, 15°, and 30° rotation to left and right in the direction of prerotation and away from it. They reported that the subjects were 20 to 25% weaker while exerting in the direction of prerotation and 30% stronger when exerting in opposite direction (Table 14.10). Kumar (1997) also reported the isokinetic torques of these subjects when they rotated their trunks from neutral to left and right, and when the trunk was rotated to the neutral posture from prerotated (left to right) postures, at angular velocities of 10°, 20°, and 40° per second. In isokinetic mode the trunk rotation from neutral to asymmetric positions produced lesser torques compared with torques from rotated to neutral position. The torque-producing capability declined with increasing velocity of the activity (Table 14.11).

The mean axial rotation torques with their standard deviations among 59 males and 43 female subjects in isometric and isokinetic modes are presented in Table 14.12 for males and Table 14.13 for females. An analysis of variance of axial rotation values to left and right showed no significant difference. For all future analyses these were pooled. With this pooled data, a multivariate ANOVA showed a strong effect of the factors of gender and activity. Their interaction was only marginally significant. The follow-up univariate analysis of variance revealed that, for peak as well as average isometric torques, the factors of gender and activity (angle of trunk rotation) were different. Their interaction, however, was not. For both peak and average torques, the normal male and females groups were different from each other. A multiple comparison for peak and average isometric torques at five different angles of trunk rotation for males revealed that axial rotation strength at 20° of trunk flexion was significantly different from all others. Strengths at 15° were significantly different from those at 0° and 5°, and strengths at 10° were significantly different from those of 0° of trunk rotation. Among females the isometric strengths at 20°

TABLE 14.10

Isometrical Axial Rotation Torque of the Trunk among Normal Young Subjects

Gender	Direction of Rotation	Peak Torque (Nm)		Time at Peak Torque (sec)		Average Torque (Nm)		Torque Area (Nm.sec)	
		Mean	SD	Mean	SD	Mean	SD	Mean	SD
Male	0 → L	110	43	3.0	1.0	81	32	483.6	194.2
	0 → R	106	37	2.9	1.3	77	29	463.0	166.0
	15L → L	90	33	3.1	1.2	69	26	402.9	146.3
	15R → R	95	35	2.9	1.2	71	27	421.4	155.7
	30L → L	77	27	3.0	1.4	57	22	335.3	125.7
	30R → R	79	27	2.9	1.2	59	18	352.7	111.7
	15L → R	122	42	3.3	1.1	93	31	547.4	189.1
	15R → L	119	45	3.0	1.1	90	35	524.9	197.2
	30L → R	131	41	3.6	1.1	97	31	581.0	181.4
	30R → L	132	41	3.3	1.2	100	35	596.3	207.7
Female	0 → L	65	21	2.7	0.9	49	16	288.3	98.8
	0 → R	67	23	3.2	1.1	50	17	308.4	105.9
	15L → L	55	20	2.0	1.1	42	17	248.0	97.5
	15R → R	57	19	2.5	1.4	44	14	258.8	86.3
	30L → L	45	17	2.2	0.9	35	15	204.6	84.9
	30R → R	47	19	2.1	1.0	35	14	210.8	83.7
	15L → R	75	22	2.7	1.2	58	18	339.0	101.1
	15R → L	73	22	2.7	1.3	56	18	339.2	111.4
	30L → R	85	24	3.1	1.4	65	20	288.3	122.8
	30R → L	78	23	2.92	1.0	61	19	366.8	110.8

Note: 0 = neutral position; L = left, R = right; 15L = 15° prerotation to the left; 15 R = 15° prerotation to right; 30L = 30° to the right.

From Kumar, 1997.

of trunk rotation were significantly different from 0° and 5° rotated trunk posture. Also, 15° of axial rotation was significantly different from the neutral posture.

The peak and average isometric axial rotation strength among normal males ranged between 17 and 24% of the reference normalizing values. The drop from the neutral posture to the 20° axially rotated trunk was only approximately 5%. The average strength remained between 80 and 87% of the corresponding peak isometric strength. From these results, it would appear that the rotational strength is approximately one-fifth to one-quarter of the extension strength, but this strength can be sustained through the range of posture in which tests were done. The extension and axial rotation ratios for the peak and average torques were 1:0.25 and 1:0.30, respectively, for normal males. Among the female sample, this strength ranged 10 to 13% of the normalizing reference torque. The decline in strength from 0° to 20° was of the order of only 2%. The average isometric strengths were between 83 and 91% of the corresponding peak isometric strength. The extension to axial rotation ratios for the peak and average isometric strengths were 1:0.24 and 1:0.25. The corresponding ratios with the flexor torques were 1:0.38 for both peak and average strengths. The normal female isometric axial rotation strength was between 52 and 57% of the corresponding values of the normal male group.

The isokinetic axial rotation strengths, peak as well as average, were found to be significantly affected by the group and the condition of the activity in the analysis of variance. However, the interactions between these factors were not significant. The multiple comparison of the different conditions among normal males revealed that all conditions were significantly different from the others for both peak axial rotation torque and the trunk rotation angle. Among normal females, however, the 0° and 5° of trunk rotation were

TABLE 14.11

Isokinetic Axial Rotation of the Trunk among Normal Young Adults

Gender	Direction of Rotation Direction	Speed (/sec)	Peak Torque (Nm) Mean	SD	Angular Displacement at Peak Torque (°) Mean	SD	Time At Peak Torque (Sec) Mean	SD	Total Angular Displacement (°) Mean	SD	Total Time (Sec) Mean	SD	Average Torque (Nm) Mean	SD	Torque Area (Nm/S) Mean	SD
Male	N → L	10	63.7	17.0	-22.7	7.3	0.7	0.3	-60.7	12.7	2.8	0.6	41.7	10.8	116.9	36.2
		20	53.5	16.3	-28.3	8.9	0.5	0.2	-69.4	9.3	1.6	0.3	34.3	10.5	52.0	13.3
		40	42.2	11.9	-30.6	7.7	0.3	0.1	-78.6	6.4	1.0	0.2	26.5	8.1	25.5	7.4
	N → R	10	63.0	19.5	23.7	5.0	0.6	0.3	61.2	10.8	2.9	0.5	41.8	13.6	120.5	37.0
		20	52.2	17.6	31.0	8.0	0.5	0.2	70.3	8.3	1.5	0.2	35.5	11.3	54.8	15.9
		40	41.0	11.1	32.4	6.8	0.3	0.1	76.2	8.3	1.0	0.2	25.3	7.8	24.5	5.6
	L → N	10	79.3	25.5	-7.6	7.4	0.9	0.4	65.6	10.6	2.3	0.4	54.6	18.5	125.0	49.3
		20	58.2	23.3	-5.2	9.4	0.6	0.2	69.5	10.6	1.4	0.3	35.7	14.0	48.2	19.5
		40	39.0	18.7	-7.2	7.0	0.3	0.1	72.9	10.3	1.0	0.2	23.6	13.7	20.6	8.3
	R → N	10	72.4	25.9	11.5	9.1	0.8	0.3	-65.0	9.0	2.4	0.4	47.6	17.7	112.6	40.5
		20	53.1	17.5	8.9	7.9	0.5	0.2	-70.1	7.6	1.2	0.2	34.1	11.8	43.4	14.1
		40	37.3	17.1	7.3	8.4	0.4	0.1	-75.3	12.3	1.0	0.2	22.5	12.5	20.2	7.9
Female	N → L	10	44.7	12.2	-19.1	7.6	0.7	0.3	-63.1	12.7	3.1	0.6	29.8	8.2	91.8	28.3
		20	35.6	10.5	-26.1	7.5	0.6	0.2	-69.6	12.7	1.6	0.3	23.7	6.8	38.0	12.2
		40	27.8	8.1	-27.3	6.8	0.4	0.1	-74.7	13.2	1.25	0.4	15.9	4.8	17.6	4.7
	N → R	10	45.7	12.8	22.7	5.7	0.7	0.2	61.0	12.5	3.1	0.5	29.6	8.6	91.6	30.4
		20	35.2	10.2	28.2	8.2	0.5	0.1	67.5	10.2	1.7	0.3	22.9	6.4	37.9	11.0
		40	25.3	8.1	32.2	9.9	0.4	0.1	74.8	12.6	1.2	0.4	15.9	4.8	17.6	4.7
	L → N	10	51.6	19.1	-11.2	6.6	0.8	0.2	63.5	2.4	2.4	0.6	34.4	12.8	82.4	36.8
		20	39.9	16.2	-5.4	10.8	0.6	0.2	70.4	1.4	1.4	0.3	25.1	10.8	34.5	14.2
		40	24.6	8.3	-7.7	11.6	0.4	0.1	77.4	1.1	1.1	0.3	14.9	4.8	15.2	4.2
	R → N	10	53.3	17.5	12.5	10.3	0.8	0.4	-62.4	2.4	2.4	0.3	34.4	11.6	81.7	32.1
		20	37.8	16.3	12.7	9.6	0.5	0.2	-69.3	1.5	1.5	0.4	24.0	9.8	35.1	17.7
		40	23.9	8.8	7.6	11.1	0.5	0.6	-75.6	1.4	1.4	1.3	13.9	4.4	15.9	4.5

Note: N = Neutral; L = left; R = right.

From Kumar, 1997.

TABLE 14.12

Torque Produced in Axial Rotation by the Normal Males (Newton-Meters)

| | | Trunk Posture in Degrees of Axial Rotation | | | | | | | | |
| | | 0° | | 5° | | 10° | | 15° | | 20° | |
Rotation Direction	Parameter	M	SD	M	SD	M	SD	M	SD	M	SD
Left	Peak isometric	79	26	78	27	73	25	70	23	66	20
	Average isometric	66	23	65	24	62	22	59	21	56	17
	Isokinetic	14	9	23	11	34	12	42	15	45	17
Right	Peak isometric	81	27	79	26	75	25	73	23	68	21
	Average isometric	67	24	65	22	64	22	61	21	57	19
	Isokinetic	11	6	19	11	31	12	40	16	47	18

From Kumar, 1991a, University of Alberta, Edmonton.

TABLE 14.13

Torque Produced in Axial Rotation by the Normal Females (Newton-Meters)

| | | Trunk Posture in Degrees of Axial Rotation | | | | | | | | |
| | | 0° | | 5° | | 10° | | 15° | | 20° | |
Rotation Direction	Parameter	M	SD	M	SD	M	SD	M	SD	M	SD
Left	Peak isometric	42	16	41	13	40	12	38	12	36	11
	Average isometric	36	14	35	12	34	10	32	10	31	10
	Isokinetic	12	7	20	9	27	10	30	11	30	11
Right	Peak isometric	45	14	44	14	42	13	40	12	37	11
	Average isometric	38	13	37	12	35	11	33	11	31	9
	Isokinetic	8	6	16	8	23	9	28	9	30	10

From Kumar, 1991a, University of Alberta, Edmonton.

significantly different from all other trunk postures for the torque as well as trunk angle. In addition, the isokinetic rotary torque at 10° was different from 20° of trunk flexion. The isokinetic torque in axial rotation for males ranged from 4 to 14% of the normalizing reference among males and 4 to 9% among females. The values were always lowest for the neutral posture and steadily increased to 20° of trunk flexion. From this it would appear that 20° was not a significant trunk rotation. In terms of the relationship of the isokinetic strength with the isometric strength, it represented 16, 29, 49, 59, and 66% at 0°, 5°, 10°, 15°, and 20° of trunk rotation among males. For females these proportions were 30, 46, 66, 75, and 81% at the neutral and progressively rotated postures. The extension–rotation and flexion–rotation ratios for male for isokinetic strength were 1:0.05 and 1:0.3, respectively. These ratios for females were 1:0.09 and 1:0.37. However, these ratios between the peak and average isokinetic extensor torque and peak and average isokinetic axial rotation torques were 1:0.21 and 1:0.20, respectively, for males. The flexion–axial rotation ratios for peak and average torques among males were 1:0.36 and 1:0.38, respectively. The average isokinetic axial rotation torques among males constituted 75% of the peak value. Among females, the extension–axial rotation ratios for peak and average isokinetic torques were 1:0.21 and 1:0.20, respectively. The flexion–axial rotation ratios were 1:0.31 and 1:0.29 for the peak and average isokinetic strengths, respectively. The average isokinetic axial strength constituted 72% of its peak value.

A comparison between the isometric and isokinetic strength initially for neutral posture revealed that, for both peak and average strengths, the factors of group, mode, and their interaction had significant effect. In both sexes the isometric and isokinetic strengths were significantly different from each other. Comparing the isokinetic strengths to isometric at all postures measured (Figure 14.14), the analysis of variance revealed that the factors of

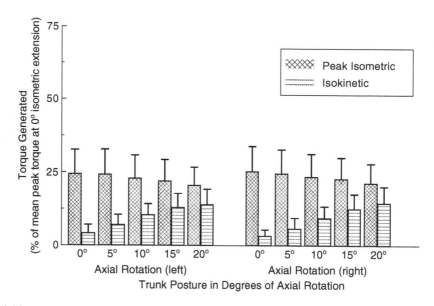

FIGURE 14.14

A comparative depiction of isometric and isokinetic axial rotation strength to left and right among males at common trunk postures. (From Kumar, 1991a, University of Alberta, Edmonton.)

group, mode, and condition all had significant effect. The multiple comparison of the peak isometric torque and the isokinetic torque at the trunk postures of measurement for the normal male sample revealed that the isokinetic torques at 0°, 5°, and 10° of trunk rotation were significantly different from all others. Isokinetic torques at 15° and 20° and isometric torque at 20° of trunk rotation were significantly different from isometric rotation torques at 0°, 5° and 10° of trunk rotation. The multiple comparison of female values showed that isokinetic rotation torque at 0° and 5° were significantly different from all others. The isokinetic axial rotation strength at 10° of trunk rotation was significantly different from all isometric axial rotation strength but not isokinetic. The comparative strength for the activities of extension, flexion, lateral flexion, and axial rotation in isokinetic mode was similar to those in isometric mode for males and females (Figure 14.15). The ratio between extension and other activities were 1:0.64, 1:0.49, and 1:0.21 for males; and 1:0.68, 1:0.55 and 1:0.21 for females.

14.4.5 Combined Motions

The human trunk is a complex organ, with 17 vertebrae in the thoracolumbar region. The intervertebral joint between any two vertebrae has six degrees of freedom of motion. Intervertebral joints can rotate about and translate along each of the three primary axes. Any occupational activity, if it is not entirely symmetrical, will be subjected to a combination of different motions. Occupational activities may involve various degrees of flexion, rotation, and lateral flexion. Measurements of strength in these combined plane motions are likely to be more beneficial than just the planar motion. However, there is no testing device currently available which is capable of providing these motions. Perhaps, it is due to the this reason that there are virtually no papers published in the scientific literature with the exceptions of Kumar (1991b), Kumar and Garand (1992), Kumar et al. (1998, 1999). These authors tested 14 normal and healthy males and 24 normal and healthy females with mean ages of 23 years (males) and 21 years (females). All subjects were

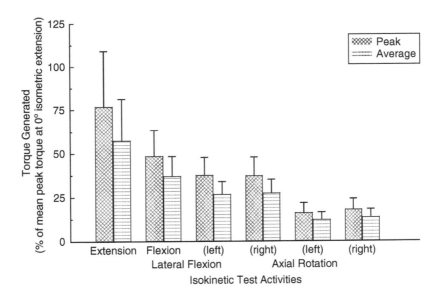

FIGURE 14.15

A comparison of peak and average strengths of flexion/extension, lateral flexion, and axial rotation in isokinetic mode. (From Kumar, 1991a, University of Alberta, Edmonton.)

required to perform 18 isometric and 18 isokinetic tasks for a total of 36 conditions. The isometric tasks consisted of flexion and extension rotations at 20°, 40°, and 60° of trunk flexion with 20°, 40°, and 60° of trunk rotation. The isokinetic tasks consisted of flexion–rotation and extension–rotation in 20°, 40°, and 60° rotation planes with angular velocities of 15, 30, and 60°/s. For each of the activities, the subjects were instructed to provide their maximal voluntary contraction without jerking. The subjects exerted their flexor or extensor torque for isometric efforts in the plane of rotation from the initially fixed flexed trunk position. For isokinetic flexion, the subjects started from an upright prerotated position flexing at one of three predetermined and adjusted velocities. In contrast, for isokinetic extension, the subjects were placed in prerotated fully flexed posture; from there they extended to reach the upright posture at one of the three predetermined and preadjusted velocities.

The isometric and isokinetic trunk functional rotation was measured using the Static Dynamic Strength Tester (SDST) (Figure 14.8 and Figure 14.9). This device has been tested for reliable velocity control for up to 150°/s (Kumar, 1996a). The isometric tests were done with equipment in a locked position. The force exerted was measured by the load cell of the SDST (Interface Model SM 500, 500 lb with a natural frequency of 1.5 KHz). The load cell of the SDST was connected with the chest harness of the subject through a flexible airplane steel cable. The harness consisted of adjustable, snugly fitting, rigid bands with an eyehook both at the front and back for connecting the cable.

The posture-stabilizing platform consisted of a horizontal circular plate mounted with two oil pipes as uprights (Figure 14.16). Slid over these uprights were three height-adjustable upholstered pads for comfortable stabilization. These were fitted with adjustable belts to fasten the lower extremities of the subject who stood on the circular plate. This plate was mounted on a rectangular base plate such that it could revolve like a turntable on rigidly installed castors but could be clamped in any position. Using this mechanism, the platform could be arranged in 20°, 40°, and 60° of rotation planes bilaterally. The rectangular plate could be slid on an iron track and locked in any position. This platform was rigidly attached to the SDST, leveled with it at one end and fixed to the wall

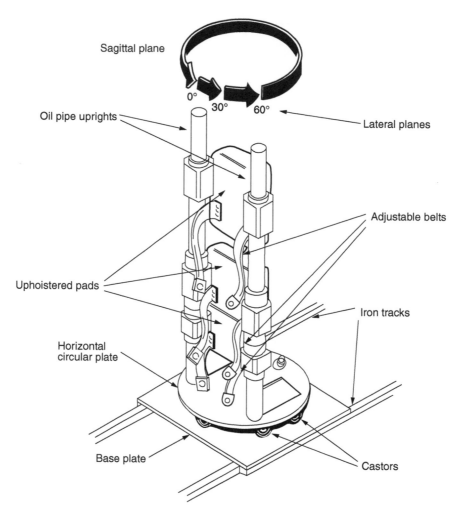

FIGURE 14.16
Posture-stabilizing platform for flexion–rotation and extension–rotation testing. (From Kumar, et al., 1999.)

on the other to achieve a rigid fixation. A high-precision potentiometer was mounted on the hip stabilizer. The shaft of this potentiometer was connected with the shoulder harness for a continuous measurement of motion at the hip.

An iron frame was built with rectangular tubing and a position-adjustable pulley system to allow orientation of the load cell and cable system in the desired rotation plane (Figure 14.17). The means and standard deviations of peak strength, average strength, time of maximum torque, and torque–time product for 18 isometric activities of males and females are presented in Table 14.14 and Table 14.15. The means and standard deviations of peak and average torques, angular deviation at peak torque, time of maximum torque, and torque–time product for 18 isokinetic activities among males and females are presented in Table 14.16 and Table 14.17. The general observations indicate that males were signifi-cantly stronger than females in each activity ($p < .001$). The isometric strengths were invariably greater than their corresponding isokinetic activities ($p < .01$). For flexion activ-ities, the subjects were strongest in isometric mode at 20° or 40° of trunk flexion, but for extension they were always strongest at 60° of flexed spinal posture. Although there was a trend for reduction in strength with increasing rotation, these differences were neither consistent nor large. Similarly, the differences between flexion and extension were not

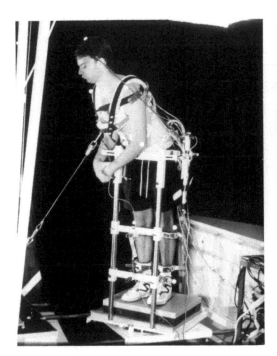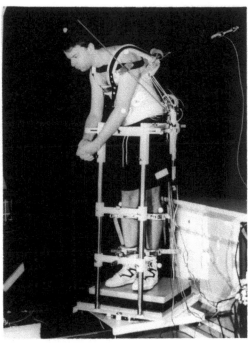

FIGURE 14.17
Subject in the flexion–rotation and extension–rotation frame for combined motion testing. (From Kumar et al., 1998.)

TABLE 14.14

Isometric Flexion–Rotation and Extension–Rotation Strength of the Male Experimental Sample

Task	Trunk Flexion (°)	Rotation Angle (°)	Max. Torque (Nm)		Average Torque (Nm)		Time of Max. Torque (S)		Torque–Time Area (Nm S)	
			Mean	SD	Mean	SD	Mean	SD	Mean	SD
Flexion–rotation	20	20	197.7	28.3	151.7	32.4	3.7	0.6	624.9	143.3
		40	19.9	19.6	139.1	24.3	3.8	0.3	568.7	115.6
		60	183.1	12.6	142.1	17.2	3.8	0.7	593.3	92.7
	40	20	205.4	32.2	164.3	27.7	4.2	0.2	684.1	111.2
		40	174.0	22.7	140.8	23.7	3.3	0.7	582.2	113.6
		60	171.6	26.3	132.5	25.0	4.0	0.7	553.8	102.9
	60	20	157.9	36.5	125.4	31.1	3.8	0.4	512.4	143.4
		40	152.3	31.4	124.1	22.7	3.9	0.4	508.5	77.8
		60	136.1	39.4	107.1	29.6	3.8	0.8	446.5	124.6
Extension–rotation	20	20	197.6	59.0	145.0	45.2	4.0	0.5	600.0	172.0
		40	211.1	47.5	164.8	39.0	3.9	0.5	685.1	169.4
		60	170.0	50.7	132.6	33.7	4.1	0.2	559.0	155.3
	40	20	219.3	65.7	159.9	54.0	3.9	0.9	674.2	222.0
		40	200.8	69.8	144.4	52.3	3.8	0.7	597.6	204.3
		60	204.4	46.3	155.3	34.1	3.7	0.5	626.9	126.7
	60	20	238.3	72.5	186.3	53.0	3.8	0.6	773.2	217.0
		40	250.9	51.2	197.7	38.6	3.6	0.9	809.7	155.9
		60	234.5	53.8	187.2	43.1	3.7	1.0	788.7	167.2

From Kumar et al. 1998, *Ergonomics*, 41(6), 835–852.

TABLE 14.15

Isometric Flexion–Rotation and Extension–Rotation Strength of the Female Experimental Sample

Task	Trunk Flexion (°)	Rotation Angle (°)	Max. Torque (Nm)		Average Torque (Nm)		Time of Max. Torque (S)		Torque–Time Area (Nm S)	
			Mean	SD	Mean	SD	Mean	SD	Mean	SD
Flexion–rotation	20	20	107.5	23.8	86.9	22.1	2.8	1.1	356.3	96.5
		40	113.4	26.1	90.0	21.8	3.0	0.9	367.4	98.9
		60	107.2	24.9	85.3	22.6	3.4	0.8	352.9	96.8
	40	20	111.0	29.9	89.3	26.1	3.5	0.8	380.6	103.6
		40	105.3	28.2	84.9	23.8	3.5	1.0	355.1	98.1
		60	111.9	28.1	90.0	23.5	3.1	1.2	372.1	104.1
	60	20	92.5	28.2	75.2	24.1	3.5	0.6	305.5	93.2
		40	91.9	30.7	75.5	26.2	3.4	1.0	319.0	113.3
		60	85.7	24.0	69.9	21.2	3.7	0.8	289.3	92.3
Extension–rotation	20	20	121.3	30.1	94.9	26.4	3.4	0.8	394.9	121.3
		40	109.5	27.4	86.4	22.6	3.3	1.0	354.0	100.8
		60	106.4	24.3	82.9	21.6	3.5	1.0	341.6	94.1
	40	20	125.2	30.4	99.2	26.7	3.2	0.9	402.6	119.8
		40	111.4	25.8	88.4	24.9	3.7	0.8	374.7	110.7
		60	122.0	28.1	95.9	23.2	3.2	0.8	393.1	98.8
	60	20	135.1	36.4	107.1	32.2	3.4	1.0	442.7	138.5
		40	134.3	32.7	104.9	24.9	3.5	0.8	430.2	119.6
		60	121.3	27.9	95.6	22.8	3.1	1.1	403.5	94.9

From Kumar et al. 1998, *Ergonomics*, 41(6), 835–852.

TABLE 14.16

Isokinetic Flexion–Rotation and Extension–Rotation Strength of the Male Experimental Sample

Task	Plane of Rotation (°)	Angular Velocity (°)	Max. Torque (Nm)		Average Torque (Nm)		Angular Deviation at Max. Torque (°)		Time of Max. Torque (S)		Torque–Time Area (Nm S)	
			Mean	SD	Mean	SD	Mean	SD	Mean	SD	Mean	SD
Flexion–rotation	20	15	160.2	4.7	99.0	1.5	48.6	6.6	2.1	0.4	651.1	23.6
		30	130.2	29.7	70.1	15.8	46.8	6.6	0.9	0.3	212.2	70.0
		60	113.0	52.7	51.7	23.8	50.4	8.2	0.7	0.2	88.7	40.8
	40	15	168.5	11.9	100.0	12.7	46.4	1.0	1.5	0.5	660.2	12.3
		30	124.2	41.3	61.9	16.2	43.8	8.8	0.7	0.3	192.9	62.3
		60	101.0	41.7	47.6	18.1	50.4	7.3	0.6	0.2	79.0	31.6
	60	15	139.3	3.9	79.6	4.8	49.8	3.9	1.3	0.1	491.9	36.6
		30	101.1	35.0	52.4	14.2	48.0	6.9	0.8	0.2	151.6	50.0
		60	103.4	39.5	44.4	12.6	47.3	7.2	0.5	0.1	70.6	20.9
Extension–rotation	20	15	168.7	24.7	112.1	7.2	43.1	0.5	2.0	0.9	760.3	17.8
		30	160.5	39.8	86.8	19.3	38.1	10.9	0.8	0.5	262.2	67.7
		60	132.5	28.8	65.2	19.0	37.0	9.0	0.5	0.1	108.1	33.6
	40	15	155.2	4.6	106.6	9.7	41.9	4.9	1.8	0.7	668.9	71.4
		30	149.7	39.0	85.1	25.0	43.8	13.8	0.7	0.3	251.0	80.8
		60	129.6	34.9	59.8	20.4	38.5	8.9	0.5	0.2	101.9	34.6
	60	15	161.0	38.7	99.6	17.2	44.3	14.1	1.0	0.2	559.4	82.8
		30	144.3	33.4	82.4	24.7	44.9	10.6	0.7	0.3	241.9	79.8
		60	118.5	33.9	56.5	20.7	42.3	12.3	0.4	0.1	87.4	29.2

From Kumar et al. 1998, *Ergonomics*, 41(6), 835–852.

TABLE 14.17

Isometric Flexion–Rotation and Extension–Rotation Strength of the Female Experimental Sample

Task	Plane of Rotation (°)	Angular Velocity (°)	Max. Torque (Nm)		Average Torque (Nm)		Angular Deviation at Max. Torque (°)		Time of Max. Torque (S)		Torque–Time Area (Nm/S)	
			Mean	SD	Mean	SD	Mean	SD	Mean	SD	Mean	SD
Flexion–rotation	20	15	99.8	20.2	57.6	13.6	46.7	9.7	1.4	0.5	362.6	112.7
		30	97.2	25.5	53.5	15.0	49.5	8.1	0.9	0.2	159.3	54.2
		60	82.0	29.7	39.7	14.9	53.5	5.8	0.6	0.1	65.3	25.5
	40	15	105.1	24.2	60.3	17.7	47.0	8.8	1.5	0.8	384.7	138.8
		30	90.6	24.3	50.5	14.3	49.2	7.9	0.9	0.3	145.2	51.3
		60	79.7	28.9	39.5	14.9	54.3	6.8	0.7	0.1	63.0	27.8
	60	15	89.4	20.6	52.6	14.8	46.2	7.6	1.0	0.4	322.3	116.5
		30	86.9	25.6	47.4	14.6	49.6	7.4	0.8	0.2	130.9	45.9
		60	71.3	24.7	34.9	12.5	52.8	10.9	0.6	0.1	55.5	24.1
Extension–rotation	20	15	114.5	29.3	73.7	18.9	49.5	6.6	1.5	0.9	485.3	86.0
		30	103.7	26.9	59.2	16.1	42.3	9.3	1.0	0.3	197.9	64.0
		60	88.8	27.7	43.6	14.8	39.4	10.5	0.6	0.1	80.0	27.0
	40	15	112.5	37.8	72.9	17.6	48.7	11.4	1.7	1.0	489.7	107.0
		30	101.0	24.3	59.8	13.5	47.9	8.7	0.9	0.2	196.6	50.0
		60	89.9	28.2	44.4	14.9	37.5	7.1	0.7	0.2	79.4	28.1
	60	15	101.1	36.1	64.9	17.6	52.4	10.3	1.4	0.8	453.6	96.8
		30	86.9	27.2	51.0	17.5	45.7	8.8	0.9	0.3	156.5	51.6
		60	83.2	24.0	42.4	14.4	41.7	8.5	0.6	0.1	75.7	21.7

striking except in a 60° flexed posture. With a progressive increase in the velocity of the isokinetic activities, there was a progressive decrease in the torque production.

14.4.5.1 Isometric Strength

The means and standard deviations of the peak and average torques for males and females are presented in Table 14.14 and Table 14.15. In flexion–rotation, the torque generated at 20° and 40° of trunk flexion were generally comparable in both genders. There was a significant drop at 60° of flexion compared to the other two settings. At each of the trunk flexion settings, increasing trunk rotation was associated with marginal decline in torque production at all trunk flexion levels. At every setting there was a significant difference between peak and average values in both genders ($p < .01$). There was also a significant difference between males and females for each of the corresponding conditions ($p < .01$). The peak torque generally occurred late in the trial. In isometric extension rotation, the magnitude of the peak torque continued to increase progressively with increasing amount of trunk flexion in both sexes. This was a general trend of marginal decrease in torques at most trunk flexion angles with increasing rotation. However, this was neither invariable nor large. The maximal extension–rotation torques were invariably higher than the maximal flexion torques. The peak extension–rotation torques also occurred late in the trial period. The torque–time products reflected the same pattern as the peak and average torques.

14.4.5.2 Isokinetic Strength

The mean peak and average isokinetic torques along with corresponding standard deviations are presented in Table 14.16 and Table 14.17. The isokinetic torques were invariably significantly lower than those of the isometric torques for all conditions ($p < .01$). This decline was generally in the order of 20 to 30%. The largest differences were observed in extension–rotation at 60° trunk flexion and 60° of trunk rotation. In these conditions the decrement ranged in the order of 35 to 50%. In flexion–rotation in isokinetic mode, the peak torque declined significantly with increasing rotation of the trunk among males but not among females. Among females there was also a pattern of decline, but this was not so large in magnitude. However, in extension–rotation, the progressive decline in torque with increasing rotation, though present, was not so pronounced. The differences between the peak and average torques in isokinetic conditions were significantly greater than those encountered in isometric conditions ranging from 33 to 55%. The reduction in peak torque with increasing velocity of the activity was more pronounced in men than in women. The time of the peak torque was greater with slower velocity of activity, and it progressively decreased with increasing velocity.

14.4.5.3 *Statistical Results*

A multiple analysis of variance revealed significant main effects due to the gender and mode and significant interactions between these variables for both peak and average torques ($p < .001$). The majority of the variance was accounted for by gender. However, the two variables together explained 50% of the variance for the peak torque and 62% of the variance for the average torque ($p < .001$). The results of follow-up univariate analyses of variance for peak as well as average torques revealed that the isometric mode was significantly different from that of isokinetic ($p < .001$). Similar results were found in both genders separately.

Follow-up multiple analyses of variances for the isometric peak and average torques were also done. Gender, task (flexion–rotation vs. extension–rotation), and the plane of rotation had significant main effects on the peak torque. The trunk flexion did not have a significant effect. There were significant two-way interactions between gender and task, and task and trunk angle, and a three-way interaction between gender, task, and trunk flexion. For average isometric torques, only gender and tasks had significant main effects. It was found that the plane of rotation and the trunk flexion did not have any effect on the average torque. Significant two-way interactions between gender and task, task and trunk flexion, and a significant three-way interaction between gender, task, and trunk angles were found. Follow-up univariate analyses of variances revealed that the peak isometric torque was not affected by trunk rotation at all. Gender significantly affected the peak as well as average torques in both flexion–rotation and extension–rotation ($p < .001$). Similarly, the task (flexion–rotation vs. extension–rotation) significantly affected isometric peak as well as average torques. The degree of the trunk flexion in which these tasks were performed also significantly affected both the peak and average torques. The multiple comparisons showed that 20° and 60° trunk flexions were always significantly different ($p < .05$), and for flexor tasks, the 60° trunk flexion was also significantly different from that of 40° flexion ($p < .05$). For isokinetic peak and average torques, the multiple analyses of variance revealed that gender, task, rotation, and speed all had significant main effects ($p < .01$, except peak rotation — $p < .02$). There was also a significant two-way interaction between gender and speed for average isokinetic torque ($p < .01$). Follow-up univariate analyses of variances showed that rotation did not have a significant effect on peak or average torque, but speed did ($p < .001$). There was also a significant two-way interaction between speed and gender ($p < .001$).

14.5 Lifting Strength

Although lifting has been a cause for grief since time immemorial, significant research was not undertaken in this area until the 1960s. However, a rapid growth followed in the 1970s, 1980s, and 1990s, which seems to be stemming to some extent from the beginning of the twenty-first century. Whether it is influenced by lack of meaningful control of low back pain/injury problems, or if it is affected by reduction in resources due to reallocation to rapidly advancing fields of biology and technology, or both is hard to say. In any case, literature in the area of lifting is extensive and diverse. The purpose of this section is not to capture exhaustive information in the field, but to focus only on one aspect of lifting — lifting strength. This section is also not going to be a "how-to" guide for manual materials-handling tasks, for which there are books written. For information of the latter variety, the readers are referred to Chaffin et al. (1977), NIOSH (1981), Ayoub and Mital (1989), Mital et al. (1993), Waters et al. (1993), NIOSH Application Manual (1993), Frymoyer et al. (1991), Nordin et al. (1997), and several others books in the field. The topic of lifting strength here will be dealt with under two categories: (a) biomechanical, and (b) psychophysical. Additionally, this section will not deal with models for spinal load prediction, but will restrict itself entirely to strength production. The metabolic variables will not be discussed.

14.5.1 Biomechanical Strength

This category refers largely to the measured forces during the act of lifting. There is more than one way to classify biomechanical strength. One way would be to consider isometric, isokinetic, and isoinertial separately. Another may be to consider the field divided into arm lift, stoop lift, squat lift and isoinertial lift. Since the latter represents a functional grouping affecting different sets of muscles more, this is the one we will use here.

14.5.1.1 Arm Lift

One of first studies reporting arm-lift strength as separate was conducted by Chaffin et al. (1977, 1978) and published by NIOSH. In this study, the authors measured arm-lift strength along with leg and torso strength, with a view to use these strength values to predict occupational injuries. To that effect, the authors had developed the "Occupational Health Monitoring and Evaluation System" (OHMES). The stated objective of the project was "to evaluate whether knowledge of a person's isometric strength can predict the risk of later injury and illness when the person is placed on jobs having various degrees of manual materials handling." The authors tested 551 industrial subjects from six different plants. They reported a mean arm-lift strength for males (443) 85.8 lb (SD 28.6 lb) and for females (108) mean arm-lift strength 44.9 lb (SD 17.6 lb). Arm-lift strength was roughly normally distributed. They found that the arm-lift strength was controlled by the body weight of the subjects, which was increased by the height and deceased by the age. Yates et al. (1980) reported static lifting strength of the elbow and shoulder. They studied nine male and nine female subjects using a spring-loaded Stoelting grip dynamometer secured to the base of the strength-testing device. The dynamometer was connected with the subject by means of a cable and nonstretchable belt. The force generated electrical output, which was displayed digitally and recorded. Maximum isometric lifting strength was

TABLE 14.18

Arm Lift Isometric Strength Reported by Yates et al. (1980)

Vertical Height (cm)	Horizontal Distance					
	18 cm		36 cm		51 cm	
	Male	Female	Male	Female	Male	Female
Male 190	177 (59)		169 (44)		122 (26)	
Female 170		66 (34)		54 (23)		41 (17)
134 cm	293 (77)	104 (58)	274 (63)	95 (40)	203 (39)	62 (27)
81 cm	553 (126)	217 (94)	489 (123)	173 (61)	251 (30)	118 (45)

Reported by Yates et al. 1980, *Ergonomics*, 23(1), 37–47.

From Kumar et al. 1988, *J. Biomech.*, 21, 1, 35–44.

measured using a metal tray 49.6 cm long, 39.5 cm wide, and 12.2 cm deep. The tray was connected in series, through its center, with a chain and cable to the dynamometer. The strength was measured in 12 positions defined by three horizontal (18, 36, 51 cm) and four vertical positions (33, 81, 134, and 190 cm) for men. The highest point for women was reduced to 172 cm. Other than the vertical height of 33 cm, the nine locations were arm lift at knuckle, elbow, and just above head heights. The data obtained is presented in Table 14.18. It was interesting to note that in the higher two heights female values were 33% of males, and at the lower height (81 cm) it represented 50% of the male sample. The authors emphasized the difficulties in deriving a general predictor of lifting capacity. Not only did the predictors vary between genders, but also substantial variation occurred from position to position.

In an isometric mode of arm-lift strength, Mamansari and Salokhe (1996) reported it to be 420 N for males and 280 N for females. The effect of posture on strength-generating capability is well established (Caldwell, 1959; Kumar, 1991b; Kumar and Garand, 1992; Kumar et al., 1991; Vink et al., 1992). Keeping this in mind, Mital and Faard (1990) studied the effect of angle of the stronger arm relative to the frontal plane. They measured the arm strength in coronal plane (0°) and 30°, 60°, 90°, 120°, and 150° of displacement in horizontal plane to the contralateral side. They found that as the arm moved to sagittal plane (90°) the strength continually increased, and beyond that point it decreased.

In a series of studies, Kumar (1987, 1991b, 1994a) and Kumar et al. (1988) reported their findings. Kumar et al. (1988) and Kumar (1991b) designed and fabricated two devices for isometric and isokinetic lifting strength measurements, arm lift among them. The details of these devices are presented in the respective publications. Kumar et al. (1988) tested 10 young males and 10 young females. None of the subjects had back pain or any musculoskeletal disorders. These 20 subjects performed a maximal isometric arm lift at their knuckle height in an upright posture and isokinetic arm lifts that started at knuckle height, and data acquisition was terminated at the shoulder height. The isokinetic lifts were performed at 20, 60, and 100 cm per second handle velocity in sagittal plane. The peak isometric and isokinetic strengths are shown in Figure 14.18. The peak static strength was invariably higher than all isokinetic strengths ($p < .01$). The peak isokinetic strength kept on decreasing with the increasing velocity of lift and occurred at progressively higher vertical displacement and later during the lift cycle for both genders (Table 14.19). The strength values of females were significantly lower than those of males. After reaching its peak, the static strength stayed high throughout the five-second test period. Dynamic strength, on the other hand, gradually reached peak, with the time of the peak progressively shifting away from the start with increasing speed of activity. The factors that influenced the arm lift are presented in a summary ANOVA table (Table 14.20).

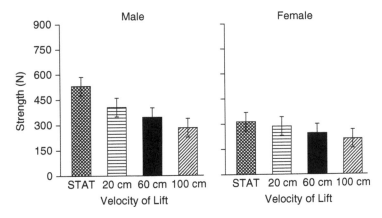

FIGURE 14.18
Isometric and isokinetic arm-lift strength of male and female subjects. The isokinetic lifts were performed at 20, 60, and 100 cm per second. (From Kumar, et al. 1988, *J. Biomech.*, 21, 1, 35–44.)

TABLE 14.19

Mean Peak Dynamic Arm-Lift Strengths, Their Corresponding Vertical Hand Location, Activity Duration, and Percent of Activity

			Duration at Which Peak Strength Occurred			
Gender		Tasks	Mean Peak Str. (N)	Mean Vertical Location (cm)	Mean Activity Duration (s)	Mean Time of Peak Occurrence (% Activity Duration)
Males	Arm	stat.	521	82.7	4.5	58
		slow	399	95.1	3.2	17
		md	332	116.1	0.9	49
		fast	275	114.0	0.7	49
Females	Arm	stat.	296	76.2	4.5	60
		slow	266	89.1	3.4	17
		md	221	95.3	1.4	34
		fast	192	103.7	1.2	53

From Kumar et al. 1988, *J. Biomech.*, 21, 1, 35–44.

TABLE 14.20

Summary Table of Analysis of Variance Showing the Factors Affecting Arm Lift Strength

		Arm Lift	
Source	DF	F ratio	p <
Speed	3	36.06	.00
Gender	1	77.64	.00
Speed-gen.	3	5.02	.00
Percentile	10	53.57	.00
Percentile speed	30	9.73	.00
Percentile gender	10	4.34	.00
Percentile speed-gen.	30	0.52	.83

From Kumar et al. 1988, *J. Biomech.*, 21, 1, 35–44.

TABLE 14.21

Mean Peak Strength during Arm Lift (N)

Mode	Sex	Stat	Sagittal Plane			30° Lateral Plane			60° Lateral Plane		
			0·5R	0·75R	1R	0·5R	0·75R	1R	0·5R	0·75R	1R
Isometric	Male	X	590	295	184	483	274	165	385	206	130
	N = 20	SD	140	67	34	148	66	37	101	55	32
		Min	331	199	116	318	165	112	228	134	87
		MAX	823	420	256	838	389	248	604	134	222
	Female	X	261	182	118	249	160	105	204	130	88
	N = 18	SD	87	63	33	100	45	33	75	40	23
		Min	99	97	71	101	94	63	109	73	60
		MAX	481	359	207	521	287	191	366	250	146
Isokinetic	Male	X	311	225	162	284	214	156	258	171	129
	N = 20	SD	87	52	28	69	58	31	69	32	29
		Min	197	146	13	168	133	108	162	116	89
		Max	499	391	217	465	394	245	479	226	201
	Female	X	184	139	101	177	145	98	160	122	81
	N = 18	SD	39	33	26	44	29	27	40	31	22
		Min	123	95	69	112	85	65	103	80	55
		Max	249	208	165	256	189	176	253	186	129

Male standard posture	Female standard posture	Stat = Statistical variable
X 331	X 170	R = Reach
SD 85	SD 44	X = Mean
Min 195	Min 68	SD = Standard Deviation
Max 452	Max 249	

From Kumar, 1991b, *Appl. Ergon.*, 22(5), 317–328.

Kumar (1991b) conducted a study to determine arm strength values (contour) for isometric and isokinetic efforts around the human trunk. Thirty-eight normal young adults (20 male, 18 female) performed a total of 19 tasks. These tasks consisted of one self-selected optimum posture with upright stance and elbows flexed at 90°, designated as standard posture for isometric test. In addition, isometric strength was done in sagittally symmetrical, 30° lateral, and 60° lateral planes at half reach, three-quarters reach, and full reach of the distances at knuckle height. The isokinetic tests were done between knuckle height and shoulder height in postures identical to isometric strength testing. The sequence of these tasks was randomized. The results for male and female samples are presented in Table 14.21 and Table 14.22 and in Figure 14.19 and Figure 14.20. The peak strength in standard posture was invariably lower than the peak strength at half reach in isometric condition in all three planes in both sexes, with the exception of one condition among females (60° lateral plan, half-reach isometric). The peak and average arm-lift strengths of males were significantly higher than those of the females ($p < .01$) and ranged between 44 and 71%. The arm-lift strength values normalized against standard posture, peak isometric, and peak isokinetic strengths are presented in Table 14.23. For both sexes, the isometric strength was significantly higher than the isokinetic strength ($p < .01$). The peak and average strengths in the sagittal plane were invariably higher than those of asymmetric postures, with one exception among females. With increasing reach distance, the strength declined significantly for all conditions among both genders ($p < .01$). The peak and average arm-lift strengths of females as percentages of males are presented in Table 14.24. The ANOVA showed that the gender and mode of lifting, in addition to affecting peak and average strength individually ($p < .01$), had significant two-way and three-way interactions ($p < .01$). All strength values were intercorrelated ($p < .010$). The developed regression equations were significant ($p < .01$) for predicting peak and average strengths based

TABLE 14.22

Mean Average Strength during Arm Lift (N)

Mode	Sex	Stat	Sagittal Plane			30° Lateral Plane			60° Lateral Plane		
			0·5R	0·75R	1R	0·5R	0·75R	1R	0·5R	0·75R	1R
Isometric	Male	X	465	237	151	386	219	135	282	160	103
	N = 20	SD	109	51	26	105	50	29	68	44	26
		Min	279	166	99	235	140	962	178	68	73
		Max	705	344	202	596	309	201	418	224	175
	Female	X	210	147	92	205	128	83	166	105	72
	N = 18	SD	64	50	26	78	36	26	60	32	19
		Min	77	77	53	85	70	53	85	58	48
		Max	355	281	158	399	221	148	287	193	118
Isokinetic	Male	X	236	180	137	219	174	130	200	140	110
	N = 20	SD	61	40	24	59	45	28	51	27	30
		Min	153	118	92	135	110	93	142	91	63
		Max	377	283	183	366	299	222	352	175	182
	Female	X	141	109	81	137	114	80	126	98	62
	N = 18	SD	25	29	22	28	22	24	28	26	19
		Min	100	70	54	93	65	49	83	64	38
		Max	176	172	141	200	161	152	188	156	110

Male standard posture	Female standard posture	Stat = Statistical variable
X 331	X 170	R = Reach
SD 85	SD 44	X = Mean
Min 195	Min 68	SD = Standard Deviation
Max 452	Max 249	

From Kumar, 1991b, *Appl. Ergon.*, 22(5), 317–328.

on anthropometric characteristics and sagittal plane strengths. They accounted for between 63 and 89% of all variance (Table 14.25).

14.5.1.2 Stoop and Squat Lifts

Stoop and squat lifts have been studied frequently and by many authors. Since the interest in lifts arises as a result of the association between lifting and low back pain/injury, the concern has been which of the two methods is superior in terms of being less hazardous. The hazard has most frequently been associated in the literature with the compressive load, which may precipitate the low back pain/injury. Lumbosacral compression is a universally accepted mode of failure of the intervertebral disk. Yet in the vast majority of cases, we all have exerted our maximal voluntary contraction without any consequence to the disk. Further it is also widely believed that 90% of the low back cases are not due to nerve root compression resulting in radiculopathy. Even though only about 10% of the low back problems are discogenic, it appears that 100% of the scientific argument is focused on the disk as the source of the problem. The entire biomechanical modeling effort is based on this premise. Despite vigorous activity and significant sophistication in modeling efforts, we are no closer to solving the problem than we were half a century ago, when everything started. Could we be barking up the wrong tree? Clearly we cannot generate forces greater than our maximal strength, and is it not a common observation that people hurt themselves when they indulge in maximal activity? By comparison, it is commonly known anecdotally that people get hurt doing the job that they have always done. How can this observation and the argument of compressive load as the injury-precipitating hazard be reconciled? Perhaps we need to return to the proverbial drawing board.

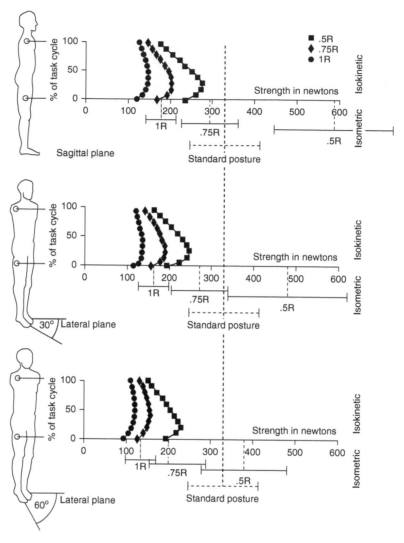

FIGURE 14.19
Isometric arm-lift strength in standard posture and at half, three-quarters, and full reach distances; and isokinetic strength contours at three reach distances of male subjects in sagittal, 30°, and 60° lateral planes. (From Kumar, 1991b, *Appl. Ergon.*, 22(5), 317–328.)

In any case, the compression hazard argument may be described as follows. A load acting on a moment arm, when lifted, has to be overcome by the contraction of back muscles (primarily erectores spinae group), which act on a very short lever arm. Thus, the force to be generated by the back muscles is multiplied by the ratio of the two lever arms. The erector spinae lever arm being 6 cm (Kumar, 1988), if the load is lifted at 30 cm distance from the fulcrum (lumbosacral disk) the muscles will have to generate a force and a lumbosacral compression 5 times greater than the load. While this is the genesis of all biomechanical back modeling efforts, with the incorporation of multiple muscles and asymmetry of activities the models have advanced significantly. In the absence of any other convincing and appropriate variable, it has served as an objective tool to compare different circumstances. Using this logic, Park and Chaffin (1974) compared stoop and squat lifts for their lumbosacral compressive load in an identical task. A load of 15.5 kg when held 38 cm away from ankle at a height of 38 cm generated a compressive load of

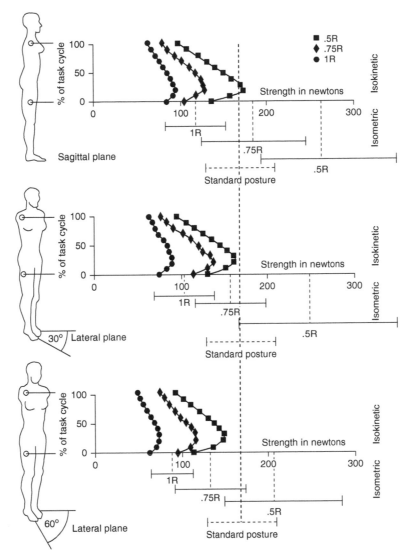

FIGURE 14.20
Isometric arm-lift strength in standard posture and at half, three-quarters, and full reach distances; and at three reach distances; and isokinetic strength contours of female subjects in sagittal, 30°, and 60° lateral planes. (From Kumar 1991b, *Appl. Ergon.*, 22(5), 317–328.)

183.3 kg at the lumbosacral disk for stoop lift, and 278.5 kg for squat lift. This was due to a longer lever arm of the load in squat posture (stoop lift 35 cm, squat lift 50.9 cm), because the load could not be brought between the knees. Given such a possible mechanical disadvantage, squat lifts are considered more hazardous because people are able to generate more force in this posture.

One of the first studies reporting stoop and squat lift strengths was by Chaffin et al. (1977) and was followed by an updated Chaffin et al. (1978). The measurements of these strengths were done with an objective of using them in screening workers during preemployment strength testing. The authors reported, from 551 industrial subjects drawn from six different plants, that the mean stoop-lift strength for males was 122.4 lbs (54.8) and for females 59.9 lbs (31). By comparison, in squat lifts the males produced a mean force of 211.8 lbs (76.5) and the females a mean force of 93.8 (44.4). The squat-lift strength values

TABLE 14.23

Peak Arm-Lift Strength Normalized against the Standard Posture(SP), Peak Isometric Strength (PIMS) and Peak Isokinetic Strength (PIKS)

Mode	Sex	NF	Sagittal Plane			30° Lateral Plane			60° Lateral Plane		
			0·5R	0·75R	1R	0·5R	0·75R	1R	0·5R	0·75R	1R
Isometric	Male	SP	165	85	52	132	79	48	97	57	38
		PIMS	100	52	31	79	48	29	58	34	23
		PIKS	178	92	56	141	85	52	104	61	39
	Female	SP	157	107	66	157	90	62	124	76	53
		PIMS	100	68	42	100	57	39	79	48	34
		PIKS	143	97	60	142	82	56	113	69	49
Isokinetic	Male	SP	93	68	46	84	63	45	78	48	39
		PIMS	56	41	28	51	38	27	47	29	24
		PIKS	100	73	50	91	68	48	83	52	42
	Female	SP	107	85	62	110	87	62	96	79	48
		PIMS	68	54	39	69	55	39	61	50	31
		PIKS	97	77	57	100	78	56	87	72	44

Note: NF = Normalizing factor; R = Reach.

From Kumar, 1991b, *Appl. Ergon.*, 22(5), 317–328.

TABLE 14.24

The Peak and Average Arm-Lift Strengths of Females as Percentage of Males

Strength Category	Test Conditions								
	Sagittal Plane			30° Lateral Plane			60° Lateral Plane		
	0·5R	0.75R	1R	0.5R	0.75R	1R	0.5R	0.75R	1R
Peak isometric	44	61	64	51	58	63	53	63	67
Average isometric	45	62	61	53	58	61	59	65	70
Peak isokinetic	59	61	62	62	67	63	62	71	63
Average isokinetic	60	60	59	62	65	61	63	70	56

Note: R = Reach.

From Kumar, 1991b, *Appl. Ergon.*, 22(5), 317–328.

were significantly higher than that of the stoop lift. Pytel and Kamon (1981) carrying on a similar objective, reported a study where they measured stoop and squat dynamic lifting strength of 10 male and 10 female subjects along with maximal acceptable load for lifting. They used a minigym for their study. Their findings are presented in Table 14.26. They also observed that the peak dynamic strength closely resembled the actual maximum acceptable lifts. On this basis, they suggested that such measured dynamic strength could be used as a predictor for lifting capacity. However, it appears that the concurrence of maximum dynamic strength and maximum acceptable lift indicates that the duration of shift may not have been given the consideration it deserves.

The first results from Cybex were reported by Mayer et al. (1985) among normal as well as low back pain patients. Keeping with the theme of the book, the clinical data will be omitted. They reported results from 125 normal subjects (62 male and 63 female). These subjects were placed into the Cybex prototype sagittal trunk strength tester and appropriately stabilized. Extension and flexion stops were set. The subjects exerted for flexion or extension, as the case was, at 30, 60, 90, and 120°/s isokinetic speeds. The authors reported that the ratio of extension to flexion was invariably greater than one, and usually in the range of 1.15 to 1.3 for both genders. However, these strength values were not for

TABLE 14.25

Linear Regression Equations for the Peak Arm-Lift Strengths in Working Space ($p < .001$)

No.	Strength	Equation	MR	R²
1.	Met-0·5R-SP	= 217·6 + 0·73 (SP Peak) + 116·5 (sex) = 0·84 (Met-0·75R-SP Peak)	0.94	0.889
2.	Met-0·75R-SP	= 1.42 + 0.23 (SP Peak) 0.67 (weight) − 0.75 (reach) + 1.2 (Met-1R-SP Peak)	0.94	0.884
3.	Met-1R-SP	= −50.2 + 0.044 (SP Peak) + 0.23 (weight) − 3.3 (R)+ 1.67 (height) + 0.40 (Met- 0·75R-SP Peak)	0.94	0.881
4.	Met-0·5R-LP1	= −115.66 + 0.51 (SP Peak) + 0.71 (Met-0·75R-SP Peak) + 1.4 (Kin-1R-SP)	0.93	0.874
5.	Met-0·75R-LP1	= 133.4 + 0.16 (SP Peak) + 1.85 (weight − 4.02 (R) + 1.52 (age) + 0.37 (Met-0·75R-SP Peak) + (0.19 Kin-0·5R-SP Peak)	0.94	0.878
6.	Met-1R-LP1	= 142.7 + 0.01 (SP Peak) + 1.52 (weight) − 3.2 (R)+ 0.37 (Met-0·75R-SP Peak)	0.90	0.828
7.	Met-0·5R-LP2	= −31.3 + 0.56 (SP Peak) + 1.12 (Met-1R-SP Peak)	0.911	0.829
8.	Met-0·75R-LP2	= 12.53 + 0.101 (SP Peak) + 0.55 (Met-0·75R-SP Peak)	0.91	0.837
9.	Met-1R-LP2	= 114.67 + 2.1 (weight) − 2.8 (R) − 0.07 (SP Peak) − 112 (Met-0·5R-SP Peak)	0.85	0.728
10.	Kin-0·5R-SP	= 21.5 + 0.39 (SP Peak) + 0.7 (Kin-0·5R-SP Peak)	0.85	0.734
11.	Kin-0·75R-SP	= 6.55 + 0.04 (SP Peak) + 0.18 (Met-1R-SP Peak) + 0.62 (Kin-1R-SP Peak) + 0.24 (Kin-0·5R-SP Peak)	0.88	0.784
12.	Kin-1R-SP	= 44.8 + 0.08 (SP Peak) + 0.27 (weight) + 1.4 (age) − 24.1 (sex) + 0.18 (Met-0·75R-SP Peak) + 0.27	0.93	0.870
13.	Kin-0·5R-LP1	= 14.4 + 0.05 (SP Peak) + 0.77 (Kin-0·75R-SP Peak) + 0.25 (Kin-0·5R-SP Peak)	0.91	0.837
14.	Kin-0·75R-LP1	= 40.75 + 0.101 (SP Peak) + 0.63 (Kin-0·75R-SP Peak)	0.81	0.670
15.	Kin-1R-LP1	= 58.8 + 0.004 (SP Peak) + 0.92 (weight) = 1.42 (R) + 0.19 (Met-0·75R-SP Peak) + 0.37 (Kin-1R-SP Peak)	0.88	0.790
16.	Kin-0·5R-LP2	= −4.07 + 0.11 (SP Peak) + 0.30 (Met-1R-SP Peak) + 0.44 (Kin-0·75R-SP Peak) + 0.24 (Kin-1R-SP Peak)	0.91	0.835
17.	Kin-0·75R-LP2	= 159.6 − 0.04 (SP Peak) + 1.15 (height) + 0.10 (Met-0·5R-SP Peak) + 0.37 (Kin-1R-SP Peak)	0.83	0.698
18.	Kin-1R-LP2	= 11.97 + 0.05 (SP Peal) + 0.29 (Met-1R-SP Peak) + 0.20 (Kin-0·75R-SP Peak)	0.81	0.670

Note: MR = Multiple regression; R² = Variance; Met = Isometric; R = Reach; SP = Standard posture; Kin = Isokinetic; LP1 = 30 Lateral Plane; LP2 = 60 Lateral Plane.

From Kumar, 1991a, University of Alberta, Edmonton.

lifting strength, and hence do not add a great deal to the discussion. Rhumann and Schmidtke (1989) reported maximum isometric forces from a large industrial sample from which data of 1245 subjects ranging in age from 35 to 65 years was reported. The results are presented in Figure 14.21. Sanchez and Grieve (1992) studied and reported the measurement and prediction of isometric lifting strength in symmetrical and asymmetrical postures. They measured the static lifting strengths of nine men and nine women at six different heights, ranging from only just above the floor to just above the head, at two horizontal reaches from the mid-ankles position in the sagittal plane and also at 45° and 90° to the right for two-handed exertions and at 45° and 90° to each side for one-handed exertions, for a total of 96 postures. A second, different group of subjects (nine male and nine females) were required to do 20 two-handed and 40 one-handed postures intermediate to those of the first group. A third group (eight males and eight females) was used to determine maximum possible reach at which lifting strength was zero at the same heights and planes as the first group. The authors expressed strength normalized to body weight and reach normalized to the stature in regression equation. The gender did not affect prediction at all. They suggested that these prediction equations could be used to develop isodyne contours for an individual in any chosen plane. The authors also found

TABLE 14.26

Means and Standard Deviations of Maximal Dynamic Lift (MDL), Maximal Acceptable Lift (MAL), and Peak Forces for Dynamic Lift Strength (DLS), Dynamic Back Extension Strength (DBES) and Dynamic Elbow Flexion Strength (DEFS) Performed at Two Speeds

		Strength (N)	
Test	Speed (Ms^{-1})	Women (*n* =10)	Men (*n* = 10)
MDL		250 ± 54	544 ± 109
MAL		55 ± 24 (22)[a]	120 ± 35 (22)
DLS	0.73	379 ± 95 (152)	601 ± 129 (110)
DBES }		315 ± 87 (140)	540 ± 101 (99)
DEFS		167 ± 33 (67)	323 ± 55 (58)
DLS	0.97	260 ± 99 (104)	398 ± 113 (73)
DBES }		210 ± 95 (84)	339 ± 102 (62)
DEFS		120 ± 38 (48)	233 ± 38.5 (43)

[a] The numbers in parentheses represent the mean strength as a percentage of MDL.

From Pytel and Kamon 1981, *Ergonomics*, 24(9), 663–672.

that some individuals did not conform to the prediction, thus limiting the usefulness of this research. Furthermore, information based on such a small number of subjects could also pose some limitation.

The next group of authors investigated lifting strengths both in isometric and isokinetic modes (Garg and Beller, 1994; Hattori et al., 1996; Hazard et al., 1992; Kumar, 1991; Kumar and Garand, 1992; Mostardi et al., 1992). Karwowski and Mital (1986) presented isometric and isokinetic strengths lifting in teamwork. They investigated activity of isometric and isokinetic strengths for teams of two and three males. Their sample consisted of six students. They tested the isometric strength evaluations for four standard measures consisting of arm, stoop, leg, and composite strength for isometric lifting strength. They used the values of the isokinetic lifting strength and isokinetic back extension strength for the isokinetic strength. They reported that the team strengths were significantly lower than the respective sums of individual strengths of team members ($p < .01$) except the leg strength for a team of two and arm strength for a team of three. They found that the isometric strengths were 94% of the sum of corresponding individual strengths for the team of two males. For the team of three males, the team isometric strength averaged only 90% of the sum of individual strengths with the exception of arm strength. The isokinetic strengths for the teams of two and three males account for only 68 and 58% of the team totals. Through the results presented here the authors emphasized that the team strengths are not additive of the members, and as such, it must be appropriately adjusted in such circumstances.

Both Mostardi et al. (1992) and Hazard et al. (1992) looked at the isokinetic lift strength from an occupational injury perspective. In a prospective study, Mostardi et al. (1992) studied 171 nurses, employed in a tertiary care facility, for their lifting strength using a Cybex Lift task machine. They also had this sample fill out an epidemiologic questionnaire. The average peak force for isokinetic lift was 63.8 kg at lift speed of 30.5 cm per second, and 59.1 kg at a lift speed of 45.7 cm per second. These nurses were followed up two years later. Job-related low back injury was reported by 16 nurses. Analysis revealed that none of the strength or epidemiological variables had any significant effect on injuries, and they did not explain any variance at all. The study concluded that in this occupation, where loads are heavy and postures are variable, prior low back strength is a poor predictor

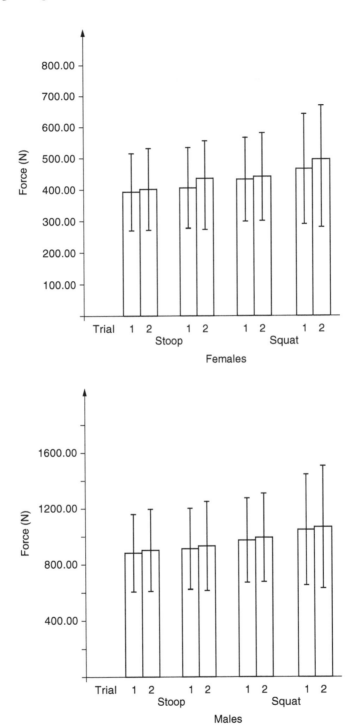

FIGURE 14.21
Stoop and squat lift strength in two consecutive trials from female and male subjects. (Modified from Ruhmann and Schmidtke, 1989, *Ergonomics*, 32, 865–879.)

of the outcome. Interestingly, in a similar population with 161 nursing aides, Kumar (1990) had reported a strong relationship between pain/injury and the cumulative load suffered by the individual. Thus, lifetime exposure played a significant role ($p < .01$). Hazard et al.

(1992) tested the validity of such strength measures, primarily on the subjects' effort during the evaluation. They used 21 men and 23 women, all asymptomatic, who underwent isometric, isokinetic, and isoinertial lift tests at their 50 and 100% effort levels. There was a significant difference in values for isokinetic force/distance curve variation, isoinertial peak force, weight ratios, and peak force–weight differences, and for heart rates in all three modes ($p < .01$). The authors reported that the only effort indices with better than 60% accuracy in identifying maximal and submaximal efforts were isokinetic force/distance curve variation (74%) and isoinertial peak force. They further reported that, for all effort indices, differences between 50 and 100% efforts were similar in magnitudes, confirming the perceptual and effort production capability reported by Kumar and Simmonds (1994), Kumar et al. (1994a) and Kumar et al. (1997).

Garg and Beller (1994) compared the isokinetic lifting strength with static strength and maximum acceptable weights for lifting. Their study was conducted to glean the effect of speed of lifting and the box size on these variables. A small sample of nine male college students lifted three different boxes, 250, 380, and 510 mm wide, from floor to bench height of 80 cm using a free-style lifting technique once every five minutes. They also measured the isometric lifting strength at the origin of the lift. Isokinetic lifting strength was measured at 41, 51, and 60 cm per second with boxes attached to the load cell. The authors reported a progressive decrease in isokinetic lifting strength with both an increase in the box width and an increase in the velocity of lift, as also reported by Kumar et al. (1988). Between these two variables, there was significantly greater effect of lifting speed than the box width on the isokinetic lifting strength. Hattori et al. (1996) investigated the effect of postural asymmetry of 45° right, 90° right, and 90° left with respect to sagittal plane in stoop and squat lifting modes on isometric and dynamic lifting. The authors reported a significant effect of posture asymmetry on isometric and dynamic peak lifting strengths, a finding consistent with Kumar (1990) and Kumar and Garand (1992).

Kumar et al. (1988) reported isometric and isokinetic stoop lifting strength in isometric and isokinetic modes at the lifting velocities of 20, 60, and 100 cm per second. They designed and fabricated a unique device for controlling the speed and guiding the path vertically upwards. The details of the device are provided in Kumar et al. (1988). The authors used 20 normal young adults (10 male, 10 female) in the study, who performed their maximal stoop lift efforts in isometric and three isokinetic velocities as stated before. The mean peak strengths of the sample are presented in Table 14.27. As reported for arm-

TABLE 14.27

Mean Peak Dynamic Strengths, Their Corresponding Vertical Hand Location, Activity Duration, and the Percent of the Activity

Gender		Tasks	Mean Peak Str. (N)	Mean Vertical Location (cm)	Mean Activity Duration (s)	Mean Time of Peak Occurrence (% Activity Duration)
					Duration At Which Peak Strength Occurred	
Males	Back	stat.	726	5.0	4.5	45
		slow	672	19.3	4.0	18
		md	639	33.9	1.3	44
		fast	597	42.4	0.8	51
Females	Back	stat.	503	5.0	4.5	41
		slow	487	15.0	4.2	19
		md	432	35.6	1.7	48
		fast	436	42.2	1.4	63

From Kumar et al. 1988, *J. Biomech.*, 21, 1, 35–44.

lift strength before, the peak isometric strength was invariably higher and significantly different from the peak isokinetic strength ($p < .01$). The peak dynamic strength decreased progressively as the lifting velocity increased and occurred at progressively higher vertical displacement and later during the lift cycle. The peak strengths of females were significantly lower than those of males, ranging from 69 to 73% for the corresponding activities ($p < .01$). Subsequently, Kumar (1990) and Kumar and Garand (1992) presented isometric and isokinetic lifting strength at different reach distances in symmetrical and asymmetrical planes to develop strength contours in the human work space. Thirty normal and healthy young adults (18 male, 12 female) were required to stoop and squat lift or exert against a standardized instrumented handle in relevant postures. The isokinetic lifts were done at a linear velocity of 50 cm per second of hand displacement between the floor and the knuckle heights. These isokinetic lifts were done on a specially designed and fabricated static and dynamic strength tester that did not constrain the path of the lift (Figure 14.8 and Figure 14.9). The isometric stoop lifting strengths were exerted in two standardized postures with 60° and 90° hip flexion. The isometric squat lifting efforts were also exerted in two standardized postures of 90° and 135° of knee flexion. All lifts were performed at half, three-quarters, and full horizontal reach distances on sagittally symmetrical, 30° left lateral and 60° left lateral planes. The isometric stoop and squat lifts were also measured in self selected optimum postures. These 56 experimental conditions were tested in random order. The mean peak and average lifting strength for stoop and squat lift efforts are presented in Table 14.28 through Table 14.31, respectively. The peak isometric strength recorded in self-select optimum posture, as proposed by Chaffin et al. (1978), was lower than the isometric strength recorded at half reach in sagittal plane at 90° hip flexion for both males and females in sagittal plane. However, this standard posture value was greater than all other strengths recorded. A similar relationship in average strength values was also observed. The peak and average standard posture strengths for squat lift for both male and females were significantly greater than all other 27 peak and 27 average strength values in both genders.

TABLE 14.28

Mean Peak Stoop-Lifting Strength (N)

Mode	Sex	Stats	Sagittal Plane			30° Lateral Plane			60° Lateral Plane		
			0·5R	0·75R	1R	0·5R	0·75R	1R	0·5R	0·75R	1R
Isometric	Male	Mean	647	342	206	508	313	202	440	227	151
Hip = 60	Female	SD	177	64	89	147	97	21	108	59	25
		Mean	319	216	139	277	218	141	230	161	106
		SD	33	46	25	39	42	41	65	28	28
Isometric	Male	Mean	697	370	208	580	318	194	381	230	142
Hip = 90	Female	SD	238	105	47	185	110	71	186	63	48
		Mean	323	254	155	321	222	155	224	162	128
		SD	51	48	11	64	45	25	40	26	35
Isokinetic	Male	Mean	542	294	233	505	299	207	334	233	168
	Female	SD	215	93	47	184	118	72	130	82	47
		Mean	296	250	202	272	237	162	243	174	135
		SD	121	104	78	122	84	56	103	70	39

Male Standard Posture
Mean 668
SD 273

Female Standard Posture
Mean 317
SD 126

Note: R = Reach, SD = Standard Deviation.

From Kumar and Garand, 1992, *Ergonomics*, 35, 35–44.

TABLE 14.29

Mean Average Stoop-Lifting Strength (N)

Mode	Sex	Stats	Sagittal Plane			30° Lateral Plane			60° Lateral Plane		
			0·5R	0·75R	1R	0·5R	0·75R	1R	0·5R	0·75R	1R
Isometric	Male	Mean	517	272	168	396	253	158	331	180	121
Hip = 60	Female	SD	191	85	40	154	89	57	139	52	38
		Mean	264	181	114	231	181	113	190	131	84
		SD	34	40	19	38	37	31	54	26	17
Isometric	Male	Mean	560	287	164	460	255	149	303	183	113
Hip = 90	Female	SD	176	70	35	151	91	46	99	63	37
		Mean	273	206	126	266	185	125	180	124	99
		SD	46	39	11	52	34	14	27	19	23
Isokinetic	Male	Mean	359	192	142	328	190	139	205	138	105
	Female	SD	136	43	36	131	60	22	79	37	16
		Mean	189	166	134	183	159	108	159	109	89
		SD	84	79	45	94	67	35	70	46	24

Male standard posture
Mean 521
SD 211

Female standard posture
Mean 259
SD 95

Note: R = Reach, SD = Standard Deviation.

From Kumar and Garand, 1992, _Ergonomics_, 35, 35–44.

TABLE 14.30

Mean Peak Squat Lift Strength (N)

Mode	Sex	Stats	Sagittal Plane			30 Lateral Plane			60 Lateral Plane		
			0·5R	0·75R	1R	0·5R	0·75R	1R	0·5R	0·75R	1R
Isometric	Male	Mean	384	290	199	361	273	185	323	233	162
Hip = 60	Female	SD	97	90	79	99	78	54	75	87	42
		Mean	175	137	93	186	129	88	141	114	72
		SD	59	34	18	60	40	15	41	38	14
Isometric	Male	Mean	409	281	187	404	262	188	311	226	164
Hip = 90	Female	SD	117	78	43	150	106	63	100	91	49
		Mean	185	117	91	182	114	81	147	110	86
		SD	63	33	18	44	28	13	40	26	18
Isokinetic	Male	Mean	411	320	215	416	269	215	296	236	182
	Female	SD	202	74	33	193	77	37	76	90	45
		Mean	232	184	155	225	176	155	186	143	111
		SD	91	69	31	59	49	23	36	43	42

Male standard posture
Mean 79
SD 264

Female standard posture
Mean 324
SD 141

Note: R = Reach, SD = Standard Deviation.

From Kumar and Garand, 1992, _Ergonomics_, 35, 35–44.

Across one isokinetic and two isometric efforts for stoop lifts at half reach, the females represented 46 to 54% of male strength capability. The female strengths as percent of their male counterparts are presented in Table 14.32. The strength contour of the isokinetic stoop and squat strengths in the workplace are presented in Figure 14.22 through 14.25. It was found that the variables of gender, mode of lifting (isometric — two postures, isokinetic stoop and squat) the plane of lifting (sagittal, 30° lateral, 60° lateral), and the reach at which the lifts were performed were all highly significant in affecting both peak and average strengths ($p <$ .01). Thus in any injury control or job design efforts, all these variables must be taken into

TABLE 14.31

Mean Average Squat Lift Strength (N)

Mode	Sex	Stats	Sagittal Plane 0·5R	0·75R	1R	30° Lateral Plane 0·5R	0·75R	1R	60° Lateral Plane 0·5R	0·75R	1R
Isometric	Male	Mean	303	236	167	286	218	150	252	183	127
Hip = 60	Female	SD	86	77	76	82	65	47	64	75	39
		Mean	142	114	78	152	103	70	110	89	56
		SD	48	26	17	49	27	11	36	31	12
Isometric	Male	Mean	320	225	154	325	215	157	225	185	130
Hip = 90	Female	SD	91	68	43	121	87	58	88	78	38
		Mean	150	97	77	151	95	65	117	88	68
		SD	56	27	16	37	25	13	33	21	12
Isokinetic	Male	Mean	273	228	153	279	187	154	197	163	120
	Female	SD	152	54	30	161	61	32	53	63	32
		Mean	148	125	107	145	116	108	121	100	76
		SD	68	54	34	52	39	24	28	37	30

Male standard posture
Mean 541
SD 217

Female standard posture
Mean 259
SD 107

Note: R = Reach, SD = Standard Deviation.

From Kumar and Garand, 1992, *Ergonomics*, 35, 35–44.

TABLE 14.32

The Peak and Average Stoop and Squat Lift Strength of Females as Percentage of the Corresponding Values for Males

Lift	Strength Category	Test Conditions Sagittal Plane 0·5R	0·75R	1R	30° Lateral Plane 0·5R	0·75R	1R	60° Lateral Plane 0·5R	0·75R	1R
Stoop	Peak isometric hip = 60	49	63	67	54	69	69	52	70	70
	Avg. isometric hip = 60	51	66	67	58	71	71	57	72	69
	Peak isometric hip = 90	46	68	74	55	69	79	58	70	90
	Avg. isometric hip = 90	48	71	76	57	72	83	59	67	87
	Peak isokinetic	54	86	86	53	79	78	72	74	80
	Avg. isokinetic	52	86	94	55	83	77	77	78	84
Squat	Peak isometric hip = 60	45	47	46	51	47	47	43	48	44
	Avg. isometric hip = 60	46	48	46	53	47	46	43	48	44
	Peak isometric hip = 90	45	41	48	45	43	43	47	48	52
	Avg. isometric hip = 90	46	43	50	46	44	41	45	47	52
	Peak isokinetic	56	57	72	54	65	72	62	60	60
	Avg. isokinetic	54	54	69	51	62	70	61	61	63

Note: Stoop standard posture: peak = 4, average = 49; squat standard posture: peak = 45, average = 47.

From Kumar and Garand, 1992, *Ergonomics*, 35, 35–44.

account. It is therefore only appropriate to design for appropriate spatial segments. If one compared the standard posture static strength with the dynamic strength (Figure 14.22 through 14.25), one could see the large magnitude of differences. In cases of inappropriate matching of workers, they could be subjected to loads greater than or closer to their limits. Even a small mismatch can accumulate over a number of years to make a difference from innocuous one-time stress to a significant cumulative load hazard (Kumar, 1990). Based on the foregoing data, Kumar (1995) developed predictive equations for strengths through multiple regression to predict strengths as a function of reach, posture, and velocity of lift. The combinations of factors are presented in Figure 14.26 and Figure 14.27. All regression

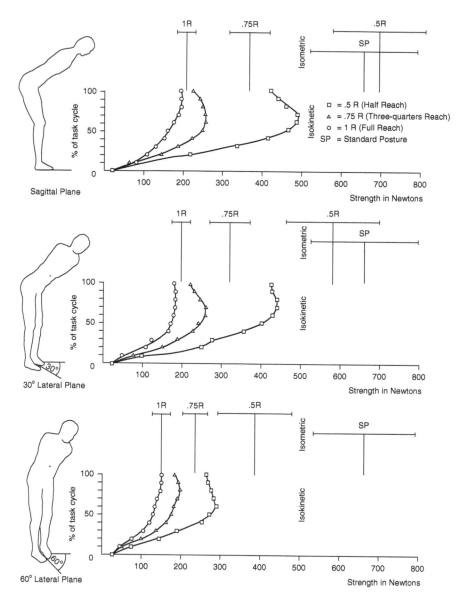

FIGURE 14.22
Isometric stoop lift strength in standard posture and at half, three-quarters, and full reach distances, and isokinetic strength contours at three reach distances of male subjects in sagittal, 30°, and 60°, lateral planes. (From Kumar and Garand, 1992, *Ergonomics*, 35, 35–44.)

equations (108) were significant ($p < .01$), and more than 70% of variance in lifting strength was accounted for by anthropometric variables and sagittal plane strength values (Table 14.33 through Table 14.36). Such an established relationship allows one to predict human lifting strength capabilities for industrial application based on simple anthropometric and strength characteristics.

14.5.1.3 *Isoinertial Lifts*

Isoinertial technique was proposed by Kroemer (1983), as described in the first chapter. It endeavors to determine the maximum amount of weight that a person can lift through

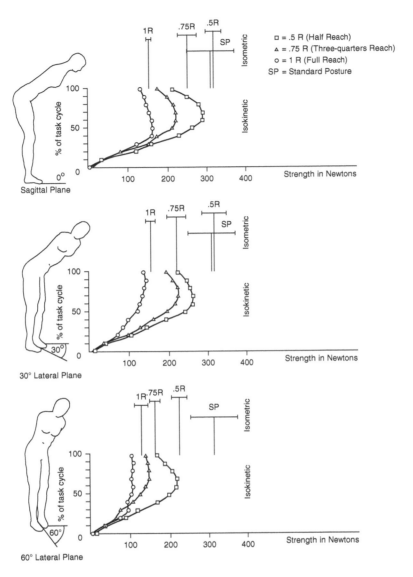

FIGURE 14.23
Isometric stoop lift strength in standard posture and at half, three-quarters, and full reach distances, and isokinetic strength contours at three reach distances of female subjects in sagittal, 30°, and 60° lateral planes. (From Kumar and Garand, 1992, *Ergonomics*, 35, 35–44.)

a range of motion. There is a specific regime associated with this methodology, which involves discrete incremental weight adjustment, the details of which are described in Chapter 8. It has been reported by Stevenson (1990) that this technique is quite sensitive to the protocol employed. He reported that changes in protocol contrasts significantly affected scores of lifting tasks. The author found that, as the subjects changed their style of lifting from set style to free style to ergonomically designed protocols, their lifting capacity changed from 37.8 to up to 58.9 kg, respectively, for males, and from 19.6 to 33.2 to 41.8 kg for females. This is really no different than what has been reported regarding stoop and squat lifts (Kumar and Garand, 1992). However, for some reason this technique has been used for industrial and return-to-work protocols. Mayer et al. (1988) developed a test based on this technique and called it Progressive Isoinertial Lifting Evaluation (PILE).

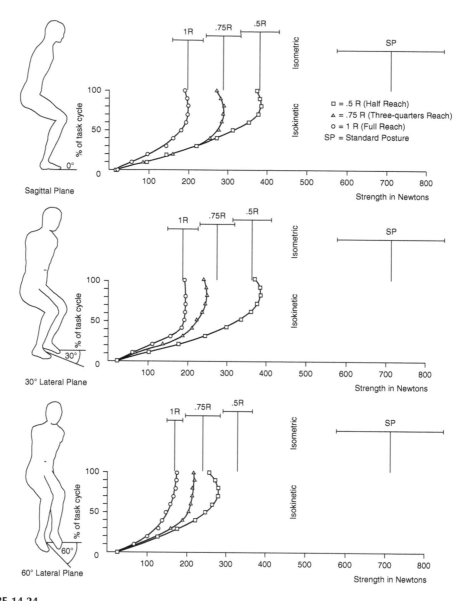

FIGURE 14.24
Isometric squat lift strength in standard posture and at half, three-quarters, and full reach distances, and isokinetic strength contours at three reach distances of male subjects in sagittal, 30°, and 60° lateral planes. (From Kumar and Garand, 1992, *Ergonomics*, 35, 35–44.)

They had male and female subjects lifting weight boxes, [the weight of which was progressively increased in fixed steps (males 10 lbs, females 5 lbs)], from floor to waist level or from waist to shoulder level at a rate of one lift every 5 seconds for a period of 20 seconds before the next increment. The test was terminated when the first of the following three conditions was achieved: (a) psychophysical end-point, a voluntary termination of test by the subject due to fatigue, discomfort, or inability to complete; (b) aerobic end-point, when 85% of age-determined maximum heart rate was achieved; or (c) safety end-point, a predetermined anthropometric safe limit, roughly 55 to 60% of body weight was reached. The authors collated the following results: (a) maximum weight lifted, (b) endurance time to discontinuation, (c) the final and target heart rate, (d) total work, and

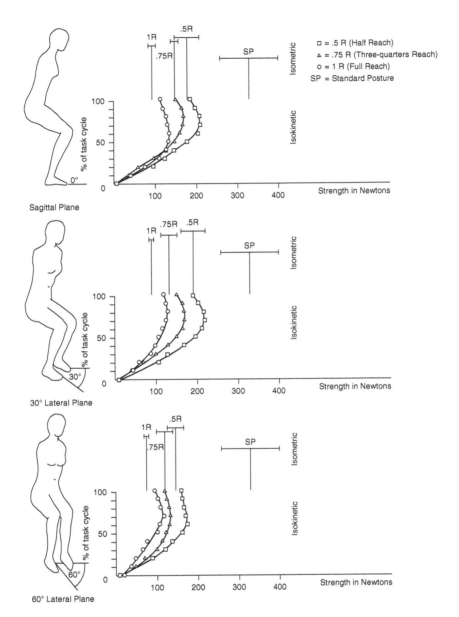

FIGURE 14.25
Isometric squat lift strength in standard posture and at half, three-quarters, and full reach distances, and isokinetic strength contours at 3-reach distances of female subjects in sagittal, 30°, and 60° lateral planes. (From Kumar and Garand, 1992, *Ergonomics*, 35, 35–44.)

(e) power consumption. The authors suggested that, since distance and endurance time were known, calculations of work and power consumption could be made and normalized against anthropometric factors. The authors suggested that this could be an effective screening tool for lifting capacity of workers.

14.5.1.4 *Psychophysical Strength*

Psychophysical measurements are about measuring physical variables (in this case, strength) conditioned by psychological perception and cognition. In case of strength, the subject will initially perceive the magnitude or intensity of work and then determine the

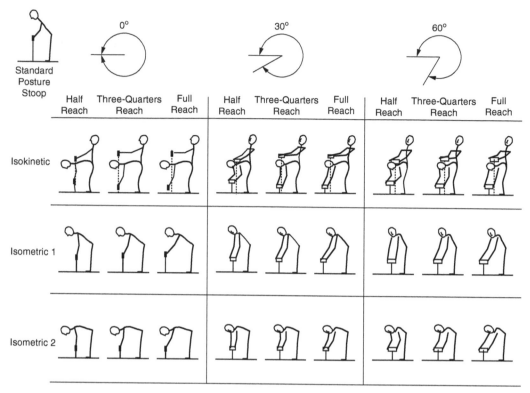

FIGURE 14.26
Pictorial depiction of postural variables in stoop lift. (From Kumar, 1995.)

comparative level in respect to the cognitive reference points provided before. This methodology derives its theoretical base from Stevens (1960), who stated that the sensation is a power function of stimulus and that it follows a mathematical relationship. Based on this principle, one can set up the desired parameters for worker perception, and vary one of the multiple variables of the task, and record the observation provided by the workers. It is for this reason that the instruction provided to the worker is the critical component of the protocol, and its consistency is extremely important. Since the psychophysical determination requires little resources, it is relatively easy to launch an investigation. Generally, people have some boxes and lead shots for lifting, after choosing the task variables for the experiment. Since the task variables are many, this field has been a fertile ground for research results. Studies have been presented since the 1950s. For example, both Emanuel et al. (1956) and Switzer (1962) reported that the maximum acceptable weight for lifting decreased with an increase in height to which the weights were lifted. However, considerable attention has been paid to workers' capacity for lifting for the duration of the shift or the entire day in order to set lifting standards for injury control.

Snook (1978) integrated the results of seven studies he conducted in his Annual Society Lecture for the Ergonomics Society, U.K. These results are presented in Table 14.37 for male and female industrial workers, respectively. These values are based on task performance, with no postural restriction or technique specifications. The instruction to subjects was to provide their maximal effort that they could sustain through the day, day after day, without getting overheated, excessively tired, or interfering with their family and social lives. However, they were informed that they would be paid based on the amount of weight they lift. What is not known is whether the workers are capable of exercising the objective balance between safety and income. As a matter of fact, Kumar and Mital

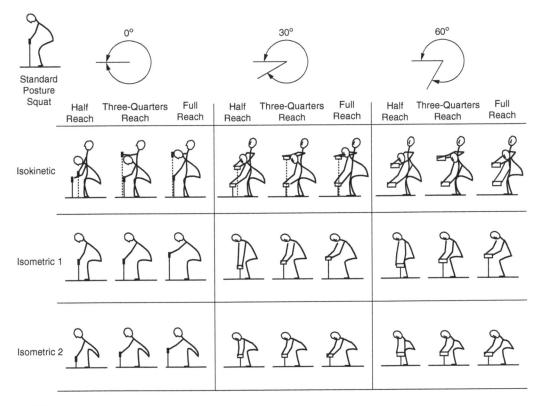

FIGURE 14.27
Pictorial depiction of postural variables in squat lift. (From Kumar, 1996d.)

(1992) indicated the possibility of psychological conditioning of the worker's own values and considerations, thus delinking, to a variable extent, the concern for safety and income. To make matters worse, income is immediate, and safety may be nebulous and hard to gauge. The latter concern also arises from the study of Kumar et al. (1994), where the industrial subjects were consistently shown to underestimate their lifting efforts, exerting harder than they needed to. However, these psychophysical tables are widely used by industries.

Ayoub et al. (1978) presented their results of the study that determined the lifting capacities of male and female industrial workers. The authors studied 73 male and 73 female subjects while they lifted weights at six different height levels at the frequencies of 2, 4, 6, and 8 lifts per minute. The height levels used were: (a) floor to knuckle, (b) floor to shoulder, (c) floor to vertical arm reach, (d) knuckle to shoulder, (e) knuckle to vertical arm reach, and (f) shoulder to vertical arm reach. These lifts were performed at three different horizontal reach distances: (a) 30 cm, (b) 38 cm, and (c) 46 cm. The results published by Ayoub et al. (1978) are presented in Table 14.38. Mital (1984) determined lifting capacity for 8- and 12-hour shifts in industrial subjects. He studied 37 male and 37 female industrial subjects in the age group 18 to 61 years, who were used to determine maximum acceptable lifts for these two shift periods. Mital investigated 36 job combinations of 4 lifting frequencies (1, 4, 8, and 12 lifts per minute), three height levels to which loads were lifted (floor to knuckle, knuckle to shoulder, and shoulder to reach), and three box sizes (38.4, 45.72, and 60.96 cm long in sagittal plane). In their balanced incomplete factorial design experiment, each subject did at least nine of the 36 job combinations. Furthermore, Mital (1984) divided his subjects in three age groups: (a) under and up to

TABLE 14.33

Stoop Lift (Peaks): Regression Equations

No.	Dependent Variable	Equation	R^2	$p <$
1	Isokinetic peak strength at half reach in sagittal plane (S.P.)	–991.5 + 18.6 [Reach] + 1.4 [Isokinetic Stoop-Lift Strength At Full Reach (S.P.)]	.742	.001
2	Isokinetic peak strength at 3/4 reach in S.P.	124.8 +.7 [isokinetic stoop-lift strength at full reach (S.P.)]	.539	.001
3	Isokinetic peak strength at full reach in S.P.	–7.5 +.7 [isometric stoop-lift strength at 3/4 reach (S.P.) with hip at 60°] -.6 [isometric squat-lift strength at full reach (S.P.) with knee at 90°] +.4 [isokinetic stoop-lift strength at 3/4 reach (S.P.)]	.884	.001
4	Isokinetic peak strength at half reach in 30° lateral plane (3P.)	–713.6 + 4.9 [body height] +.6 [isokinetic stoop-lift strength at half reach (S.P.)]	.838	.001
5	Isokinetic peak strength at 3/4 reach in 3P	64.5 +.02 [isometric stoop-lift strength with hip at 60° at 3/4 reach (S.P.)] +.9 [isokinetic stoop-lift strength at full reach (S.P.)]	.741	.001
6	Isokinetic peak strength at full reach in 3P	–29.6 + 2.0 [body weight] +.3 [isokinetic stoop-lift strength at 3/4 reach (S.P.)]	.720	.001
7	Isokinetic peak strength at half reach at 60° lateral plane (6P)	–48.6 +.7 [isometric stoop-lift strength with hip at 60° at 3/4 reach (S.P.)] -.5 [isometric squat-lift strength with knee at 90° at full reach (S.P.) +.7 [isokinetic stoop-lift strength at 3/4 reach (S.P.)]	.877	.001
8	Isokinetic peak strength at 3/4 reach in 6P	–105.5 + 3.0 [body weight +.5 [isokinetic stoop-lift strength at full reach (S.P.)]	.782	.001
9	Isokinetic peak strength at full reach in 6P	–206.1 + 2.1 [body height]	.401	.01
10	Isometric peak strength at half reach with hip at 60° in S.P.	–41.9 +.3 [standard squat peak] + 1.5 [isometric stoop-lift strength with hip at 60° at 3/4 reach (S.P.)]	.920	.001
11	Isometric peak strength at 3/4 reach with hip a 60° in S.P.	35.3 +.4 [isometric stoop-lift strength with hip at 60° at half reach (S.P.)] +.3 [isokinetic stoop-lift strength at full reach (S.P.)] -.05 [standard squat peak]	.911	.001
12	Isometric peak strength at full reach with hip at 60° in S.P.	74.8 - 6.1 [gender] +.4 [isometric squat-lift strength with knee at 90° at full reach (S.P.)] +.2 [isometric stoop-lift strength with hip at 60° at 3/4 reach (S.P.)]	.891	.001
13	Isometric peak strength at half reach with hip at 60° in 3P	2.8 -.05 [standard squat peak] +.6 [isometric stoop-lift strength with hip at 60° at half reach (S.P.)] +.4 [isometric squat-lift strength with knee at 90° at 3/4 reach (S.P.)]	.944	.001
14	Isometric peak strength at 3/4 reach with hip at 60° in 3P	40.0 +.03 [standard squat peak] +.8 [isometric stoop-lift strength with hip at 60° at 3/4 reach (S.P.)] +.1 [isokinetic squat-lift strength at half reach (S.P.)] -.1 [isokinetic stoop-lift strength at half reach (S.P.)]	.941	.001
15	Isometric peak strength at full reach with hip at 60° in 3P	41.4 +.04 [standard squat peak] +.7 [isometric stoop-lift strength with hip at 60° at 3/4 reach (S.P.)] -.3 [isokinetic squat-lift strength at 3/4 reach (S.P.)]	.903	.001
16	Isometric peak strength at half reach with hip at 60° in 6P	–38.5 +.1 [standard squat peak] + 1.3 [isometric stoop- lift strength with hip at 60° at 3/4 reach (S.P.)] +.3 isokinetic squat-lift at half reach (S.P.)] –.7 [isokinetic squat-lift strength at 3/4 reach (S.P.)]	.957	.001

(continued)

TABLE 14.33 (CONTINUED)

Stoop Lift (Peaks): Regression Equations

No.	Dependent Variable	Equation	R^2	$p <$
17	Isometric peak strength at 3/4 reach with hip at 60° in 6P	80.3 +.1 [body weight] +.5 [isometric squat-lift strength with knee at 90° at 3/4 reach (S.P.)]	.670	.001
18	Isometric peak strength at full reach with hip at 60° in 6P	–27.8 +.5 [body weight] +.7 [isometric stoop-lift strength with hip at 60° at full reach (S.P.)]	.732	.001
19	Isometric peak strength at half reach with hip at 90° in S.P.	286.1 +.06 [standard squat peak] - 108.5 [gender] +.7 [isometric stoop-lift strength with hip at 60° at half reach (S.P.)	.918	.001
20	Isometric peak strength at 3/4 reach with hip at 90° in S.P.	124.1 -.1 [standard squat peak] +.5 [isometric stoop-lift strength with hip at 60° at half reach (S.P.)	.880	.001
21	Isometric peak strength at full reach with hip at 90° S.P.	71.8 - 5.2 [gender] +.7 [isometric stoop-lift strength with hip at 60° at full reach (S.P.)]	.673	.001
22	Isometric peak strength at half reach with hip at 90° in 3P	59.7 -.08 [standard squat peak] +.7 [isometric stoop-lift strength with hip at 60° at half reach (S.P.)] +.3 isokinetic squat-lift strength at half reach (S.P.)]	.898	.001
23	Isometric peak strength at 3/4 reach with hip at 90° in 3P	9.8 +.02 [standard squat peak] +.9 [isometric stoop-lift strength with hip at 60° at full reach (S.P.)] +.3 [isokinetic squat-lift strength at half reach (S.P.)]	.738	.001
24	Isometric peak strength at full reach with hip at 90° in 3P	72.3 +.05 [standard squat peak] +.6 [isometric stoop-lift strength with hip at 60° at 3/4 quarters reach (S.P.)] -.4 [isokinetic squat-lift strength at 3/4 reach (S.P.)]	.861	.001
25	Isometric peak strength at half reach with hip at 90° in 6P	–26.4 +.1 [standard squat peak] + 1.5 [isometric stoop-lift strength with hip at 60° at full reach (S.P.)]	.784	.001
26	Isometric peak strength at 3/4 reach with hip at 90° in 6P	52.6 +.03 [standard squat peak] +.5 [isometric stoop-lift strength with hip at 60° at 3/4 reach (S.P.)] +.2 [isokinetic squat-lift strength at half reach (S.P.)] -.3 [isokinetic squat-lift strength at 3/4 reach (S.P.)]	.937	.001
27	Isometric peak strength at full reach with hip at 90° in 6P	60.3 +.3 [isometric stoop-lift strength with hip at 60° at 3/4 reach (S.P.)]	.407	.01

From Kumar, 1995.

29 years, (b) between 30 and 39, and (c) above 39 years of age. All experimental boxes were fitted with handles at 12.7-cm height above the bottom of the box. A free-style lifting technique was used in all trials in the experiment. The subjects initiated their shifts either with low or high weights in the box, which was lifted according the task criteria provided to them. They were asked to assume an eight-hour shift and make as many adjustments as necessary to arrive at the maximum acceptable they felt they could lift comfortably without feeling tired, overheated, and exhausted. The subjects were subsequently asked to assume an extra 4-hour work period (4 hours overtime, having just completed an 8-hour shift, and no rest allowances were provided in between, and the activity continued. During this extended phase, the subjects were asked to make additional adjustments if required. This protocol was carried out with all task combinations for each subject. These trials were conducted in a random order. The maximum acceptable weights for various lifts obtained from these experiments are presented in Table 14.39 through Table 14.42. The percentage decreases in the maximum acceptable weight for lifting with increasing frequency, box size, and height of the lifts are presented in Table 14.43. The physiological data that the author collected during these trials showed 23 and 24% increase in energy expenditure in

TABLE 14.34

Stoop Lift (Averages): Regression Equations

No	Dependent Variable	Equation	R^2	$p <$
1	Isokinetic average strength at half reach in sagittal plane (S.P.)	58.3 - 1.5 [Reach] +.7 [Isokinetic Stoop-Lift Strength At Half Reach (S.P.)]	.963	.001
2	Isokinetic average strength at 3/4 reach in S.P.	−9.5 +.7 [isokinetic stoop-lift strength at 3/4 reach (S.P.)]	.941	.001
3	Isokinetic average strength at full reach in S.P.	46.7 +.4 [isokinetic stoop-lift strength at full reach (S.P.)	.837	.001
4	Isokinetic average strength at half reach in 30° lateral plane (3p)	−294 +.5 [isokinetic stoop-lift strength at half reach (S.P.) + 1.9 HT	.741	.001
5	Isokinetic average strength at 3/4 reach in 3P	51.4 -.04 [isometric stoop-lift strength with hip at 60° at 3/4 reach (S.P.)] +.6 [isokinetic stoop-lift strength at full reach (S.P.)]	.674	.001
6	Isokinetic average strength at full reach in 3P	−13.8 + 1.4 [body weight] +.2 [isokinetic stoop-lift strength at 3/4 reach (S.P.)]	.559	.001
7	Isokinetic average strength at half reach in 60° lateral plane	−5.6 -.06 [isometric stoop-lift strength with hip at 60° in 3/4 reach (S.P.)] +.09 [isometric squat-lift strength with knee at 90° at full reach (S.P.)] +.9 [isokinetic stoop-lift strength at full reach (S.P.)]	.841	.001
8	Isokinetic average strength at 3/4 reach in 6P	−44.6 + 1.6 [body weight] +.3 [isokinetic stoop-lift strength at full reach (S.P.)]	.612	.001
9	Isokinetic average strength at full reach in 6P	73.3 +.05 [isokinetic stoop-lift strength at half reach (S.P.)]	.283	.04
10	Isometric average strength at half reach with hip at 60° in S.P.	−29.6 -.04 [standard squat peak] +.6 [isometric stoop-lift strength with hip at 60° at half reach (S.P.)] +.4 [isometric stoop-lift strength with hip at 60° at 3/4 reach (S.P.)]	.991	.001
11	Isometric average strength at 3/4 reach with hip at 60° in S.P.	−19.5 +.03 [standard squat peak] +.8 [isometric stoop-lift strength with hip at 60° at 3/4 reach (S.P.) +.3 [isometric stoop-lift strength with hip at 60° at full reach (S.P.)] -.09 [isometric stoop-lift strength with hip at 60° at half reach (S.P.)]	.990	.001
12	Isometric average strength at full reach with hip at 60° in S.P.	6.1 + 3.4 [gender] +.4 [isometric stoop-lift strength with hip at 60° at full reach (S.P.)] +.1 [isometric stoop-lift strength with hip at 60° at 3/4 reach (S.P.)] +.1 [isometric squat-lift strength with knee at 90° at full reach (S.P.)]	.979	.001
13	Isometric average strength at half reach with hip at 60° in 3P	31.9 -.09 [standard squat peak] +.6 [isometric stoop-lift strength with hip at 60° at half reach (S.P.)]	.877	.001
14	Isometric average strength at 3/4 reach with hip at 60° in 3P	45.5 +.03 [standard squat peak] +.8 [isometric stoop-lift strength with hip at 60° at 3/4 reach (S.P.)] -.2 [isokinetic stoop-lift strength at half reach (S.P.)]	.915	.001
15	Isometric average strength at full reach with hip at 60° in 3P	15.0 +.04 [standard squat peak] +.4 [isometric stoop-lift strength with hip at 60° at 3/4 reach (S.P.)] +.4 [isometric stoop-lift strength with hip at 60° at full reach (S.P.)] -.3 [isokinetic squat-lift strength at 3/4 reach (S.P.)]	.929	.001
16	Isometric average strength at half reach with hip at 60° in 6P	10.6 +.1 [standard squat peak] +.8 [isometric stoop-lift strength with hip at 60° at 3/4 reach (S.P.)] +.3 [isokinetic squat-lift strength at half reach (S.P.)] -.6 [isokinetic squat-lift strength at 3/4 reach (S.P.)]	.939	.001

(continued)

TABLE 14.34 (CONTINUED)

Stoop Lift (Averages): Regression Equations

No	Dependent Variable	Equation	R²	p <
17	Isometric average strength at 3/4 reach with hip at 60° in 6P	71.8 +.04 [standard stoop peak] +.3 [isometric squat-lift strength with knee at 90° at 3/4 reach (S.P.)]	.660	.001
18	Isometric average strength at full reach with hip at 60° in 6P	–23.5 +.3 [body weight] +.6 [isometric stoop-lift strength with hip at 60° at full reach (S.P.)]	.790	.001
19	Isometric average strength at half reach with hip at 90° in S.P.	45.0 +.02 [standard squat peak] +.7 [isometric stoop-lift strength with hip at 60° at half reach (S.P.)]	.909	.001
20	Isometric average strength at 3/4 reach with hip at 90° in S.P.	43.6 +.00 [standard squat peak] +.4 [isometric stoop-lift strength with hip at 60° at 3/4 reach (S.P.)] +.5 [isometric stoop-lift strength with hip at 60° at full reach (S.P.)]	.908	.001
21	Isometric average strength at full reach with hip at 90° in 3P	43.5 + 2.1 [gender] + 0.575 [isometric stoop at full reach]	.852	.001
22	Isometric average strength at half reach with hip at 90° in 3P	71.3 -.08 [standard squat peak] +.7 [isometric stoop-lift strength with hip at 60° at half reach (S.P.)]	.862	.001
23	Isometric average strength at 3/4 reach with hip at 90° in 3P	125 -.07 [standard squat peak] +.4 [isometric stoop-lift strength with hip at 60° at half reach (S.P.)] -.2 [isokinetic stoop-lift strength at half reach (S.P.)]	.755	.001
24	Isometric average strength at full reach with hip at 90°° in 3P	63.5 -.00 [standard squat peak] +.4 [isometric stoop-lift strength with hip at 60° at 3/4 reach (S.P.)] -.1 [isokinetic stoop-lift strength at half reach (S.P.)]	.787	.001
25	Isometric average strength at half reach with hip at 90° in 6P	–18.4 +.1 [standard squat peak] + 1.3 [isometric stoop-lift strength with hip at 60° at full reach (S.P.)]	.803	.001
26	Isometric average strength at 3/4 reach with hip at 90° in 6P	35.4 -.04 [standard squat peak] +.1 [isometric stoop-lift strength with hip at 60° at half reach (S.P.)] +.2 [isokinetic squat-lift strength at half reach (S.P.)]	.863	.001
27	Isometric average strength at full reach with hip at 90° in 6P	44.5 +.2 [isometric stoop-lift strength at 3/4 reach (S.P.)]	.473	.001

From Kumar, 1995.

males and females, respectively, for 12-hour shifts, even though the magnitude of the weight declined.

Garg and Badger (1986) investigated maximum acceptable weights for asymmetric lifting. Using 13 male college students and three different boxes, they recorded psychophysically determined maximum acceptable weight from floor to 81-cm high table in sagittal plane and 30°, 60°, and 90° asymmetrical plane. They found that the maximum acceptable weights for asymmetric planes were significantly lower than that obtained in sagittal plane ($p < .01$). These decreases increased with an increase in the magnitude of asymmetry ($p < 0.01$) (Table 14.44). The box size did not significantly affect the maximum acceptable weight. They suggested correction factors of 7, 15, and 22% for maximum acceptable weights at 30°, 60°, and 90° asymmetry.

A psychophysical study of the effects of load and frequency upon selection of workload in repetitive lifting was published by Nicholson and Legg (1986). In their study, eight male soldiers repeatedly lifted and lowered a box to and from a platform at 40% stature for 10 minutes and adjusted either the load or the lifting frequency or a combination of both in order to select a workload that they considered to be maximum acceptable load for a period

TABLE 14.35

Squat Lift (Peaks): Regression Equations

No	Dependent Variable	Equation	R^2	$p<$
1	Isokinetic peak strength at half reach in sagittal plane (S.P.)	−1788.6 + 12.5 [Body Height]	.501	.001
2	Isokinetic peak strength at 3/4 reach in S.P.	−161.9 +.1 [standard squat peak] + 2.1 [reach] + 1.3 [isokinetic squat-lift strength at full reach (S.P.)]	.914	.001
3	Isokinetic peak strength at full reach in S.P.	49.2 +.8 [body weight] +.3 [isokinetic squat-lift strength at 3/4 reach (S.P.)]	.866	.001
4	Isokinetic peak strength at half reach in 30° lateral plane (3P)	34.9 +.1 [standard squat peak] +.8 [isokinetic squat-lift strength at half reach (S.P.)]	.903	.001
5	Isokinetic peak strength at 3/4 reach in 3P	−30.0 -.04 [standard squat peak] +.2 [isometric stoop-lift strength at half reach (S.P.)] +.9 [isokinetic squat-lift strength at full reach (S.P.)]	.787	.001
6	Isokinetic peak strength at full reach in 3P	91.5 +.04 [standard squat peak] +.2 [isometric stoop-lift strength at 3/4 reach (S.P.)] +.09 [isokinetic squat-lift strength at half reach (S.P.)]	.901	.001
7	Isokinetic peak strength at half reach in 60° lateral plane (6P)	−392 + 2.5 [body height] + 1.2 [isokinetic squat-lift strength at full reach (S.P.)]	.715	.001
8	Isokinetic peak strength in 3/4 reach in 6P	60.2 +.1 [standard squat peak] +.3 [isokinetic squat-lift strength at half reach (S.P.)]	.820	.001
9	Isokinetic peak strength at full reach in 6P	−40.6 + 2.2 [body weight] +.2 [isokinetic squat-lift strength at half reach (S.P.)]	.670	.001
10	Isometric peak strength at half reach with knee at 90° in S.P.	282.8 − 119.7 [gender] +.6 [isometric stoop-lift strength with hip at 60° at 3/4 reach (S.P.)]	.770	.001
11	Isometric peak strength at 3/4 reach with knee at 90° in S.P.	−102.8 + 4.1 [body weight] + 1.0 [isometric squat-lift strength with knee at 90° at full reach (S.P.)] −.3 [isokinetic squat-lift strength at 3/4 reach (S.P.)]	.943	.001
12	Isometric peak strength at full reach with knee at 90° in S.P.	−49.6 +.04 [standard stoop peak] +.3 [isometric squat-lift strength with knee at 90° at 3/4 reach (S.P.)] +.9 [isometric stoop-lift strength with hip at 60° at full reach (S.P.)] -.2 [isokinetic stoop-lift strength at full reach (S.P.)]	.961	.001
13	Isometric peak strength at half reach with knee at 90° in 3P	30.6 + 2.2 [gender] +.9 [isometric squat-lift strength with knee at 90° at half reach (S.P.)] +.2 [isokinetic squat-lift strength at half reach (S.P.)] −.3 [isokinetic squat-lift strength at 3/4 reach (S.P.)]	.948	.001
14	Isometric peak strength at 3/4 reach with knee at 90° in 3P	18.4 - 9.8 [gender] +.6 [isometric squat-lift strength with knee at 90° at 3/4 reach (S.P.)] +.1 [isometric stoop-lift strength with hip at 60° at 1/2 reach (S.P.)]	.953	.001
15	Isometric peak strength at full reach with knee at 90° in full reach	−59.1 + 1.7 [body weight] +.6 [isometric squat-lift strength with knee at 90° at full reach (S.P.)]	.925	.001
16	Isometric peak strength at half reach with knee at 90° in 6P	143 - 65.7 [gender] +.2 [isokinetic squat-lift strength at half reach (S.P.)] +.6 [isometric squat-lift strength with knee at 90° at 3/4 reach (S.P.)]	.926	.001
17	Isometric peak strength at 3/4 reach with knee at 90° in 6P	9.6 -.5 [body weight] +.4 [isometric squat-lift strength with knee at 90° at half reach (S.P.)] +.5 [isometric squat-lift strength with knee at 90° at full reach (S.P.)]	.902	.001
18	Isometric peak strength at full reach with knee at 90° in 6P	50.8 - 37.5 [gender] +.7 [isometric stoop-lift strength with hip at 60° at full reach (S.P.)]	.873	.001
19	Isometric peak strength at half reach with knee at 135° in S.P.	237.4 - 96.6 [gender] +.9 [isometric squat-lift strength with knee at 90° at 3/4 reach (S.P.)]	.806	.001

(continued)

TABLE 14.35 (CONTINUED)

Squat Lift (Peaks): Regression Equations

No	Dependent Variable	Equation	R^2	$p<$
20	Isometric peak strength at 3/4 reach with knee at 135° in S.P.	46.2 - 32.5 [gender] +.7 [isometric squat-lift strength with knee at 90° at 3/4 reach (S.P.)] +.1 [isometric squat-lift strength with knee at 90° at half reach (S.P.)]	.958	.001
21	Isometric peak strength at full reach with knee at 135° in S.P.	136.4 - 45.2 [gender] +.5 [isometric squat-lift strength with knee at 90° at full reach (S.P.)]	.914	.001
22	Isometric peak strength at half reach with knee at 135° in 3P	−117.6 +.2 [standard squat peak] + 1.9 [isometric stoop-lift strength with hip at 60° at full reach (S.P.)]	.887	.001
23	Isometric peak strength at 3/4 reach with knee at 135° in 3P	−46.4 +.05 [standard stoop peak] +.6 [isometric squat-lift strength with knee at 90° at 3/4 reach (S.P.)] +.5 [isometric stoop-lift strength with knee at 90° at 3/4 reach (S.P.)] -.1 [isokinetic stoop-lift strength at half reach (S.P.)]	.956	.001
24	Isometric peak strength at full reach with knee at 135° in 3P	17.0 - 12.1 [gender] +.5 [isometric squat-lift strength with knee at 90° at full reach (S.P.)] +.2 [isometric squat-lift strength with knee at 90° at half reach (S.P.)]	.912	.001
25	Isometric peak strength at half reach with knee at 135° in 6P	−69.8 + 1.9 [body weight] +.1 [standard squat peak] +.6 [isometric squat-lift strength with knee at 90° at 3/4 reach (S.P.)]	.889	.001
26	Isometric peak strength at 3/4 reach with knee at 135° in 6P	−16.9 +.2 [body weight] +.8 [isometric squat-lift strength with knee at 90° at 3/4 reach (S.P.)]	.843	.001
27	Isometric peak strength at full reach with knee at 135° in 6P	61.2 +.4 [isometric stoop-lift strength with hip at 60° at 3/4 reach (S.P.)] - 31.8 [gender]	.882	.001

From Kumar, 1995.

of 1 hour. They reported that when the soldiers adjusted the lifting frequency, the mean maximum acceptable weight was 94.5 kg·m·min^{-1}, which was significantly greater (27%; $p < 0.05$) than when the load alone was adjusted 74.3 kg·m·min^{-1}). A control of both variables together produced a maximum acceptable weight for lifting of 76.5 kg·m·min^{-1}, which was similar to and insignificantly different from the load alone adjustment. Mital et al. (1986) stated that repetitive dynamic strength determined psychophysically was a better measure of an individual's lifting capability. Yates and Karwowski (1987) found that the maximum acceptable load for lifting in seated posture was significantly lower than in the standing posture. They recommended a 16% downward adjustment for seated lifting.

14.6 Summary and Conclusions

Physical strength is an essential trait through which we mediate industrial activities. Since the trunk is centrally involved in manual lifting and most cases of materials handling, it is the human organ most frequently injured. These injuries have proven to be expensive from personal, societal, and industrial point of views. Any strategy of control has to appropriately incorporate the trunk and lifting strength characteristics of the workforce.

The trunk is strongest in extension, followed by flexion, lateral flexion, and axial rotation. The axial rotation represents approximately 25% of the extension strength. Also, the trunk is least stable in rotation and suffers a loss of contribution from 50% of trunk muscles and

TABLE 14.36

Squat Lift (Averages): Regression Equations

No	Dependent Variable	Equation	R^2	$p<$
1	Isokinetic average strength at half reach in S.P.	−147.3 + 2.8 [Body Height] − 5.8 [Reach] +.7 [Isokinetic Squat-Lift Strength At Half Reach (S.P.)]	.978	.001
2	Isokinetic average strength at 3/4 reach in S.P.	−59.1 +.02 [standard squat peak] + 1.1 [reach] +.6 [isokinetic squat-lift strength at 3/4 reach (S.P.)]	.954	.001
3	Isokinetic average strength at full reach in S.P.	−45.7 +.01 [standard stoop peak] +.5 [reach] +.9 [isokinetic squat-lift strength at full reach (S.P.)] −.05 [isokinetic stoop-lift strength at half reach (S.P.)]	.958	.001
4	Isokinetic average strength at half reach in 3P	−21.0 +.05 [standard squat peak] +.7 [isokinetic squat-lift strength at half reach (S.P.)]	.910	.001
5	Isokinetic average strength at 3/4 reach in 3P	39.7 +.11 [isometric stoop at half reach] +.19 [isokinetic squat at half reach]	.821	.001
6	Isokinetic average strength at full reach in 3P	88.0 +.1 [standard squat peak]	.746	.001
7	Isokinetic strength at half reach in 6P	−387.8 +.2.9 [body height] +.2 [isometric stoop-lift strength with hip at 60° at 3/4 reach (S.P.)]	.715	.001
8	Isokinetic average strength at 3/4 reach in 6P	42.7 +.05 [standard squat peak] +.2 [isokinetic squat-lift strength at half reach (S.P.)]	.759	.001
9	Isokinetic average strength at full reach in 6P	46.7 +.04 [standard stoop peak] +.1 [isokinetic squat-lift strength at half reach (S.P.)]	.640	.001
10	Isometric average strength at half reach with knee at 90° in S.P.	−31.2 + 11.9 [gender] +.8 [isometric squat-lift strength with knee at 90° at half reach (S.P.)]	.973	.001
11	Isometric average strength at 3/4 reach with knee at 90° in S.P.	13.4 +.8 [isometric squat-lift strength with knee at 90° at 3/4 reach (S.P.)] −.3 [body weight]	.989	.001
12	Isometric average strength at full reach with knee at 90° in S.P.	−6.1 +.00 [standard stoop peak] +.9 [isometric squat-lift strength with knee at 90° at full reach (S.P.)] −.03 [isokinetic squat-lift strength at half reach (S.P.)]	.992	.001
13	Isometric average strength at half reach with knee at 90° in S.P.	−81.9 + 19.9 [gender] +.5 [isometric squat-lift strength with knee at 90° at half reach (S.P.) +.7 [isometric stoop-lift strength with hip at 60° at full reach (S.P.)]	.877	.001
14	Isometric average strength at 3/4 reach with knee at 90° in 3P	3.4 − 4.9 [gender] +.5 [isometric squat-lift with hip at 60° at 3/4 reach (S.P.)] +.1 [isometric stoop-lift strength with hip at 60° at half reach (S.P.)]	.957	.001
15	Isometric average strength at full reach with knee at 90° in 3P	−55.1 + 1.4 [body weight] +.5 [isometric squat-lift strength with knee at 90° at full reach (S.P.)]	.923	.001
16	Isometric average strength at half reach with knee at 90° in 6P	70.8 +.04 [standard stoop peak] − 34.7 [gender] +.4 [isometric squat-lift strength with knee at 90° at 3/4 reach (S.P.)] +.1 [isokinetic squat-lift strength at half reach (S.P.)]	.944	.001
17	Isometric average strength at 3/4 reach with knee at 90° in 6P	2.9 − 1.1 [body weight] +.6 [isometric squat-lift strength with knee at 90° at 3/4 reach (S.P.)] +.2 [isometric stoop-lift strength with hip at 60° at 3/4 reach (S.P.)]	.903	.001
18	Isometric average strength at full reach with knee at 90° in 6P	20.7 − 25.3 [gender] +.6 [isometric stoop-lift strength with hip at 60° at full reach (S.P.)]	.835	.001
19	Isometric average strength at half reach with knee at 135° in S.P.	57.2 − 32.8 [gender] +.7 [isometric squat at 3/4 reach] +.2 [isokinetic squat at half reach]	.890	.001
20	Isometric average strength at 3/4 reach with knee at 135° in S.P.	11.3 − 16.2 [gender] +.6 [isometric squat-lift strength with knee at 90° at 3/4 reach (S.P.)] +.1 [isokinetic squat-lift strength at half reach (S.P.)]	.959	.001

(continued)

TABLE 14.36 (CONTINUED)

Squat Lift (Averages): Regression Equations

No	Dependent Variable	Equation	R^2	$p<$
21	Isometric average strength at full reach with knee at 135° in S.P.	84.9 - 26.8 [gender] +.5 [isometric squat-type strength with knee at 90° at full reach (S.P.)]	.895	.001
22	Isometric average strength at half reach with knee at 135° in 3P	−.4 +.09 [standard squat peak] +.5 [isometric squat-lift strength with knee at 90° at 3/4 reach (S.P.)] +.2 [isometric stoop-lift strength with hip at 60° at 1/2 reach (S.P.)] +.2 [isokinetic squat-lift strength at half reach (S.P.)] -.3 [isokinetic squat-lift strength at 3/4 reach (S.P.)]	.980	.001
23	Isometric average strength at 3/4 reach with knee at 135° in 3P	−22.5 +.1 [standard stoop peak] +.4 [isometric squat-lift strength with knee at 90° at 3/4 reach (S.P.)] +.3 [isometric stoop-lift strength with hip at 60° at 3/4 reach (S.P.)] +.08 [isokinetic squat-lift strength at half reach (S.P.)] -.3 [isokinetic squat-lift strength at 3/4 reach (S.P.)]	.912	.001
24	Isometric average strength at full reach with knee at 135° in 3P	−9.3 - 4.2 [gender] +.5 [isometric squat-lift strength with knee at 90° at full reach (S.P.)] +.2 [isometric squat-lift strength with knee at 90° at half reach (S.P.)]	.912	.001
25	Isometric average strength at half reach with knee at 135° in 6P	−46.9 + 1.3 [body weight] +.1 [standard stoop peak] +.5 [isometric squat-lift strength with knee at 90° at 3/4 reach (S.P.)]	.872	.001
26	Isometric average strength at 3/4 reach with knee at 135° in 6P	−54.5 + 1.2 [body weight] +.7 [isometric squat-lift strength with knee at 90° at full reach (S.P.)]	.837	.001
27	Isometric average strength at full reach with knee at 135° in 6P	−49.2 +.01 [standard squat peak] +.9 [body weight] +.3 [isometric stoop-lift strength with hip at 60° in 3/4 reach (S.P.)]	.905	.001

From Kumar, 1995.

loss of support from 50% of annular fibers. Isometric strength was greater than isokinetic strength in all cases. The strength continually declines with increasing postural asymmetry and increasing velocity of motion.

Lifting strength is a composite of hands, arms, shoulders, trunk, and hip strengths. Load being lifted generally has a varying demand on each these joints. However, due to the greatest mechanical disadvantage at spinal joints, which also the harbor most vulnerable structures, the injuries are precipitated at these sites. The lifting strength capability is mode, posture, technique, and speed dependent. Isometric strength is always greater than isokinetic. With increasing postural asymmetry, increasing velocity, and increasing mechanical disadvantage, the strength value observed progressively declines. The isoinertial technique commonly used in industry likely does not account for repetition and duration of the task. Similarly, the psychophysical techniques so far have not been designed to strike an optimum balance between the worker's safety and his/her income. Unfortunately, we do not yet have a rigorously determined and validated "gold standard." Therefore, wider consideration and caution are advised while using any system.

TABLE 14.37

Maximum Acceptable Weight of Lift for Males (kg)

Width (A)	Distance (B)	Percent (C)	Floor to Knuckle Ht. One Lift Every						Knuckle to Shoulder Ht. One Lift Every						Shoulder to Arm Ht. One Lift Every					
			5	9 s	14	1 min	5	8 h	5	9 s	14	1 min	5	8 h	5	9 s	14	1 min	5	8 h
75	76	75	10	14	15	18	25	29	12	16	18	17	21	24	9	12	14	16	20	23
		50	13	17	19	22	30	36	14	19	21	21	27	30	11	15	18	20	25	28
		25	16	20	23	26	36	42	17	22	25	26	32	36	13	18	21	24	29	33
	51	75	11	14	16	19	26	31	13	17	19	20	24	27	10	14	15	18	22	25
		50	14	18	20	23	31	37	15	20	23	24	30	34	12	17	19	22	28	31
		25	16	21	24	27	37	44	18	24	27	29	36	40	14	20	23	27	33	37
	25	75	13	17	19	21	29	34	15	20	22	23	28	32	11	16	18	21	26	30
		50	16	21	23	26	35	42	18	24	27	28	35	40	14	20	22	26	33	37
		25	19	25	28	31	42	50	21	28	32	34	42	47	17	24	27	31	39	44
49	76	75	12	15	17	21	28	34	12	16	18	17	21	24	9	12	14	16	20	23
		50	15	19	21	26	35	42	14	19	21	21	27	30	11	15	18	20	25	28
		25	17	23	26	31	42	50	17	22	25	26	32	36	13	18	21	24	29	33
	51	75	12	16	18	22	30	35	13	17	19	20	24	27	10	14	15	18	22	25
		50	15	20	22	27	37	43	15	20	23	24	30	34	12	17	19	22	28	31
		25	18	24	27	32	44	52	18	24	27	29	36	40	14	20	23	27	33	37
	25	75	14	18	21	24	33	39	15	20	22	23	28	32	11	16	18	21	26	30
		50	18	23	26	30	41	49	18	24	27	28	35	40	14	20	22	26	33	37
		25	21	28	31	36	49	59	21	28	32	34	42	47	17	24	27	31	39	44
36	76	75	13	17	20	23	31	37	13	17	19	18	23	26	9	13	15	17	21	24
		50	17	22	25	29	39	46	15	20	23	23	29	32	11	16	19	21	27	30
		25	20	27	30	34	47	55	18	23	26	28	34	39	14	19	23	26	32	36
	51	75	14	18	20	24	32	38	13	18	20	21	26	29	10	15	16	19	24	27
		50	17	23	26	30	40	48	16	21	24	26	32	36	13	18	20	24	30	34
		25	21	28	31	36	49	57	19	25	28	31	39	44	15	22	25	29	36	40
	25	75	16	21	24	27	37	43	16	21	24	24	30	34	12	17	19	23	28	32
		50	20	27	30	34	46	54	19	25	28	31	38	43	15	22	24	28	35	40
		25	25	32	36	40	55	65	22	29	33	37	45	51	18	26	29	34	42	48

			1	2	3	4	5	6	7	8	9	10	11	12	13	14	15	16	17	18
75	76	75	14	12	10	9	9	5	15	14	11	11	11	8	20	17	13	11	10	8
		50	15	13	11	10	9	6	18	16	13	12	12	9	23	20	14	13	12	9
		25	17	15	12	11	10	6	20	18	14	13	13	10	26	22	16	14	13	10
	51	75	15	14	11	11	10	6	17	15	12	12	12	9	21	18	13	12	11	8
		50	17	15	12	12	11	6	20	18	14	13	13	10	24	20	15	14	13	9
		25	19	17	13	13	12	7	22	20	16	14	14	11	27	23	17	16	15	10
	25	75	18	16	13	13	11	7	20	18	15	14	14	11	24	20	15	13	13	10
		50	20	18	14	14	13	8	23	21	17	15	15	12	27	23	17	15	15	11
		25	22	20	16	15	14	8	26	23	19	17	17	13	31	26	19	17	17	12
49	76	75	14	12	10	9	9	5	15	14	11	11	11	8	24	20	15	12	12	9
		50	15	13	11	10	9	6	18	16	13	12	12	9	27	23	17	14	14	10
		25	17	15	12	11	10	6	20	18	14	13	13	10	31	26	20	16	16	11
	51	75	15	14	11	11	10	6	17	15	12	12	12	9	25	21	16	14	14	10
		50	17	15	12	12	11	6	20	18	14	13	13	10	28	24	19	16	16	11
		25	19	17	13	13	12	7	22	20	16	14	14	11	32	27	22	19	18	12
	25	75	18	16	13	13	11	7	20	18	15	14	14	11	28	23	16	15	16	12
		50	20	18	14	14	13	8	23	21	17	15	15	12	32	27	19	17	18	13
		25	22	20	16	15	14	8	26	23	19	17	17	13	36	30	21	20	21	14
36	76	75	15	13	11	10	9	6	17	14	12	12	12	9	26	22	17	13	13	9
		50	16	15	12	11	10	6	19	17	14	13	13	10	30	25	19	15	15	10
		25	18	16	13	12	11	7	21	19	15	14	14	11	34	28	22	17	17	11
	51	75	17	15	12	11	10	6	19	17	13	12	12	9	27	23	19	14	15	10
		50	18	16	13	13	11	7	21	19	15	14	14	10	30	26	22	17	17	11
		25	20	18	14	14	12	8	24	21	17	15	15	11	35	30	25	19	19	12
	25	75	20	17	14	14	12	8	22	19	16	15	15	11	31	26	20	16	17	12
		50	22	19	16	15	14	8	25	22	18	16	16	12	35	30	23	18	19	13
		25	24	21	17	16	15	9	29	25	20	18	18	13	40	33	26	21	21	14

Note: Horizontal hand location is at least (15 + width/2). (a) Width of object (cm), (b) Vertical distance of lift (cm), (c) Percent of industrial population exceeding table value.

From Snook, 1978.

TABLE 14.38

Maximum Recommended Weights Based on Dynamic Strength (kg)

Height of Lift	Horz. (cm)	Freq. (Lift/Min)	Female		Male	
			25%ile	50%ile	50%ile	75%ile
Floor to knuckle	30	1	18	20	30	36
		2	17	18	28	34
		4	14	16	24	28
		6	12	14	22	28
		8	11	13	21	26
		12	9	11	18	21
	38	1	15	17	27	32
		2	11	13	26	31
		4	10	13	24	30
		6	10	12	22	25
		8	10	12	20	24
		12	9	10	15	18
	46	1	13	16	24	29
		2	11	13	23	27
		4	11	12	21	26
		6	11	12	2	23
		8	10	11	18	23
		12	8	10	14	17
Floor to shoulder	30	2	11	13	23	27
		4	12	13	22	25
		6	11	13	20	24
		8	11	13	19	23
	38	2	12	14	24	26
		4	11	13	23	27
		6	9	13	22	25
		8	10	12	21	255
	46	2	11	13	23	26
		4	11	13	22	25
		6	10	12	21	24
		8	9	11	20	25
Floor to reach	30	2	11	12	21	24
		4	10	12	20	24
		6	11	11	19	21
		8	10	11	18	20
	38	2	11	13	24	29
		4	11	12	21	25
		6	10	11	18	21
		8	10	11	15	17
	46	2	10	12	18	22
		4	9	11	18	20
		6	9	11	17	20
		8	9	10	17	20
Knuckle to shoulder	30	1	13	14	24	29
		2	12	14	23	27
		4	12	13	22	26
		6	11	13	20	24
		8	9	12	18	22
		12	9	10	15	18

(continued)

TABLE 14.38 (CONTINUED)

Maximum Recommended Weights Based on Dynamic Strength (kg)

Height of Lift	Horz. (cm)	Freq. (Lift/Min)	Female		Male	
			25%ile	50%ile	50%ile	75%ile
Knuckle to shoulder	38	1	12	13	27	31
(continued)		2	11	13	26	30
		4	12	13	24	28
		6	11	13	22	27
		8	9	11	20	24
		12	8	9	14	17
	46	1	11	13	21	25
		2	11	13	20	25
		4	10	12	19	24
		6	9	11	18	24
		8	9	10	17	21
		12	8	9	14	17
Knuckle to reach	30	2	10	13	21	26
		4	10	12	20	22
		6	10	12	18	21
		8	9	11	17	19
	38	2	11	12	24	27
		4	11	12	22	24
		6	10	11	20	22
		8	10	11	18	21
	46	2	13	14	24	28
		4	11	13	22	24
		6	11	12	20	21
		8	9	11	18	70
Shoulder to reach	30	1	11	12	23	27
		2	10	12	22	24
		4	10	11	21	25
		6	9	10	19	22
		8	7	8	15	18
		12	6	6	11	14
	38	1	10	11	20	24
		2	9	11	19	22
		4	9	10	18	22
		6	8	9	17	22
		8	6	8	15	18
		12	5	6	11	13
	46	1	11	12	20	24
		2	10	11	18	22
		4	9	10	18	21
		6	9	10	17	21
		8	8	9	15	18
		12	5	6	11	13

From Ayoub et al. 1978, DHEW (NIOSH) Report.

TABLE 14.39

Maximum Weight of Lift Acceptable to Male Industrial Workers for 12 h and Their Physiological Responses at That Weight

Lift Height	Box Size (cm)		1 30.48	1 45.72	1 60.96	4 30.48	4 45.72	4 60.96	8 30.48	8 45.72	8 60.96	12 30.48	12 45.72	12 60.96
						Frequency of Lift (lifts/min)								
Floor to knuckle	Weight (kg)	x[a]	13.33	12.02	13.05	12.20	10.56	10.66	11.89	9.39	8.59	8.19	7.54	7.34
		S	3.63	2.92	2.70	3.87	2.77	3.12	4.66	1.79	1.71	2.93	2.53	1.84
	Heart rate (beats/min)	x	81.44	72.82	74.19	88.03	88.69	92.64	102.91	101.24	100.65	100.88	109.90	107.36
		S	13.29	10.34	13.24	15.27	8.38	18.64	18.60	14.03	18.55	14.87	17.10	11.55
	Oxygen uptake (1/min)	x	0.40	0.34	0.40	0.69	0.73	0.68	0.85	0.86	0.91	0.99	0.87	0.88
		S	0.21	0.15	0.20	0.19	0.25	0.20	0.23	0.20	0.16	0.23	0.24	0.19
Knuckle to shoulder	Weight	x	12.44	12.82	11.88	10.51	11.67	10.69	9.90	9.48	8.40	8.69	9.10	7.93
		S	3.28	3.75	3.26	2.35	2.19	3.37	2.04	2.34	2.25	2.18	2.69	1.95
	Heart rate	x	83.33	80.79	78.12	95.88	94.61	89.72	98.51	91.11	91.00	97.52	106.54	98.10
		S	14.13	12.2	11.60	14.87	13.44	15.68	13.75	9.79	16.78	14.91	12.96	12.40
	Oxygen uptake	x	0.27	0.38	0.39	0.51	0.57	0.59	0.63	0.72	0.65	0.75	0.75	0.58
		S	0.13	0.20	0.11	0.18	0.19	0.23	0.19	0.25	0.16	0.19	0.21	0.12
Shoulder to reach	Weight	x	10.11	9.58	10.86	9.69	10.05	9.36	8.95	9.23	9.80	8.85	7.70	7.87
		S	2.75	1.76	3.10	3.40	3.57	1.80	2.44	1.63	3.59	2.24	1.72	1.78
	Heart rate	x	69.50	76.09	72.18	81.63	84.99	83.30	90.39	88.82	96.74	94.77	85.67	89.55
		S	11.27	8.45	15.15	12.31	13.37	16.64	11.00	10.09	16.81	20.61	10.98	13.58
	Oxygen uptake	x	0.30	0.30	0.31	0.46	0.52	0.52	0.54	0.53	0.70	0.68	0.62	0.68
		S	0.15	0.14	0.16	0.23	0.17	0.18	0.18	0.20	0.28	0.18	0.20	0.21

[a] x = mean; S = standard deviation.

From Mital, 1984a, J. Occup Accid., 5, 223–231.

TABLE 14.40

Maximum Weight of Lift Acceptable to Female Industrial Workers for 12 h and Their Physiological Responses at That Weight

Lift Height	Box Size (cm)		Frequency of Lift (Lifts/Min)											
			1			4			8			12		
			30.48	45.72	60.96	30.48	45.72	60.96	30.48	45.72	60.96	30.48	45.72	60.96
Floor to knuckle	Weight (kg)	x[a]	11.90	11.71	10.18	10.43	9.46	10.17	8.87	7.55	7.53	7.42	8.37	8.17
		S	2.27	2.38	2.06	2.73	2.37	2.67	2.44	2.10	1.00	2.16	2.64	2.47
	Heart rate (beats/min)	x	89.08	92.46	87.18	103.15	99.60	97.54	108.61	111.09	98.16	116.44	121.78	113.51
		S	7.22	10.11	9.92	15.02	17.13	14.07	19.08	8.01	10.18	17.67	14.86	12.43
	Oxygen uptake (l/min)	x	0.36	0.30	0.36	0.52	0.44	0.55	0.64	0.54	0.50	0.57	0.66	0.56
		S	0.17	0.15	0.17	0.13	0.20	0.27	0.37	0.13	0.16	0.17	0.22	0.20
Knuckle to shoulder	Weight	x	10.06	10.58	9.41	9.07	9.32	8.91	8.15	7.86	7.93	6.93	7.91	7.50
		S	1.97	2.47	2.18	1.64	2.17	2.29	1.58	1.30	2.11	0.89	1.67	2.00
	Heart rate	x	90.89	89.07	88.27	98.57	98.33	103.52	107.57	100.41	102.08	114.70	120.51	102.73
		S	12.11	9.45	13.25	9.76	10.95	12.98	12.64	8.67	14.65	12.43	13.79	16.54
	Oxygen uptake	x	0.25	0.26	0.16	0.34	0.34	0.37	0.43	0.45	0.45	0.45	0.59	0.47
		S	0.13	0.15	0.15	0.18	0.21	0.24	0.13	0.07	0.22	0.15	0.17	0.15
Shoulder to reach	Weight	x	9.38	7.92	7.66	8.57	7.98	8.03	7.93	7.93	7.57	7.33	6.46	7.15
		S	1.89	1.34	0.96	1.94	1.39	1.68	1.03	1.66	1.50	0.93	0.91	1.04
	Heart rate	x	86.80	86.65	83.07	97.58	93.58	84.12	106.20	94.96	105.83	102.44	111.27	109.67
		S	8.91	9.10	6.66	11.87	14.03	13.04	12.85	11.57	14.58	15.11	10.13	18.52
	Oxygen uptake	x	0.15	0.24	0.18	0.35	0.28	0.23	0.49	0.30	0.47	0.31	0.39	0.46
		S	0.09	0.18	0.09	0.12	0.17	0.08	0.27	0.20	0.23	0.10	0.18	0.19

[a] x = mean; S = standard deviation.

From Mital, 1984a, *J. Occup. Accid.*, 5, 223–231.

TABLE 14.41

Maximum Weight of Lift Acceptable to Male Industrial Workers for 8 h and Their Physiological Responses at That Weight

Lift Height	Box Size (cm)		Frequency of Lift (Lifts/Min)											
			1			4			8			12		
			30.48	45.72	60.96	30.48	45.72	60.96	30.48	45.72	60.96	30.48	45.72	60.96
Floor to knuckle	Weight (kg)	x[a]	16.83	15.12	16.46	15.42	14.28	13.60	14.99	12.46	11.11	10.57	10.12	9.62
		S	4.68	3.48	3.47	4.52	3.92	3.85	5.34	2.30	1.69	3.50	2.55	2.10
	Heart rate (beats/min)	x	90.72	82.36	81.04	96.68	97.88	101.75	114.85	111.15	109.60	111.74	118.03	114.87
		S	14.33	12.35	13.83	16.76	7.86	20.80	20.16	14.93	22.31	19.52	20.06	13.88
	Oxygen uptake (1/min)	x	0.47	0.46	0.49	0.83	0.79	0.81	1.03	1.03	1.08	1.20	0.96	1.08
		S	0.22	0.19	0.22	0.20	0.21	0.23	0.26	0.24	0.19	0.23	0.20	0.31
Knuckle to shoulder	Weight	x	15.46	16.27	14.86	13.56	14.47	13.51	12.96	11.89	10.63	11.31	11.88	10.35
		S	3.96	4.17	3.81	3.15	2.53	4.15	2.70	2.68	2.48	2.73	3.03	2.56
	Heart rate	x	93.32	91.96	84.76	105.72	102.62	100.48	109.80	99.86	101.73	107.37	119.20	109.32
		S	15.97	10.83	12.29	18.72	13.33	17.54	13.22	12.72	20.36	18.76	14.83	13.53
	Oxygen uptake	x	0.34	0.46	0.44	0.62	0.70	0.72	0.77	0.83	0.75	0.92	0.92	0.72
		S	0.15	0.22	0.11	0.21	0.23	0.26	0.21	0.25	0.09	0.26	0.28	0.13
Shoulder to reach	Weight	x	13.05	12.19	13.62	12.02	12.81	11.94	11.28	11.56	12.26	11.44	9.63	10.34
		S	2.94	2.49	3.74	4.13	4.06	2.18	3.02	1.97	4.24	2.74	1.97	2.47
	Heart rate	x	76.76	82.67	78.26	90.07	93.97	92.80	100.63	97.31	105.18	103.61	94.12	100.72
		S	13.08	8.30	18.76	12.27	12.53	18.36	13.15	10.90	18.85	20.67	20.67	16.46
	Oxygen uptake	x	0.37	0.39	0.37	0.57	0.64	0.60	0.70	0.64	0.83	0.79	0.77	0.82
		S	0.18	0.16	0.14	0.25	0.21	0.19	0.19	0.21	0.30	0.20	0.21	0.24

[a] x = mean; S = standard deviation.

From Mital, 1984.

TABLE 14.42

Maximum Weight of Lift Acceptable to Female Industrial Workers for 8 h and Their Physiological Responses at That Weight

Lift Height	Box Size (cm)		Frequency of Lift (lifts/min)											
			1			4			8			12		
			30.48	45.72	60.96	30.48	45.72	60.96	30.48	45.72	60.96	30.48	45.72	60.96
Floor to knuckle	Weight (kg)	x^a	13.33	12.94	11.40	11.84	10.52	11.51	10.47	9.05	8.64	8.41	9.54	9.83
		S	2.39	2.63	2.19	3.47	2.64	2.74	3.76	2.07	1.22	2.30	2.88	2.72
	Heart rate (beats/min)	x	92.23	95.19	91.97	107.94	104.19	99.75	115.06	116.55	101.76	123.04	123.75	118.36
		S	7.28	10.30	13.13	14.59	17.38	15.12	22.74	8.87	10.19	17.60	18.09	12.43
	Oxygen uptake (1/min)	x	0.40	0.35	0.39	0.56	0.51	0.60	0.71	0.63	0.59	0.60	0.79	0.62
		S	0.19	0.16	0.17	0.11	0.25	0.28	0.44	0.16	0.21	0.15	0.21	0.16
Knuckle to shoulder	Weight	x	11.37	11.69	10.45	10.30	10.38	10.34	9.52	8.90	8.93	8.07	8.34	8.40
		S	2.01	2.73	2.38	1.73	2.38	3.00	1.85	1.28	2.23	1.16	2.04	2.31
	Heart rate	x	94.07	91.34	91.59	102.44	101.57	105.71	111.90	100.80	105.79	118.98	123.62	106.77
		S	12.47	9.37	14.51	9.93	11.15	11.00	12.83	10.35	15.61	14.00	14.47	17.50
	Oxygen uptake	x	0.28	0.28	0.22	0.41	0.39	0.44	0.54	0.51	0.53	0.53	0.67	0.53
		S	0.14	0.17	0.09	0.22	0.21	0.29	0.18	0.09	0.24	0.22	0.21	0.21
Shoulder to reach	Weight	x	10.53	8.94	8.68	9.65	9.32	8.97	8.95	9.10	9.02	8.31	7.68	8.21
		S	2.05	1.61	0.95	2.09	1.35	1.82	1.15	2.14	2.39	1.03	1.16	1.26
	Heart rate	x	90.16	89.75	86.54	100.19	96.95	86.70	109.34	100.12	109.35	106.39	115.14	114.89
		S	8.96	10.21	7.33	12.13	14.82	12.78	11.36	16.28	18.28	17.19	11.19	21.55
	Oxygen uptake	x	0.18	0.27	0.22	0.40	0.35	0.28	0.54	0.38	0.55	0.37	0.47	0.53
		S	0.11	0.19	0.10	0.15	0.17	0.10	0.23	0.23	0.24	0.14	0.21	0.21

[a] x = mean; S = standard deviation.

From Mital, 1984.

TABLE 14.43

Decrement (%) in Maximum Acceptable Weight of Lift with Frequency, Box Size, and Height of Lift

Task Factor	12-hr Shift		8-hr Shift	
	Males	Females	Males	Females
Box size (cm)				
30.48	100[a]	100	100	100
45.72	94	98	94	98
60.96	92	95	92	95
Frequency (lifts/min)				
1	100	100	100	100
4	91	92	92	93
8	81	80	81	83
12	69	76	71	78
Height of lift				
Floor to knuckle	100	100	100	100
Knuckle to shoulder	97	92	96	92
Shoulder to reach	89	84	87	84

[a] Numbers rounded to the nearest integer.

From Mital, 1984.

TABLE 14.44

Maximum Acceptable Weights and Maximum Voluntary Isometric Strength Expressed as Percentage of the Corresponding Values in the Sagittal Plane

Variable	Angle from the Sagittal Plane (Degrees)	Box Size								
		Small			Medium			Large		
		Mean	SD	Range	Mean	SD	Range	Mean	SD	Range
Maximum acceptable weight (%)	0	100.0	—	—	100.0	—	—	100.0	—	—
	30	93.7	2.3	88.5–97.2	93.5	2.5	89.6–97.7	92.7	2.8	89.2–96.6
	60	84.8	6.6	66.6–91.5	85.7	2.9	81.2–90.1	85.9	4.2	80.3–93.9
	90	76.8	6.4	61.7–87.7	79.4	4.3	70.8–85.4	77.4	5.8	67.4–86.6
Maximum voluntary isometric strength	0	100.0	—	—	100.0	—	—	100.0	—	—
	30	87.0	5.6	72.7–93.3	87.8	7.1	67.6–94.4	88.9	5.9	74.1–96.5
	60	77.6	8.5	63.6–90.7	78.7	6.9	59.4–86.1	81.5	7.5	66.7–89.7
	90	69.3	10.6	46.7–88.0	68.7	9.2	48.6–81.8	69.3	10.6	47.4–82.7

From Garg and Badger, 1986, *Ergonomics, 29*(7), 879–892.

References

Addison, R. and Schultz, A. 1980.Trunk strengths in patients seeking hospitalization for chronic low back disorders, *Spine*, 5, 539–544.

Ayoub, M., Bethea, N.J., Deivanayangam, S., Asfour, S.S., Bakker, G.M., Liles, D., Mital, A., and Sherif M. 1978. Determination and modeling of lifting capacity. DHEW (NIOSH) Report.

Ayoub, M.M., Selan, J.L., and Liles, D.H. 1983. An ergonomics approach for the design of manual materials handling tasks, *Hum. Factors*, 25(5), 507–515.

Ayoub, M.M. and Mital, A. 1989. *Manual Materials Handling* (Taylor and Francis: London).

Biemborn, D.S. and Morrissey, M.C. 1988. A review of the literature related to trunk muscle performance, *Spine*, 13(6), 655–660.

Berkson, M., Schultz, A., Nachenson, A., and Anderson, G. 1977. Voluntary strengths of male adults with acute low back syndromes, *Clin. Orthop.*, 129, 84–95.

Bernard, B.P. and Fine, L.J. 1997. Musculoskeletal disorders and workplace factors, DHHS (NIOSH) Publication No. 97B141, Cincinnati, OH.

Bigos, S., Battie, M.C., and Spengler, D.M. 1991. A prospective study of work perceptions and psychosocial factors affecting the report of back injury, *Spine*, 16, 1–6.

Borg, G.A.V. 1962. Physical performance and perceived exertion, *Studia Psychologicaet Paedagogica*, 5, 62.

Bureau of Labor Statistics. 1999. Lost work time injuries and illnesses: characteristics and resulting time away from work, 1977. USDL 99-102, Washington, DC.

Caldwell, L.S. 1959. The effect of the spatial position of a control on the strength of six linear hand movements, USA: Report No. 411, Medical Research Laboratory.

Chaffin, D.B., Herrin, G.D., and Keyserling, W.M. 1978. Pre-employment strength testing. An updated position, *J. Occup. Med.*, 20(6), 403–408.

Chaffin, D.B., Herrin, G.D., Keyserling, W.M., and Garg, A. 1977. A method for evaluating the biomechanical stresses resulting from manual materials handling jobs, *AIHAJ*, 38, 662–675.

Chaffin, D.B. and Park, K.S. 1973. A longitudinal study of low-back pain as associated with occupational weight lifting factors, *AIHAJ*, , 513–525.

Copley, F.B. 1923. *Frederick W. Taylor*, Vol. 1 (New York: Harper and Bros.).

Dempsey, P., Burdof, A., and Webster, B.S. 1997. The influence of personal variables on work-related low-back disorders and implications for future research, *J. Occup. Environ. Med.* 39, 748–759.

Duncan, A.N. and Ahmed, A.M. 1991. The role of axial rotation in the aetiology of unilateral disc prolapse. An experimental and fine element analysis, 16, 1089–1098.

Ekblom, B. and Goldberg, N.A. 1971. The Influence of training and other factors on the subjective rating of perceived exertion. *Acta. Physiol. Scand.*, 83, 399–486.

Emanuel, I., Chaffee, J., and Wing, J.A. 1956. A study of human weight lifting capabilities for loading ammunition into the F-86 aircraft. U.S. Air Force, Technical Report, WADC-TR 056-367, August 1956.

Flint, M. 1955. Effect of increasing back and abdominal muscle strength on low back pain. *Res.Q.*, 29, 160–171.

Frymoyer, J. and Cats-Baril, W. 1986. Predictors of low back pain disability, *Clin. Orthop.*, 221,89–98.

Frymoyer, J., and Cats-Baril, W. 1991. An overview of the incidences and costs of low back pain, *Orthop. Clin. North Am.*, 22, 263–271.

Frymoyer, J.W., Ducker, T.B., Hadler, N.M., Kostiuk, J.P., Weinstein, J.N., and Whitecloud, T.S. 1991. *The Adult Spine, Principles and Practice* (New York: Raven Press).

Frymoyer, J.W., Pope, M.H., Clements, J.H., et al. 1980. Risk factors in low back pain: an epidemiologic study, *J. Bone Joint Surg.*, 65, 213–218.

Frymoyer, J.W., Pope, M.H., Costanza, M.C., et al. 1983. Epidemiologic studies of low back pain. *Spine*, 5, 419–423.

Gamberale, F. 1972. Perceived exertion, heart rate oxygen uptake and blood lactate in different work operations, *Ergonomics*, 15(5), 545–554.

Garg, A. and Badger, D. 1986. Maximum acceptable weights and maximum voluntary isometric strengths for asymmetric lifting, *Ergonomics*, 29(7), 879–892.

Garg, A. and Beller, D. 1994. A comparison of isokinetic lifting strength with static strength and maximum acceptable weight with special reference to speed of lifting, *Ergonomics*, 37(8), 1363–1374.

Hadler, N. 2000. Comments on the "Ergonomics Program Standards" proposed by the Occupational Safety and Health Administrations, *J. Occup. Environ. Med.*, 42, 951–969.

Hasue, M., Fujiwara, M., and Kikuchi, S. 1980. A new method of quantitative measurement of abdominal and back muscle strength, *Spine*, 5(2), 143–148

Hattori, Y., Ono, Y., Shimaoka, M., Hiruta, S., Kamijima, M., Shibata, E., Ichihara, G., Ando, S., Beatriz, M., Villaneauva, G., and Takeuchi, Y. 1996. Effects of asymmetric dynamic and isometric lifting on strength/force and rating of perceived exertion, *Ergonomics*, 39, 862–876.

Hazard, R.G., Reeves, V., and Fenwick, J.W. 1992. Lift capacity: indices of subject effort, *Spine*, 17(9), 1065–1070.

Johansen, J.G. 1986. Computed tomography in assessment of myelographic nerve root compression in the lateral recess, *Spine*, 11(5), 492–495.

Kelsey, J.L. 1984. An epidemiologic study of lifting and twisting on the job and risk for acute prolapsed lumbar intervertebral disc, *J. Orthop. Res*, 2(1), 61.6.

Keyserling, W.M. 2000. Workplace risk factors and occupational musculoskeletal disorders, Part 2: a review of biomechanical and psychophysical research on risk factors associated with upper extremity disorders, *AIHAJ*, 61(1), 39–50.

Kroemer, K.H.E. 1983. An isoinertial technique to access individual lifting capability, *Hum. Factors*, 25(5), 493–506.

Kumar, S. 1987. Arm strength at different reach distances, in *Trends in Ergonomics/Human Factors IV*, S.S. Astour, Ed. (North-Holland), pp. 623–630.

Kumar, S. 1988. Moment arms of spinal musculature determined from CT scans, *Clin. Biomech.*, 3,137–144.

Kumar, S. 1990. Cumulative load as a risk factor for back pain, *Spine*, 15(12),1311–1316.

Kumar, S. 1991a. Functional evaluation of the human back: A research report. University of Alberta, Edmonton.

Kumar, S. 1991b. Arm lift strength in work place, *Appl. Ergon.*, 22(5), 317–328.

Kumar, S. 1994a. A function evaluation of human back: isometric and isokinetic strength of trunk muscles, *Eur. J. Phys. Med. Rehabil.*, 4(3), 73–82.

Kumar, S. 1994b. A conceptual model of overexertion, safety, and risk of injury in occupational settings, *Hum. Factors*, 197–209.

Kumar, S., 1995. Prediction of lifting strength in three-dimensional space, *Appl. Ergonomics*, 26, 327–341.

Kumar, S. 1996a. Isolated planar trunk strength and mobility measurement for the normal and impaired backs: Part I — The devices, *Int. J. Ind. Ergon.*, 17, 81–90.

Kumar, S. 1996b. Trunk strength and mobility measurement for the normal and impaired backs: Part II. Protocol, software logic and sample results, *Int. J. Ind. Ergon.*, 17, 91–101.

Kumar, S. 1996c. Isolated planar trunk strength measurement in normals: Part III. Results and database, *Int. J. Ind. Ergon.*, 17, 103–111.

Kumar, S. 1996d. Spinal compression during peak isometric and isokinetic lifting, *Clin. Biomechanics*, 11, 281–289.

Kumar, S. 1997. Axial rotation strength in seated neutral and prerorated postures of young adults, *Spine*, 22, 2213–2221.

Kumar, S. 1997. The effect of sustained spinal load on intra-abdominal pressure and EMG characteristics of trunk muscles, , 40(12), 1312–1334.

Kumar, S. 2001. Theories of musculoskeletal injury causation, *Ergonomics*, 44(1), 17–47.

Kumar, S. 2002. Trunk rotation: ergonomic and evolutionary perspective, *Theor. Issues Ergon. Sci.*, 3, 235–256.

Kumar, S., Chaffin, D.B., and Redfern, M. 1988. Isometric and isokinetic back and arm lifting strengths: device and measurement, *J. Biomech.*, 21, 1. 35–44.

Kumar, S., Dufresne, R.M., and Garand, D. 1991. Effect of body posture on isometric and torque producing capability of the back, *Int. J. Ind. Ergon.*, 7, 53–62.

Kumar, S., Dufresne, R.M., and Schoort, T.V. 1995. Human trunk strength profile in lateral flexion and axial rotation, *Spine*, 20(2), 169–177.

Kumar, S. and Garand, D. 1992. Static and dynamic strength at different reach distances on symmetrical and asymmetrical planes, *Ergonomics*, 35, 35–44.

Kumar, S. and Mital, A. 1992. Margin of safety for the human back: a probable consensus based on published studies, *Ergonomics*, 35, 769–781.

Kumar, S., Narayan, Y., and Chouinart, K. 1997. Effort reproduction accuracy in pinching, gripping and lifting among industrial males, 20, 109–119.

Kumar, S., Narayan, Y., and Garand, D. 2002. Isometric axial rotation of the trunk from prerotated postures, *Eur. J. Appl. Physiol.*, 87, 7–16.

Kumar, S., Narayan, Y., Stein, R.B, et al. 2001. Muscle fatigue in axial rotation of the trunk, *Int. J. Ind. Ergon.*, 28(2), 113–125.

Kumar, S., Narayan, Y., and Zedka, M. 1998. Trunk strength in combined motions of rotations and flexion extension in normal young adults, *Ergonomics*, 41(6), 835–852.

Kumar, S. and Simmonds, M. 1994. The accuracy of magnitude production of submaximal precision and power grips and gross motor efforts, *Ergonomics*, 37(8), 1345–1353.

Kumar, S., Simmonds, M., and Lechelt, D. 1994. Maximal and graded effort perception by young females in stoop lifting, hand grip, and finger pinch activity with comparisons to males, *Int. J. Ind. Ergon.*, 13, 3–13.

Langarana, N., Lee, C., Alexander, H., and Mayott, C. 1984. Quantitative assessment of back strength using isokinetic testing, *Spine*, 9, 287–297.

Mamansari, D., and Salokhe, V. 1996. Static strength and physical work capacity of agricultural labourers in the central plain of Thailand, *Appl. Ergon.*, 27, 53–60.

Manning, D.P., Mitchell, R.G., Blanchfield, L.B. 1984. Body movements and events contributing to accidental and non-accidental back injuries, 9, 734–749.

Marras, W., Lavender, S., Leurgans, S., et al. 1993. The role of dynamic three-dimensional trunk motion in occupationally-related low back disorders, *Spine*, 18, 617–628.

Mayer, T.G., Barnes, D., Kishino, N.D., Nichols, G., Gatchel, R.J., Mayer, H., and Mooney, V. 1988. Progressive isoinertial lifting evaluation. I. A standardized protocol and normative database, *Spine*, 13(9), 993–997.

Mayer, T.G., Gatchel, R.J., Kishino, N., Keeley, J., Capra, P., Mayer, H., Barnett, J., and Mooney, V. 1985a. Objective assessment of spine function following industrial injury. A prospective study with comparison group and one-year follow-up, *Spine*, 10(6), 482–493.

Mayer, T.G., Smith, S.S., Keeley, J., and Mooney, V. 1985b. Quantification of lumbar function. Part 2: Sagittal plane trunk strength in chronic low-back pain patients, *Spine,* 10(8), 765–772.

Mayer, T.G., Smith, S.S., Kondraske, G., Gatchel R.J., Carmichael, T.W., and Mooney, V. 1985c. Quantification of lumbar function. Part 3: preliminary data on isokinetic torso rotation testing with myoelectric spectral analysis in normal and low-back pain subjects, *Spine*, 19(10), 912–920.

McGill, S. and Hoodless, K. 1990.

Mcneill, T., Warwick, D., Andersson, G., and Schultz, A. 1980. Trunk strengths in attempted flexion, extension and lateral bending in healthy subjects and patients with low back disorders, *Spine*, 5, 529–538.

Mital, A. 1983. The psychophysical approach in manual lifting — a verification study, *Hum. Factors*, 25(5), 485–491.

Mital, A. 1984a. Prediction of maximum weights of lift acceptable to male and female industrial workers, *J. Occup Accid.*, 5, 223–231.

Mital, A. and Faard, H.F. 1990. Effects of sitting and standing, reach and distance and arm orientation on isokinetic pull strengths in the horizontal plane, *Int. J. Ind. Ergon.*, 6, 241–248.

Mital, A., Karwowski, W., Mazouz, K., and Orsarh, E. 1986. Prediction of maximum acceptable weight of lift in the horizontal and vertical planes using simulated job dynamic strengths, *AIHAJ*, 47(5), 288–292.

Mital, A., Nicholson, A.S., and Ayoub, M.M. 1993. *A Guide To Manual Materials Handling* (Taylor and Francis: London).

Mostardi, R.A., Noe, D.A., Kovacik, M.W., and Porterfield, J.A. 1992. Isokinetic lifting strength and occupational injury. A prospective study, *Spine*, 17(2), 189–193.

Nachemson, A. and Lindh, M. 1969. Measurement of abdominal and back muscle strength with and without low back pain, *Scand. J. Rehab. Med.*, 1, 60–65.

National Academy of Science (2001). *Musculoskeletal Disorders and the Workplace* (Washington, DC: National Academy Press).

Newton, M. and Waddell, G. 1993. Trunk strength testing with iso machines. Part I: Review of a decade of scientific evidence, *Spine*, 18(7), 801–811.

Newton, M., Thow, M., Somerville, D., Henderson, I., and Waddell, G. 1993. Trunk strength testing with iso machines. Part 2: Experimental evaluation of the Cybex II back testing system in normal subjects and patients with chronic low back pain, *Spine*, 18(7), 812–824.

Nicholson, L.M.L.S. 1986. A psychophysical study of the effects of load and frequency upon selection of workload in repetitive lifting, *Ergonomics*, 29(7), 903–911.

Nicholson, L.M. and Legg, S.J. 1986. A psychophysical study of the effects of load and frequency upon selection of workload in repetitive lifting, *Ergonomics*, 29, 903–911.

NIOSH (National Institute of Occupational Safety and Health). Work practices guide for manual lifting. Dept of Health and Human Services 1981: Publication No. 81-122.

Nordin, M., Andersson, G.B., and Pope, M. (1997). *Musculoskeletal Disorders in the Workplace: Principles and Practice* (St. Louis: Mosby).

Park K.S. and Chaffin, D.B. 1974. A biomechanical evaluation of two methods of manual load lifting, *Am. Inst. Industrial Eng. Trans.*, 6, 105–113.

Pandolf, K.B., Cafarelli, E., Noble, B.J., and Metz, K.F. 1972. Perceptual response during prolonged work, *Percept. Mot. Skills*, 35, 975–985.

Parnianpour, M., Li, F., Nordin, M., and Kahanovitz, N. 1989. A database of isoinertial trunk strength tests against three resistance levels in sagittal, frontal, and transverse planes in normal male subjects, *Spine*, 14(4), 409–411.

Pope, M. 1998. Does manual materials handling cause low back pain? *Int. J. Ind. Ergon.*, 22, 489–492.

Pope, M.H., Bevins, T., Wilder, D.G., and Frymoyer, J.W. 1985. The relationship between anthropometric postural, muscular, and mobility characteristics of males ages 18–55, *Spine*, 19(7), 644–648.

Pope, M.H., Svensson, M., Andersson, G.B.J., Broman, H., and Zetterberg, C. 1987. The role of prerotation of the trunk in axial twisting efforts, *Spine*, 12(10), 1041–1045.

Punnett, L. 2000. Commentary on the scientific basis of the proposed Occupational Safety and Health Administration ergonomics program standards, *J. Occup. Environ. Med.*, 42, 970–981.

Pytel, J.L. and Kamon, E. 1981. Dynamic strength test as a predictor for maximal and acceptable lifting, *Ergonomics*, 24(9), 663–672.

Ralston, H.J., Inman, V.T., Strait, L.A., et al. 1974. Mechanics of human isolated voluntary muscle, 151, 612–620.

Rhumann, H. and Schmidtke, H. 1989. Human strength: measurements of maximum isometric forces in industry, *Ergonomics*, 32, 865–879.

Sanchez, D. and Grieve, D.W. 1992. The measurement and prediction of isometric lifting strength in symmetrical and asymmetrical postures, *Ergonomics*, 35, 49–64.

Schaffer, H. 1982. Back injuries associated with lifting, Bulletin 2144 (Washington, DC, Department of Labor, Bureau of Statistics), 1–20.

Schmidt, G., Amundsen, L.R., and Dostal, W.F. 1980. Muscle strength at the trunk. *J. Orthop. Sport Phys. Ther.*, 1, 165–170.

Shankland, M. and Seaver, E.C. 2002. Evolution of the bilaterian body plan: what have we learned from annelids, *Proc. Natl. Acad. Sci. U.S.A.*, 97, 4434–4437.

Smith, S., Mayer, T., Gatchel, R., et al. 1985. Qualification of lumbar function, *Spine*; 10, 757–764.

Snook, S.H. 1978. The design of manual handling tasks, *Ergonomics*, 21(12), 963–985.

Snook, S.H., Campanelli, R.A., and Hart, J.W. 1980. A study of back injuries at Pratt and Whitney Aircraft.

Statistics Canada. 1995. *Work Injuries*. Ottawa.

Stevens, S.S. 1960. The psychophysics of sensory function, *Am Sci.*, 48, 226–253.

Stevenson, J. 1990. The effect of lifting protocol on comparisons with isoinertial lifting performance, *Ergonomics*, 33(12), 1455–1469.

Suzuki, N.E.S. 1983. A quantitative study of trunk muscle strength and fatigability in the low back pain syndrome, *Spine*, 8(1), 69–74.

Switzer, S.A. 1962. Weight lifting capabilities of a selected sample of human males. Wright-Patterson AFB, OH: Aerospace Medical Research Laboratory, Report No. AP-284054.

Thorstensson, A. and Arvidson, A. 1982. Trunk muscle strength and low back pain, *Scand. J. Rehabil. Med.*, 14, 69–75.

Thorstensson, A. and Nilsson, J. 1982. Trunk muscle strength during constant velocity movements, *Scand. J. Rehabil. Med.*, 14, 61–68.

Videman, T., Nurminen, M., and Troup, J.D. 1990. Lumbar spinal pathology in cadaveric material in relation to history of back pain, occupation and physical loading, *Spine*, 15, 728–740.

Vink, P., Daaneen, A.M., Meijst, W.J., and Ligteringen, J. 1992. Decrease in back strength in asymmetric trunk postures, *Ergonomics*, 35(4), 405–416.

Waters, T., Putz-Anderson, V., Garg, A., and Fine, L.J. 1993. Application manual for the revised NIOSH lifting equation, U.S. Department of Health and Human Services, CDC, NIOSH, Cincinnati.

Waters, T., Putz-Anderson, V., Garg, A., and Fine, L.J. 1993. Revised NIOSH equation for the design and evaluation of manual lifting tasks, *Ergonomics*, 36, 749–776.

World Health Organization. 1995. Global Strategy on Occupational Health for All, WHO, Geneva.

Yates, J.W. and Karwowski, W. 1987. Maximum acceptable lifting loads during seated and standing work positions, *Appl. Ergon.*, 18(3), 239–243.

Yates, J.W., Kamon, E., Rodgers, S.H., and Champney, P.C. 1980. Static lifting strength and maximal isometric voluntary contractions of back, arm and shoulder muscles, *Ergonomics*, 23(1), 37–47).

15

Pushing and Pulling Strength

Robert O. Andres

CONTENTS

ABSTRACT Pushing and pulling actions take place in the workplace and away from the workplace on a daily basis. While many of these actions require minimal exertion, sometimes strength is needed to complete a task. As ergonomists involved in either task analysis or task design, we need to understand human capabilities to exert push and pull forces required by specific task(s). This chapter reviews the research relevant to determining human capability to exert these required forces and how these forces are determined.

15.1 Ergonomic Relevance

This chapter provides a review of human pushing and pulling research and guides task design that includes pushing and pulling exertions.

15.2 Introduction

What is the distinction between pushing and pulling vs. lifting? On a mechanical basis, any task that requires a resultant force (most often exerted at the hands, but not always) that includes a horizontal component involves pushing or pulling. Pushing implies that the horizontal component is directed away from the whole-body center of mass (CM — the point of an object or system which may be treated as if the entire mass of the object or system were concentrated at that point, and any external translational forces appear to act through that point – from Stramler, 1993); pulling implies a horizontal component directed towards the whole-body CM. Pure lifting or lowering tasks have resultant forces with vertical components only. From this strict definition, it can be found that most manual materials-handling (MMH) tasks involve some pushing or pulling even if they are considered primarily lifting tasks.

15.2.1 Pushing and Pulling Tasks in Manual Materials Handling

One of the most comprehensive reviews of pushing and pulling strengths relevant to MMH is contained in Ayoub and Mital (1989). Several of the investigations that provided the databases for their compilation are discussed in this chapter; however, research since their review has improved our understanding of pushing and pulling strength expression and requirements.

A common industrial MMH task that focuses on pushing and pulling is the use of carts. A Swedish study (Winkel, 1983) concerned the manual handling of food and beverage carts on wide-body airplanes. Wheeled cages are pushed and pulled around to transport mail at the Dutch Postal Service (van der Beek, 1998). Large wheeled carts are used in food distribution centers for loading food onto trucks and delivering food to stores (Pabon-Gonzalez and Andres, 1999). Other examples of tasks predominantly requiring pushing or pulling would be the pulling of fire hoses by firemen, the pulling of lines or cables by deckhands on ships, (although these tasks may have a lifting component as well), or the transport of dustbins (Jaeger et al., 1984). Quite often pushing or pulling occurs to position items for further MMH purposes — such as pulling a box to the edge of a shelf to move it closer before lifting it. Or, in an effort to minimize lifting, jobs are redesigned such that pushing or pulling replaces lifting. Examples here include the use of carts to replace carrying items and the use of mechanical lifting aids like overhead hoists to vertically raise or lower objects that otherwise would have to be lifted. While the hoists can move something up and down, the operator usually has to push or pull on the object (while attached to the hoist) or on the hoist itself to position the object at its destination (Resnick and Chaffin, 1996).

15.2.2 Other Pushing and Pulling Tasks

Pushing and pulling is also an important consideration in the operation and control of machinery. Many work vehicles involve a seated operator manipulating levers to control vehicle motions and actuate specialized mechanisms (i.e., the bucket on a front-end loader, the boom on a crane, etc.) (Dempster, 1955; Gaughran and Dempster, 1956). Dupuis et al. (1955), Murrell (1965), and Van Cott and Kincade (1972) have studied the pull force that a subject can generate on a control lever from the seated position. Bench work in factories often requires lever manipulation to control a process or to fasten or release a work piece from a fixture. Assembly lines employ levers to control equipment at particular stations. Thumbs and fingers push during assembly operations to attach fasteners, engage electrical connections, and accomplish other tasks (Wright et al., 1998). Finger pulling with pinch grips has also been studied (Imrhan and Sundararajan, 1992), because a handle compatible with a power grip is not always available. While the manipulation of control levers or buttons usually requires less strength than MMH tasks, the frequency of exertion may be dramatically greater. Therefore, MMH is not the only situation where pushing and pulling strength capability is important.

Activating pedals in a seated posture has been investigated by Slater (1949), Rees and Graham (1952), Hugh-Jones (1947), and Martin and Johnson (1952). Orlansky and Dunlop (1948) studied hand and foot forces needed for aircraft cockpit design. Extensive guidelines for seated arm strengths and pedal activation strength have been promulgated by the military (MIL-STD-1472F, 1999) and tabulated in design handbooks (e.g., Woodson et al., 1992).

15.2.3 Musculoskeletal Demands and Injury Potential

What happens when a worker does not have the required strength to complete a task requiring pushing and pulling? The most straightforward result in the case where the worker attempts to complete the task instead of giving up on it would be an overexertion type of injury to soft tissue — typically a ligament sprain or a musculotendinous strain, but sometimes involving other structures (intervertebral disks, menisci, etc.). The risk of these overexertion injuries would be expected to follow the trend found in lifting research: individuals having to use a greater percentage of their strength capability in a particular posture to complete a task are at a greater risk for overexertion. This provides the impetus for designing occupational pushing and pulling tasks well within the strength capabilities of the relevant worker population.

There are other scenarios, however, that are unique to pushing and pulling tasks among MMH tasks. An overexertion incident while pushing a loaded cart up a ramp could result in the cart coming back on the employee, causing contact injuries or worse (if the employee gets pushed into equipment or off a ledge). An overexertion injury while pulling a loaded cart may allow the cart to continue moving into the injured employee. The most extreme example may be found in an airplane cockpit scenario where the pilot needs to pull the handle or face curtain for seat ejection. If the pull force exceeds the pilot's capability, then whether an overexertion injury occurs or not is unfortunately moot. The concept of equipment or machinery controls requiring more push or pull strength than the user can muster summons up a plethora of scenarios involving injury both to the operator and to coworkers.

Biomechanical models are another method of predicting these risks. Several biomechanical models have been developed to predict strength requirements (Garg and Chaffin, 1975; Schanne, 1972) and L5/S1 disk loading (Chaffin, 1975; Garg, 1973; Park, 1973) in static or isometric conditions. Dynamic biomechanical models for lifting have been reported (Ayoub and El-Bassoussi, 1978; Freivalds et al., 1984; McGill and Norman, 1985). The doctoral research of Lee (1982) formulated a dynamic biomechanical model of cart pushing and pulling. This model predicted the slip potential and back loading for given hand forces as a function of task and subject variables. In particular, the inputs to the model included: subject anthropometry, body postures during dynamic tasks, and hand forces exerted on the cart handle. The specific model predictions were: horizontal and vertical foot forces for given hand forces and gross torso muscle and vertebral column loadings when pushing or pulling. Detailed validation testing of a modified version of this dynamic push/pull model over a wide range of subject anthropometry revealed that correlations between predicted back loading and measured back root-mean-square elec-tromyograms (RMS–EMGs) were low (Andres and Chaffin, 1991), but predictions of required slip resistance to prevent foot slip were more reasonable. Hoozemans et al. (1998) provided an extensive review of push/pull risk factors for the development of muscu-loskeletal disorders.

15.2.4   Environmental Demands That Limit Pushing and Pulling Strength

The application of resultant forces with horizontal components, whether through the hands or another region of the body (i.e., the shoulder), can create horizontal forces on the whole-body CM that either must be counteracted to maintain the position of the whole-body CM or motion of the whole-body CM will occur. These counteracting forces are usually generated by reactive forces at the feet (when the task involves standing) or by reactive forces exerted by the seat on a seated worker.

The friction characteristics between the worker's shoes and the floor can limit the amount of push or pull strength that can be expressed (Redfern and Andres, 1984). The coefficient of friction between shoes and floors has been used to represent the slip-resis-tance of industrial situations (Andres and Chaffin, 1985). Mechanically, the static coefficient of friction (SCOF) is defined as the ratio between the shear force applied to the sensor (a shoe heel, sole sample, or a shoe itself) and the normal force exerted on the sensor (holding the sensor in contact with the floor with or without lubricants) at the point in time just before the sensor begins to move horizontally. In a situation where persons are standing with their feet in a fixed position while they push on an immovable object, a low SCOF will limit the amount of push force they can exert before their foot (or feet) start to slide backwards. Likewise, with a heavily loaded cart, the break-away force to start the cart moving is typically higher than the force required to keep it moving once it has started rolling. So even though it may not be immovable, the initial amount of horizontal force may create more shear force at the shoe/floor interface than the available friction can counteract, resulting in shoe slip. A different condition exists when the shoe is moving with respect to the floor while pushing or pulling on carts. Under these conditions, the SCOF is not applicable; a dynamic coefficient of friction (DCOF) expressing the ratio of shear to normal forces is required because these values can depend on relative surface velocities and specific characteristics of the interface (e.g., hydrodynamics of intervening lubricants, sensor material properties, etc.). Although not directly dealing with pushing and pulling task performance, more recent work has been completed concerning slip and fall mechanisms and coefficients of friction (Gronqvist, 1999; Hanson et al., 1999).

Another demand imposed by the working environment is the workspace available. In confined spaces, the body may be constrained to postures that do not permit the exertion of maximum force. This may be due to the lack of a sufficient support base to maintain balance, or to major joints having to operate in positions where their strength capability is decremented.

15.3   Measurement of Required Pushing and Pulling Forces

Strength measurement methods discussed elsewhere are relevant but are mostly used in the lab or the clinic as opposed to the field. Techniques for measuring the amount of force exerted when pushing or pulling range from inexpensive field techniques to elaborate and costly laboratory methods. There are advantages and disadvantages of every approach; these will be detailed next.

One of the least costly tools useful for push or pull force measurement is the spring force gauge. The minimal requirements for such a device are calibrated precision over the range of forces to be measured, ability to measure force when either pushing or pulling, and a mechanism to maintain the peak force reading. While these devices are the least expensive, the requirement for precision over a wide range (up to 120 kg) and the need to measure push forces preclude the use of fish scales or other cheap tools. Electronic digital versions of these gauges (using load cells) are available at some additional cost. The advantage of force gauges, besides their low cost, is portability; these devices are the easiest to take into the field to make on-site measurements. The primary disadvantage of the spring devices is their inability to measure dynamic forces continuously; some of the more expensive gauges have computer connectivity to avoid this shortcoming. Most importantly, these devices cannot directly measure the direction of force application.

Field measurements have also been made by instrumenting existing handles with load cells or strain gauges or by attaching instrumented frames to the item to be manipulated (Kreutzberg and Andres, 1985). The disadvantages of this approach involve the loss of portability (due to the requirement for data acquisition and storage via data logger or computer), the need for recalibration each time the system is attached to the item to be pushed or pulled, potential crosstalk between channels, and possible modification of the task. An example of the latter would be the situation where the instrumented handle(s) are mounted to a handle that already exists, changing the spatial relationships to the point that different torques or forces are expressed during the task than what would be seen at the actual handle.

Laboratory systems range from instrumented handles to force plates, or a combination of both. Static testing of standing exertions can be successfully performed with force plates alone if the body segments between the point(s) of force application and the feet are stationary. Instrumented handles in the laboratory are subject to the same limitations as in the field except for the concern for portability. Another disadvantage of these laboratory systems is cost — the necessary instrumentation is more costly than simple force gauges. Laboratory systems that employed both instrumented handles and force platforms have been described elsewhere (Andres and Chaffin, 1991; Lee, 1982).

Assessing push/pull forces in the field during construction tasks provided an opportunity to compare handheld digital force gauge readings with an instrumented frame attached to a concrete hopper (Hoozemans et al., 2001). When forces were applied simultaneously to the force gauge and the instrumented frame there were no significant differences between the mean force or the maximum force values during either pushing or pulling. Subjects were also asked to recreate the forces they had just exerted on the hopper by exerting forces with the hand-held force gauge; these simulations revealed that the simulated values were significantly lower than the actual measured values except for mean pull forces. All of the construction workers over-estimated the exerted forces. These authors found that the use of a simple force gauge to assess push/pull forces in the workplace was justified. The approach did not allow an assessment of the direction of force exertion, however, which is still essential to understanding joint loading and traction requirements.

Measurements of seated push and pull forces have typically been performed in the laboratory using load cells for registering static exertions. Depending on the purpose of the study, subjects are constrained by seat belts, shoulder harnesses, and other straps to minimize movement of the torso or other body segments (Stobbe, 1982).

15.4 Review of Research on Pushing and Pulling Strength

Studies of human pushing and pulling force capacity generally fall into 3 categories. Biomechanical investigations measure these forces either directly or indirectly during specific task performance. Psychophysical research follows established protocols to estimate human capacity to push or pull to meet particular criteria (usually the acceptable level of loading that can voluntarily be exerted at a given frequency for a given duration). Finally, the metabolic demands of pushing and pulling tasks have been included within psychophysical studies or directly examined with physiological studies. Table 15.1 summarizes much of the following research for comparison or task design purposes.

15.4.1 Biomechanical Research

15.4.1.1 Static or Isometric Push/Pull Strength Capability

Pushing and pulling in the standing position will be discussed first, followed by seated pushing and pulling. Finally, cart pushing and pulling involving realistic demands in industry where static strength is an issue are discussed.

Dempster (1958) performed one of the first studies of static pull forces during standing. Kroemer measured maximal isometric pushing forces in 65 different positions and assessed the effects of varied foot friction during pushing (Kroemer, 1969; Kroemer and Robinson, 1971). In some conditions, his subjects were allowed to brace themselves against an external structure. Ayoub and McDaniel (1974) had their subjects either push or pull on a horizontal bar mounted at different heights based on either the subject's reach height or shoulder height. Foot distance from the vertical projection of the handle to the floor was varied also. Forces were exerted with arms extended at the elbows and either the rear leg (pushing) or the front leg (pulling) extended at the knee. A high-friction floor was employed, but there was no bracing against external objects. Bar height ranged from 30–80% reach height for males and 60–100% shoulder height for females. Foot distance during pulling ranged from 30–70% reach height for males and 50–90% shoulder height for females; for pushing, the corresponding male and female foot distances ranged from 1–30% reach height and 10–30% shoulder height, respectively. Pushing forces increased for both males and females when the foot distance increased (except at the highest handle heights for males). At the smallest foot distances there were only minimal differences between male and female push capability; the strength differences emerged where there was a greater foot distance advantage. The calculated efficiency during these pushing tasks was the ratio of the horizontal force to the resultant force; the greatest efficiency was found at a bar height of approximately 70 to 80% shoulder height and dropped off most markedly for the lowest foot distance as it varied from this height. Further testing to determine how long the subjects could maintain their maximum force before dropping off 10% was conducted to derive the optimum position. Maximum fatigue was found when foot distance was between 70 and 80% shoulder height, and the lower the bar height, the longer the sustention time. They concluded that the best position for pushing tasks occurs when force is applied at approximately 70% shoulder height (or 50% reach height) and a rear foot distance of 100% shoulder height (or 70% reach height). For pulling tasks, the pulling forces increased as the bar height decreased to 30% reach height for males (40% reach height for females), while the foot distance was at the 10% reach height. Their results removed the effects of arm strength by keeping the elbows locked. They also

TABLE 15.1

Review of Research on Pushing and Pulling

Authors	Dependent Variables	Subjects	Conditions	Significant Findings	Reference Data Format
Ayoub and McDaniel (1974)	Isometric push and pull strength	35 male, 11 female	Bar heights and foot distances varied	Pushing forces increased for both males and females when the foot distance increased Maximum fatigue found when foot distance was between 70–80% shoulder height Best position for pushing occurs when force is applied at 70% shoulder height and a rear foot distance of 100% shoulder height Pulling forces increased as bar height decreased to 40% reach height	Graphs and one table of the best configuration
Warwick, Novak, Schultz, and Berkson (1980)	Maximum voluntary isometric strengths	29 male	Handle at shoulder or knee height Feet in one of 5 positions with respect to handle (straight or twisting to either side) Both hands or one hand at a time Force exertion directions varied	Largest forces exerted when the work piece was directly in front and both hands used Use of single hand or bending/twisting decreased exertion force	Tables
Chaffin, Andres, and Garg (1983)	Maximal one-handed and two-handed push and pull forces and associated postures	3 male, 3 female	Three handle heights Two force directions One-handed and two-handed Two stances	Males stronger than females in this study Two-handed pushes and pulls exceeded one-handed exertions Postural changes explained the diminished strength when pushing or pulling on a high handle	Tables and body pictographs of mean body postures for maximum force generation
Daams (1993)	Isometric force exertions and associated postures	Expt. 1 – 3 male, 2 female Expt. 2 – 10 male, 10 female	Type of posture (free, standard, or functional) Force direction Handle heights varied	Highest forces exerted with freely chosen posture Forces and postures reproducible over short (2–3 days) and long (10 months) terms	Tables, charts, and body pictographs of postures

(continued)

TABLE 15.1 (CONTINUED)

Review of Research on Pushing and Pulling

Authors	Dependent Variables	Subjects	Conditions	Significant Findings	Reference Data Format
Fothergill, Grieve, and Pheasant (1992)	One-handed maximal pulling forces	16 female, 14 male	Four handle types Two handle heights Freely chosen grip type and posture	Hook grip or power grip used on bar and handle exceeded pull forces with precision grips on knobs Mechanical stress concentrations caused by handle features caused pain and discomfort Force exertion on well designed handles was dependent on body posture	Tables, graphs, and sketches of typical grip types
Voorbij and Steenbekkers (2001)	Isometric forces for pushing and pulling (and others)	750 male and female, segmented by decade	Fixed handle height Two-handed pushing in a free position One-handed pulling in a free position	Little decline in strength between 20 and 55 years of age Decrease in strength with age is similar in men and women Pushing and pulling strengths were correlated with gripping and twisting forces	Graphs with fit curves and tables
Kumar, Narayan, and Bacchus (1995)	Two-handed maximal pull-push isometric and isokinematic forces	20 male and 20 female	3 Handle heights 3 Different lateral planes Subjects strapped into support frame up to hips	Female subjects weaker than males in all experimental conditions Pulling strengths higher than pushing strengths in all conditions Strength declined for both genders and all handle heights with increasing postural asymmetry	Tables
Lee, Chaffin, Herrin, and Waikar (1991)	L5/S1 compressive loading during cart pushing and pulling	4 male, 2 female	3 Different pushing and pulling forces 3 Different handle heights 2 Moving speeds	Subject body weight affected low back compressive loads more in pulling than in pushing Middle handle height better for pushing and high handle height better for pulling to reduce low back loading Compressive forces greater for pulling than for pushing Compressive forces increased as cart speed increased	Graphs, regression equations

Study	Measures	Subjects	Variables	Results	Output
De Looze, van Greuningen, Rebel, Kingma, and Kuijer (2000)	Net joint torques in shoulder and lumbosacral joints; Force exertion against stationary bar while walking on treadmill	8 males	Steady pace in pushing (0.75 m/s) and pulling (0.5 m/s); Target forces varied; Handle heights varied	Exerted forces were more horizontal at the higher handle height and at the higher required force level; Shoulder torques hardly affected by handle height level; L5/S1 torques were low for pushing and high for pulling	Charts and graphs
Resnick and Chaffin (1996)	Peak push and pull hand forces; Peak hand velocities; Psychophysical ratings of perceived exertion	7 males, 3 females	Load levels varied; 2 Target sizes for destination; 2 Friction levels in the articulating joint of the material handling device; Sagittally symmetric vs. asymmetric postures	Symmetric pushing and pulling yielded the highest ratings of perceived exertion with the heaviest load and the highest friction in the articulations; Peak push and pull forces increased when the torso twisted as opposed to the symmetric task; Psychophysical responses did not reflect increased loading in the low back	Tables and graphs
Snook and Ciriello (1991)	Psychophysically acceptable push and pull forces	12 female, 10 male (Expt.1); 12 females (Expt. 2); 12 female, 10 male (Expt. 3); 6 males (Expt. 4)	Task frequency varied; Distance varied; Height varied; Duration varied	Extension and completion of tables of maximum acceptable forces for push and pull for males and females	Tables
Shoaf, Genaidy, Karwowski, Waters, and Christensen (1997)	Load capacity values on the basis of psychophysical, biomechanical, and physiological criteria	NA	Multipliers for vertical distance from floor, traveled distance, frequency of exertion, age group, body weight, and task duration	Load capacity values are lower than psychophysically established limits	Tables of multiplier values, and prediction equations of pushing and pulling exertion capacities

demonstrated that heavier subjects were capable of exerting greater pushing and pulling forces.

Maximum voluntary strengths of 29 healthy male subjects were measured during the performance of 120 activities including pushing or pulling in four different horizontal directions (Warwick et al., 1980). Eight different body positions were tested; two were symmetric in the sagittal plane, while the other six involved varying amounts of twisting (determined by foot positions with respect to the work piece). Subjects exerted forces with the right hand, the left hand, and with both hands from each position on a work piece located at shoulder height (142 cm) or at knee height (60 cm). Pushing and pulling were forward or backward or to the left and to the right. Mean force magnitudes over all subjects for all positions ranged from 74 to 386 N. The largest forces were generally exerted when the subjects were in the anterior position (facing the work piece) and used both hands. The use of a single hand or the need to bend and/or twist to reach the work piece generally decreased the maximum exertion force. These authors concluded that the details of body configuration can significantly influence maximum voluntary strengths.

In an extension of the study by Ayoub and McDaniel (1974), Chaffin et al. (1983) measured push and pull forces in unrestricted (but known) postures chosen by a group of subjects of varied anthropometry. Subjects were permitted to lift their heels off the floor and unlock elbows and knees; two stances were evaluated: feet symmetrically placed beside each other, and one foot in front of the other. Subjects tried different positions until they felt they had reached a position that maximized their strength. Three fixed handle heights were tested (68, 109, and 152 cm) for both one-handed and two-handed pushes and pulls. Six subjects (3 male, 3 female) participated in the experiment. Males were significantly stronger than females for all tasks. Pushing and pulling strengths were not significantly different when the feet were kept symmetrically located beside each other, but when one foot was placed in front of the other, the pushing strength was significantly greater than the pulling strength for males. Male strength was greatly diminished when using the high handle as compared to the low handle. Females had smaller pull force reductions with increasing handle height, but the push force did not vary significantly with increasing handle height. Two-handed pushes and pulls exceeded the forces exerted with one-handed pushes and pulls. Postural changes explained the diminished strength when pushing or pulling on a high handle. Subjects stood more erect, limiting the turning moments about the feet necessary to produce a large handle force. In the low handle position, however, the subjects (particularly the males) leaned forward more for pushing and used a deep squat backward lean for pulling, increasing the turning moment about the pivot foot. The disadvantage of the low handle height, however, was shown to be greater compressive loading on the low back. Mean body postures were presented such that task designers could lay out a workspace for maximum push or pull force exertion.

Static force exertion in standing positions characterized as standardized, functional, and freely chosen postures found that the forces were reproducible for all postures (Daams, 1993). However, the magnitude of the forces differed considerably. Standardized postures constrained the subjects to stand upright, pulling or pushing with a pronated hand, the elbow in 90° of flexion, and either one foot 30 cm in front of the other, or feet together. Handle height was either elbow or shoulder height. The only constraint on the free posture was that the preferred hand exerted the force. The functional postures were derived from the free postures from the first experiment, with three fixed handle heights (0.7, 1.3, and 1.7 m) tested. Results revealed that the highest forces could be exerted with the freely chosen posture, as opposed to either the standardized or functional postures. The author proposed that all standing static strength testing allow freely chosen postures for realistic assessment of force capability and standing room required.

Maximal pull forces as a function of handle type were acquired by Fothergill et al. (1992). Subjects pulled with one hand on four different handles at either 1.0 or 1.75 m height. Handle types were a handle, two shapes of knobs, and a steel bar. The floor provided high traction, and subjects were allowed to freely choose their posture. The precision grips used on the knobs yielded lower pull forces than the hook grip or power grip used on the bar and the handle. The prominent corners or ridges on the handle and one of the knobs caused pain and discomfort, whereas the other knob was smooth and provided a poor grip purchase. The poorer the handle design, the less effect handle height had on the force which could be exerted; force exertion on well designed handles was more dependent on body posture.

Changes in isometric pushing and pulling strength with age were investigated by Voorbij and Steenbekkers (2001). The pushing task involved freestanding two-hand pushes on a T-shaped handle instrumented with strain gauges. The pulling task was also a freely chosen posture with one hand pulling on an H-shaped instrumented handle. A total of 750 subjects ranging in age from 20 to over 80 participated, both males and females. It was unclear from their methodology whether handle heights were fixed or also freely chosen. They fitted curves to their data, omitting the data for subjects 30 to 50 years old, by means of asymmetric transition functions formed from a non-linear equation based on a logistic-dose-response curve. Results were presented in equation form and in graphs depicting strength vs. age for each gender. Pushing and pulling forces both decreased with age, similar to grip forces. Male push/pull strength was greater than female strength, but after age 50 the drop-off in the strength curve for males was steeper than for females.

Push and pull force guidelines have been published in MIL-STD 1472F (1999). Some of what were termed push and pull forces, however, were exerted vertically (i.e., pull up, push up, etc.), and so related more directly to isometric lifting strengths. One-handed and two-handed vertical push and pull strengths were tabulated for both standing and seated positions (Table XXV in the document) for 5th and 95th percentile males and females.

Isometric forces exerted on hand controls while seated are of interest for vehicle and process control design. Hand brakes, boom swing levers, and aircraft controls all require pushing and pulling from the seated position. Comprehensive tables for maximal forces produced by male college students pushing or pulling on handles at various orientations and elbow postures, or pushing or pulling on aircraft control sticks or control wheels were presented in Woodson et al. (1992). The data were apparently extracted from Morgan et al. (1963). Maximum strength was presented for 5th, 50th, and 95th percentile males. One unique aspect of these data was the presentation of results when the handle was not simply pushed or pulled (away and towards the body, respectively), but also when exertions in the horizontal plane were directed to the left or the right. Results were presented for each arm acting alone; a statistical comparison of the strength data from the different upper limbs was not presented. Testing of push and pull forces on aircraft control sticks was performed at fixed horizontal locations of the control with respect to the seat reference point (SRP) and the distance from the mid-plane of the body (lateral distance). Testing was again performed, exerting force to the right and the left with the right arm in addition to strict pushing and pulling. In general, exertion forces were greater for fore and aft horizontal exertions than for lateral (right or left) exertions. Maximum forces were generated when the shoulder and elbow joints were in their respective optimal positions for isolated joint isometric exertions. There were too much data presented to allow any other global statements; the reader is referred to the handbook (Woodson et al., 1992) if seated one-hand push and pull strength capability is sought.

Forces exerted on pedals while seated have also been investigated. Hertzberg and Burke (1971) determined the mean maximum brake-pedal forces exerted by 100 Air force pilots in two leg positions and various brake pedal angles. Forces were greatest when the foot angle on the pedal was between 15 and 35 degrees from vertical. Other factors that affected pedal force capability included the angular relationship between the seat back and the seat pan, distances between the seat and the pedal, and the angle and size of the pedal. Force capability also depended on whether the operator was expected to use only ankle flexion or full-leg extension (Woodson et al., 1992). For example, suggested limits (based on 5th percentile male strength) for automobile brakes would be 265 N, for typical accelerator pedal operation the suggested range is 44-89 N, and for aircraft rudder pedals the limits would be 89 N for ankle operation and 666 N for full-leg operation.

Static strength was measured on stationary carts in a Swedish study (Winkel, 1983) concerning the manual handling of food and beverage carts on wide-body airplanes. Several recommendations were made about cart configuration and loading as a result of this study. A German group studied the load on the spine during the transport of dustbins (Jaeger et al., 1984). These authors utilized a simple static model of L5/S1 torques, and they also measured the EMG activity of back, leg, and hand muscles. However, no validation of their model was offered, and the EMG information of the back muscles was never related to the L5/S1 torque predictions.

A food distribution company had changed the material handling for their route drivers from lifting boxes singly onto dollies for delivery to off-loading large carts pre-loaded at the warehouse. Field and laboratory studies were performed to determine the biomechanical stresses when pulling and pushing this large (weighing up to 560 kg) cart with one fixed wheel and three wheels that swiveled freely (Pabon-Gonzalez and Andres, 1999). Trials were completed for three different conditions as follows: (a) straight pull with cart's wheels aligned straight, (b) straight pull with cart's wheels perpendicular, and (c) side pull with cart's wheels aligned straight (field) or maneuver to the subject's left with cart wheels aligned straight (laboratory). In both studies the forces required to start the cart's motion from a stationary position were determined. These were found to be the highest exertion forces required when performing the pull/push motion. A sagittal plane biomechanical model during single support (Andres and Chaffin, 1991), assuming that link and whole body center of gravity accelerations were negligible, was used to calculate the resultant hand forces that were compared to the spring gauge measurements. No significant differences between the measured and predicted hand forces were found. Laboratory and field results were similar. Back compression forces did not exceed the Back Compression Design Limit (criterion back compressive force representing only a nominal risk of low back injury) of 3430 N, but the results suggested that caster wheel alignment was a major factor affecting biomechanical stresses in the body of the subjects pulling these carts. A hand tool was designed to aid in realigning the caster wheels in the direction of cart travel. By swiveling the caster wheels in the direction of cart travel, the pulling forces could be decreased by as much as 40%.

15.4.1.2 *Dynamic Push/Pull Strength Capability*

Dynamic strength development implies exertion over a range of motion, either of a joint system or of a measurement device. Dynamic strength relates to the expression of this force vs. time within this motion range while moving at a known velocity (if the velocity is fixed, then this is termed isokinematic strength). Asmussen et al. (1965) presented results demonstrating the decrement in maximum pull force as the speed of movement increased during shortening (concentric) exertions; conversely the maximum pull force increased as the lengthening speed increased (eccentric contraction). These results confirmed the force–velocity properties of skeletal muscle. The question becomes: what are the implications of dynamic pushing and pulling exertion capacity in relevant activities? Laboratory testing is discussed first, followed by dynamic push/pull exertions in industrial tasks.

Both isometric and isokinematic push and pull strength were assessed by Kumar et al. (1995) at three handle heights, (50, 100, and 150 cm) and three body orientations (sagittal, 30-degree lateral, and 60-degree lateral planes). The isokinematic trials were at 50 cm/s of linear handle velocity. A modified static dynamic strength tester with vertical handle orientation measured force during the exertions; peak and average forces were derived for each exertion. Subjects were constrained in a stabilizing platform with their feet together and their legs and hips secured by straps to the frame; this platform was rotated to secure the subjects in the respective orientation for symmetric or asymmetric exertions. The female subjects were weaker than the males in all of the experimental conditions, ranging from a 1% decrement at the high handle height in the 60-degree plane to a 29% decrement at the medium handle height in the sagittal plane. Pulling strengths were higher than pushing strengths in all conditions. The greatest discrepancy occurred with isometric exertions in the sagittal plane (a 25 to 30% decrement). Strength values also declined for both genders for all handle heights with increasing postural asymmetry. Maximal strengths were always found in the isometric conditions, declining with the dynamic (isokinematic) conditions; this effect was most notable for pulling as opposed to pushing. Maximal strengths were always found at the middle handle height, with higher pull forces measured at the high handle height than the low handle height, and the reverse situation for pushing (i.e., higher push forces at the low height vs. the high height). Handle height effects were most pronounced in the sagittal plane.

Most investigations of dynamic forces exerted in pushing and pulling involve carts. Unless the subject is constrained appropriately, instrumented handle forces will not reflect dynamic strength of any particular joint system. Just because these instrumented handles register force variations with time does not mean that maximum dynamic push or pull forces are being measured. Wheeled cart systems typically present their greatest resistance to motion when they are stationary; the breakaway force to overcome inertia and start the cart moving usually exceeds the amount of force required to keep the cart in motion. This phenomenon is partially explained by the inertia of the loaded cart greatly exceeding the rolling friction in properly functioning wheels; in situations where the wheel bearings become worn, the rolling friction increases and the forces to keep a cart moving approach the force required to start it moving. Requirements for strength when moving carts also exist when stopping a cart that is already in motion. When the arms are locked at the elbows the dynamic strength expressed during cart pushing or pulling reflects trunk and lower limb exertions while moving in the gait pattern dictated by the intended cart path. If the elbows are not locked, the upper limbs may also move cyclically with the gait cycle, spreading the dynamic loading among more joint systems. In most situations cart motion is intended to be smooth and controlled, essentially minimizing the dynamics to avoid overexertion. However, since carts are often selected to replace other dynamic manual material-handling tasks (lifting, lowering, and carrying), a review of cart pushing and pulling tasks follows.

Lee et al. (1991) investigated the effects of handle height on low back loading during cart pushing and pulling. Horizontal hand forces were established by adjusting cart resistance on a cart simulator moving on rails. Three hand force levels (98, 196, and 294 N), three handle heights (66, 109, and 152 cm), and two cart speeds (1.8 and 3.6 km/h) were tested with six subjects, and a dynamic biomechanical model predicted L5/S1 compressive loading. The maximum horizontal force of 294 N was selected because subjects could not maintain a constant speed of 1.8 km/h above this force level; cart speed effects were only tested at the middle handle height. The higher speed testing only took place at the middle handle height because subjects slipped or could not keep the cart moving at the other handle heights. Low back compressive forces increased as the hand force increased for both pushing and pulling. These compressive forces decreased when pulling as the handle height increased, but were not affected when handle height was changed for pushing tasks. Compressive forces at L5/S1 were significantly greater for pulling instead of pushing regardless of handle height and horizontal hand force. As cart speed increased, maximum compressive forces increased. Although this study did not directly address maximum dynamic push/pull strength, the limitation of horizontal hand force to 294 N to achieve a cart speed of 1.8 km/h because of safety concerns provides some relevant information.

A sequence of experiments was designed to test the biomechanical predictions from Lee's (1982) dynamic push/pull model. Static calibration experiments derived regression equations relating predicted erector spinae and rectus abdominis muscle force to measured RMS-EMGs during gradual ramps of push or pull force (Andres, 1986). The dynamic testing took place with a cart simulator offering low or high resistance (88 and 127 N, respectively), three handle heights (50, 100, and 150 cm), two walking cadences (60 and 100 steps per minute), and two directions (pushing and pulling). Hand force amplitude parameters for both horizontal and vertical coordinates (average, maximum, and minimum force) were all dependent on the individual subject. This indicated that individual strategies have an important role in pushing and pulling hand forces. For the vertical hand forces, the height of the handle significantly affected average and maximum amplitudes with the greatest forces (downward) exerted on the lowest handle. Males exerted greater vertical forces in this study than females, which was reflected also as the heavier subjects exerting greater vertical forces. Subject height, handle height, and direction of exertion were significant factors in the horizontal hand force amplitudes. Females exerted higher horizontal hand forces than males, which was reflected also by the finding that taller subjects exerted less horizontal hand force than shorter subjects. Heavier subjects exerted larger horizontal hand forces also. Horizontal hand force amplitudes were greater for pulls than for pushes (Andres, 1989). As previously discussed, the validation experiments found that the predictions of required slip resistance to prevent foot slip were better than the predictions of low back loading (Andres and Chaffin, 1991).

Resnick and Chaffin (1995) investigated the effects of inertial load when using a wheeled cart. Higher inertial loads (up to 450 kg) required higher peak hand forces (up to 500 N for strong males). Peak velocities were higher for inertial loads of 45 kg vs. inertial loads of 450 kg.

An investigation by De Looze et al. (2000) compared a simulated cart push/pull task (pulling or pushing on a fixed bar while walking on a treadmill) with actual cart pushing or pulling on the treadmill. They documented similar cyclical patterns of resultant hand forces and force directions for the simulations and the actual cart handling tasks. Nine conditions were tested for both pushing and pulling: target forces of 15, 30, and 45% of each subject's total body weight for handle heights of 60, 70, or 80% of shoulder height for pushing or 50, 60, or 70% of shoulder height for pulling. A constant velocity of 0.75 m/s was employed for the pushing tasks; 0.50 m/s was the constant velocity for the pulling tasks. Once again, maximum dynamic push or pull forces were not directly assessed. The direction of force exertion was affected by the horizontal force level and the handle height. Exerted forces were more horizontal at the higher required force levels and at the higher handle height — more so for pushing than for pulling. Shoulder torques were hardly affected by handle height and only moderately affected by horizontal force levels, because the line of action of the hand force ran only slightly below the shoulder joint axis of rotation. L5/S1 torques were low for pushing, because the upper body gravity vector counteracted the reactive torque caused by the hand forces. Low back torques were greater when pulling, as found by others also. This study confirmed the need for different guidelines for maximum push or pull force at different handle heights.

Resnick and Chaffin (1996) investigated push and pull forces applied to an articulated-arm material-handling device with loads ranging from 0 to 68 kg supported by the device. Subjects pushed the device and load forward to a target and then pulled backward to the starting position once every 15 s for five trials. They obtained Borg CR-10 (Borg, 1982) ratings to assess perceived exertion. When performing symmetric pushing or pulling, subjects had the highest ratings of perceived exertion with the heaviest load when the 25 N static hand force resistance was activated. When the task was modified to require twisting of the torso while pushing and pulling, mean peak push and pull forces increased above those seen for the symmetric task. For these twisting tasks, however, the Borg CR-10 results were not significantly greater than for the symmetric tasks; the psychophysical approach in this case did not reflect increased loading in the low back, although biomechanically demonstrated.

Workers at the Dutch Postal Service push and pull wheeled cages to transport mail back and forth from the sorting area to the mail train. A study aimed at determining the effects of gender and body weight on exerted forces was completed (van der Beek et al., 1998). Three subject groups of four each (females weighing 50–64 kg, females weighing 65–75 kg, and males weighing 65–75 kg) pushed and pulled cages ranging from 130 to 550 kg. There was a measuring frame attached to the cage to register forces in three dimensions, although it was not clear whether there were fixed handles on the cages or not. The peak force during cage acceleration, the mean force during the entire 11-m push or pull, and the peak force during the deceleration phase were all acquired. The mean forces over the entire action ranged from 49 N (push) and 56 N (pull) for the 130 kg cage up to approximately 135 N for both pushing and pulling of the 550 kg cage. Peak acceleration forces to set the 550 kg cage in motion exceeded 400 N for pulling, and were 367 N for pushing. Deceleration forces for the 550 kg cage were 308 N (push) and 346 N (pull). Male workers generally exerted higher forces than females; exertion forces increased with cage weight, but there was a significant interaction between subject group and cage weight. The results suggested that women and men adopted different working techniques: the women used slightly more time to complete the tasks, thus spreading out the impulse loads.

15.4.2 Psychophysical Research

Psychophysical methodologies essentially give the subject control of a particular variable (such as force or frequency) while the experimenters control all other task variables. The subjects then monitor their own feelings of fatigue or exertion level and adjusts the variable they control to meet a specified criterion (i.e., perform this task for 8 h). Details of experimental designs and procedures can be found in Ciriello et al. (1990).

Pushing and pulling forces have been measured while the subject walked on a treadmill (Snook et al., 1970) with different handle heights and adjustable resistance. Strindberg and Peterson (1972) used psychophysical methods to study force perception while pushing trolleys. These studies began to approach more realistic dynamic simulations of actual industrial situations, but the risks inherent in these tasks were not described.

Snook and co-workers have developed guidelines for design of manual material-handling tasks over years of testing at the Liberty Mutual Research Center. Guidelines for maximum acceptable forces for pushing and pulling were included in the report of the results of seven studies (Snook, 1978). A subsequent publication integrated the results of four additional experiments with the previous results to develop a revised, more comprehensive set of guidelines (Snook and Ciriello, 1991). Pushing and pulling tasks were simulated by having subjects walking on a treadmill push or pull on a stationary bar instrumented with load cells. The subjects varied the treadmill resistance. Pushing and pulling forces were measured for both the initial exertion and for the sustained exertion. Pushing and pulling distances ranged from 2.1 m to 61.0 m, and exertion frequencies ranged from once every 6 s to once every 8 h. Three handle heights were evaluated for males (64, 95, and 144 cm) and for females (57, 89, and 135 cm). Results were presented for industrial population percentages capable of exerting force at a particular level (10, 25, 50, 75, and 90% of either the male or female industrial population). The greatest push forces were 735 N and 402 N for the strongest males and females, respectively, at the middle handle heights for one push every 8 h with a 2.1-m push distance. The lowest maximal push forces were found in the weakest individuals required to exert sustained forces once every 2 min for 61.0 m of travel at any handle height (males = 68.6 N, females = 39.2 N). For pulling tasks, the greatest maximal forces were found in the strongest subjects' initial forces performing one lift every 8 h over 2.1 m at the lowest handle heights (males = 676 N, females = 421 N). The lowest maximal forces were sustained forces for the 61.0 m pulls performed one every 2 min at the highest handle height for the weakest males (58.8 N) and at the lowest handle height for the weakest females (39.2 N). As can be seen from these data, maximum acceptable dynamic pushing or pulling strength changed by a factor of 10 depending on the conditions under which the exertions were performed.

A further study (Ciriello et al., 1993) found that the maximum acceptable initial and sustained forces of pull trended 13 and 20% lower, respectively, than the maximum initial and sustained forces of push, although the results were not significant. However, the subjects exceeded the recommended physiological criteria for 8 hours. All testing was done with the handle at a height midway between knuckle and elbow height on their treadmill setup.

More recently Ciriello et al. (1999) did comparison testing of pushing on their treadmill setup vs. pushing on a high-inertia cart. The cart system had pneumatic tires that were constrained inside aluminum channels on either side of the runway; cart weight was adjusted by a pump system that either off-loaded or on-loaded water in a storage tank on the cart. Instrumented handles were mounted on the cart horizontally (requiring prone hand positions). Maximum acceptable initial and sustained forces of pushing on the high-inertia cart were significantly higher than pushing forces on the treadmill system. The authors speculated that the subjects' perceived exertion of pushing a cart with high mass may be lower than pushing against the constant resistance of the treadmill system because of the momentum of the cart once it has started moving. However, only 8 male subjects participated in the experiment, so the results need confirmation in further testing.

15.4.3 Physiological Research

The energy cost of pushing and pulling tasks has rarely been studied by itself. One exception was the study of Hansson (1968), which found that the energy expenditure of pushing wheelbarrows was higher the smaller and softer the wheel, and that a 2-wheel cart was more efficient than a wheelbarrow with a single wheel. Most investigations that acquire energy expenditure data during pushing and pulling tasks have done so to validate psychophysical criteria. For instance, Snook and Ciriello (1991) reported that their previous experiments (Snook, 1978) violated accepted the physiological criterion of 33% maximum oxygen uptake rate for an 8-hr day established by NIOSH (1981) for the high-frequency tasks.

15.5 Implications for Task Design Involving Pushing and Pulling

The wide variety of tasks that involve pushing and pulling exertions in the workplace or outside the work environment, juxtaposed against the wide range of tasks for which strength data has been gathered in the laboratory or in the field, makes it challenging to design tasks within the strength capacity of the exposed population. All of the respective research approaches — biomechanical, physiological, and psychophysical — have inherent limitations and lack extensive epidemiological validation (Dempsey, 1998). Probably the most used criteria for task design are the psychophysical data from Snook, Ciriello, and coworkers at the Liberty Mutual Research Center.

An approach modeled after the NIOSH Lifting Equation (Waters et al., 1993) that combines biomechanical, physiological, and psychophysical considerations into prediction equations for lowering, pushing, pulling, and carrying tasks has been proposed by Shoaf et al. (1997). The pushing and pulling models predict the pushing or pulling capacity by starting with the maximum force acceptable to a specified percentage of the worker population depending on the type of force (initial or sustained). This maximum acceptable force is then decremented by multipliers (≤ 1.0) that account for the following factors: the vertical distance from the floor to the hands, the traveled distance, the frequency of push or pull, the age group, the body weight, and the task duration. The maximum acceptable forces at given travel distances and handle heights were based on Snook (1978) and Snook and Ciriello (1991); the multipliers for age and body weight were derived from biomechanical data reported in Genaidy et al. (1993); and the task duration multiplier was based on a physiological study performed by Asfour et al. (1991). Although their biomechanical analysis found that base weights were lower than those established by psychophysical testing for lowering, lifting, and carrying tasks, the psychophysical data yielded more conservative base weights than the biomechanically derived forces for pushing and pulling tasks. Their comprehensive models for pushing and pulling, therefore, were based on the psychophysical results described previously. However, they modified the frequency-discounting factors to reflect physiological limitations at higher frequencies of pushing or pulling.

This set of prediction equations can be helpful for tasks involving two-handed pushing or pulling in the standing position with symmetrical postures. Realistically, however, several of the tasks previously mentioned above would not fit into these limitations. One-handed tasks, seated tasks, and asymmetrical tasks all necessitate referral back to the original research or to summary tables presented in handbooks (e.g., *Human Factors Design Handbook* by Woodson et al., 1992).

A table of horizontal push and pull forces exertable intermittently or for short durations by males was developed for MIL-STD-1472D (Table XXIV in the document). Horizontal forces applied via hands, one shoulder, or the back were limited to 100 N with low traction (0.2 < COF < 0.3), 200 N with medium traction (COF approx. 0.6), or 300 N with high traction (COF > 0.9). One-handed pushing when braced against a vertical wall (510–1525 mm from and parallel to the push panel) was limited to 250 N; the limit increased to 500 N if both hands, one shoulder, or the back applied the force if the feet were anchored, and up to 750 N if the force was applied with the back while the feet were braced.

15.6 Future Directions

As more females perform manual work and as the working population ages, the need for demographically relevant strength data increases. Further biomechanical studies are necessary to provide a clearer understanding of the interplay between gender and anthropometry when exerting pushing or pulling forces. Development of strength criteria based on 5th percentile female strength capabilities can follow these studies — allowing the majority of the population to work within their strength capability. Finally, the generation of models to predict maximum acceptable push and pull forces is a promising development. However, epidemiological studies to examine the validity of model predictions will still be needed in the future.

15.7 Summary

A wide variety of investigations of pushing and pulling strength have been completed since the 1940s. Table 15.1 presents descriptions of some of these studies, so that researchers can compare findings to other studies, and so that task designers can refer to the citation that includes results in the format desired (i.e., tables, graphs, posture depictions, prediction equations).

Biomechanical evaluations of static and dynamic push/pull strength and psychophysical investigations of practical pushing and pulling tasks have led to the following conclusions:

- Seated hand forces and pedal forces depend on joint positions and muscle groups utilized.
- Free-position standing push/pull forces are greater than forces exerted in constrained standing postures unless bracing can be used.
- Two handed push/pull strength is greater than one hand strength.
- Heavier subjects have greater push/pull strength than lighter subjects when exertion depends on traction.
- Symmetrical (two-handed mid-sagittal plane) push/pull exertion forces are greater than asymmetrical exertion forces.
- Push/pull strength decreases with age, similar to other human static strengths.

For pushing and pulling carts, the following conclusions have been supported:

- Pulling generally creates greater loading on the low back than pushing.
- Wheel alignment with respect to the direction of exertion affects hand forces.
- Greater hand forces are required to move carts with greater inertia.

The reader is referred to the cited research or to Ayoub and Mital (1989) for databases of strength values.

References

Andres, R.O. 1986. Low back compressive loads during gradual ramps of push or pull force, *J. Biomechanics*, 19, 469 (abstract).

Andres, R.O. 1989. A Revised Final Report on a Biodynamic Model of Industrial Pushing and Pulling Tasks, NIOSH Contract 210-81-3104.

Andres, R.O. and Chaffin, D.B. 1985. Ergonomic analysis of slip-resistance measurement devices, *Ergonomics*, 28, 1065–1080.

Andres, R.O. and Chaffin, D.B. 1991. Validation of a biodynamic model of pushing and pulling, *J. Biomech.*, 24, 1033–1045.

Asfour, S., Khalil, T., Genaidy, A., Akcin, M., Jomoah, I., Koshy, J., and M. Tritar. 1991. Ergonomic injury control in high frequency lifting tasks Final Report, NIOSH Grant Nos. 1 R01 OH02591-01 and 5 R01 OH0259-02, NIOSH, Cincinnati, OH.

Asmussen, E., Hansen, O., and Lammert, O. 1965. The relation between isometric and dynamic muscle strength in man. Communications from the Testing and Observations Institute of the Danish National Association for Infantile Paralysis, No. 20.

Ayoub, M. and El-Bassoussi, M. 1978. Dynamic biomechanical model for sagittal plane lifting activities, in *Safety in Manual Materials Handling*, C.G. Drury, Ed. (Cincinnati, OH: DHEW (NIOSH) Publication No. 78-105), pp. 88–95.

Ayoub, M. and McDaniel, J. 1974. Effects of operators' stance on pushing and pulling tasks, *AIIE Trans.*, 6, 185–195.

Ayoub, M.M., and Mital, A. 1989. *Manual Materials Handling*. (Taylor and Francis, London).

Borg, G. 1982. Psychophysical bases of perceived exertion. *Med. Sci. Sports Exerc.*, 14, 377–381.

Chaffin, D.B. 1975. On the validity of biomechanical models of the low-back for weight lifting analysis, in *ASME Proceedings*, 75-WA-Bio-1, (New York: American Society of Mechanical Engineers), pp. 1–13.

Chaffin, D.B., Andres, R.O., and Garg, A. 1983. Volitional postures during maximal push/pull exertions in the sagittal plane, *Hum. Factors*, 25, 541–550.

Ciriello, V.M., McGorry, R.W., Martin, S.E., and Bezverkhny, I.B. 1999. Maximum acceptable forces of dynamic pushing: comparison of two techniques, *Ergonomics*, 42, 32–39.

Ciriello, V.M., Snook, S.H., and Hughes, G.J. 1993. Further studies of psychophysically determined maximum acceptable weights and forces, *Hum. Factors*, 35, 175–186.

Ciriello, V.M., Snook, S.H., Blick, A.C., and Wilkinson, P.L 1990. The effects of task duration on psychophysically-determined maximum acceptable weights and forces, *Ergonomics*, 33, 187–200.

Daams, B.J. 1993. Static force exertion in postures with different degrees of freedom, *Ergonomics*, 36, 397–406.

De Looze, M.P., van Greuningen, K., Rebel, J., Kingma, I., and Kuijer, P. 2000. Force direction and physical load in dynamic pushing and pulling, *Ergonomics*, 43, 377–390.

Dempsey, P.G. 1998. A critical review of biomechanical, epidemiological, physiological and psychophysical criteria for designing manual materials handling tasks, *Ergonomics*, 41, 73–88.

Dempster, W.T. 1955. *Space Requirements of the Seated Operator*, WADC Technical Report 55-159, University of Michigan, Ann Arbor, MI.

Dempster, W.T. 1958. Analysis of the two-handed pulls using free body diagrams, *J. Appl. Physiol.*, 13, 469–480.

Dupuis, H., Preuschen, R., and Schulte, B. 1955. *Zweckmäbige Gestaltung des Schlepper-Führerstandes*, Max Planck Institut für Arbeitsphysiologie, Dortmund, Germany.

Fothergill, D.M., Grieve, D.W., and Pheasant, S.T. 1992. The influence of some handle designs and handle height on the strength of the horizontal pulling action, *Ergonomics*, 35, 203–212.

Freivalds, A., Chaffin, D.B., Garg, A., and Lee, K.S. 1984. A dynamic biomechanical evaluation of lifting maximum acceptable loads, *J. Biomech.*, 17, 251–262.

Garg, A. 1973. The Development of the Validation of a Three-Dimensional Hand Force Capability Model, unpublished Ph.D. thesis, University of Michigan, Department of Industrial and Operations Engineering, Ann Arbor, MI.

Garg, A., and Chaffin, D.B. 1975. A biomechanic computerized simulation of human strength, *AIIE Trans.*, March, pp. 1–15.

Gaughran, G., and Dempster, W.T. 1956. Force analysis of horizontal two-handed pushes and pulls in the sagittal plane, *Hum. Biol.*, 28, 67–92.

Genaidy, A., Waly, S., Khalil, T., and Hidalgo, J. 1993. Compression tolerance limits of the lumbar spine for the design of manual materials handling operations in the workplace, *Ergonomics*, 36, 415–434.

Gronqvist, R. 1999. Slips and falls, in *Biomechanics in Ergonomics*, S. Kumar, Ed. (London: Taylor and Francis), pp. 351–371.

Hanson, J.P., Redfern, M.S., and Mazumdar, M. 1999. Predicting slips and falls considering required and available friction, *Ergonomics*, 42, 1619–1633.

Hansson, J.E. 1968. *Work Physiology as a Tool in Ergonomics and Production Engineering*, Al-Rapport 2, Ergonomi och Produktionsteknik, National Institute of Occupational Health, Stockholm.

Hertzberg, H., and Burke, F. 1971. Foot forces exerted at various aircraft brake pedal angles, *Hum. Factors*, 13, 445–456.

Hoozemans, M., Van Der Beek, A.J., Frings-Dresen, M., and Van Der Molen, H. 2001. Evaluation of methods to assess push/pull forces in a construction task, *Appl. Ergon.*, 32, 509–516.

Hoozemans, M., Van Der Beek, A.J., Frings-Dresen, M., Van Dijk, F., and Van Der Woude, L.H.V. 1998. Pushing and pulling in relation to musculoskeletal disorders: a review of risk factors, *Ergonomics*, 41, 757–781.

Hugh-Jones, P. 1947. The effect of limb position in seated subjects on their ability to utilize the maximum contractile force of limb muscles, *J. Physiol.*, 105, 322–344.

Imrhan, S.N., and Sundararajan, K. 1992. An investigation of finger pull strengths, *Ergonomics*, 35, 289–299.

Jaeger, M., Luttmann, A., and Laurig, W. 1984. The load on the spine during the transport of dustbins, *Appl. Ergon.*, 15, 91–98.

Kreutzberg, K.L. and Andres, R.O. 1985. Field study methods for applying a biodynamical model to industrial push/pull tasks, *J. Biomech.*, 18, 234 (abstract).

Kroemer, K.H.E. 1969. *Push Forces Exerted in 65 Common Work Positions*, Aerospace Med. Res. Lab. Tech. Report, USAF AMRL-TR-68-143, Wright Patterson Air Force Base, OH.

Kroemer, K.H.E. and Robinson, D.E. 1971. *Horizontal Static Forces Exerted by Men Standing in Common Working Positions on Surfaces of Various Tractions*, USAF AMRL-TR-70-114, Wright Patterson Air Force Base, OH.

Kumar, S., Narayan, Y., and Bacchus, C. 1995. Symmetric and asymmetric two-handed pull-push strength of young adults, *Hum. Factors*, 37, 854–865.

Lee, K.S. 1982. Biomechanical Modeling of Cart Pushing and Pulling, unpublished Ph.D. dissertation, University of Michigan, Ann Arbor, MI.

Lee, K.S., Chaffin, D.B., Herrin, G.D., and Waikar, A.M. 1991. Effect of handle height on lower-back loading in cart pushing and pulling, *Appl. Ergon.*, 22, 117–123.

Martin, W.R. and Johnson, E.E. 1952. *An optimum range of seat positions as determined by exertion upon a foot pedal*, AMRL Report No. 86, 1–9.

McGill, S.M., and Norman, R.W. 1985. Dynamically and statically determined low back moments during lifting, *J. Biomech.*, 18, 877–886.

MIL-STD-1472F, 1999. *Department of Defense Design Criteria Standard, Human Engineering*.

Morgan, C.T. et al. Eds. 1963. *Human Engineering Guide to Equipment Design*. (New York: McGraw-Hill).

Murrell, K. 1965. *Human Performance in Industry*, (New York: Reinhold Publishing).

NIOSH 1981. Work practices guide for manual lifting, DHHS (NIOSH) Publication No. 81-122, Cincinnati, OH.

Orlansky, J. and Dunlop, J.W. 1948. *The Human Factor in the Design of Stick and Rudder Controls for Aircraft*, Biomechanical Office of Naval Research, 1–77.

Pabon-Gonzalez, M., and Andres, R.O. 1999. Biomechanical evaluation of cart pushing and pulling operations for a distribution center, in *Advances in Occupational Ergonomics and Safety* 3rd ed. (Amsterdam: G.C.H. Lee, IOS Press), pp. 111–115.

Park, K. 1973. A Computerized Simulation Model of Postures During Manual Materials Handling, unpublished Ph.D. thesis, University of Michigan, Department of Industrial and Operations Engineering, Ann Arbor, MI.

Redfern, M.S. and Andres, R.O. 1984. The analysis of dynamic pushing and pulling: required coefficients of friction, *Proceedings of the 1984 International Conference on Occupational Ergonomics*, Toronto, Canada, May 7–9.

Rees, J.E., and Graham, N.E. 1952. The effect of backrest position on the push which can be exerted on an isometric foot pedal, *J. Anat.*, 86, 310–319.

Resnick, M.L. and Chaffin, D.B. 1995. An ergonomic evaluation of handle height and load in maximal and submaximal cart pushing, *Appl. Ergon.*, 26, 173–178.

Resnick, M.L. and Chaffin, D.B. 1996. Kinematics, kinetics, and psychophysical perceptions in symmetric and twisting pushing and pulling tasks, *Hum. Factors*, 38, 114–129.

Schanne, F.J. 1972. A Three-Dimensional Hand Force Capability Model for the Seated Operator, unpublished Ph.D. thesis, University of Michigan, Ann Arbor, MI.

Shoaf, C., Genaidy, A., Karwowski, W., Waters, T., and Christensen, D. 1997. Comprehensive manual handling limits for lowering, pushing, pulling and carrying activities, *Ergonomics*, 40, 1183–1200.

Slater, J.K.W. 1949. An Aspect of Attitude and Physical Control in Agricultural Tractors, unpublished thesis, Durham University.

Snook, S.H. 1978. The design of manual handling tasks, *Ergonomics*, 21, 963–985.

Snook, S.H. and Ciriello, V.M. 1991. The design of manual handling tasks: revised tables of maximum acceptable weights and forces, *Ergonomics*, 34, 1197–1213.

Snook, S.H., Irvine, C.H., and Bass, S.F. 1970. Maximum weights and workloads acceptable to male industrial workers, *AIHA J.*, 31, 579–586.

Stobbe, T. 1982. The Development of a Practical Strength Testing Program in Industry, unpublished Ph.D. dissertation, University of Michigan, Ann Arbor, MI.

Stramler, J.H. 1993. *The Dictionary for Human Factors/Ergonomics*, (Boca Raton, FL: CRC Press).

Strindberg, L. and Peterson, N. 1972. Measurement of force perception in pushing trolleys, *Ergonomics*, 15, 435–438.

Van Cott, H.P. and Kincade, R.G. 1972. *Human Engineering Guide to Equipment Design, Revised Edition*. (Washington, DC: American Institutes for Research).

van der Beek, A.J., Kluver, B.D.R., Frings-Dresen, M.W., and Burdorf, A. 1998. Pushing and pulling of wheeled cages: exerted forces and energetic workload, in *Advances in Occupational Ergonomics and Safety*, S. Kumar, Ed. (Amsterdam: IOS Press), pp. 339–342.

Voorbij, A.I.M. and Steenbekkers, L.P.A. 2001. The composition of a graph on the decline of total body strength with age based on pushing, pulling, twisting and gripping force, *Appl. Ergon.*, 32, 287–292.

Warwick, D., Novak, G., Schultz, A., and Berkson, M. 1980. Maximum voluntary strengths of male adults in some lifting, pushing and pulling activities, *Ergonomics*, 23, 49–54.

Waters, T., Putz-Anderson, V., Garg, A., and Fine, L. 1993. Revised NIOSH equation for the design and evaluation of manual lifting tasks, *Ergonomics*, 36, 749–776.

Winkel, J. 1983. On the manual handling of wide-body carts used by cabin attendants in civil aircraft, *Appl. Ergon.*, 14, 186–168.

Woodson, W.E., Tillman, B., and Tillman, P. 1992. *Human Factors Design Handbook, Second Edition*. (New York: McGraw-Hill).

Wright, U., Leonard, J., Neigebauer, D., Fox, R., and Peacock, B. 1998. A case study: thumb and index finger push forces common to assembly processes, in *Advances in Occupational Ergonomics and Safety*, S. Kumar, Ed. (Amsterdam: IOS Press), pp. 390–393.

16

Design Applications of Strength Data

Karl H. E. Kroemer

CONTENTS

ABSTRACT The designer who wants to consider operator strength when developing a new process or device must decide *which* strength values to use. "Maximal" user strength usually determines the structural integrity of the object, so that even the strongest operator cannot break a handle or a pedal. "Minimal" user strength is that expected from the weakest operator, which still yields the desired result (such as successfully operating a brake handle or pedal) under the worst circumstances. Actual strength applications will fall within that range between minimum and maximum.

16.1 Ergonomic Relevance

Body strength is among the many variables in which people differ from each other. Information on strength is therefore of great importance to the designer of tools, equipment, and work tasks. Proper design for strength requires two basic decisions: first, *what* database to use, and second, *how* to use the numbers — specifically, *which* percentile value(s) to select.

16.2 Introduction

Taking into account human variability is the hallmark of ergonomics. People are not all the same, and everybody changes during the course of life. Overtly, persons differ from each other in gender and age, size and strength.

People do not come assembled of parts that are all average, nor in any other single percentile value. While politicians and journalists like to invoke the illusory "average person," we may neither simply design for such a phantom nor for the similarly unreal "5th" or "95th percentile person." Yet as the following text shows, proper use of strength data allows for design of tasks and devices that "fit" human strength capabilities.

16.3 Design for Variability

In biomechanics as well as in anthropometry and physiology (for instance, see Kroemer, 1998, 2002; Kroemer et al., 1997, 2001; Kuczmarski et al., 2002) diversity in body strength comes into view primarily in four aspects: interindividual variability, intraindividual variability, secular changes, and variability in measurements.

16.3.1 Interindividual Variability

Interindividual variability means that persons differ from each other. Data describing a population sample usually stem from a "cross-sectional" study in which every subject is measured at (about) the same moment in time. This means that people of different age, health, and fitness are included in the sample set. The interindividual strength data in this book and the compilations found in most textbooks describe the results of cross-sectional studies.

16.3.2 Intraindividual Variability

Intraindividual variability reflects that body and segment size, strength, and endurance change within a person with age, nutrition, exercise, health, and fatigue. "Longitudinal" studies, which observe an individual over years and decades, show such intraindividual changes. (Because of time and cost constraints, the literature contains only few longitudinal studies.) Most, but not all, variations within a person from infancy to old age follow this scheme: during childhood and adolescence, most body descriptors such as stature and muscle strength start out small and then increase rapidly. From the 20s into the 50s, little change occurs in general, but stature begins to decline. From the 50s or 60s on, strength and stature diminish, while other dimensions — for example body weight or bone circumference — often increase.

16.3.3 Secular Changes

Europeans, North Americans, and Japanese people are, on average, larger (although possibly neither stronger nor more able to do demanding work) than their ancestors. Also, life expectancy has greatly increased, so the portion of aged people in the workforce and in the overall population is much larger now than just a few decades ago. The probable reasons for this secular development are improved nutrition, hygiene, and health care, all of which have allowed persons to achieve more of their genetically determined potential than was possible generations ago.

16.3.4 Variability in Measurements

Measured/reported data can vary due to differences in selecting population samples, measuring at different points in time, using dissimilar techniques in measuring, storing the measured numbers, and applying statistical treatments to the raw data. Of these sources of diversity, unwanted *variability in measurements* usually precludes use of such data. In contrast, *intraindividual variability* is the most common reason to seek descriptive information and to apply it in human-engineered designs.

16.4 Basic Statistics

Like most ergonomic information, data on muscular strength are usually reported by the basic statistics *mean* (*m*) (often called *average*), *standard deviation* (*SD*), and *sample size* (*N*). These simple statistical descriptors fully portray "normal" (Gaussian) distributions of numbers that show the characteristic bell-shaped pattern symmetrical to the average.

Fortunately for the designer, many ergonomic data — anthropometrics in particular — occur in reasonably normal distributions. However, muscle strength data (as well as other physiological and psychological information) may not turn out to be normally distributed, although this is often carelessly assumed.

- If the data distribution is indeed nonnormal, the designer wishing to use that information still has easy ways to do so. Mathematical routines are able to use nonparametric statistical procedures, or to transform the data. Another practical solution is to estimate (interpolate) data points of interest, such as certain percentile values. Such estimation is often simple to do, as discussed below.

- If the distribution is Gaussian, a fast way for evaluating data diversity is by dividing the standard deviation in question by its average. This yields the *coefficient of variation*, CV. In most cross-sectional studies of adult body dimensions, the CV is in the neighborhood of 5%, but in many strength data it is up to 25% or even larger.

We may want to determine how one variable *y*, arm strength for example, changes with another variable *x*, say grip strength. For doing this, we regularly assume that *y* varies directly with *x*. The correlation coefficient *r* describes such linear relation. (But few people ever check the propriety of that assumption of a linear regression.) The *coefficient of determination* is the square of the correlation coefficient: R^2 indicates the proportion of variation in the dependent variable *y* that is due to the scatter in variable *x*.

These remarks just provide a brief overview of some statistical procedures important to the human-factors engineer. Textbooks on statistics contain more detailed treatments of the topic.

16.5 The "Average Person" and Other Phantoms

Actually, relationships among body descriptors vary a great deal. Values of the coefficient of correlation, *r*, may lie anywhere between negative and positive "one" (indicating a *perfect* relation), including "zero" (meaning *no* relation). It is delusive to attempt expressing "all" body characteristics as related to one basic denominator. In the past, too many designers used a misleading scheme which relied on stature (standing body height) to predict other heights and such body dimensions as breadths, depths, and weight, even muscular strength, as a given percentage of stature. The most often invoked ghost was the "average person" who not only was 50th percentile in height but average all over. Two other notorious phantoms have either all 5th or all 95th percentile values. Of course, designs for these figments hardly fit actual users.

Some engineers, even physiologists and physicians, are still willing to express body attributes in terms of stature (probably because height is easily measured), even when a proper logical, empirical, or statistical correlation does not exist. For example, body weight is often compared to stature in spite of the fact that the correlation between these two variables is only around 0.5 at best (Kroemer et al., 1997, 2001).

The lack of persistent correlations not only makes single-percentile templates of body size illusory, but also renders similar models of body strength highly questionable. The literature demonstrates that neither stature nor body weight are reliable predictors of muscular strength. Even within compilations of strength data, coefficients of correlations vary widely (see, for instance, Daams, 2001; Kroemer, 1974, 1999a,b; Kroemer et al., 1971, 1997; Norris and Wilson, 1995; Peebles and Norris, 1998, 2000; Steenbekkers and Beijster-veldt, 1998; University of Nottingham, 2002; and chapters in this book). This means that as a rule, the human-factors engineer must carefully ascertain that an available set of strength data does describe the capabilities of the intended user group. Predicting strength capacities from an existing compilation is a risky enterprise, because strength varies among body parts (for example, grip strength is not a good predictor of leg strength), and strength differs from person to person. Furthermore, exerted strength depends on the extant biomechanical and emotional situation (Daams, 1993; Kroemer, 1974; Marras, 1999).

16.6 Proper Design Procedures

When designing a process or device for human strength, the ergonomist has two basic tasks:

1. Identify the critical strength demand(s) and determine the corresponding available body strength(s).
2. Match the demands to the users' capabilities.

The following text treats these two ergonomic tasks.

16.6.1 Identify the Critical Strength

This book presents examples of human strength exertions by hand, foot, or the whole body. These compilations show the orders of magnitude of body strength that various persons have exerted. Other data sets have been compiled recently, for example, by Norris and Wilson (1995), Salvendy (1997), Steenbekkers and Beijsterveldt (1998), Peebles and Norris (1998, 2000), Karwowski and Marras (1999), Marras (1999), Konz and Johnson (2000), Phillips (2000), Daams (2001), Kroemer et al. (2001, 2004 under review), and University of Nottingham (2002). While these data indicate "orders of magnitude" of forces and torques, design applications of these exact numbers require great caution because they were measured on various subject groups under widely varying circumstances.

16.6.1.1 Factors That Determine Actual Strength

The actual amounts of strength that people do generate depend decidedly on several factors. These include gender, age, and fitness, but *situational factors* often also strongly affect the amount of body strength that a person will or can apply:

- Motivation
- Skill and experience
- Application with suitable body sites, such as one foot or both feet, shoulder, hand or hands
- Ability to brace the body against a support structure
- Body posture, especially the use of strong muscles at advantageous leverage
- Body motion

16.6.1.2 Static Strength Exertions

Body motion, or lack thereof, is a major situational factor to determine the force, torque, or impulse transmitted from the body to an external object (Kroemer, 1999b). When the body does not move, the muscles involved in this "static" strength exertion remain at constant (unchanging) length. In physiological terms, this called an *isometric* muscle contraction (because the muscle sarcomeres neither shorten nor become lengthened). In physics terms, all forces acting within the system are in equilibrium, as Newton's First Law requires.

The static condition is theoretically simple and experimentally well controllable. It allows rather easy measurement of muscular effort. Therefore, most of the information available on "body strength" describes outcomes of static (isometric) testing. Accordingly, most of the tables on body segment strength in the human engineering and physiologic literature contain static data.

16.6.1.3   Dynamic Strength Exertions

Dynamic muscular efforts are much more complicated to describe than static contractions and are much more difficult to control in experiments. In dynamic activities, muscle length changes and therefore involved body segments move. This results in displacement. The time derivatives (velocity, acceleration, and jerk) of displacement are of importance both for the muscular effort (as discussed extensively in several chapters of this book) and for the external effect: for example, change in velocity determines impact and force, as per Newton's Second Law.

16.6.1.4   Classifications of Strength Exertions

To overcome the complexities in definition and experimental control of dynamic muscle exertions, various new classification schemes for independent and dependent experimental variables can be developed. The system described by Marras et al. (1993) includes the traditional static (isometric) and dynamic conditions. Instead of keeping muscle length constant, one may choose to control the rate at which muscle length changes (velocity) as an independent variable. When velocity is a constant value, one speaks of an *isokinematic* muscle strength measurement. (Note that this *isovelocity* condition is often mislabeled "isokinetic.")

Other tests keep the amount of muscle tension (force) to a constant value. In such an *isoforce* test, mass properties and displacement (and its time derivatives) are likely to become controlled independent variables, and repetition a dependent variable. This *isotonic* condition is, for practical reasons, often combined with an isometric condition, such as in holding a load motionless. (Note that some older texts labeled the examples of lifting or lowering a weight "isotonic." This is physically and physiologically false.)

For the sake of experimental control, all of these tests impose certain protocols that make the procedures somewhat artificial, different from everyday activities. One exception is the *isoinertial* test, where the external mass is constant, while the conditions at the muscle are not experimentally controlled. This procedure often serves to determine lifting ability.

Marras et al. (1993) labeled the most general case of motor performance measurement "free dynamic." This is the most realistic condition, but it allows little experimental control because the subject has the free choice of the independent variables force and displacement (and its time derivatives) to achieve the desired effect, say, to push a heavy object or throw a ball.

16.6.2   Design to Accommodate Strength Capabilities

Designing for human body strength (in terms of the force, torque, work, power, or impulse that a person can exert) involves a number of decisions:

16.6.2.1   Is the Strength Exertion Static or Dynamic?

- If it is *static*, we employ information about isometric strength capabilities.
- If the motion is very slow, especially when eccentric, data on isometric strength measured at points on the path may yield a reasonable estimate of the maximally possible exertion.
- If it is *dynamic*, we need strength information that is specific to the kind of exertion. Isometric data do not estimate fast exertions well, especially if they are concentric and of the ballistic/impulse type, such as throwing or hammering. A special experiment may be necessary to measure the dynamic strength capability in demand.
- As a rule, strength exerted in motion is of lower magnitude than static strength measured in positions located on the path of motion. Physiologic and biomechanical texts (Astrand and Rodahl, 1986; Marras et al., 1993; Winter 1990) provide guidance for dynamic measurements.

16.6.2.2   Is a Maximal or a Minimal Strength Exertion Critical for the Design?

- *Maximal* user output usually determines the structural strength of the object, so that even the strongest operator may not break it. The chosen design value is above the highest possible strength application.
- *Minimal* user output is an exertion expected from the weakest operator in the most disadvantageous situation that still yields the desired result, such as successfully operating a handle or brake pedal, or moving a heavy object.

16.6.3   Design Strategies

It is hard to imagine even one activity or object that could be properly devised for the average of a data set: whatever is designed for a mean value is too little for half the users, and too much for all the others, as logic posits and experience confirms. Employing the average as a design guide invariably cuts off half the users. (Yet, while the mean of a distribution is very unlikely to serve as a criterion for design, indeed it is a useful *statistical tool*, as shown below.)

To be practical, the ergonomist must determine a single number that establishes upper or lower design dimensions, or both. (See below for statistical procedures that apply to relevant data sets.) Then, these design values serve in either of two design strategies:

16.6.3.1   "Min OR Max" Design

Design for a minimal value — for the feeble operator, for the situation that makes for a weak exertion, the lowest limit

or

Design for the maximal value — for the brute exertion under the most favorable condition, for the uppermost limit

16.6.3.2   *"Min AND Max" Design*

In many applications, lower and upper values *both* determine the appropriate design. A typical example is a brake pedal in a car that a weak driver can use to safely stop the vehicle, while a strong operator cannot break it even when stomping on it during an emergency.

Some U.S. military procurement requirements state that the range from 5th to 95th percentile of soldiers must be accommodated by equipment. This is a laudable example of setting "Min *AND* Max" limits, but unfortunately, the military documents commonly lack a definition of the crucial characteristic(s) for which these limits shall be fixed — clearly, one can not assume soldiers to be of exclusively of either 5th or 95th percentiles.

16.6.4   How to Determine Critical Design Values

16.6.4.1   *Calculating Data*

If a set of strength data has a "normal" (Gaussian) distribution, we can fully explicate it using just two descriptors. The *mean (m)* (same as "average") identifies the central point of the distribution. Fifty percent of the data lie below and the other 50% above the mean (average), which is therefore called the *50th percentile* (short: *p50*) as well. The other key descriptor is the *standard deviation (SD)*, which specifies the spread of the data about the mean. The numerical value of the standard deviation is larger when the data scatter widely than when they cluster close to the mean. The *coefficient of variation (V = SD/m)* expresses this in relative terms.

We can easily calculate the location of any (percentile) point of the distribution using mean and standard deviation. Simply multiply the standard deviation *SD* by a factor *k*, selected from Table 16.1. Then add the product to the mean *m*:

$$p = m + k \times SD$$

If the desired percentile is above the 50th percentile, the factor *k* has a positive sign, and the product $k \times SD$ is added to the mean *m*; if the p-value is below average, *k* is negative and therefore the product $k \times SD$ is subtracted from the mean.

Examples:

> 1st percentile is at $m - 2.33 \times SD$ with $k = -2.33$ (see Table 16.1)
> 5th percentile is at $m - 1.64 \times SD$ with $k = -1.64$
> 10th percentile is at $m - 1.28 \times SD$ with $k = -1.28$
> 50th percentile is at m with $k = 0$
> 60th percentile is at $m + 1.28 \times SD$ with $k = 1.28$
> 95th percentile is at $m + 1.64 \times SD$ with $k = 1.64$

16.6.4.1.1   *Determining a Cut-Off Point (Single Percentile): "Min OR Max"*

1. Select the desired percentile value *p*.
2. Determine the associated *k* value from Table 16.1.
3. Calculate the *p* value from $p = m + k \times SD$. (Note that *k* and hence the product may be negative.)

TABLE 16.1

Percentile Values with Their k Factors

Below Mean				Above Mean			
Percentile	Factor k	Percentile	Factor k	Percentile	Factor k	Percentile	Factor k
0.001	-4.25	25	-0.67	50	0	76	0.71
0.01	-3.72	26	-0.64	51	0.03	77	0.74
0.1	-3.09	27	-0.61	52	0.05	78	0.77
0.5	-2.58	28	-0.58	53	0.08	79	0.81
1	-2.33	29	-0.55	54	0.10	80	0.84
2	-2.05	30	-0.52	55	0.13	81	0.88
2.5	-1.96	31	-0.50	56	0.15	82	0.92
3	-1.88	32	-0.47	57	0.18	83	0.95
4	-1.75	33	-0.44	58	0.20	84	0.99
5	-1.64	34	-0.41	59	0.23	85	1.04
6	-1.55	35	-0.39	60	0.25	86	1.08
7	-1.48	36	-0.36	61	0.28	87	1.13
8	-1.41	37	-0.33	62	0.31	88	1.18
9	-1.34	38	-0.31	63	0.33	89	1.23
10	-1.28	39	-0.28	64	0.36	90	1.28
11	-1.23	40	-0.25	65	0.39	91	1.34
12	-1.18	41	-0.23	66	0.41	92	1.41
13	-1.13	42	-0.20	67	0.44	93	1.48
14	-1.08	43	-0.18	68	0.47	94	1.55
15	-1.04	44	-0.15	69	0.50	95	1.64
16	-0.99	45	-0.13	70	0.52	96	1.75
17	-0.95	46	-0.10	71	0.55	97	1.88
18	-0.92	47	-0.08	72	0.58	98	2.05
19	-0.88	48	-0.05	73	0.61	99	2.33
20	-0.84	49	-0.03	74	0.64	99.5	2.58
21	-0.81	50	0	75	0.67	99.9	3.09
22	-0.77					99.99	3.72
23	-0.74					99.999	4.25
24	-0.71						

Note: Any percentile value p can be calculated from the mean m and the standard deviation SD (normal distribution assumed) by $p = m + k \times SD$.

16.6.4.1.2 Determining the Upper and Lower Limits (Percentiles) of a Range: "Min AND Max"

1a. Select upper percentile p_{max}.

1b. Find related k_{max} value in Table 16.1.

1c. Calculate upper percentile value $p_{max} = m + k_{max} \times SD$.

2a. Select lower percentile p_{min}. (Note that the two percentile values need not be at the same distance above and below the mean m; that is, the range does not have to be "symmetrical to the average." Also, *both* values may be above the mean, or below.)

2b. Find related k_{min} value in Table 16.1.

2c. Calculate lower percentile value $p_{min} = m + k_{min} \times SD$.

3. Determine the range $R = p_{max} - p_{min}$.

A note of caution: When using the formula $p = m + k \times SD$ to calculate percentiles at either tail of a distribution, the result may not reflect data outliers. These are actual strength scores that are extremely low or high. Such outliers can result from unusual conditions, for example if a person is injured or intensely motivated. (The proverbial case is the mother lifting a car to free her pinned child.) Such extreme values may not have been included in the original set of measurements, or they can vanish from sight by the "smoothing" of the data that occurs automatically as part of statistical procedures to fit entries into a normal distribution. If extremely low or high strength applications are expected, the designer may select uncommonly large values for the k factor used in the calculating formula or apply sound judgment to estimate (instead of calculate) the critical strength score.

16.6.4.2   Estimating Data

There are instances were no experimentally obtained strength data are at hand that apply to the design task. If no measurements are feasible, then we have to "guesstimate" the design value. Such approximation has merit if done carefully, for example by examining and comparing related data. However, the responsible human-factors engineer should thoroughly document the reasons that lead to the estimated result.

Then there are cases were applicable data exist, but they do not follow a normal distribution. We may be able to "normalize" the compilation mathematically and then apply the formulas listed above to calculate percentile values. Another solution is to use nonparametric statistics to calculate the desired data points.

If a curve graphically represents strength scores, this offers yet a different way to determine given percentile points, without use of any formal calculation. In this case, we can simply measure, cut, count, or similarly estimate from the graph the desired percentile values. This procedure works whether the distribution is normal, skewed, binomial, or in any other form.

Whichever method is employed to establish critical design value, the human-factors engineer should ascertain on a sample of the intended user population that a new design is indeed operable under all foreseeable circumstances.

16.7   Future Directions

Currently, standardized experimental procedures are lacking by which to record dynamic strength scores. Such experimental procedures are needed. Part of the task is to develop suitable measuring instruments and means to control subject motivation.

Computer-aided design tools must avoid the past mistake of relying on average values, but instead prompt the engineer to select proper strength criteria, such as minimal and/ or maximal limits.

16.8   Summary

The literature contains many sources of information on human body strength. However, the ergonomist should be cautious when applying the data if they stem from persons other than the prospective users and reflect circumstances different from those during actual use.

The engineer, who wants to consider operator strength when developing a new process or device, has to make a series of decisions. These follow from answers to the following questions:

- Is the use static or dynamic? If static, information about isometric strength capabilities applies. If dynamic, special experimental controls are appropriate, concerning, for example, displacement and its time derivatives. The operator's motivation for exertion and prevailing situational conditions can strikingly influence strength outcomes.

- Is a maximal or a minimal strength exertion the critical design factor? "Maximal" user strength usually determines the structural integrity of the object, so that even the strongest operator cannot break a handle or a pedal. The design value is set, with a safety margin, above the highest perceivable strength application. "Minimal" user strength is that expected from the weakest operator, which still yields the desired result, such as successfully operating a brake handle or pedal, under the worst circumstances.

The "range" of expected operator strength exertions lies, obviously, between the considered minimum and maximum. "Average" user strength is usually of no design value.

Statistically, measured strength data often follow a normal distribution. This facilitates reporting them in terms of averages (means) and standard deviations. These numeric descriptors allow the use of common statistical techniques (described in this chapter) to determine selected percentiles, which represent the design points of special interest.

Yet, certain body strength data do not fall into a normal distribution. However, this is not of great concern, because usually the data points of design interest are the extremes. These can be extracted from any type of distribution, either by calculation or by estimation.

In any case, it is prudent to test the newly designed products or processes by having actual users operate them under realistic conditions.

References

Astrand, P.O. and Rodahl, L. 1986. *Textbook of Work Physiology*, 3rd ed. (New York: McGraw-Hill).

Daams, B.J. 1993. Static force exertion in postures with different degrees of freedom, *Ergonomics, 36,* 397–406.

Daams, B.J. 2001. Push and pull data (pp. 299–316), Torque data (pp. 334–342), in *International Encyclopedia of Ergonomics and Human Factors*, W. Karwowski, Ed. (London, UK: Taylor and Francis).

Karwowski, W. and Marras, W.S. (Eds.) 1999. *The Occupational Ergonomics Handbook*. (Boca Raton, FL: CRC Press).

Konz, S. and Johnson, S. 2000. *Work Design: Industrial Ergonomics*. 5th ed. (Scottsdale, AZ: Holcomb Hataway).

Kroemer, K.H.E. 1974. *Designing for Muscular Strength of Various Populations*, AMRL-Technical Report-72-46, Aerospace Medical Research Laboratory, Wright-Patterson AFB, OH.

Kroemer, K.H.E. 1998. Relating muscle strength and its internal transmission to design data, in *Advances in Occupational Ergonomics and Safety*, S. Kumar, Ed. (Amsterdam, NL: IOS Press), pp. 349–352.

Kroemer, K.H.E. 1999a. Assessment of human muscle strength for engineering purposes: Basics and definitions, *Ergonomics, 42,* 74–93.

Kroemer, K.H.E. 1999b. Human strength evaluation, Chapter 11 in *The Industrial Ergonomics Handbook*, W. Karwowski and W.S. Marras, Eds. (Boca Raton, FL: CRC Press), pp. 205–227.

Kroemer, K.H.E. 2002. Ergonomics, Chapter 13 in *Fundamentals of Industrial Hygiene*, 5th ed., B.A. Plog, Ed. (Itasca, IL: National Safety Council), pp. 357–418.

Kroemer, K.H.E. 2004 (under review). *Extra-Ordinary Ergonomics*. (Santa Monica, CA: Human Factors and Ergonomics Society).

Kroemer, K.H.E., Kroemer, H.J., and Kroemer-Elbert, K.E. 1997. *Engineering Physiology: Bases of Human Factors/Ergonomics*, 3rd ed. (New York, NY: Van Nostrand Reinhold–Wiley).

Kroemer, K.H.E., Kroemer, H.B. and Kroemer-Elbert, K.E. 2001. *Ergonomics: How to Design for Ease and Efficiency*, 2nd. ed. (Upper Saddle River, NJ: Prentice-Hall/Pearson Education).

Kroemer, K.H.E. and Robinson, D.E. 1971. *Horizontal Static Forces Exerted by Men Standing in Common Working Positions on Surfaces of Various Tractions*, AMRL-Technical Report-70-114, Aerospace Medical Research Laboratory, Wright-Patterson AFB, OH.

Kuczmarski, R.J., Ogden, C.L., Guo, S.S., Grummer-Strawn, L.M., Flegal, K.M., Mei, Z., Wei, R., Curtin, L.R., Roche, A.F., and Johnson, C.L. 2002. *2000 CDC Growth Charts for the United States: Methods and Development*, DHHS Publication No. PHS 2002-1696, Vital and Health Statistics, Series 11, No. 246, Department of Health and Human Services, Hyattsville, MD.

Marras, W.S. 1999. Occupational biomechanics, Chapter 10 in *The Industrial Ergonomics Handbook*, W. Karwowski and W.S. Marras, Eds. (Boca Raton, FL: CRC Press), pp. 167–204.

Marras, W.S., McGlothlin, J.D., McIntyre, D.R., Nordin, M., and Kroemer, K.H.E. 1993. *Dynamic Measures of Low Back Performance*. (Fairfax, VA: American Industrial Hygiene Association).

Norris, B. and Wilson, J.R. 1995. *Childata: The Handbook of Child Measurements and Capabilities — Data for Design Safety*, DTI/Pub 1732/2k/6/2.96 AR, Department of Trade and Industry, London, UK.

Peebles, L. and Norris, B. 1998. *Adultdata: The Handbook of Adult Anthropometric and Strength Measurements — Data for Design Safety*, DTI/Pub 2917/3k/6/98/NP, Department of Trade and Industry, London, UK.

Peebles, L. and Norris, B. 2000. *Strength Data*, DTI/URN 00/1070, Department of Trade and Industry, London, UK.

Phillips, C.A. 2000. *Human Factors Engineering*. (New York: Wiley).

Steenbekkers, L.P.A., and Beijsterveldt, C.E.M., Eds. 1998. *Design-Relevant Characteristics of Ageing Users*. (Delft, NL: Delft University Press).

University of Nottingham. 2002. *Strength Data for Design Safety, Phase 2*, DTI URN 01/1433, Department of Trade and Industry, London, UK.

Winter, D.A. 1990. *Biomechanics and Motor Control of Human Movement*, 2nd ed. (New York: Wiley).

17

Electromyography and Muscle Force

Carolyn M. Sommerich and William S. Marras

CONTENTS

ABSTRACT A key interest in occupational biomechanics is understanding how and to what extent joints are loaded, passive tissues are stressed, and muscles are activated under different working conditions. This chapter provides an examination of the means by which electromyographic signals can be used to estimate muscle force, and in turn, how those muscle force estimates are used in biomechanical models to estimate joint loads during work. The chapter includes brief reviews of appropriate EMG data collection techniques and muscle physiology relevant to EMG sampling, as well as references to more detailed reviews. Practical use of this information for work design assessment purposes is discussed.

17.1 Ergonomic Relevance

Used correctly, electromyography (EMG) can be used in conjunction with other tools to improve understanding of muscle force and joint loading in response to work task configurations. This can facilitate improved designs of new work tasks or modification of existing tasks that promote reduced exposure of workers to unfavorable loading conditions.

17.2 Introduction

EMG can be a valuable tool for the assessment of musculoskeletal behavior during the exertion of force. It is a measure of the depolarization of muscle fibers, and when the EMG signal is treated properly can be used as an indirect measure of muscle activity, fatigue, or force.

Muscle force and strength are not synonymous terms. Strength can be considered "the capacity to produce force or torque with voluntary muscle contraction" (Gallagher et al., 1998). Operationally, strength is usually defined as the maximum torque that can be generated about a joint, or the maximum amount of torque or force that is demanded by a work task (Chaffin and Park, 1973). Both of these strength determinations refer to forces or torques measured external to the body, though they are in fact the resultant output of internal forces actively generated by muscles acting about the joints that they cross, as well as internal forces stemming from the stretching of ligaments and other passive tissues associated with those same joints. In determining whether or not someone has enough strength to perform a specific task, each involved joint must generate a requisite amount of torque, so that the body (the system of joints working together) produces the external force required for the task. A work task can be characterized by the percentage of the working population that has the strength (the force-generating capacity) to perform it.

From a biomechanical standpoint, there is clearly a direct connection between the internal muscle forces and passive tissue forces that combine to produce joint moments (torques) and, in turn, the external force that is imparted by the human body to a work object (tool, piece of machinery, part, etc). However, quantifying internal muscle forces associated with measured external forces continues to be a challenge. Only a few studies have directly measured and linked external forces imparted by the body and internal muscle forces (Dennerlein et al., 1998; Heckathorne and Childress, 1981), because invasive methods or other special circumstances are required (Figure 17.1). More commonly, the link between internal and external forces is a mathematical model which produces estimates of internal muscle forces based on measured external force or torque, kinematic information about the body during the exertion, and numerous assumptions about the

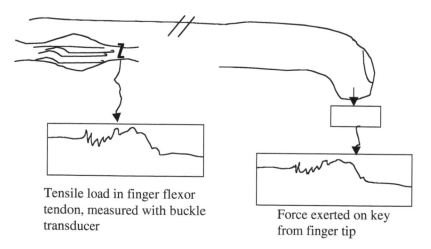

Tensile load in finger flexor
tendon, measured with buckle
transducer

Force exerted on key
from finger tip

FIGURE 17.1

Depiction of an example of a rare situation wherein both muscle tension and external force are directly measured.

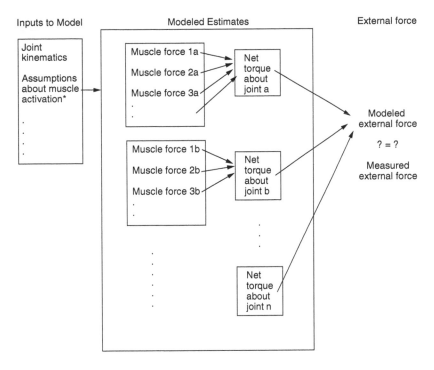

FIGURE 17.2
Depiction of mathematical model that produces estimates of muscle forces and joint torques. Modeled torques are compared with the measured torque (or force) to determine how well a system of joints is modeled. *Note*: Assumptions are often based on a researcher's ideas about what the body may be trying to optimize. This provides the opportunity to reduce the number of muscles for which muscle force must be estimated.

behavior and force contributions of the passive tissues and muscles (Figure 17.2). Some assumptions about muscle behavior and force can be eliminated with the use of EMG, the technique of measuring muscle activity (Figure 17.3). Models that incorporate EMG data may be referred to as "biologically assisted" or "EMG-assisted" models. When conditioned properly and combined with other muscle state and trait information (such as relative length and physiological cross-sectional area, respectively), an EMG signal can be used to estimate force generated within a particular muscle. However, it is important to have an appreciation for the complex nature of the relationship between an EMG signal sampled from a muscle and the force generated within that muscle.

This brief review of EMG and muscle force will discuss some of the key factors that need to be considered when attempting to assess force within a muscle based upon an EMG signal. Practical uses of this information for work design assessment purposes will also be discussed.

17.3 Factors Affecting the EMG–Force Relationship

17.3.1 Nature of the EMG–Force Relationship: Isometric Exertion

It is important to acknowledge the stochastic nature of EMG signals, which precludes the establishment of an instantaneous, one-to-one relationship between EMG amplitude and

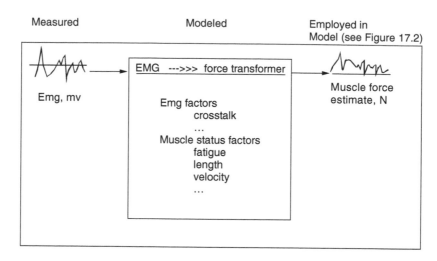

FIGURE 17.3

Depiction of EMG signal recording and transformation of that signal into an estimation of the force (tension) developed in the muscle, based on consideration of the factors known to affect the EMG–force relationship.

muscle force. However, a fundamental assumption of EMG-assisted models is that a relationship between the EMG signal and muscle force can be quantitatively characterized. In order to make this assumption, raw EMG signals must be processed, as described in Section 17.3.2.1, in order to remove the moment-to-moment variation in the signal. In general, studies of EMG and muscle force relationships measure muscle activity and isometric joint torque (which must serve as a surrogate for muscle force). Assumptions must be made regarding the stability of the contribution of the muscle of interest to the net joint torque over the range of torque produced. Activity in the muscle fibers within the detection volume of the EMG electrodes is assumed to be representative of the activity of the muscle as a whole. Both linear and curvilinear relationships have been reported by various researchers in describing the relationship between the EMG signal and muscle force (joint torque) during the simplest of situations, an isometric exertion. Lippold (1952), Moritani and deVries (1978), Yoo et al. (1979), Wang et al. (2000), and Karlsson and Gerdle (2001) have all demonstrated linear relationships between surface EMG activity and voluntary isometric joint torque. Zuniga and Simons (1969), Vredenbregt and Rau (1973), Komi and Vitasalo (1976), and Madeleine et al. (2001) measured EMG proportional to the square of the isometric joint torque. It has also been demonstrated that a single muscle can produce a linear or nonlinear EMG–force relationship at the tendon, depending upon the function performed (Solomonow et al. 1987) or upon the type of electrode employed (Moritani and deVries, 1978). However, in order to simplify this issue, it is common for EMG-assisted models to employ a linear EMG–force relationship. In addition, it is assumed that this relationship is valid for those motions wherein the time delay between the onset of myoelectric activity and muscular contractile force is minimal (i.e., smooth exertions).

17.3.2 Muscle Force Estimates

In order to derive muscle force from an EMG signal, the signal must be processed to adjust for the various factors that can modulate the EMG–force relationship. A condensed listing of the factors that affect the relationship between EMG and muscle force can be organized into four categories (refer to Table 17.1). The purpose of this section is only to provide a basic introduction to these factors. More extensive explanations can be found in other sources, including Soderberg (1991), De Luca (1997), and Farina et al. (2002).

In spite of the number of factors that have been shown to affect the EMG–force relationship, research has shown that muscle force can be estimated from a valid, suitably processed EMG signal (Marras and Granata, 1997; McGill and Norman, 1986). The general form of that relationship is:

$$\text{Force}_j = [\ \text{EMG}_j(t)/\text{EMGMax}_j\ ] \times \text{MaxForce}_j \times \text{CrossArea}_j \times g(\text{Length}_j) \times f(\text{Vel}_j) \quad (1)$$

Where:

Force_j: Force produced in muscle j, N

$\text{EMG}_j(t)$:   Processed EMG amplitude from muscle j at time t, arbitrary units

EMGMax_j: Maximum amplitude of processed EMG signal from muscle j while performing a maximum isometric exertion; may also show as EMGRef_j, for a reference contraction that is other than a maximum; arbitrary units

MaxForce_j: Capacity of muscle j, N/cm^2 (aka stress or specific tension)

CrossArea_j: Physiological cross-sectional area of muscle j, cm^2

$g(\text{Length}_j)$:   Coefficient for length modulation (change in fiber/muscle length from resting/neutral length)

$f(\text{Vel}_j)$:   Coefficient for velocity modulation (speed of change in length)

Factors in Table 17.1 and their influence on, or roles in, Equation 1 are presented in the next four subsections of the chapter.

17.3.2.1 *EMG Collection and Processing Methods*

The first step in estimating the force produced by a specific muscle is to collect a valid EMG signal from the muscle (provides $\text{EMG}_j(t)$ in Equation 1). In the time domain, myoelectric data represent muscle activity and are used to calculate relative muscle force. EMG signals are often collected from pairs of Ag/Ag Cl surface electrodes affixed to the skin after it is prepared in such a way that skin impedance is sufficiently reduced. Electrode pairs should be placed in parallel with the muscle fibers. A single pair of electrodes cannot be assumed to represent the activity in a whole muscle if the muscle is divided into multiple parts (such as the deltoid) or if the muscle is very wide (such as the trapezius). In those cases, multiple pairs of electrodes would be required to provide an accurate assessment of the muscle's activity.

For a given pair of electrodes, the center-to-center spacing of the pair and the size of the electrodes determine the detection volume, which introduces one stage of filtering that occurs in the EMG signal collection process. Electrodes should not straddle an innervation site/zone (Jensen et al., 1993; Queisser et al., 1994) or be placed too close to the muscle's tendon or the muscle's edge. If placed too close to the edge of the muscle, the electrodes will be close enough to nearby muscles to detect their myoelectric activity, which will introduce crosstalk into the signal from the muscle of interest. All of these placement problems will adversely affect the EMG signal, and in turn affect the EMG–force relationship. Interested readers can learn more about crosstalk and volume conduction of EMG signals in Farina et al. (2002).

Typical recommendations for EMG signal collection include sampling at a collection rate that satisfies the Nyquist rule (typically about 1000 Hz), high-pass filtering at 10 or 20 Hz, and low-pass filtering at 500 or 1000 Hz. More information about appropriate methods of EMG data collection can be found at the end of each issue of the *Journal of Electromyography and Kinesiology*, and in Hermens et al. (1997). Signal processing includes full-wave rectification and smoothing via a sliding window filter. Normalization of EMG signals to a reference value ($[EMG_j(t)/EMGMax_j]$ in Equation 1) is recommended, in order to neutralize some of the effects related to electrode placement, and in order to allow for comparison of data between different people. Normalizing to a maximum, as opposed to another reference level, provides an estimate of the proportional effort present in the muscle.

17.3.2.2 Muscle Physiology

17.3.2.2.1 Trait Factors

These factors represent the more stable characteristics of the muscle of interest and include the fiber type composition and distribution, quantity and orientation of muscle fibers, motor unit efficiency, and the cross-sectional area of the muscle. All of the items in the list except for the last one affect the capacity of the muscle fibers or motor units to generate force (represented as MaxForce, in Equation 1). Muscle force capacity is reported to be between 30 and 100 N/cm^2 (Reid and Costigan, 1987). In some modeling efforts, a single capacity value is arbitrarily assumed for all muscles being modeled. However, a more precise estimate can be computed by comparing model-computed muscle-generated joint moments with measured, applied moments about a joint. In order to satisfy the equations of dynamic equilibrium, model-estimated muscle-generated moments must equal the measured moment. EMG-assisted models often determine muscle capacity by choosing a value that satisfies this condition (Marras and Granata, 1997). Muscle force per unit area can be variable from one person to the next and is dependent upon a person's conditioning and natural ability (referred to as "efficiency" in Table 1). However, the capacity predicted for a given individual is assumed to be constant in the absence of fatigue.

Cross-sectional area is the other key muscle trait factor listed in Table 17.1. It appears in Equation 1 as CrossArea$_j$. The normalized and modulated EMG data from a given muscle can be multiplied by the muscle's cross-sectional area to account for the relative force-generating capacity of the muscle. It has been demonstrated that maximum muscle force is proportional to the cross-sectional area for fusiform muscles (Close, 1972). Therefore, scaling the EMG by muscle area provides larger muscles with greater modeled force-generating capacity. In order to account for muscle fiber orientation, Brand (1985) and Chao et al. (1989) suggested utilizing a modified physiologic cross-sectional area: muscle fiber length measured with the muscle in the resting position divided by the muscle volume.

17.3.2.2.2 State Factors

These factors are related to the momentary condition of the muscle and include the extent of muscle fatigue, the rate at which the muscle fibers are being stimulated and which fibers are currently being stimulated, the relative length of muscle (fibers), and the velocity of muscle (fiber) contraction/stretch. Fatigue will affect MaxForce$_j$ and EMG$_j$(t) in Equation 1; recruitment and stimulation rate will affect EMG$_j$(t).

Relative myoelectric activities are multiplied by a unitless function of length, g(Length$_j$) to account for the effect of the relative length of the muscle (fiber) on the muscle's force-generating capacity. The length coefficient incorporates the physiologic length–strength relation and may also be used to account for artifacts due to the variation in the myoelectric potential picked up by the surface electrode. In an effort to enhance their EMG-assisted biomechanical model of the spine, Marras and Granata (1997) empirically derived an

TABLE 17.1

Factors That Affect the EMG–Force Relationship

EMG Collection and Processing Methods	Muscle Physiology (Trait and State Factors)	Interaction between EMG Methods and Muscle Physiology	Force Assessment
Electrodes (type, configuration) EMG signal filtering (explicit and other) EMG sampling rate (comply with Nyquist rule) EMG signal processing	Trait: Fiber type composition and distribution, quantity and orientation of muscle fibers, motor unit efficiency, cross-sectional area of muscle State: Extent of fatigue, rate and recruitment control levels, relative length of muscle (fibers), velocity of muscle contraction/ stretch (fibers)	Electrodes: Orientation relative to muscle fibers, thicknesses and types of tissue between electrodes and muscle of interest, distance from active muscle fibers, distance to other muscles, location of electrodes relative to innervation zone	Type of exertion (extent of isolation, static v. dynamic) Choice of joint and muscle (complexity of muscle arrangement)

equation that described the modulating effect of muscle length on the EMG–force relationship (Equation 2a). Functional coefficients were determined by minimizing the average variation in predicted gain as a function of length. The length modulation factor employed the instantaneous length of muscle j, determined from the anthropometry coefficients and kinematic inputs (Equation 2a). This empirically determined relationship also agrees with the theoretically derived expanded form of the length-modulation factor used in another EMG-assisted model shown in Equation 2b (McGill and Norman, 1986).

$$g(Length_j) = -3.2 + 10.2\ Length_j - 10.4\ Length^2_j + 4.6\ Length^3_j \qquad (2a)$$

$$g(Length_j) = SIN\ (\pi \times (Length_j) - 0.5) \qquad (2b)$$

where

$$Length_j = muscle\ length_j / resting\ muscle\ length_j$$

Relative muscle activity must also be multiplied by a velocity coefficient, $f(Vel_j)$, to account for the physiologic force–velocity relation, and which may also account for associated EMG artifact. Bigland and Lippold (1954) demonstrated that increased muscle contraction velocity yields increased myoelectric activity without a corresponding increase in force output. Most EMG–force relationship models compute this coefficient by minimizing the average variation of muscle capacity predicted by the model as a function of velocity. The velocity coefficient should include the time-dependent contraction velocity of each muscle, j, determined from anthropometry as well as kinematic data. As an example, Equation 3 is the empirically derived expression of the effect of velocity on the EMG–force relationship in the study by Marras and Granata (1997). The coefficients for this modulation factor approximate the theoretical force–velocity relationship represented in the Hill equation (Hill, 1938) and are appropriate for smooth, nonballistic muscle contraction.

$$f(Vel_j) = 1.2 - 0.99\ Vel_j + 0.72\ Vel_j^2 \qquad (3)$$

17.3.2.3   Interaction Between EMG Methods and Muscle Physiology

Placement of the EMG electrodes relative to the muscle of interest and nearby muscles can impact the strength of the EMG signal, as well as the very validity of it. Most of the key points of these interactions were mentioned previously in the subsection on EMG collection and processing methods. Another factor that can have a significant influence on the detected EMG signal is the thickness of the layer of adipose tissue that lies between the skin and the muscle of interest. Farina et al. (2002) reported results of their EMG simulation model that showed a dual, negative effect of adipose tissue on the EMG signal. The signal of interest detected from electrodes appropriately positioned directly over the muscle of interest was increasingly attenuated as the thickness of the adipose tissue layer increased. At the same time, signals from other muscles (crosstalk) were shown to be amplified by thicker layers of adipose tissue.

17.3.2.4   Force Assessment

In estimating muscle force from an EMG signal, it is important to consider the type of exertion that is performed. If an exertion can be attributed to a particular muscle, or if the muscle works with other synergists but its contribution to the exertion remains consistently proportional throughout the range of effort, then the EMG–force relationship is more predictable. As mentioned previously, in dynamic exertions the speed and direction of muscle length change must be accounted for (as $f(Vel_j)$ in Equation 1).

17.3.3   Measurement Requirements

Measurements required to employ an EMG-assisted estimate of muscle force include time-domain EMG, exertion kinetics, and kinematics. Maximum exertion EMG levels and subject anthropometry are needed to calibrate and format the dynamic data for use in the model mechanics. For some muscles, cross-sectional area of muscles may be computed from regression equations based on the subject's trunk depth and breadth (Marras et al. 2001).

Muscle velocity can be computed from dynamic measures of joint movement collected from either goniometers or motion imaging systems. Kinematic data are used to: (a) describe motion as a function of time, (b) determine the muscle force and moment vector directions, and (c) modulate muscle EMG values to account for muscle length and velocity effects.

17.4   Using the EMG–Force Relationship in a Biologically Assisted Model

Biomechanical models typically describe mathematically the forces imparted to a joint or system of joints by the muscles (and possibly other soft tissues) that affect the joint. Among the many assumptions that must be made in developing a biomechanical model are those regarding the lines of action of the muscles, lengths of moment arms, and the proportional contribution made by muscles that act synergistically about the joint. Acquiring and incorporating person-specific information, such as EMG, anthropometric, or muscle imaging data can improve the estimations from the model. Such models have been referred to in this chapter as "biologically assisted," or more specifically, "EMG-assisted" models.

Biologically assisted models use information collected from the body to determine which internal force-generating structures are active and use this information to "drive" or simulate activity in the biomechanical model. The EMG-assisted model provides a framework from which the implications of muscle forces can be appreciated. An EMG-assisted model integrates the EMG–force relationship with the geometric framework of the musculoskeletal system. Such a model would typically describe the relationship of the articulating surfaces relative to one another, so that the axis of rotation of the joint could be described. In addition, the locations of the muscles/tendons (active force sources) and ligaments (passive force sources) relative to the axis of rotation are specified in the model. Muscle locations are typically described through the origins and insertions of the muscles relative to the frame of reference of the system. EMG-assisted models have been particularly appealing for understanding spine loading, because most biomechanical models of the torso are indeterminate. The mechanical advantage of the internal force generators relative to the axis of rotation of the joint can be described through a model of the trunk mechanics. EMG-assisted models have been developed for describing trunk loading by McGill and Norman (1986), Thelen et al. (1994), Granata and Marras (1995), and van Dieen and Kingma (1999).

17.4.1 Example of Using the EMG–Force Relationship to Understand Trunk Mechanics

As an example of how the EMG–force relationship can be used to understand spine loading, we describe an EMG-assisted model that has been under development in the Biodynamics Laboratory for the past decade. Figure 17.4 represents the physical embodiment of the model. The mechanics of this model can be visualized as two "plates" that can be allowed to move relative to one another. Based upon MRI data (Jorgensen et al. 2001), muscle origins were assigned a three-dimensional location relative to the spinal axis along one of the "plates" coplanar with the iliac crest. Muscle insertions are located coplanar with the 12th rib. Muscle forces were represented as vector quantities between their two endpoints. Hence, these plates represent the attachment points of the muscle on the pelvis and thorax. The muscles were represented by vectors between these plates that change their orientation as the trunk moves dynamically (Figure 17.4). Using this approach, muscle orientations and lengths change throughout a movement, accounting for a muscle's changing mechanical advantage throughout the task. In addition, this representation affords the opportunity to change the relative angles between the iliac and thoracic planes. Pelvic tilt or asymmetric bends of the trunk could, therefore, be simulated. Muscle origins and insertions were dynamically located via Euler rotation of the anatomically defined three-dimensional coordinates relative to trunk motion measured by a goniometer. Muscle vector directions, lengths, and velocities were continuously determined from the instantaneous positions and motions of the muscles. As discussed previously, time- and position-dependent force vectors significantly affect the predicted trunk moments and forces generated by the musculature by allowing the vector direction to move throughout an exertion.

Spinal loading (compression, lateral shear, and anterior–posterior (A–P) shear forces) are calculated from the vector sum of validated muscle forces. Moments generated by the muscles about the spinal axis are predicted from the summed vector products combining dynamic tensile forces of each muscle, j, and respective moment arms.

$$\text{Net Muscle Moment} = \text{Sum}_{j=1-n} (\text{moment arm}_j \times \text{Force}_j) \qquad (4)$$

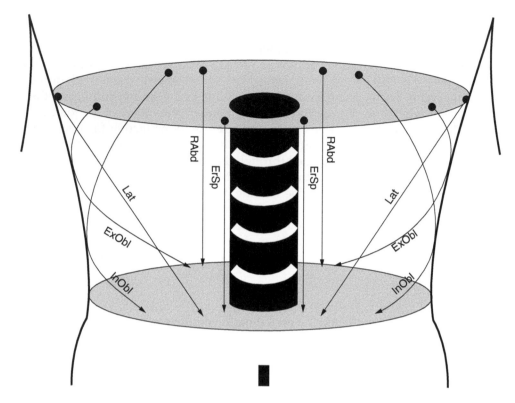

FIGURE 17.4
Depiction of the underlying physical model structure of the EMG-assisted model of Marras and Granata (1997). Muscle origins and insertions are shown relative to their mechanical advantage.

Model output consists of spinal compression, lateral shear force, and A–P shear force that are displayed as a function of time. Measured and predicted trunk moments are compared and serve as a measure of how well the model is simulating trunk mechanics. Correlations between predicted and measured moment profiles (R^2) serve as a measure of model performance and indicate how well the model accounts for the variability in the dynamic moment. A high correlation implies the model accounts for a high degree of variability in the simulation of dynamic spinal loading.

The model has been incorporated into a Windows™ environment that allows the user to analyze the modeled lift in various ways. In this manner, associations between the EMG-related measures and other measurements (e.g., force production) can be evaluated. An example of this model is shown in Figure 17.5.

This model has been used to assess various work situations. One study used this model to evaluate differences in spine loading as workers lift loads from different levels (layers) of a pallet. Figure 17.6 summarizes the results of this study for spine compression (Marras et al. 1999). Two advantages of using EMG-assisted models can be gleaned from this summary. First, there is significant variability in spine loading associated with each condition. EMG-assisted models, such as the one employed here, are particularly sensitive to these individual differences, and can document the range of spine loading values expected due to the task. Second, even though the load mass lifted under all conditions was the same, large differences in spine loading were observed when subjects lifted from different layers of the pallet. Hence, this example emphasizes the significance of considering factors

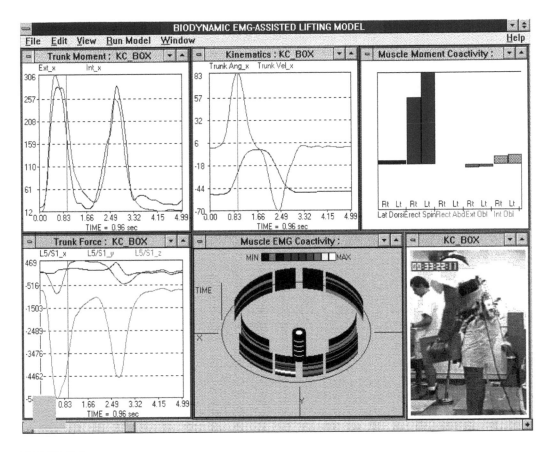

FIGURE 17.5

Windows™ environment for the EMG-assisted model of Marras and Granata (1997), used to assess spinal loading during materials-handling tasks. Displayed in the boxes in the first row, left to right, are continuous displays, over the course of a materials-handling task, of external (measured) and internal (model-predicted) moments about an axis passing through the sagittal plane (box labeled Trunk Moment); trunk flexion and velocity in the sagittal plane (box labeled Kinematics); and instantaneous muscle moments for five pairs of trunk muscles (box labeled Muscle Moment Coactivity; also see Figure 17.4 for muscles). Displayed in the boxes in the second row, left to right, are continuous displays of the model's estimates of compression and shear loads at L5/S1, an alternative display of the muscle activity, and a video clip of the materials handling task.

such as muscle length and muscle velocity when assessing the activities of the muscles and subsequent internal forces imposed upon the body.

17.5 Conclusions

Biomechanical models can provide insight into the loading of joints within the human body during task performance and the level of exertion and contribution to that joint loading by the muscles that cross the joint. Effects of task method, tool, and workstation design can be studied through such models. However, development of such models requires incorporation of many assumptions. Some key assumptions regarding the degree

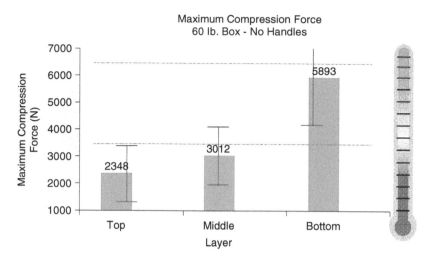

FIGURE 17.6

Mean spine compression estimated from an EMG-assisted biomechanical model as 10 experienced warehouse workers lifted 27.3-kg cases from three different pallet height regions (top, middle, and bottom). These values represent the summary statistics from over 2700 individual lifts. Note that dramatically different spine compressions are experienced as subjects lifted from different height origins.

to which particular muscles contribute to joint loading may be replaced by the use of EMG data which can convey, when properly recorded and processed, the contribution of the muscle from which the signal is recorded. This allows models to provide assessment of an individual worker, which can provide insight regarding why some individuals develop musculoskeletal problems in performing certain tasks while others do not. This is a key question of biomechanists who are interested in designing work and workspaces that enhance worker productivity while improving worker safety.

References

Bigland, B. and Lippold, O.C. 1954. The relation between force velocity and integrated electrical activity in human muscles, *J. Physiol.*, 123, 214–224.

Brand, P.W. 1985. *Clinical Mechanics of the Hand.* (St. Louis: Mosby Year Book).

Chaffin, D.B. and Park, K.S. 1973. A longitudinal study of low-back pain as associated with occupational weight lifting factors, *AIHA J.*, 34, 513–525.

Chao, E.Y.S., An, K.-N., Cooney, W.P., III, and Linscheid, R.L. 1989. *Biomechanics of the Hand: A Basic Research Study* (New Jersey: World Scientific Publishing).

Close, R.I. 1972. Dynamic properties of mammalian skeletal muscles, *Physiol. Rev.*, 52, 129–197.

De Luca, C.J. 1997. The use of surface electromyography in biomechanics, *J. Appl. Biomech.*, 13, 135–163.

Dennerlein, J.T., Diao, E., Mote, C.D., Jr., and Rempel, D.M. 1998. Tensions of the flexor digitorum superficialis are higher than a current model predicts, *J. Biomech.*, 31, 295–301.

Farina, D., Merletti, R., Indino, B., Nazzaro, M. and Pozzo, M. 2002. Surface EMG crosstalk between knee extensor muscles: experimental and model results, *Muscle Nerve*, 26, 681–95.

Gallagher, S., Moore, J.S. and Stobbe, T.J. 1998. *Physical strength sssessment in ergonomics* (Fairfax, VA: American Industrial Hygiene Association Press).

Granata, K.P. and Marras, W.S. 1995. An EMG-assisted model of trunk loading during free-dynamic lifting, *J. Biomech.*, 28, 1309–1317.

Heckathorne, C.W. and Childress, D.S. 1981. Relationships of the surface electromyogram to the force, length, velocity, and contraction rate of the cineplastic human biceps, *Am. J. Phys. Med.*, 60, 1–19.

Hermens, H.J., Hägg, G. and Frekiks, B. 1997. European applications of surface electromography, *Proceedings of the Second General SENIAM Workshop*, Roessingh Research and Development, Enschede, Netherlands.

Hill, A. 1938. The heat of shortening and the dynamic constants of muscle, *Proc. R. Soc. Biol. Sci.*, 126, 136–195.

Jensen, C., Vasseljen, O. and Westgaard, R.H. 1993. The influence of electrode position on bipolar surface electromyogram recordings of the upper trapezius muscle, *European J. Appl. Physiol. Occup. Physiol.*, 67, 266–273.

Jorgensen, M.J., Marras, W.S., Granata, K.P. and Wiand, J.W. 2001. MRI-derived moment-arms of the female and male spine loading muscles, *Clin. Biomech.*, 16, 182–193.

Karlsson, S. and Gerdle, B. 2001. Mean frequency and signal amplitude of the surface EMG of the quadriceps muscles increase with increasing torque — a study using the continuous wavelet transform, *J. Electromyogr. Kinesiol.*, 11, 131–140.

Komi, P.V. and Vitasalo, J.H. 1976. Signal characteristics of EMG at different levels of muscle tension, *Acta Physiol. Scand.*, 96, 267–276.

Lippold, O. 1952. The relation between integrated action potentials in the human muscle and its isometric tension, *J. Physiol.*, 117, 492–499.

Madeleine, P., Bajaj, P., Sogaard, K. and Arendt-Nielsen, L. 2001. Mechanomyography and electromyography force relationships during concentric, isometric and eccentric contractions, *J. Electromyogr. Kinesiol.*, 11, 113–121.

Marras, W.S. and Granata, K.P. 1997. The development of an EMG-assisted model to assess spine loading during whole-body free-dynamic lifting, *J. Electromyogr. Kinesiol.*, 7, 259–268.

Marras, W.S., Granata, K.P., Davis, K.G., Allread, W.G. and Jorgensen, M.J. 1999. Effects of box features on spine loading during warehouse order selecting, *Ergonomics*, 42, 980–996.

Marras, W.S., Jorgensen, M.J., Granata, K.P. and Wiand, B. 2001. Female and male trunk geometry: Size and prediction of the spine loading trunk muscles derived from MRI, *Clin. Biomech.*, 16, 38–46.

McGill, S.M. and Norman, R.W. 1986. Partitioning of the l4–l5 dynamic moment into disc, ligamentous, and muscular components during lifting, *Spine*, 11, 666–678.

Moritani, T. and deVries, H.A. 1978. Reexamination of the relationship between the surface integrated electromyogram (IEMG) and force of isometric contraction, *Am. J. Phys. Med.*, 57, 263–277.

Queisser, F., Bluthner, R., Brauer, D. and Seidel, H. 1994. The relationship between the electromyogram–amplitude and isometric extension torques of neck muscles at different positions of the cervical spine, *Eur. J. Appl. Physiol. Occup. Physiol.*, 68, 92–101.

Reid, J.G. and Costigan, P.A. 1987. Trunk muscle balance and muscular force, *Spine*, 12, 783–786.

Soderberg, G.L. 1991. Selected Topics in Surface Electromyography for Use in the Occupational Setting: Expert Perspectives, Pub. No. 91–100, DHHS/CDC/NIOSH.

Solomonow, M., Baratta, R., Zhou, B.H., Shoji, H. and D'Ambrosia, R.D. 1987. The EMG–force model of electrically stimulated muscles: dependence on control strategy and predominant fiber composition, *IEEE Trans. Biomed. Eng.*, 34, 692–703.

Thelen, D.G., Schultz, A.B., Fassois, S.D. and Ashton-Miller, J.A. 1994. Identification of dynamic myoelectric signal-to-force models during isometric lumbar muscle contractions, *J. Biomech.*, 27, 907–919.

Van Dieen, J.H. and Kingma, I. 1999. Total trunk muscle force and spinal compression are lower in asymmetric moments as compared to pure extension moments, *J. Biomech.*, 32, 681–687.

Vredenbregt, J. and Rau, G. 1973. Surface electromyography in relation to force, muscle length and endurance, in JE Desmedt (ed.), *New Developments in Electromyography and Clinical Neurophysiology*, Vol. 1. (Basel, Switzerland: S. Karger).

Wang, K., Arima, T., Arendt-Nielsen, L. and Svensson, P. 2000. EMG–force relationships are influenced by experimental jaw-muscle pain, *J. Oral Rehabil.*, 27, 394–402.

Yoo, J.H., Herring, J.M. and Yu, J. 1979. Power spectral changes of the vastus medialis electromyogram for graded isometric torques (i), *Electromyogr. Clin. Neurophysiol.* 19, 183–197.

Zuniga, E.N. and Simons, D.G. 1969. Nonlinear relationship between averaged electromyogram potential and muscle tension in normal subjects, *Arch. Phys. Med. Rehabil.* 50, 613–620.

18

Myoelectric Manifestations of Muscle Fatigue

Roberto Merletti, Dario Farina, and Alberto Rainoldi

CONTENTS

ABSTRACT The issue of myoelectric manifestations of muscle fatigue during isometric voluntary or electrically elicited muscle contractions is addressed in this chapter. Global manifestations are described by the "fatigue plot," a graphic representation of the fatigue-related changes reflected by surface EMG variables such as amplitude, spectral characteristic frequencies, and muscle fiber conduction velocity. Different definitions of "myoelectric manifestations of muscle fatigue" are presented and discussed. The concept of "fatigue" at low contraction levels sustained for long time, when the motor unit pool is not stable, is also discussed. The evidence of correlation between the "fatigue indexes" and the muscle fiber constituency is presented. Expected future developments of the field and applications in ergonomics, sport, space, and rehabilitation medicine are outlined. A short appendix provides the basic concepts of signal spectrum and spectral characteristic frequencies.

18.1 Introduction

Muscle fatigue is a subjective feeling whose quantitative assessment is very complex, not unique and controversial. In this chapter a possible and frequently adopted approach to fatigue assessment is described. The approach is based on the analysis of the surface EMG signal. This signal reflects physiological phenomena evolving in the muscle during a sustained isometric contraction. In a dynamic contraction, many confounding factors intervene in the surface EMG generation/detection process, the main ones being signal nonstationarity, highly variable motor unit pool, relative shifts between muscle and detection electrodes, and movement-related artifacts. As a consequence, sustained isometric contractions provide a preferable bench test to assess the properties and degree of fatigability of a muscle. On the other hand, fatigue assessment during daily activities, sport performances, occupational activities, and dynamic contractions in general requires great attention and care, competence, and full awareness of the limitations of the signal processing approaches. This experience and awareness are needed to avoid falling in the many traps of the techniques.

Indexes of myoelectric manifestations of fatigue are defined on the basis of the time evolution of the surface EMG signal features during the contraction under investigation. In this chapter, we describe techniques and applications for the assessment of muscle fatigue by surface EMG signal analysis, we indicate the potentials and limitations of each technique, as well as its repeatability, and we describe future research topics to be addressed.

Muscle fatigue is often associated with an event or condition, such as the inability to further perform a task or sustain an effort, and therefore is somehow related to mechanical performance. Another description deals with the inability to reach the same initial level of maximal voluntary contraction (MVC) force (the force generation capacity), again related to an event or time instant associated with a reduced mechanical performance. These definitions indirectly imply that fatigue is absent before a specific time, whereas it is present after it.

The definition of "myoelectric manifestations of muscle fatigue," based on EMG, accounts for the physiological changes taking place in a muscle before the inability to perform sets in as "mechanical manifestation of muscle fatigue." This definition is clearly explained in Basmajian and De Luca (1985), and introduces the concept of fatigue as an analog function of time, which starts evolving from the beginning of the contraction. The evolution may be fast or slow, depending on the effort performed, and leads eventually to mechanically detectable changes of performance. Many factors contributing to this evolution affect the surface EMG signal and can be detected through it. Unless otherwise specified, this is the concept of fatigue that will be adopted in this chapter.

There are many potential sites of fatigue in the neuromuscular system: the motor cortex, the excitatory drive and the control strategies of the spinal (upper) and the α (lower) motoneurons, the motoneuron conduction properties, the neuromuscular transmission, the sarcolemmal excitability and conduction properties, the excitation–contraction coupling, the metabolic energy supply, and the contraction mechanisms. They can be grouped under the headings of (a) central fatigue, (b) fatigue of the neuromuscular junction, and (c) muscle fatigue. All these factors directly or indirectly affect the EMG signal in ways that are very difficult to unscramble, especially because the information obtained from the surface EMG signal is usually related to a large group of motor units (MUs). However, recent developments, described in this chapter, provide insight in many of these mechanisms.

18.2 Assessment of Muscle Fatigue: Voluntary and Electrically Elicited Contractions

Fatigue itself is not a physical variable and can be "measured" only through the measurement of physical variables associated with it. These variables are affected by noise or include random components, which are particularly relevant in voluntary contractions, due to the stochastic nature of the signal. For this reason, assessment of fatigue becomes an estimation problem. Physical variables that can be estimated are force or torque (current value or MVC value), power, angular velocity of a joint, variables associated to single MUs, such as firing rates, conduction velocity, or activation timing, or variables associated to the global EMG signal detected above a muscle, such as amplitude and spectral estimates, degree of MU synchronization, or muscle fiber conduction velocity (CV) estimates.

The association between mechanical and myoelectric variables requires great caution and awareness of the different phenomena affecting the two sets of variables. In most cases, voluntary efforts activate agonist and antagonist muscle groups, not individual muscles. They often involve coactivation of antagonist muscles for the purpose of limb stabilization, and it is possible that the net force acting over a joint is near zero while agonist and antagonist muscles are both active and "fatiguing". While the resulting force or torque is the algebraic summation of the contributions of different muscles, the detected EMG signal predominantly reflects the activity of the muscle(s) underneath the electrodes. A variation of the sharing of force contributions among synergic and antagonist muscles may leave the net force or torque unchanged but may redistribute the EMG signals, decreasing those of some muscles and increasing those of others, thereby changing their myoelectric manifestations of muscle fatigue (Farina et al., 1999). Furthermore, even within the same muscle, the MU pool may not remain constant during a constant-force sustained isometric contraction (see the following section). New MUs may be additionally recruited, therefore altering mechanical or myoelectric manifestations of fatigue (Gazzoni et al., 2001) and bringing up the question, "What is the structure we are estimating the fatigue of?" As a consequence, a definition of fatigue (or of some specific aspect of it) becomes important in order to describe and understand what is going on within the muscle and its controlling system. A general definition does not yet exist, and research is focusing on specific conditions, mainly isometric constant-force contractions with a (presumably) stable MU pool.

Isometric constant-force contractions are performed as "bench tests" to observe differences between muscles or between the results of tests performed on the same muscle at different times. The force or torque level is sustained for some time at a fraction of the MVC value. In this way, it is believed that the results obtained from different individuals performing similar relative efforts could be compared. Indeed, the same force or torque would imply a rather different relative effort for a weight lifter or an old lady; therefore performing a test with the same relative effort seems more reasonable. However, the second approach has a major drawback, because fatigue depends on the rate of metabolite removal, which in turn depends on blood flow. At a certain level of contraction, blood flow is blocked by intramuscular pressure, and the muscle becomes ischemic. Myoelectric manifestations of muscle fatigue are affected by this condition (Merletti et al., 1984), which probably depends on absolute, rather than relative, contraction level.

Specific experimental protocols have been designed to reduce and isolate the many factors influencing the surface EMG signal features during fatigue. A very useful method to investigate fatigue implies the use of selective electrical stimulation of a nerve branch or of the motor point of a muscle (Merletti et al., 1992). The purpose of this approach is to "disconnect" the muscles from the central nervous system (CNS) and activate only one (or a portion of one) muscle at a time at a controlled frequency and with a MU pool that is more likely to be stable. This approach may be considered as highly unphysiological, but we should not forget that our knowledge of the system under investigation is limited and can be increased only by testing it in controlled conditions with the goal of answering one question at a time. During electrically elicited contractions, the factors of variability due to crosstalk, synchronization and all the central control strategies can be ruled out. On the other hand, the variability due to the critical location and pressure of the stimulation electrodes is added and implies poor interexperiment repeatability.

18.3 Myoelectric Manifestations of Muscle Fatigue in Isometric Voluntary Contractions

In 1912 Piper (1912) observed a progressive "slowing" of the EMG during isometric voluntary sustained contractions. Given the random nature of voluntary EMG, this "slowing" cannot easily be quantified in the time domain. It is easier to describe it in the frequency domain using spectral characteristic frequencies, such as the mean or median frequency (indicated as MNF or f_{mean} and MDF or f_{med} and defined in the Appendix of this chapter) of the power spectral density function (indicated in the following simply as power spectrum), as suggested in the early works of Chaffin, De Luca, Kadefors, and Lindstrom (Chaffin, 1973; De Luca, 1984; Kadefors et al., 1968; Lindstrom and Magnusson, 1977; Lindstrom et al., 1970). This approach has the advantage of being applicable equally well to the almost stochastic signals generated during voluntary contractions and to the almost deterministic signals generated during electrically elicited contractions. MNF is the centroid frequency of the power spectrum (i.e., the moment of order one, the line of the center of gravity) while MDF is the 50th percentile of the power spectrum (i.e., the frequency value splitting the spectrum in two parts of equal power). See Appendix for further details.

Muscle fiber conduction velocity (CV) is known to decrease during sustained isometric constant-force contractions (Arendt-Nielsen and Mills, 1985; De Luca, 1984; Merletti et al., 1990), and as a consequence, the cross-correlation function between two signals detected from adjacent electrode pairs widens, while its peak shifts to the right, and the power spectrum of each of the two signals compresses to the left. The power spectrum is usually scaled by a factor greater than that predictable by the changes of CV. In addition, the power spectrum may undergo a change of shape. These observations strongly suggest that other factors affect myoelectric manifestations of muscle fatigue in addition to CV decrements (Merletti et al., 1990). Myoelectric manifestations of muscle fatigue are a multifactorial phenomenon, involving a number of physiological processes that evolve simultaneously; among these are the changes in shape and width of the action potential of the most superficial fibers (the deep ones have a smaller effect), the different degree of MU synchronization, and the spreading of CV values (Farina et al., 2002b).

Consider a segment (epoch) of the signal generated at time 1 by a single MU as $x_1(l_1)$, where l_1 is the local time within epoch 1, and a second signal epoch $x_2(l_2) = hx_1(kl_1)$ generated at some later time 2 by the same MU and scaled in amplitude by a factor h and in time by a factor k as a consequence of the change of CV by a factor k ($k < 1$ means slowing). If these are the only changes, the magnitude of the Fourier transform $X_2(f)$, the power spectrum $P_2(f)$, and the autocorrelation function $\Phi_{22}(\tau)$ of $x_2(l_2)$ are associated to those of $x_1(l_1)$ by the following relations:

$$X_2(f) = \frac{h}{k} X_1\left(\frac{f}{k}\right) \tag{1}$$

$$P_2(f) = \frac{h^2}{k^2} P_1\left(\frac{f}{k}\right) \tag{2}$$

$$\Phi_{22}(\tau) = \frac{h^2}{k} \Phi_{11}(k\tau) \tag{3}$$

As a consequence, the EMG "variables" MDF (f_{med}), MNF (f_{mean}), lag of the first zero of the autocorrelation function τ_0, average rectified value (ARV), and root mean square value (RMS) of $x_2(l_2)$ (at time t), are related to those of $x_1(l_1)$ (at time 0) by the relations:

$$\frac{f_{med}(t)}{f_{med}(0)} = \frac{f_{mean}(t)}{f_{mean}(0)} = k \tag{4}$$

$$\frac{\tau_0(t)}{\tau_0(0)} = \frac{1}{k}; \quad \frac{ARV(t)}{ARV(0)} = \frac{h}{k}; \quad \frac{RMS(t)}{RMS(0)} = \frac{h}{\sqrt{k}} \tag{5}$$

where t indicates the time interval between epoch 1 (time = 0) and epoch 2 (time = t).

During a sustained isometric contraction the value k changes, in subsequent epochs, as a function of time, starting from the initial value equal to 1. The current value of k could therefore be chosen as a fatigue index. Unfortunately, not all the MUs have the same value of k. In addition, their motor unit action potentials (MUAPs) may change shape, and other phenomena, such as a change of MU pool, or a change in the shape of the sources or in their degree of synchronization, may further affect the interference EMG signal. As a consequence, Equations 1 through 5, which account only for amplitude and frequency scaling factors resulting from changes of CV of a single MU, no longer hold. Different EMG variables show different sensitivity to these changes; therefore, different values of k and h are obtained, depending on the variable used for their estimate (CV, MNF, MDF, ARV, RMS). In particular, different values of k obtained from $f_{med}(t)/f_{med}(0)$ and from $f_{mean}(t)/f_{mean}(0)$ indicate a change of spectral shape. The plots of these variables, normalized with respect to their initial values (Equations 4 and 5), vs. time are referred to as "fatigue plots" (Merletti et al., 1991). This type of plot allows us to compare percent changes of the different variables during an isometric constant-force contraction. The association of this plot with physiological events is not trivial, but the approach is useful to outline differences among muscles or individuals and to identify the ongoing processes by testing if Equations 1 through 3 are holding or not. An example of fatigue plots obtained from the biceps of two young healthy individuals is presented in Figure 18.1 and shows that the two individuals respond differently to the same (relative) effort. Although both subjects sustain the same effort (in percent of MVC) without any force decrement for the required time, subject 8 shows much greater myoelectric manifestations of muscle fatigue than subject 1. The reason may be attributed to a number of factors that include different fiber constituency, different degrees of circulation occlusion, different degrees of MU synchronization, and different spreading of CV values.

While Figure 18.1 compares multiple measurements obtained from two subjects belonging to the same group (young healthy males), Figure 18.2 shows a sample result from two groups of subjects (young and elderly) whose different fatigue behavior was reported by Merletti et al. (2002). In agreement with histological findings showing higher percentage of type I fibers in elderly subjects, these authors demonstrated significantly lower myoelectric manifestations of muscle fatigue (associated with lower MVC values) in the elderly group.

Although the fatigue plot provides considerable information concerning the evolving situation of the neuromuscular system, a more compact representation could be desirable to facilitate comparisons between subjects or muscles. If a graph of a variable shows a linear pattern, the slope of the regression line would provide an intuitive fatigue index. If a pattern is curvilinear, the parameters of an exponential or polynomial regression could be used as indexes. This implies the choice of a regression curve, a fact that introduces a subjective factor. In addition, a negative slope (for example, of CV or MNF) would be associated with "positive" fatigue. A way to improve this representation was proposed by Merletti et al. (1991), who suggested the use of the regression-free "area ratio," defined in Figure 18.3, which provides a positive fatigue index between 0 and 1 (or 0% and 100%) for decreasing patterns and a negative index for increasing patterns. This index may either be defined for the entire contraction or as a function of time. While, in the case of linear patterns, this approach does not provide advantages with respect to a regression line, in the case of curvilinear behavior, it avoids dealing with a wealth of coefficients of regression functions. A disadvantage of this index is its sensitivity to the initial value required to

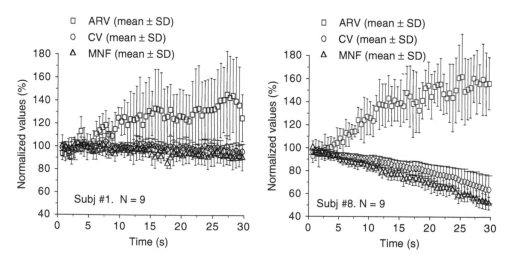

FIGURE 18.1

Fatigue plot obtained from two subjects, each repeating nine times (three times per day for three days) an isometric voluntary contraction of the biceps brachii sustained for 30 s at 70% of the maximal voluntary contraction (MVC) effort. The differences between the subjects are obvious. For clarity, only three EMG variables are shown. (Redrawn from Rainoldi et al. 1999, *J. Electromyogr. Kinesiol.*, 9(2), 105–119. With permission.)

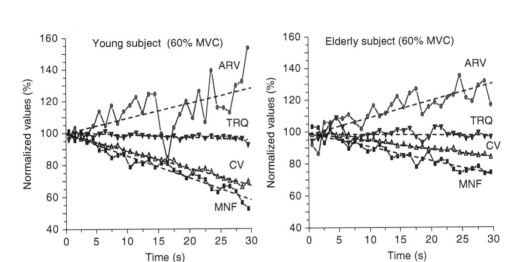

FIGURE 18.2

Comparison of fatigue plots obtained at 60% MVC from (a) a young and (b) an elderly subject. Three EMG variables (ARV, CV, MNF) as well as the torque signal are reported as a function of time with the respective regression lines. Each graph is normalized with respect to the intercept of the corresponding regression line. (Redrawn from Merletti et al. 2002, *Muscle Nerve*, 25(1), 65–76. With permission.)

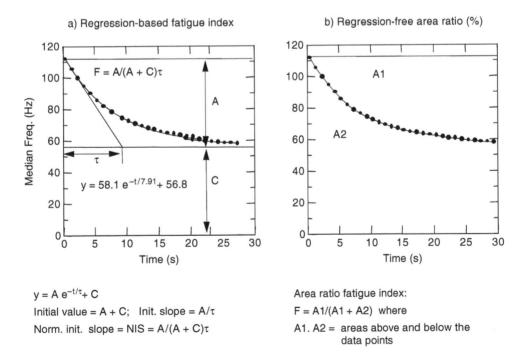

FIGURE 18.3

Definition of two fatigue indexes. (a) Normalized initial slope of an exponentially interpolated pattern. (b) Definition of the area–ratio fatigue index as A1/(A1+A2); this ratio may be computed vs. time. (Redrawn from Merletti et al. 1998, *Muscle Nerve*, 21(2), 184–193. With permission.)

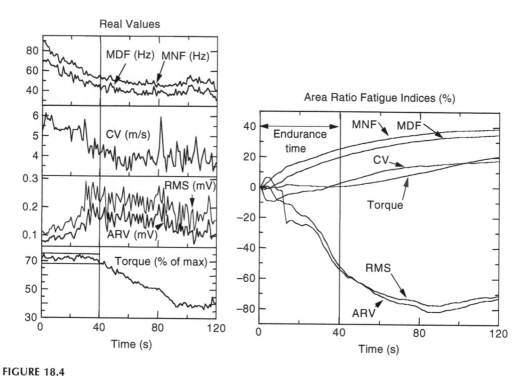

FIGURE 18.4

Example of area–ratio fatigue indexes for six EMG variables computed during a voluntary isometric contraction of the tibialis anterior muscle of a healthy subject. The subject was requested to match a torque target set at 70% MVC until exhaustion. The endurance time was defined as the interval during which the torque was maintained within ± 5% of the target (40 s). Exhaustion time was 120 s. Myoelectric manifestations of muscle fatigue begin at the beginning of the contraction and predict endurance time.

define the rectangle area. Of course, an "area ratio" index may be computed for each EMG variable, therefore providing a "fatigue vector" and allowing clustering of behaviors on the basis of the length, direction, and time evolution of such vector in the space defined by the EMG variables considered. An example of this fatigue plot based on the area ratio index is given in Figure 18.4 for a voluntary isometric contraction at 70% MVC of the tibialis anterior muscle of a healthy subject. The subject was required to try to match the target even beyond the endurance time (when the contraction level became 100% MVC).

Other indexes of fatigue have been proposed by Lo Conte and Merletti (1995) and Farina et al. (2002b). In particular, the first index is based on the cumulative power functions (the integral of the normalized power density function, monotonically increasing from 0 to 1) of two EMG signals detected in subsequent epochs of the same contraction and is of interest because it provides an estimate of the spectral scaling between the signals and an additional number indicating the degree of shape change of the power spectrum. A band, similar to a confidence interval, can also be associated to the fatigue curve describing spectral compression. The second index, based on Recurrence Quantification Analysis, provides indicators that are more sensitive to fatigue than the "traditional" spectral variables, but are not yet fully assessed.

Finally, fatigue indexes related to single MUs have been recently defined and are based on more advanced methods of signal detection and processing described in (Farina et al., 2000, 2001b).

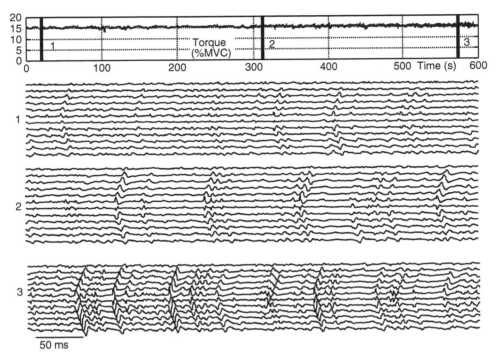

FIGURE 18.5
Upper panel: plot of the elbow torque during an isometric constant-force contraction of the elbow at 15% MVC sustained for 10 min. Panel 1, 2, and 3 depict the raw differential EMG signal detected with a linear electrode array placed on the muscle and corresponding to points 1, 2, and 3 of panel 1. Interelectrode distance is 10 mm. The progressive recruitment of new MUs is evident. (Redrawn from Gazzoni et al. 2001, *Acta Physiol. Pharmacol. Bulg.*, 26(1–2), 67–71. With permission.)

The examples provided in Figure 18.1 through 18.4 show estimates of EMG variables computed during relatively high-level contractions (60–70% MVC), when almost all the available MUs are likely to be recruited and their pool is stable. It is interesting to ask if the same behavior could be expected at lower contraction levels sustained for longer time when the active MU pool may not be stable. Figure 18.5 shows three multichannel linear EMG array recordings (0.5 s each) at the beginning, middle, and end of an isometric contraction of a biceps brachii sustained for 600 s in isometric conditions at 15% MVC. Details about the linear array technique can be found in Farina et al. (2000, 2001a,b) and Gazzoni et al. (2001). The typical "signatures" of MUAPs are quite clear, and the appearance of larger MUAPs after some time is evident. The CNS is therefore recruiting new MUs to maintain (by visual feedback) the required constant force. Figure 18.6 shows the plots of MNF, CV, and ARV for the entire signal of Figure 18.5. The expected progressive decrement of MNF and CV takes place and continues for about 350–400 s and is then followed by an increase, presumably due to the recruitment of "fresh" MUs. This experiment (Gazzoni et al., 2001) underlines the difficulty of defining fatigue when the pool of active MUs is not constant.

An important tool for understanding the role of various phenomena in myoelectric manifestations of fatigue is the simulation of such phenomena by means of mathematical models. This topic will not be addressed in this chapter, but the interested reader may find application examples in (Farina and Merletti, 2001; Farina et al., 2002a,c).

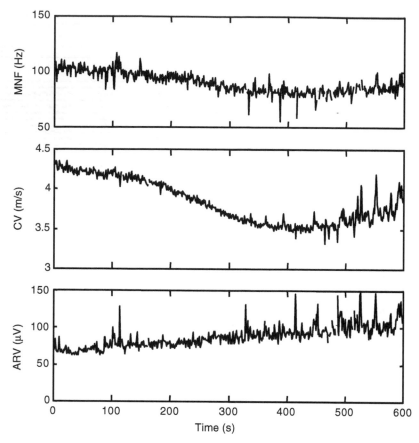

FIGURE 18.6
Estimates of three global EMG variables during the contraction described in Figure 18.5. The change in trend of MNF and CV after about 350 s is due to the recruitment of "fresh" MUs. Such a recruitment is evident in the raw signals depicted in Figure 18.5. (Redrawn from Gazzoni et al. 2001, *Acta Physiol. Pharmacol. Bulg.*, 26(1–2), 67–71. With permission.)

18.4 Myoelectric Manifestations of Muscle Fatigue in Isometric Electrically Elicited Contractions

The response of a muscle to an electrical stimulus is referred to as a "compound action potential" (CAP), or M-wave, and is the summation of the synchronized contributions provided by the activated MUs. The stimulation frequency is externally controlled, and the degree of synchronization is 100%. Stimuli spaced by at least 100 ms (f < 10 Hz) produce single twitches, whereas trains at least 20 Hz produce tetanic contractions. Higher frequencies result in progressively faster manifestations of muscle fatigue without substantial force increase. During an isometric electrically elicited contraction, force is not constant and shows an initial increase (potentiation) followed by a decrease (mechanical fatigue). The M-wave evolves in time, showing changes of amplitude and width (reflected by the amplitude and spectral variables) as well as conduction velocity (Merletti et al., 1992). Myoelectric manifestations of muscle fatigue seem more closely correlated to the number of applied stimuli than to their frequency (Merletti et al., 1992). At frequencies

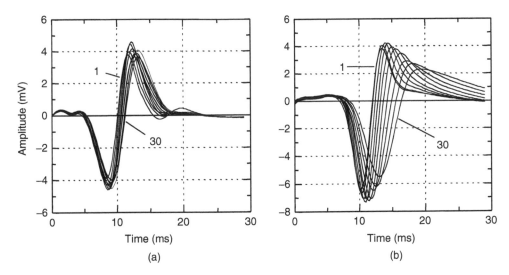

FIGURE 18.7
Sequence of M-waves elicited in the vastus medialis muscle of two healthy male subjects during supramaximal stimulation applied to the muscle's motor point at 30 Hz for 30 s. Each wave is the average of the 30 responses elicited in 1 s. One wave every 3 s is depicted. Subject *a* shows limited myoelectric manifestations of muscle fatigue whereas subject *b* shows more intense manifestations. (Redrawn from Merletti et al. 1998, *Muscle Nerve*, 21(2), 184–193. With permission.)

above 30 to 35 Hz the analysis of the elicited signal becomes difficult because of the overlapping (or truncation by the artifact suppression hardware) of the M-waves.

Because of the quasi-deterministic nature of the EMG signal, during a sustained isometric electrically elicited contraction the patterns of the EMG variables depicted in the fatigue plot show fluctuations that are much smaller than in the case of voluntary contractions. As indicated in Figure 18.7 and 18.8, differences in myoelectric manifestations of fatigue between individuals (same muscle) or muscles are more clearly evident than during voluntary contractions, although intrasubject repeatability may be lower because of critical factors such as position and pressure of the stimulation electrode(s) on the skin.

Different fatigue plot patterns may be identified in the same muscle of normal subjects, as indicated in Figure 18.8, suggesting the capability of the technique to discriminate among different peripheral fatigue mechanisms or muscle structures.

18.5 Myoelectric Manifestations of Muscle Fatigue in Dynamic Contractions

Sustained isometric contractions provide a useful bench test to assess the properties and degree of fatigability of a muscle. On the contrary, fatigue assessment during daily activities, sport performances, occupational activities, and dynamic contractions in general requires great attention and care, competence, and proper signal processing approaches.

It has been recently shown that geometrical factors, among which are the position of EMG electrodes with respect to the innervation and tendon zones and the fiber inclination with respect to the skin plane, play a major role in determining surface EMG variables and fatigue indexes (Li and Sakamoto, 1996; Merletti and Roy, 1996; Rainoldi et al., 2000; Roy et al., 1986). In a dynamic contraction, these confounding and unpredictable variables

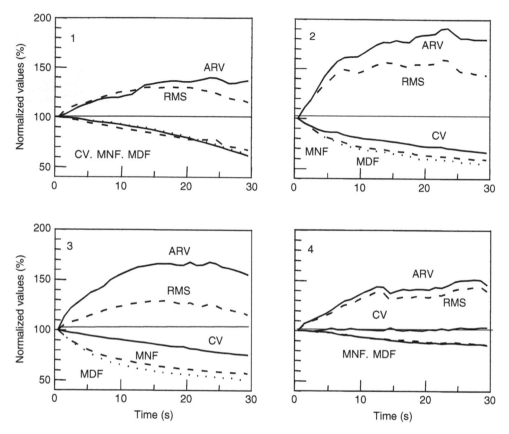

FIGURE 18.8

Fatigue plots showing myoelectric manifestations of muscle fatigue of the electrically stimulated vastus medialis muscle in four healthy male subjects. Case 1 shows the same rate of change for CV, MNF, and MDF, suggesting that Equations 1 through 5 apply for this muscle. Case 2 shows the frequently observed pattern of MDF and MNF changing more than CV, a pattern even more evident in case 3. Case 4 shows a relatively uncommon pattern of decrement of MDF and MNF with no change of CV. Electrical stimulation amplitude was supramaximal in all cases; that is, it was increased slightly above the value producing the maximal M-wave.

change continuously and strongly affect the surface EMG generation/detection process, introducing changes that may incorrectly be attributed to physiological factors.

One of the main problems when dealing with signals acquired during dynamic muscle contractions is the high degree of nonstationarity of the signal (see Appendix). EMG signals, indeed, may suddenly change their spectral properties during a dynamic task, and this may be difficult to investigate with classic spectral techniques. Recently, methods for time-frequency analysis of surface EMG signals detected during dynamic conditions have been proposed. Most of them are based on the Cohen's class of time-frequency distributions (Bonato, 2001) or wavelets (Karlsson et al., 2000). These methods have been shown to provide better results than classic short-time Fourier transform, although there are still no modeling studies showing real improvements in practical applications.

Apart from the spectral estimation problem, many issues related to the dynamic condition should be considered, in addition to those that affect results in isometric contractions. During movement, the muscle changes its size and moves with respect to the electrodes located over the skin. This phenomenon depends on the specific muscle and subject (Rainoldi et al., 2000). The relative shift of the muscle with respect to the skin over the dynamic range of the joint ranges from a few millimeters, as in the upper trapezius muscle or gastrocnemii muscles (Farina et al., 2001a, 2002d), up to 3 centimeters, as in semitendinosus muscle (Farina et al., 2001a). Measurements during cyclic contractions (Bonato et al., 2001) may be less sensitive to this problem, in particular if CV of individual MUs is estimated (with electrode arrays) instead of global spectral variables.

18.6 Fiber Typing and Myoelectric Manifestations of Muscle Fatigue

Myoelectric manifestations of muscle fatigue are affected by (and therefore reflect) muscle fiber constituency. One way to distinguish between the two main MU types is their mechanical response. Fast MUs produce force twitches characterized by short time to peak, while slow fibers need more time to produce the maximum. Fast (Type II) MUs usually have larger fibers and higher CV values, whereas slow (Type I) MUs usually show the opposite. The first work aimed at assessing this correlation was done by Hopf et al. (1974). These authors electrically evoked single twitches in human biceps brachii muscle, estimating both the contraction times (defined as the time from the onset of the deflection to the peak) and the muscle fiber CV (using invasive needle technique). A negative correlation between contraction times and muscle fiber CV was found. Further evidence was provided by Andreassen and Arendt-Nielsen (1987).

The work by Kupa et al. (1995) provides a clear demonstration, on an animal model, of the correlation between variations of the EMG signal power spectrum and the percentage of fiber types in a muscle. EMG signals were recorded *in vitro* from neuromuscular preparations of rat soleus, extensor digitorum longus, and diaphragm muscles during electrically elicited contractions. Fibers from the three rat muscles were then typed as slow oxidative, fast oxidative glycolytic, and fast glycolytic, respectively. MDF initial values and rates of change during the contraction were positively correlated with the percentage of fast glycolytic fiber within the muscle (Figure 18.9). These findings were obtained in a very particular laboratory condition, where the whole muscle was extracted with a nerve segment, free of subcutaneous layers, entirely stimulated by electrical pulses, and the electrodes were placed directly on its surface. Thus, the strong correlation found in rat muscles may not be reachable with noninvasive methods in humans because of several factors that modify the surface EMG variables. Mannion et al. (1998) demonstrated the correlation between the rate of change of MNF and the distribution of fiber types in the erector spinae muscle, providing a confirmation of the work of Kupa et al. (1995) in the case of human muscles. Their results are depicted in Figure 18.10.

The EMG variables presumably most affected by fiber type distribution are MNF (or MDF) and CV. They are related, although not exclusively, to the pH decrease due to the increment of metabolites produced during fatiguing contractions. Lactic acid accumulation, in fact, determines a decrease of muscle fiber CV (and, as a consequence, of MNF) (Brody et al., 1991). However, the works of Linssen et al. (1990, 1991) on McArdle's disease patients, demonstrated that a CV reduction was present during biceps brachii contraction at 80% MVC, although the muscles could not produce lactic acid. In the same direction, Sadoyama et al. (1988) showed a strong correlation between fiber type compositions and CV estimated during maximal voluntary knee extensions, which lasted for 2 to 3 s, from a group of sprinters and one of distance runners (Figure 18.11).

Modifications of fiber type composition due to diseases and/or adaptation can be monitored by the EMG spectral modification, as shown in recent works (Linssen et al., 1990, 1991; Pedrinelli et al., 1998). In the very particular case of a congenital myopathy,

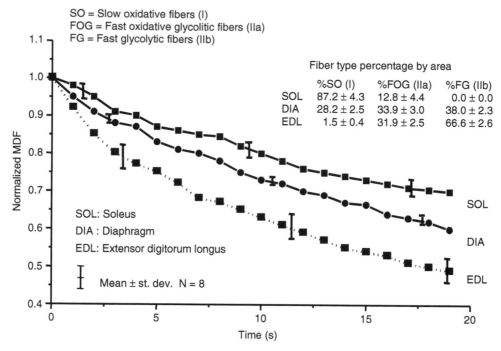

FIGURE 18.9

Normalized values of MDF plotted for the soleus (SOL), diaphragm (DIA) and extensor digitorum longus (EDL) muscles of eight Wistar rats. The neuromuscular preparations were stimulated *in vitro* at 40 Hz for 20 s. Data for each muscle are normalized with respect to their initial values. The rate of change of MDF in the three muscles is correlated with the percentage of fast glycolytic fiber. SO = Slow oxidative fibers (Type I); FOG = Fast Oxidative Glycolitic fibers (Type IIa); FG = Fast Glycolytic fibers (Type IIb). (Redrawn from Kupa et al. 1995, *J. Appl. Physiol.*, 79(1), 23–32. With permission.)

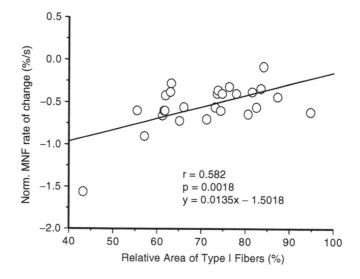

FIGURE 18.10

Correlation between relative area of the muscle occupied by Type I fiber and the rate of change of MNF during a 60% MVC contraction recorded from the erector spinae in 31 subjects. (Redrawn from Mannion et al. 1998, *Spine*, 23(5), 576–584. With permission.)

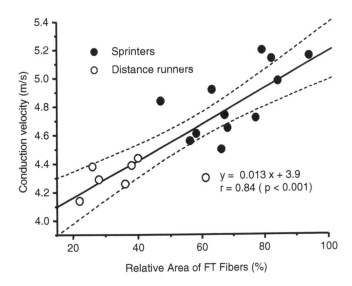

FIGURE 18.11
Correlation between muscle fiber conduction velocity and the ratio of fast type fiber areas in the vastus lateralis.
Filled circles are for sprinters, and open circle are for distance runners. The two separate clusters confirm the
hypothesis that the two groups have opposite fiber type distributions and suggest that the conduction velocity
as a suitable variable for the noninvasive estimate of the fast type fiber percentage in the studied muscle.
(Redrawn from Sadoyama et al. 1988, *Eur. J. Appl. Physiol.*, 57, 767–771. With permission.)

where a generalized congenital predominance (95 to 100%) of Type I fibers is found in all
skeletal muscles, initial values and rate of change of CV, estimated from vastus lateralis
muscle, were found significantly lower compared to the control group of healthy subjects
(Linssen et al., 1991). According to these findings, the lactic acid production was found to
be lower in pathological subjects than in the control group. In subjects with peripheral
vascular disease, CV initial values from the tibialis anterior muscle were found to correlate
with Type II fiber diameter, while MNF initial values correlated both with Type II fiber
diameter and with their percentage in the muscle (Pedrinelli et al., 1998). With their
findings, Linssen et al. (1990, 1991), Pedrinelli et al. (1998), and Mannion et al. (1998)
provided a further confirmation of the hypothesis of Sadoyama et al. (1988), indicating
that myoelectric manifestations of fatigue are related to the percentage area of Type II
fibers in the muscle cross section.

Further indirect confirmation comes from the work of Falla et al. (2003), where the
myoelectric manifestations of fatigue of neck muscles were found to be greater in a group
of whiplash patients with respect to the control group, in agreement with the hypothesis
that the patient group had a greater percentage of more fatigable Type II fibers, as found
in biopsies (Uhlig et al., 1995). Further indirect evidence in the same direction comes from
the work of Merletti et al. (2002), who observed that myoelectric manifestations of muscle
fatigue in the biceps of elderly subjects are lower than in young subjects. Autoptic findings
available from the literature indicate that Type II fibers decrease in number and size with
age, a very likely explanation for the above observation.

Further work is necessary to develop a noninvasive tool, alternative to the biopsy
technique, for noninvasive fiber type estimation. The information provided by the myoelectric manifestations of fatigue seems to be a promising way to reach the objective.

18.7 Repeatability of Estimates of EMG Variables and Fatigue Indexes

Assessment of the repeatability of EMG variables is of considerable relevance for the clinical daily use of surface EMG. To reach this goal it is necessary (a) to define a standard so that data recorded in different laboratories and by different operators are comparable and (b) to define the minimum change of the observed variables that may be related to physiological/pathological variations and not to random fluctuations due to experimental noise. Several works on repeatability of surface EMG variables and fatigue indexes on different muscles, such as elbow flexors, quadriceps, back, and respiratory muscles, can be found in the literature of the last 15 years (Bilodeau et al., 1994; Daanen et al., 1990; Dedering et al., 2000; Kollmitzer et al., 1999; Lariviere et al., 2002a ; Linssen et al., 1993; Merletti et al., 1995, 1998; Ng and Richardson, 1996; Rainoldi et al., 1999, 2001; Van Dieen and Heijblom, 1996). The issue of repeatability is associated with the problem of standardization, proper electrode positioning, and the correct use of statistical tools and has a dramatic importance in the validation of EMG-based fatigue assessment. The efforts toward standardization promoted by the European Concerted Action "Surface EMG for Non-Invasive Assessment of Muscles" (SENIAM) (Hermens et al., 1999) provide an important contribution and should be considered with great attention by EMG clinical researchers.

The issue of repeatability can be addressed considering two different and complementary aspects. The first aspect concerns the reliability (within days) and constancy (between days) of the repeated measure (Viitasalo and Komi, 1975), which are important in order to assess the measure precision (described by the standard error of the mean, SEM) and the variability due to repeated trials and electrode repositioning. The second concerns the capability of the EMG variables to distinguish between different subjects. This capability is in general assessed using the Intraclass Correlation Coefficient (ICC) or the Fisher test; both lead to similar results (Rainoldi et al., 2001). When the between-subject variability is comparable to (or less than) the within-subject variability, the degree of repeatability defined by the ICC becomes meaningless (Rainoldi et al., 1999, 2001). When very little variation in the EMG variable estimates across subjects and trials is found, the measure might not be able to detect individual differences within homogenous groups (e.g., healthy subjects). In this case the EMG variable might be suitable to provide a reference range for the homogenous group of subjects (i.e., normative data), but the ICC would not be the appropriate indicator to use.

18.8 Applications in Ergonomics, Occupational, Sport, Space, and Rehabilitation Medicine

18.8.1 Ergonomics and Occupational Medicine Applications

There is a wide spectrum of either known or strongly suspected work-related disorders affecting the neuromuscular and musculoskeletal systems. Some are referred to as Cumulative Trauma Disorders (CTD) or Repetitive Strain Injuries (RSI). They cause neck, shoulder, arm, and back pain, carpal tunnel syndrome, and a variety of other problems with extremely high economical and social costs. Muscle fatigue is now an accepted fundamental concept in ergonomics. It is a basic principle that fatigue should be avoided, and therefore proper gaps should be introduced when some subjective or objective index exceeds a threshold. This implies continuous monitoring of EMG as an indicator of muscle activity. It is likely that, in the near future, EMG electrode arrays, painted or glued on the skin or embedded in garments, will classify and monitor the "Cinderella" motor units identified by Hägg (1991) or the trapezius or erector spinae bursts of activity during the working day. It is also likely that specific neuromuscular tests will be carried out periodically at critical working places to detect early signs of work-related disorders. Monitoring endurance and myoelectric manifestations of muscle fatigue will certainly be an important part of such tests.

18.8.2 Sport and Space Medicine

Modifications of muscle conditions and performance may be desired in some cases (e.g., sports) and undesired in others (e.g., permanence in microgravity environment). The assessment of effectiveness of a particular type of sport training or of a particular countermeasure to reduce muscle wasting in space is of obvious importance. Surface EMG variables are becoming indexes of performance as important as endurance, maximal voluntary force, and explosive power in exercise physiology, sport, and space medicine.

18.8.3 Rehabilitation Medicine

The pervasive concepts of evidence-based medicine are difficult to apply in the rehabilitation field because of the great intersubject variability and subjective assessment of outcome. In the search for objective indicators of changes, EMG variables play a very important role. In particular, the simplicity of the technique makes it suitable for home applications under supervision through phone, TV, or Web systems. Together with cardiovascular monitoring, neuromuscular remote assessment may become an important application of telemedicine.

18.9 Future Perspectives and Challenges

18.9.1 Understanding Myoelectric Manifestations of Muscle Fatigue by Means of Computer Models

With the exception of muscle fiber CV, the variables extracted from the surface EMG signal are not directly related to any physiological quantity. For example, MNF does not reflect any frequency physiologically generated by the muscle. Rather, these variables are sensitive to a large number of interacting physiological quantities. The most important challenge has been (and still is) the analysis of the relationships between surface EMG variables and the physiological phenomena under study. Due to the high complexity of the EMG generation and detection system, predictions usually have been based on simple models of the problem. With the increment in computational speed and power and the advancement of research in the field, sophisticated models and array signal processing software are becoming available, providing powerful tools for EMG interpretation.

This will certainly be a major line of research in the future. Together with the technical improvements in two-dimensional array detection systems, this research will likely allow monitoring fatigue of individual MUs in superficial skeletal muscles.

18.9.2   Understanding Central Nervous System Control Strategies and Central Fatigue

Myoelectric manifestations of muscle fatigue reflect both central and peripheral phenomena. A clinically reliable tool for separating the two contributions is not yet available. However, the technological progress anticipated in the previous paragraph will allow surface EMG decomposition and, therefore, the estimation of firing rate of individual MUs, the identification of recruitment and derecruitment time, and the estimation of the degree of MU synchronization. These are all central factors that may be identified with the combined use of more sophisticated technology and interpreted with the use of more accurate models.

18.9.3   Understanding Dynamic Fatigue

Dynamic EMG is very difficult to interpret because of the many confounding factors and the high signal nonstationarity. A future challenge is the definition of fatigue indexes appropriate for dynamic conditions where the MU pool is highly variable.

18.9.4   Understanding Pathologies and Estimating Fiber Type Composition

Specific tests based on surface EMG will likely be developed in the coming years for monitoring the evolution of pathologies and for estimating changes in the distribution of fiber types in a muscle.

18.9.5   Understanding and Solving the Crosstalk Problem

Crosstalk depends on many anatomical factors, such as the subcutaneous layer thickness and the length of the fibers, as well as many detection system parameters such as the spatial filter used, the interelectrode distance, and the size and location of the electrodes (Farina et al., 2002e). The separation of electrical contributions from muscles placed at different distances from the electrodes is a task that will likely be solved in the near future through appropriate signal processing procedures and the use of multiple electrode systems and EMG modeling.

18.9.6   Understanding "Difficult" Muscles

Not all muscles are well suited for surface EMG analysis; some have a structure that makes detection and interpretation of EMG signals very difficult. Among these are sphincters and pennated muscles. As indicated above, advancements in detection technology and mathematical modeling will allow the study of these muscles and of their myoelectric manifestations of fatigue.

18.9.7   Move to Clinical Routine Application

The efforts to improve EMG signal processing and interpretation techniques described above are in the perspective of strong clinical applications. To date, many works are available in the literature in different field such as back pain (Kumar and Narayan, 1998; Lariviere et al., 2001, 2002b, Moseley et al., 2002; Roy and Oddsson, 1998; Roy et al., 1998), whiplash injury (Falla et al., 2003), neurological diseases (Roeleveld and Stegeman, 2002;

Rossi et al., 1996), sport applications (Gerdle et al., 1991; Komi and Tesch, 1979; Sadoyama et al., 1988), and ergonomics (Daanen et al., 1990; Hägg, 1991).

International research projects and networks of laboratories are facing the challenge of increasing the degree of standardization of the EMG technique and of EMG signal interpretation in order to increase its clinical use according to the so-called "evidence based medicine."

18.9.8 Education and Training

A thorough understanding of EMG signals requires a solid background in physiology, biophysics, neuropathology, electronics, signal processing, and modeling. This collection of competences is difficult to find among either biomedical engineers or life scientists in the research environment. In the clinical environment, this expertise is absent. On the other hand, the future "EMG expert" will need these competences and will have to be familiar with the concepts of motor unit and electrode impedance, Fourier transform and actin-myosin complexes, estimation of conduction velocity, and histology of Type I and Type II motor units. Training this professional figure will be an interesting challenge in the coming years.

18.10 Summary

The issue of myoelectric manifestations of muscle fatigue during isometric constant-force contractions is addressed in this chapter. Myoelectric manifestations of muscle fatigue may concern either an individual MU (single-MU fatigue) or the EMG generated in the detection volume during either voluntary or electrically elicited contractions (global fatigue). The latter assumes a constant motor unit pool and must be regarded with caution if this is not the case. It is therefore applicable at high contraction levels when all MUs are recruited. Global manifestations are described by the "fatigue plot," a graphic representation of the fatigue-related changes reflected by surface EMG variables such as amplitude, spectral characteristic frequencies, and muscle fiber conduction velocity. Although the general trend reflecting fatigue is an increase of amplitude variables (ARV, RMS), a decrease of conduction velocity (CV), and a greater decrease of spectral frequencies (MNF, MDF), they show different patterns for different muscles and individuals; such differences are likely due to different fiber constituency, different CV mean and/or variance, different degree of MU synchronization, or different degree of muscle ischemia. Current research is focusing on the separation of the contributions due to these factors. One way to separate central from peripheral factors is the study of electrically elicited contractions where all fluctuations due to motor control are eliminated and firing rate is strictly controlled.

Single-MU fatigue requires at least some level of surface EMG decomposition and the identification/classification of MUAPs using an electrode array. At this time, this approach is feasible only at low contraction levels and is under development.

An outline of future developments is provided and includes technological advances (electrode arrays, detection in dynamic conditions, etc.) and modeling advances providing support to the unraveling of the many factors contributing to the initial values and rates of change of the surface EMG variables. Rapidly developing applications in ergonomics, occupational, sport, space, and rehabilitation medicine are outlined. The need to introduce

"EMG experts" with multidisciplinary competence in the clinical environment is under-lined. Such need cannot be overestimated, because the EMG methodology can be easily applied in the wrong way and the results are susceptible to misinterpretation by untrained operators.

Appendix: Basic Concepts of EMG Spectral Analysis

A signal is the representation of a physical variable vs. another variable, usually time or space. Since computers accept as input only electrical signals, other signals (angles, positions, pressures, forces, etc.) are converted into electrical voltages or currents by appropriate transducers. The myoelectric (or electromyographic, EMG) signal is the voltage distribution over the surface of the skin measured with respect to a reference electrode placed in a "neutral" position, away from the sources of such a signal. EMG is changing as a function of space (at any given time instant, every point on the skin surface has a different voltage) and of time (every point on the skin surface has a voltage evolving in time). The EMG signal is detected between a point over the muscle and a reference (monopolar detection). Linear combinations of monopolar signals, obtained from an electrode array of at least two contacts, provide spatial filters. The weights of the combination can be chosen to obtain special purposes. The simplest and most common version combines two monopolar signals with weights +1 and −1 to form the differential montage with the purpose to reduce (theoretically eliminate) common mode voltages generated by the power line or remote sources in the muscle. When we refer to the EMG, we mean either the monopolar signal or the output of a spatial filter.

The French mathematician J. Fourier demonstrated that any periodic signal can be described as a sum of sinusoids that are referred to as "harmonics." The first harmonic (fundamental harmonic) has the same frequency as the periodic signal. The other harmonics have frequencies that are multiples of the fundamental. The sum may have few, many, or infinite terms, depending on the complexity of the signal being analyzed. Figure 18.12a shows the result of the summation of four harmonics that generate the signal $s(t)$ at the bottom left of the figure (signal synthesis). This process may be seen in the opposite direction as the decomposition of $s(t)$ into the four harmonics (Fourier series expansion). Each harmonic is identified by an amplitude, a phase, and its order (i.e., its frequency). Figure 18.12b shows the "spectrum" of the peak amplitudes of the sinusoids, and Figure 18.12c show the "spectrum" of the power contributed by each component. It is obvious that the signal to be analyzed must contain at least one full cycle of the lowest frequency of interest (fundamental harmonic). A DC component does not count as a harmonic and is considered at zero frequency.

This concept can be applied to nonperiodic signals, such as a segment of music or the EMG, with a conceptual extension that is acceptable because it has no practical consequences. Consider a segment of EMG having a duration of $\frac{1}{4}$ s (250 ms). This will be referred to as a signal "epoch." For the purpose of the analysis, we may consider this signal as the first cycle of a periodic signal whose first harmonic has a frequency of 4 Hz, the second 8 Hz, the third 12 Hz, and so on for as many harmonics as is required to reach a good signal representation (about 100). The "spectrum" will therefore have about 100 harmonics of nonnegligible amplitude, 4 Hz apart from each other. Alternatively, if the epoch is 1 s long, there will be about 400 harmonics, 1 Hz apart from each other. This implies a higher frequency resolution because, with respect to the previous example, there

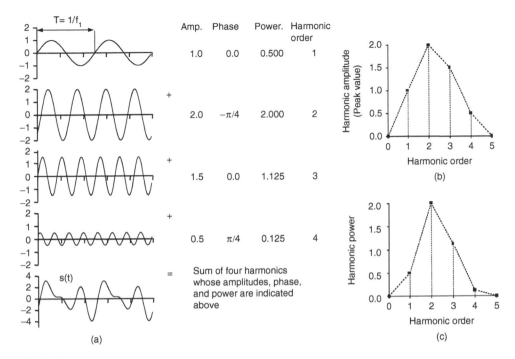

FIGURE18.12

Primer on Fourier analysis. (a) Four sine waves of frequencies that are multiples of the first are added to form the signal s(t). It can be shown that any signal s(t) can be expressed as a sum of a sufficient number of sine waves. (b) Plot of the "amplitude" spectrum of signal s(t). (c) Plot of the "power" spectrum of signal s(t).

are more "spectral points" (400 instead of 100) within the same "frequency bandwidth," but a lower time resolution, since it is not possible to discriminate from the spectrum events or changes of signal structure taking place within a single epoch. If the spectra of the example (four spectra per second obtained from 0.25 s epochs and one from a 1.00 s epoch) are, statistically speaking, the same, the signal is said to be stationary within 1 s. If they are not the same, it means that there are ongoing changes, such as a trend in some feature, and the signal is not stationary. If the changes progress slowly, the signal is considered quasi-stationary. This is the case of surface EMG, which can be considered quasi-stationary within epochs up to 1–2 s during isometric constant-force contractions.

The sinusoids used to describe the EMG power (or amplitude) spectrum are a mathematical tool and neither reflect nor describe physiological phenomena. In particular, their frequency should not be confused with the MU firing rate. The Fourier series expansion of an EMG signal has harmonics of negligible amplitude (or power) below 5–10 Hz and above 350–400 Hz. This frequency range is said to be the bandwidth of the surface EMG signal.

Power outside this range, if present, is due to electrode contact noise or interferences from other equipment and should be removed by an operation referred to as "filtering," which removes harmonics above and/or below a frequency value and is described in Figure 18.13.

Figure 18.13a shows an example of an original signal whose high frequency components are removed by a low-pass filter (Figure 18.13b) or whose low frequency components are removed by a high-pass filter (Figure 18.13c), while the effect of a band-pass filter is depicted in Figure 18.13d. The latter operation removes low-frequency fluctuations and high-frequency noise.

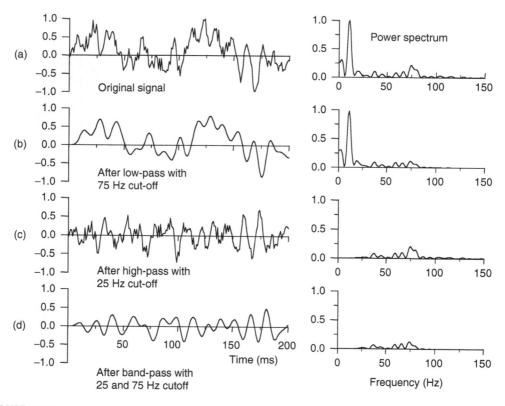

FIGURE 18.13

Example of signal filtering. (a) Original signal and its power spectrum, (b) result of low-pass filtering with 75 Hz cut-off, (c) result of high-pass filtering with 25 Hz cut-off, (d) result of band-pass filtering with band 25 to 75 Hz.

The concept of spectrum is associated to the problem of sampling. Sampling is the process of reading signal values at equal time intervals for the purpose of converting them into digital (binary) form for computer storage. If a signal contains significant harmonics up to the frequency value f_M, sampling must take place at a frequency higher than $2 \cdot f_M$, after the noise components (if any) above f_M have been filtered out. If this criterion (known as Nyquist principle) is met, no information is lost as a consequence of the sampling process, and the original signal can be exactly reconstructed. If the criterion is not met, forms of ambiguity will arise. They are known as "aliasing." One example is given in Figure 18.14 where two signals are depicted. Signal 1 in Figure 18.14a is sampled at a frequency above the Nyquist rate and, although the lines connecting the sampled values (signal 2) do not match signal 1, it is possible, with a proper algorithm, to fully reconstruct signal 1. Figure 18.14b shows a case of ambiguity where the same samples could be coming from either signal 1 (properly sampled) or from signal 2 (under-sampled). If proper filtering had been applied, signal 2 would have been ruled out.

The main reason for performing spectral analysis of either the voluntary or electrically elicited EMG signal is to monitor and quantify spectral changes due to myoelectric manifestations of muscle fatigue. These manifestations reflect a "slowing" of the signal due to a number of physiological phenomena and factors discussed in this chapter. A way to quantify these changes is to describe them by means of spectral characteristic frequencies such as the centroid frequency (the center-of-gravity line), also called mean frequency

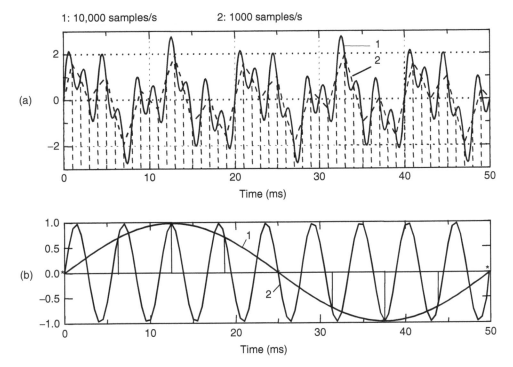

FIGURE 18.14
Examples of signal sampling. (a) Sampling at a frequency higher than the Nyquist rate: the signal can be reconstructed exactly, and no information is lost. (b) A special example showing ambiguity. Signals 1 and 2 are sampled at the same time instant and generate the same sequence of samples, but one of them (signal 2) is sampled below Nyquist rate.

(MNF), or the 50th-percentile line or median frequency (MDF) which splits the spectrum into two parts of equal power (equal area). Figure 18.15 shows the time course of MNF of a biceps muscle during an isometric contraction sustained for 90 s. Ninety epochs of 1 s are defined: 90 spectra are computed and plotted in a three-dimensional plot (only four are depicted in Figure 18.15 for clarity); the centroid line is defined for each spectrum; and the value of MNF is displayed vs. time. The vertical hatching of the spectra schematically depicts the harmonics. A clear progressive "compression" of the spectra is evident during the 90 s and is quantitatively described by the time course of MNF, which decreases from 130 Hz to 60 Hz. In turn, the MNF plot may be further reduced to a few numbers such as the initial value and the initial rate of change or the "area ratio." The last two are often used (as well as other values) as "fatigue indexes" and used for quantitative comparisons of muscle behavior and fatigability.

Both MNF and MDF have been widely used in literature: the first can be estimated with smaller statistical errors; the second is less sensitive to wideband noise. Jointly they provide additional information: if they decrease by the same percentage the spectrum is scaled without change of shape; if MDF is decreasing more than MNF the spectrum is becoming more skewed.

The Fourier analysis is not the only way to estimate these characteristic frequencies. A frequently used alternative method is based on the "autoregressive modeling" technique. For most practical purposes the two methods, both widely used, are equivalent (Farina and Merletti, 2000).

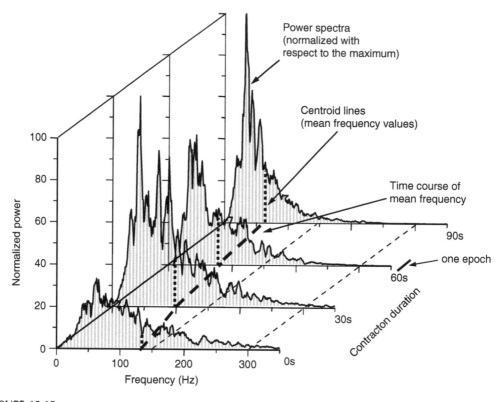

FIGURE 18.15

Example of time course of the power spectrum of a real EMG signal during a 90 s contraction of a biceps brachii. For simplicity only four of the 90 spectra are displayed to show the time course of MNF.

Acknowledgment

Preparation of this chapter was partially supported by the European Shared Cost Project "Neuromuscular assessment in the Elderly Worker" and by grants of the Regional Administration of Piemonte, Italy.

References

Andreassen, S. and Arendt-Nielsen, L. 1987. Muscle fibre conduction velocity in motor units of the human anterior tibial muscle: a new size principle parameter, *J. Physiol*, 391, 561–571.

Arendt-Nielsen, L. and Mills, K.R. 1985. The relationship between mean power frequency of the EMG spectrum and muscle fibre conduction velocity, *Electroencephalogr. Clin. Neurophysiol.*, 60(2), 130–134.

Basmajian, J.V. and De Luca, C.J. 1985. *Muscles Alive* (Baltimore: Williams & Wilkins).

Bilodeau, M., Arsenault, A.B., Gravel, D., and Bourbonnais, D. 1994. EMG power spectrum of elbow extensors: a reliability study, *Electromyogr. Clin. Neurophysiol.*, 34(3), 149–158.

Bonato, P. 2001. Recent advancements in the analysis of dynamic EMG data, *IEEE Eng. Med. Biol. Mag.*, 20(6), 29–32.

Bonato, P., Roy, S.H., Knaflitz, M., and De Luca, C.J. 2001. Time-frequency parameters of the surface myoelectric signal for assessing muscle fatigue during cyclic dynamic contractions, *IEEE Trans. Biomed. Eng.*, 48(7), 745–753.

Brody, L.R., Pollock, M.T., Roy, S.H., De Luca, C.J., and Celli, B. 1991. pH-induced effects on median frequency and conduction velocity of the myoelectric signal, *J. Appl. Physiol.*, 71(5), 1878–1885.

Chaffin, D.B. 1973. Localized muscle fatigue — definition and measurement, *J. Occup. Med.*, 15(4), 346–354.

Daanen, H.A., Mazure, M., Holewijn, M., and van der Velde, E.A. 1990. Reproducibility of the mean power frequency of the surface electromyogram, *Eur. J. Appl. Physiol. Occup. Physiol.*, 61(3–4), 274–277.

De Luca, C.J. 1984. Myoelectrical manifestations of localized muscular fatigue in humans, *Crit. Rev. Biomed. Eng.*, 11(4), 251–279.

Dedering, A., Roos, A.F., Hjelmsater, M., Elfving, B., Harms-Ringdahl, K., and Nemeth, G. 2000. Between-days reliability of subjective and objective assessments of back extensor muscle fatigue in subjects without lower-back pain, *J. Electromyogr. Kinesiol.*, 10(3), 151–158.

Falla, D., Rainoldi, A., Merletti, R., and Jull, G. 2003. Myoelectric manifestations of sternocleidomastoid and anterior scalene muscle fatigue in chronic neck pain patients, *Clin. Neurophysiol.*, 114(3), 488–495.

Farina, D., Cescon, C., and Merletti, R. 2002a, Influence of anatomical, physical, and detection-system parameters on surface EMG, *Biol. Cybern.*, 86(6), 445–456.

Farina, D., Fattorini, L., Felici, F., and Filligoi, G. 2002b, Nonlinear surface EMG analysis to detect changes of motor unit conduction velocity and synchronization, *J. Appl. Physiol.*, 93(5), 1753–1763.

Farina, D., Fortunato, E., and Merletti, R. 2000. Noninvasive estimation of motor unit conduction velocity distribution using linear electrode arrays, *IEEE Trans. Biomed. Eng.*, 47(3), 380–388.

Farina, D., Fosci, M., and Merletti, R. 2002c. Motor unit recruitment strategies investigated by surface EMG variables, *J. Appl. Physiol.*, 92(1), 235–247.

Farina, D., Madeleine, P., Graven-Nielsen, T., Merletti, R., and Arendt-Nielsen, L. 2002d. Standardising surface electromyogram recordings for assessment of activity and fatigue in the human upper trapezius muscle, *Eur. J. Appl. Physiol.*, 86(6), 469–478.

Farina, D. and Merletti, R. 2000. Comparison of algorithms for estimation of EMG variables during voluntary isometric contractions, *J. Electromyogr. Kinesiol.*, 10(5), 337–349.

Farina, D. and Merletti, R. 2001. Effect of electrode shape on spectral features of surface detected motor unit action potentials, *Acta Physiol. Pharmacol. Bulg.*, 26(1–2), 63–66.

Farina, D., Merletti, R., Indino, B., Nazzaro, M., and Pozzo, M. 2002e. Surface EMG crosstalk between knee extensor muscles: experimental and model results, *Muscle Nerve*, 26(5), 681–695.

Farina, D., Merletti, R., Nazzaro, M., and Caruso, I. 2001a. Effect of joint angle on EMG variables in leg and thigh muscles, *IEEE Eng. Med. Biol. Mag.*, 20(6), 62–71.

Farina, D., Merletti, R., Rainoldi, A., Buonocore, M., and Casale, R. 1999. Two methods for the measurement of voluntary contraction torque in the biceps brachii muscle, *Med. Eng. Phys.*, 21(8), 533–540.

Farina, D., Muhammad, W., Fortunato, E., Meste, O., Merletti, R., and Rix, H. 2001b. Estimation of single motor unit conduction velocity from surface electromyogram signals detected with linear electrode arrays, *Med. Biol. Eng. Comput.*, 39(2), 225–236.

Gazzoni, M., Farina, D., and Merletti, R. 2001. Motor unit recruitment during constant low force and long duration muscle contractions investigated with surface electromyography, *Acta Physiol. Pharmacol. Bulg.*, 26(1–2), 67–71.

Gerdle, B., Henriksson-Larsen, K., Lorentzon, R., and Wretling, M.L. 1991. Dependence of the mean power frequency of the electromyogram on muscle force and fibre type, *Acta Physiol. Scand.*, 142(4), 457–465.

Hägg, G. 1991. Static work loads and occupational myalgia — a new explanation model, in *Electromyographical Kinesiology*, H.D. Anderson PA, Danoff JV, Eds. (Elsevier Science Publisher), pp. 141–144.

Hermens, H.J., Freriks, B., Merletti, R., Stegeman, D., Blok, J., Rau, G., Disselhorst-Klug, C., and Hägg, G. 1999. *European recommendations for Surface Electromyography* (Enschede, NL: Roessing Research Development) ISBN 90-75452-15-2.

Hopf, H.C., Herbort, R.L., Gnass, M., Gunther, H., and Lowitzsch, K. 1974. Fast and slow contraction times associated with fast and slow spike conduction of skeletal muscle fibres in normal subjects and in spastic hemiparesis, *Z. Neurol.*, 206(3), 193–202.

Kadefors, R., Kaiser, E., and Petersen, I. 1968. Dynamic frequency analysis of myo-potentials, *Electroencephalogr. Clin. Neurophysiol.*, 25(4), 402–403.

Karlsson, S., Yu, J., and Akay, M. 2000. Time-frequency analysis of myoelectric signals during dynamic contractions: a comparative study, *IEEE Trans. Biomed. Eng.*, 47(2), 228–238.

Kollmitzer, J., Ebenbichler, G.R., and Kopf, A. 1999. Reliability of surface electromyographic measurements, *Clin. Neurophysiol.*, 110(4), 725–734.

Komi, P.V. and Tesch, P. 1979. EMG frequency spectrum, muscle structure, and fatigue during dynamic contractions in man, *Eur. J. Appl. Physiol. Occup. Physiol.*, 42(1), 41–50.

Kumar, S. and Narayan, Y. 1998. Spectral parameters of trunk muscles during fatiguing isometric axial rotation in neutral posture, *J. Electromyogr. Kinesiol.*, 8(4), 257–267.

Kupa, E.J., Roy, S.H., Kandarian, S.C., and De Luca, C.J. 1995. Effects of muscle fiber type and size on EMG median frequency and conduction velocity, *J. Appl. Physiol.*, 79(1), 23–32.

Lariviere, C., Arsenault, A.B., Gravel, D., Gagnon, D., and Loisel, P. 2001. Effect of step and ramp static contractions on the median frequency of electromyograms of back muscles in humans, *Eur. J. Appl. Physiol.*, 85(6), 552–559.

Lariviere, C., Arsenault, A.B., Gravel, D., Gagnon, D., and Loisel, P. 2002a. Evaluation of measurement strategies to increase the reliability of EMG indices to assess back muscle fatigue and recovery, *J. Electromyogr. Kinesiol.*, 12(2), 91–102.

Lariviere, C., Arsenault, A.B., Gravel, D., Gagnon, D., Loisel, P., and Vadeboncoeur, R. 2002b. Electromyographic assessment of back muscle weakness and muscle composition: reliability and validity issues, *Arch. Phys. Med. Rehabil.*, 83(9), 1206–1214.

Li, W. and Sakamoto, K. 1996. The influence of location of electrode on muscle fiber conduction velocity and EMG power spectrum during voluntary isometric contraction measured with surface array electrodes, *Appl. Hum. Sci*, 15(1), 25–32.

Lindstrom, L. and Magnusson, R. 1977. Interpretation of myoelectric power spectra: a model and its application, *Proc. IEEE*, 65, 653–662.

Lindstrom, L., Magnusson, R., and Petersen, I. 1970. Muscular fatigue and action potential conduction velocity changes studied with frequency analysis of EMG signals, *Electromyography*, 10(4), 341–356.

Linssen, W.H., Jacobs, M., Stegeman, D.F., Joosten, E.M., and Moleman, J. 1990. Muscle fatigue in McArdle's disease. Muscle fibre conduction velocity and surface EMG frequency spectrum during ischaemic exercise, *Brain*, 113(Pt 6), 1779–1793.

Linssen, W.H., Stegeman, D.F., Joosten, E.M., Binkhorst, R.A., Merks, M.J., Ter Laak, H.J., and Notermans, S.L. 1991. Fatigue in type I fiber predominance: a muscle force and surface EMG study on the relative role of type I and type II muscle fibers, *Muscle Nerve*, 14(9), 829–837.

Linssen, W.H., Stegeman, D.F., Joosten, E.M., Van't Hof, M.A., Binkhorst, R.A., and Notermans, S.L. 1993. Variability and interrelationships of surface EMG parameters during local muscle fatigue, *Muscle Nerve*, 16(8), 849–856.

Lo Conte, L.R. and Merletti, R. 1995. Advances in processing of surface myoelectric signals: Part 2, *Med. Biol. Eng. Comput.*, 33(3 Spec No), 373–384.

Mannion, A.F., Dumas, G.A., Stevenson, J.M., and Cooper, R.G. 1998. The influence of muscle fiber size and type distribution on electromyographic measures of back muscle fatigability, *Spine*, 23(5), 576–584.

Merletti, R., Farina, D., Gazzoni, M., and Schieroni, M.P. 2002. Effect of age on muscle functions investigated with surface electromyography, *Muscle Nerve*, 25(1), 65–76.

Merletti, R., Fiorito, A., Lo Conte, L.R., and Cisari, C. 1998. Repeatability of electrically evoked EMG signals in the human vastus medialis muscle, *Muscle Nerve*, 21(2), 184–193.

Merletti, R., Knaflitz, M., and De Luca, C.J. 1990. Myoelectric manifestations of fatigue in voluntary and electrically elicited contractions, *J. Appl. Physiol.*, 69(5), 1810–1820.

Merletti, R., Knaflitz, M., and De Luca, C.J. 1992. Electrically evoked myoelectric signals, *Crit. Rev. Biomed. Eng.*, 19(4), 293–340.

Merletti, R., Lo Conte, L., and Orizio, C. 1991. Indices of muscle fatigue, *J. Electromyogr. Kinesiol.*, 1, 20–33.

Merletti, R., Lo Conte, L., and Sathyan, D. 1995. Repeatability of electrically evoked myoelectric signals in human tibialis anterior muscle, *J. Electromyogr. Kinesiol.*, 5, 67–80.

Merletti, R. and Roy, S. 1996. Myoelectric and mechanical manifestations of muscle fatigue in voluntary contractions, *J. Orthop. Sports Phys. Ther.*, 24(6), 342–353.

Merletti, R., Sabbahi, M.A., and De Luca, C.J. 1984. Median frequency of the myoelectric signal. Effects of muscle ischemia and cooling, *Eur. J. Appl. Physiol. Occup. Physiol.*, 52(3), 258–265.

Moseley, G.L., Hodges, P.W., and Gandevia, S.C. 2003. External perturbation of the trunk in standing humans differentially activates components of the medial back muscles, *J. Physiol.*, 547, 581–587.

Ng, J.K. and Richardson, C.A. 1996. Reliability of electromyographic power spectral analysis of back muscle endurance in healthy subjects, *Arch. Phys. Med. Rehabil.*, 77(3), 259–264.

Pedrinelli, R., Marino, L., Dellomo, G., Siciliano, G., and Rossi, B. 1998. Altered surface myoelectric signals in peripheral vascular disease: correlations with muscle fiber composition, *Muscle Nerve*, 21(2), 201–210.

Piper, H. 1912. *Electrophysiologie* (Berlin: Menschlicher Muskeln, Springer Verlag).

Rainoldi, A., Bullock-Saxton, J.E., Cavarretta, F., and Hogan, N. 2001. Repeatability of maximal voluntary force and of surface EMG variables during voluntary isometric contraction of quadriceps muscles in healthy subjects, *J. Electromyogr. Kinesiol.*, 11(6), 425–438.

Rainoldi, A., Galardi, G., Maderna, L., Comi, G., Lo Conte, L., and Merletti, R. 1999. Repeatability of surface EMG variables during voluntary isometric contractions of the biceps brachii muscle, *J. Electromyogr. Kinesiol.*, 9(2), 105–119.

Rainoldi, A., Nazzaro, M., Merletti, R., Farina, D., Caruso, I., and Gaudenti, S. 2000. Geometrical factors in surface EMG of the vastus medialis and lateralis muscles, *J. Electromyogr. Kinesiol.*, 10(5), 327–336.

Roeleveld, K. and Stegeman, D.F. 2002. What do we learn from motor unit action potentials in surface electromyography? *Muscle Nerve*, Suppl(11), S92–S97.

Rossi, B., Siciliano, G., Carboncini, M.C., Manca, M.L., Massetani, R., Viacava, P., and Muratorio, A. 1996. Muscle modifications in Parkinson's disease: myoelectric manifestations, *Electroencephalogr. Clin. Neurophysiol.*, 101(3), 211–218.

Roy, S.H., Bonato, P., and Knaflitz, M. 1998. EMG assessment of back muscle function during cyclical lifting, *J. Electromyogr. Kinesiol.*, 8(4), 233–245.

Roy, S.H., De Luca, C.J., and Schneider, J. 1986. Effects of electrode location on myoelectric conduction velocity and median frequency estimates, *J. Appl. Physiol.*, 61(4), 1510–1517.

Roy, S.H. and Oddsson, L.I. 1998. Classification of paraspinal muscle impairments by surface electromyography, *Phys. Ther.*, 78(8), 838–851.

Sadoyama, T., Masuda, M., Miyata, H., and Katsuta, S. 1988. Fibre conduction velocity and fibre composition in human vastus lateralis, *Eur. J. Appl. Physiol.*, 57, 767–771.

Uhlig, Y., Weber, B.R., Grob, D., and Muntener, M. 1995. Fiber composition and fiber transformations in neck muscles of patients with dysfunction of the cervical spine, *J. Orthop. Res.*, 13(2), 240–249.

Van Dieen, J.H. and Heijblom, P. 1996. Reproducibility of isometric trunk extension torque, trunk extensor endurance, and related electromyographic parameters in the context of their clinical applicability, *J. Orthop. Res.*, 14(1), 139–143.

Viitasalo, J. and Komi, P. 1975. Signal characteristics of EMG with special reference to reproducibility of measurements, *Acta Physiol. Scand.*, 93, 531–539.

19

Physical Demands Analysis: A Critique of Current Tools

Troy Jones and Shrawan Kumar

CONTENTS

ABSTRACT This review summarizes currently existing peer reviewed methods of physical demands analysis. The review was performed to aid practitioners in the selection of appropriate assessment methods and highlights the relative role of the strength (force) variable. Methods of physical demands analysis which have been presented in peer-reviewed literature are critiqued and assessed. Cut points used to determine risk by each method are considered in a comparative synthesis. Currently, there is little consistency between either the cut points used or the method of risk calculation in physical demands analysis methods described in scientific literature. Significant limitations exist in all templates described. Determination of the most appropriate method for industrial application requires careful consideration of factors including the industrial population for which the template has been developed, body region(s) considered, and the mechanism(s) of injury accounted for.

19.1 Introduction and Ergonomic Relevance

The negative impact of work-related musculoskeletal injuries on the economies of industrialized countries has now been well established. In the United States alone, the National Research Council estimates that musculoskeletal disorders account for nearly 130 million health care encounters annually (National Research Council, 2001). Conservative estimates of the economic burden imposed, as measured by compensation costs, lost wages, and lost productivity, are between 45 and 54 billion dollars annually (National Research Council, 2001). Clearly the demands imposed by work tasks may exceed, in many instances, the capacities of the human system. There is currently a movement within health care systems internationally to require the investigation of job demands for the purpose of assessing the fit between the work performed and the worker. These regulations, guidelines, etc. are commonly termed "ergonomic regulations" and are central to the directed efforts addressing prevention of work-related musculoskeletal disorders. Obviously the central tenet of any effort directed at the identification of levels of risk due to physical exposure must be the assessment of physical demand imposed by the work task(s). The physical demands analysis (PDA) is used for this purpose to not only assess the strength requirements of the occupational task but also rates of repetition required, postures required, etc. Information quantified through PDA is necessary to effectively implement initial job design, preemployment worker suitability assessments, musculoskeletal injury prevention efforts, and job redesign following injury. A considerable epidemiologic knowledge base is now present, identifying the relationship between physical risk factors and incidence of musculoskeletal disorders. This epidemiologic evidence base may be used indirectly to validate those items identified as modifying risk in physical demands analysis methods. Experimental evidence has begun to determine the precise mechanisms for this injury category. Although it is not the intent of this chapter to describe these mechanisms, a presentation of current expert opinion may be found in the National Research Council reference (2001). Despite the knowledge gained so far, specific cause-and-effect relationships have not yet been established and precise cut points identifying safe exposure levels have not been determined. An examination of current methods of physical demands analysis is therefore warranted.

19.2 Theories of Musculoskeletal Injury Causation

Before an examination of the selected job demands analyses is presented, a brief review of the current theories of musculoskeletal injury causation and the state of epidemiologic evidence is indicated. Kumar (2001) has proposed four theories of musculoskeletal injury causation, which have been summarized below.

Multivariate interaction theory of musculoskeletal injury precipitation states that the precipitation of injury is based on the interaction of genetic, morphological, psychophysical, and biomechanical factors. Within each of these categories are many variables which potentiate and may effect precipitation of a musculoskeletal injury. Given the sheer number of variations in each of these categories and their interactive effects, the precise mechanisms by which injury may occur are many (Kumar, 2001).

Differential fatigue theory states that occupational injury may result from the mismatch between occupational demands and biological compatibility. This mismatch results in differential loading of active and passive tissues, potentially beyond the range of specific tissue tolerance. This may result in differential fatigue of active structures as well as lengthening of passive structures. Imbalance in load distribution may result in injury (Kumar, 2001).

Cumulative load theory states that biological tissues are viscoelastic. Biological tissues undergo degradation with repeated and prolonged usage due to the cumulative effect of loading precipitating injuries (Kumar, 2001).

Overexertion theory states that a physical exertion of force may exceed the tolerance of the musculoskeletal system or its component parts. Overexertion will be a function of force duration, posture, and motion (Kumar, 2001).

19.3 Epidemiologic Evidence Base

Numerous epidemiologic studies have now been performed examining those physical factors thought to modify risk of musculoskeletal injury. It is taken that to be valid, the physical demands analysis templates proposed must reflect those factors shown to have a causal relationship in the development of musculoskeletal disorders (MSDs). The term musculoskeletal disorder as defined by the United Sates National Institute for Occupational Safety and Health (NIOSH) refers to conditions that involve the nerves, tendons, muscles, and supporting structures of the body (U.S. Department of Health and Human Services, 1997). In 1997, NIOSH performed a detailed review of over 600 epidemiologic studies examining work-related back pain, tension neck syndrome, shoulder tendonitis, epicondylitis, carpal tunnel syndrome, and hand–arm vibration syndrome. This 1997 review concluded that at least moderate evidence has been presented that heavy or forceful work (those tasks requiring significant strength) was related to the low back, neck, elbow, and wrist disorders examined (U.S. Department of Health and Human Services, 1997). Figure 19.1 presents the relative strength of the evidence supporting a cause effect relationship between high levels of exposure to a physical factor and incidence of a MSD. The variables considered and the discounting factors (or multipliers) applied in physical demands analysis templates may be indirectly evaluated through comparison of the relative weight of the variable, in the determination of risk, with supporting epidemiologic evidence.

19.4 Physical Demands Analysis Templates

The physical demands analysis templates presented in the literature may be divided into groups according to the body region of focus, model used, and physical factors considered.

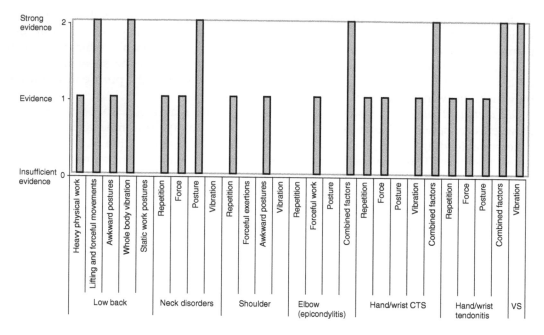

FIGURE 19.1
Epidemiological evidence of work relatedness by physical factor as presented by U.S. Department of Health and Human Services, (NIOSH). 1997. Musculoskeletal Disorders and Work Place Factors, Pub. No. 97B141, U.S. Department of Health and Human Services, National Institute for Occupational Safety and Health, Cincinnati, OH.

Additionally, the occupation group the template was developed to describe, the type of analysis, and the provision of a method to calculate risk are relevant factors. The purpose of this chapter is to discuss and compare for consistency the various criteria used to describe physical factors associated with risk of MSD.

19.4.1 General Templates

Table 19.1 presents some of the general templates (examining two or more body regions) that have been presented in literature for the purpose of identifying risk of musculoskeletal injury based on the quantification of physical factors.

19.4.1.1 Commentary

All of the templates examined under the general template category follow a model of musculoskeletal injury that states that precipitation of MSI, due to physical factors, is modulated by the elements of force (strength), posture, repetition and lack of recovery. The general templates examined may be divided into those that propose a method by which risk of musculoskeletal injury may be calculated and those that do not. Those that allow either a direct or indirect assessment of risk of musculoskeletal injury by providing cut points in risk factors assessed include those proposed or described by Hignett and McAtamney (2000), McAtamney and Corlett (1993), Karhu et al. (1977), Drury (1987), Chen et al. (1989), Corlett et al. (1979), Cote-Gill and Tunes (1989), and Fransson-Hall et al. (1995). The PLIBEL method described by Kemmlert (1995) is not included in the above classification because risk factors identified are dichotomously classified. Dichotomous classification does not facilitate determination of intervention priority, nor does it allow rehabilitation programs to reproduce critical job demands by providing detailed information. Those

TABLE 19.1

General Physical Demands Analysis Templates Examined

Template	Body Regions Examined	Physical Factors Examined	Occupation Group	Calculation of Risk	Static or Dynamic Analysis
Hignett and McAtamney (2000) REBA (Rapid entire body assessment)	• Trunk • Neck • Legs • Upper arms • Lower arms • Wrists	• Posture • Force • Coupling	Health care industry	Yes	Dynamic
McAtamney and Corlett (1993) RULA (Rapid upper limb assessment)	• Upper arm • Lower arm • Wrist • Neck • Trunk • Leg	• Posture • Force • Repetition	Data processing operations, sewing machine operations, production line packing, brick sorting and wire twisting	Yes	Dynamic
Karhu et al. (1977) OWAS (Ovako Working posture analysis system)	• Head and neck • Trunk • Upper limbs • Lower limbs	• Posture • Force	Steel, textiles, meat, mining, wood, and light metal industries	Yes	Static
Drury (1987) PDA template	• Neck • Back • Shoulder • Elbow • Forearm • Wrist	• Force (grip type) • Postural discomfort • Posture • Repetition	Shoe industry	No	Dynamic
Foreman et al. (1988) PDA template	• Whole body posture (e.g., stand, stoop, squat, walk, sit)	• Activity • Posture • Frequency and duration of activities	Health care industry (Nurses)	No	Dynamic
Chen et al. (1989) PWSI (Physical work stress index)	• Overall physiological stress	• Movement (location) • Orientation • Base posture • Hand position • external work load • Load due to imposed accelerations • Thermal environment	Lifting task, hand tool task, light assembly task	Yes	Dynamic
Priel (1974) PDA template	• Head • Shoulder • Arms • Forearms • Trunk • Thighs • Legs • Feet	• Posture	General working postures	No	Dynamic

(continued)

TABLE 19.1 (CONTINUED)

General Physical Demands Analysis Templates Examined

Template	Body Regions Examined	Physical Factors Examined	Occupation Group	Calculation of Risk	Static or Dynamic Analysis
Corlett et al. (1979) Posture targeting	• Head • Neck • Shoulder • Trunk • Wrist • Hip • Knee • Ankle	• Posture • Manual activity performed	Static posture (slides) of machine operators in the electronics industry	Yes	Dynamic
Ridd et al. (1989) ROTA (Robens occupational task analysis system)	• Undefined (description of a system that may be used with dedicated posture/ activity libraries.	• Posture • Repetition • Force • Environment • Workstation	General	No	Dynamic
Kemmlert (1995) PLIBEL (Method for the identification of musculoskeletal stress factors which may have injurious effects)	• Neck/ shoulders, upper part of back • Elbows, forearms, hands • Feet • Knees and hips • Low back	• Dichotomous, general ergonomic risk factor identification	Multiple work groups including; small enterprise, furniture manufacturing, construction, data terminals, farming	Yes	Dynamic
Wells et al. (1994) PDA template	• Hand • Wrist • Shoulder • Back	• Posture • Force through EMG (static, dynamic and peak)	Car seat cover manufacturers and electrical panel manufacturers	No	Dynamic
Cote-Gill and Tunes (1989) PDA template	• Head • Forearm • Trunk • Thigh • Knee • Ankle	• Sitting posture	Seated subjects undergoing classroom activities	No	Dynamic
Wiktorin et al. (1995) HARBO (Hands relative to the body)	• Whole body posture	• Posture of the hands relative to the body	Ceiling builder, carpet layer, railway track layer, car assembly worker	No	Dynamic
Fransson-Hall et al. (1995) PEO (portable observation method)	• Hand • Neck • Trunk • Knee	• Posture • Force	Cook, secretary, mechanic, furniture mover	No	Dynamic
Holzmann (1982) ARBAN	• Head–neck • Shoulder–arm • Trunk and Back • Leg	• Posture • Force • Static load • Vibration • Psychophysical demand	Methodology presented only	No	Dynamic

methods proposed or described (Wilktorin et al., 1995; Wells et al., 1994; Ridd et al., 1989; Foreman et al., 1988; Holzmann, 1982; Priel, 1974) present only methods by which physical factors may be recorded. This examination will be limited to a review of the cut points proposed by the various templates allowing an assessment of risk. Those general templates not allowing either a direct or indirect calculation of risk will not be included, because a review of data collection methodology is not the focus of this chapter. Table 19.2 through Table 19.5 summarize the various cut points used by each "general" template examined. The majority of templates described in this section do provide research-based justification for the cut points used. The current epidemiologic and experimental evidence base in this area may not allow precise cut points to be determined, however. This difficulty is further compounded by the inability to directly transfer cut points supported by epidemiologic research from one working population to another.

19.4.2   Low-Back Templates

Table 19.6 presents some of the lower back templates that have been presented in literature for the purpose of identifying risk of musculoskeletal injury based on the quantification of physical factors.

19.4.2.1   *Revised NIOSH Equation (Waters et al., 1993)*

The revised NIOSH equation is a multiplicative model which uses weight constants and modifier variables to arrive at an index of risk. The lifting model is constructed using the same mathematical format developed by Drury and Pfeil (1975). Biomechanical, physiological, and psychophysical data, in addition to expert opinion, is used to determine the weighting of the multiplier variables described. Low-frequency lifting (i.e., repetition rates below 4 lifts/min) is limited by biomechanical compression limits at the L5/S1 level. High-frequency limits are based on physiological calculation of energy expenditure using the model proposed by Garg et al. (1978). Maximum weight guidelines used in the equation have been set using the psychophysical data presented by Snook (1978) and revised by Snook and Ciriello (1991). Thus, an underlying assumption of the revised NIOSH equation is that the maximum acceptable weight of lift (determined psychophysically) provides an empirical measure that integrates biomechanical and physiologic sources of stress (Karwowski, 1983; Karwowski and Ayoub, 1984). The NIOSH equation may not be used to determine risk associated with tasks involving one hand, lifting while sitting or kneeling, lifting in a constrained work space, lifting temperate items, high-speed lifting (lifting that is performed in a 2- to 4-second time frame) lifting wheel barrels, or shoveling (Waters et al., 1993). Additionally, it is assumed that manual handling tasks other than lifting are minimal and do not require significant energy expenditure, especially when repetitive lifting tasks are performed. For this reason, the NIOSH assessment procedure may not be well suited to application in nonindustrial sectors, given the variability in characteristics of the load lifted, variability in lifting tasks, their frequent association with other handling tasks (trolley pushing or pulling), and finally, the presence of other risk factors for the lumbar spine (i.e., whole-body vibration) (Grieco et al., 1997). Agriculture, transport and delivery of goods, and assistance to individuals who are not self sufficient (at home or in hospital) are typical examples (Grieco et al., 1997). In these situations, although the NIOSH lifting index is useful, validated procedures for integrated exposure assessment are not yet available. Use of the 3.4 kN L5/S1 compression limit has been questioned by Leamon et al. (1994) based on the variability in observed compression tolerance limits across both epidemiologic and cadaveric studies. Considering the research used in the formation of the 3.4 kN guideline, Leamon (1994) suggests that a compression tolerance limit of 5 kN

TABLE 19.2

Posture Cut Points Used to Identify Risk of Musculoskeletal Injury by Template

Body Region	Hignett and McAtamney (2000)	McAtamney and Corlett (1993)	Karhu et al. (1977)	Drury (1987)[a]	Corlett et al. (1979)	Cote-Gill and Tunes (1989)	Fransson-Hall et al. (1995)
Hand/wrist	0–15° of flexion or extension > 15° of flexion or extension Increased risk if wrists are deviated or twisted	Neutral Flexion or extension 0–15° Flexion or extension > 15° Increased risk for any radial or ulnar deviation	Category not applicable	Flexion; 0–9, 9–23, 23–45, 45+ Extension; 0–10, 10–25, 25–50, 50+ Radial deviation; 0–3, 3–7, 7–14, 14+ Ulnar deviation; 0–5, 5–12, 12–24, 24+	Posture is recorded in 1-degree increments for joint movements in the sagittal or frontal plane	Category not applicable	Below shoulder level Above shoulder level
Forearm	Category not applicable	"Mid range of twist" "At or near end range of twist"	Category not applicable	Pronation; 0–8, 8–19, 19–39, 39+ Supination; 0–11, 11–28, 28–57, 57+	Category not applicable	Supported Unsupported	Category not applicable
Elbow	Lower arm: 60–100° of flexion < 60° flexion or >100° flexion	60–100° of flexion < 60° or > 100° flexion Increased risk if working across the midline or out to the side	Category not applicable	Flexion; 0–14, 14–36, 36–71, 71+	Category not applicable	Category not applicable	Category not applicable
Shoulder	Upper arm: 20° extension to 20° flexion >20° extension 25–45° of flexion 45–90° of flexion > 90° flexion Increased risk if arm is abducted or rotated or if shoulder is raised Decreased risk if leaning, supporting weight of arm, or if posture is gravity assisted)	20° Flexion to 20° degrees extension Flexion 20–45° or Extension > 20° Flexion 45–90° Flexion > 90° Increased risk if shoulder is elevated or if upper arm is abducted Decreased risk if the operator is leaning or the weight of the arm is supported	< 90° shoulder flexion Both arms. 90° shoulder flexion One arm > 90° flexion	Outward rotation; 0–3, 3–9, 9–17, 17+ Inward rotation; 0–10, 10–24, 24–49, 49+ Abduction; 0–13, 13–34, 34–67, 67+ Adduction; 0–5, 5–12, 12–24, 24+ Flexion; 0–19, 19–47, 47–94, 94+ Extension; 0–6, 6–15, 15–31, 31+	Posture is recorded in 1° increments for joint movements in the sagittal or frontal plane	Recorded in 15° increments from 60° extension to 90° flexion	Category not applicable

Neck	0–20° flexion > 20° flexion or in extension Increased risk if twisting or side flexed		0° flexion/extension, 0° rot, 0° side flexion > 30° flexion > 30° lateral flexion > 45° of rotation > 30° extension	Rotation; 0–8, 8–20, 20–40, 40+ Lateral bend; 0–5, 5–12, 12–24, 24+ Flexion; 0–6, 6–15, 15–30, 30+ Extension; 0–9, 9–22, 22–45, 45+	Posture is recorded in 1° increments for joint movements in the sagittal or frontal plane	Forward bent Neutral position Backward bent	Flexion > 20 degrees Rotation > 45 degrees
Trunk	Upright 0–20° flexion or extension 20–60° flexion or > 20° extension > 60° flexion Increased risk if twisting or side flexed	Sitting supported with hip/trunk angle of > 90° 0–20° flexion 20–60° flexion > 60° flexion Increased risk if side bent or twisted	0° flex/ext, 0° rot, 0° side flexion Rotation and lateral flexion (undefined rotation or lateral flexion angle) 20–30° of axial twisting (undefined rotation angle) 20–30° forward flexion (undefined hip flexion/lumbar flexion angles)	Rotation; 0–10, 10–25, 25–45, 45+ Lateral bend; 0–5, 5–10, 10–20, 20+ Flexion; 0–10, 10–25, 25–45, 45+ Extension; 0–5, 5–10, 10–20, 20+	Posture is recorded in 1° increments for joint movements in the sagittal or frontal plane	Recorded in 15° increments from 60° extension to 90° flexion	Flexion 20–60° Flexion > 60° Rotation > 45°
Hip	Legs: Bilateral weight bearing walking or sitting Unilateral weight bearing, feather weight bearing or an unstable posture Increased risk if knee(s) are between 30 and 60° of flexion Increased risk if knee(s) are > 60° flexion (not for sitting)	Leg posture: Legs and feet well supported with weight borne evenly OR If standing with body weight even on both feet and room for position change Increased risk if legs and feet are not supported or weight is unevenly balanced	Lower limbs: Standing on one leg (knee straight) with other off the floor. Standing with knees fully extended Generally normal seated posture	Category not applicable	Posture is recorded in 1° increments for joint movements in the sagittal or frontal plane	Angle between the trunk and the thigh is recorded in 15° increments from 135 to 30°	Category not applicable

(continued)

TABLE 19.2 (CONTINUED)

Posture Cut Points Used to Identify Risk of Musculoskeletal Injury by Template

Body Region	Hignett and McAtamney (2000)	McAtamney and Corlett (1993)	Karhu et al. (1977)	Drury (1987)[a]	Corlett et al. (1979)	Cote-Gill and Tunes (1989)	Fransson-Hall et al. (1995)
Knee	Category not applicable	Category not applicable	Category not applicable	Category not applicable	Posture is recorded in 1° increments for joint movements in the sagittal or frontal plane	Angle between the trunk and the thigh is recorded in 15° increments from 135 to 0° / Crossed (adducted across midline) / Uncrossed	Category not applicable
Ankle	Category not applicable	Category not applicable	Category not applicable	Category not applicable	Posture is recorded in 1° increments for joint movements in the sagittal or frontal plane	Crossed (adducted across midline) / Uncrossed	Category not applicable

Note: In general, increased risk is associated with descending categories.

[a] Postural ranges are given in % of maximal range per joint. Risk increases as % range increases.

TABLE 19.3

Repetition/Frequency (Including Determination of Static Posture) Cut Points Used to Identify
Risk of Musculoskeletal Injury by Template

Author(s)	McAtamney and Corlett (1993)	Drury (1987)
Description	One point is added to the risk calculation if the task is mainly static (held for more than 1 min) or is repeated more than 4 times per min.	Frequency of movements constituting a risk factor are not described

would allow greater discrimination between low and high risk groups. Hidalgo et al.
(1997) suggests modification of the existing physiologic criteria through consideration of
the data presented by Asfour et al. (1991) and presents lifting frequency limits based upon
task duration. Marras et al. (1999b) found that only the average weight of box and average
horizontal distance multipliers contributed significantly to the revised lifting equation
model. The authors suggest that further description of the functional nature of the multipliers may lead to higher predictive ability. Further, upon application of the revised
NIOSH equation to a database of 353 industrial jobs, it was found that, while 73% of the
high-risk jobs were correctly classified, about 25% of the jobs that had never experienced
a back injury were classified as high risk. In addition, over 66% of the medium-risk jobs
were incorrectly classified as high risk.

19.4.2.2 Lifting Model (Hidalgo et al., 1997)

The lifting model proposed by Hidalgo et al. (1997) is based on the revised NIOSH lifting
equation (i.e., it is a multiplicative model with weight constants and modifier variables)
with the following modifications. Maximum frequency of lift is calculated with respect to
task duration, and therefore the frequency multiplier is calculated considering separately
the frequency of lift and the duration of lift. Several additional modifiers are considered
in the calculation of the proposed risk index including: age, weight, and heat stress. Age
and weight modifiers were developed using the biomechanical data presented by Genaidy
et al. (1993b). The heat stress multiplier is generated from the unpublished work of Havez,
(1984). Similar to the NIOSH model, base weights are calculated using the psychophysical
data presented by Snook and Ciriello (1991) and modified using the benchmarks established by Tichauer (1978). The authors built and tested the model in two stages. First, the
model was built using psychophysical data. Second, the discounting factors of the various
variables were tested and adjusted using physiologic and biomechanical data. Discounting
factors relying on physiological data were predicted using the data presented by Garg et
al. (1978) and modified through consideration of the physiologic fatigue data presented
by Asfour et al. (1991).

19.4.2.3 Lifting Model (Grieco et al., 1997)

The lifting model proposed by Grieco et al. (1997) is a multiplicative model based on the
revised NIOSH lifting equation. Proposed modifications are directed at enabling exposure
assessment, associated with manual handling tasks, in Italy. Two discounting factors in
addition to those proposed by Waters et al. (1993) are described. Guidelines for the manual
materials-handling activities of pushing, pulling, and carrying are also described. One-
arm lifting is discounted by a factor of 0.6, and if lifting is carried out by two or more
operators, always in the same workplace, the weight lifted is divided by the number of
operators and discounted by a further factor of 0.85. Guidelines for the manual materials-
handling tasks of pushing, pulling, and carrying are based solely on the psychophysical

TABLE 19.4

Force Cut Points Used to Identify Risk of Musculoskeletal Injury by Template

Author(s):	Hignett and McAtamney (2000)	McAtamney and Corlett (1993)	Karhu et al. (1977)	Drury (1987)[a]	Chen et al. (1989)[b]	Corlett et al. (1979)[c]	Fransson-Hall et al. (1995)
Manual handling/ undefined	< 5 kg 5–10 kg > 10 kg Increased risk if there is shock or rapid build up of force	No resistance or less than 2 kg intermittent load or force 2–10 kg intermittent load or force 2–10 kg static load or repeated load or force 10 kg or more static load or 10 kg or more repeated loads or forces or shock or forces with rapid buildup	Less than 10 kg Between 10 and 20 kg Greater than 20 kg	Grip type: Power grip Finger tip pinch Pulp pinch Lateral pinch	External load: 0–0.5 kg 0.5–5 kg 5–20 kg Acceleration: Zero Slight Moderate Heavy	Manual activities: crank, strike, push, pull, hold, squeeze, twist, and wipe. Weight of object.	Manual handling: 1–5 kg 6–15 kg 16–45 kg > 45 kg unknown force

[a] Increasing levels of risk during grip are not clearly identified, forces are measured for each grip.

[b] Force required to perform task is a variable in the calculation of physiologic load.

[c] Force required may be indirectly determined via activity variables marked dichotomously and weight recorded.

TABLE 19.5

Additional Factors, Description, and Cut Points Used to Identify Risk of Musculoskeletal Injury by Template

Author(s)	Hignett and McAtamney (2000)	Drury (1987)	Chen et al. (1989)
Description	Activity score is used to modify risk (elevate) if any of the following are observed; 1 or more body parts are static (held for longer than 1 minute) Action causes rapid large range changes in posture or an unstable base. Repeated small range actions, e.g., repeated more than 4 times per minute (not including walking) Coupling modifier is used to elevate risk any of the following are observed; Hand hold acceptable but not ideal or coupling is acceptable via another part of the body Hand hold not acceptable although possible Awkward, unsafe grip, no handles Coupling is unacceptable using other parts of the body	Postural discomfort is assessed psychophysically using the body discomfort scale and the general discomfort scale	Additional factors considered are; movement (location), orientation, base posture, hand position, external work load, load due to imposed accelerations, and thermal environment. Hand orientation is given relative to the "box" bordered superiorly, laterally, and distally by the arms when the shoulders are flexed to 90°, and the level of the waist inferiorly. Left and right hand position are recorded relative to the "box" in four categories. In box Edge of box Outside of box Outside of box in two planes Location (movement) of the worker is recorded relative to the work station in four categories: Primary work space Meters from primary workplace: 5–10 m Meters from primary work place; 10–50 m Meters from primary work place: > 50 m Orientation of the worker is recorded in relation to the primary work place in four categories: Forward Right Left Backward Thermal load is recorded in relation to four categories: 20–25°C 25–30° or 15–20°C 30–35° or 0–15° > 35° or < 0° Postural base is recorded in relation to four categories: Lying Sitting Leaning Standing

TABLE 19.6

Low Back Physical Demands Analysis Templates Examined

Template	Physical Factors Examined	Tasks Considered	Calculation of Risk	Single and/or Multitask Assessment
Waters et al. (1993) Revised NIOSH equation	• Frequency multiplier • Coupling multiplier • Asymmetric multiplier • Distance multiplier • Vertical Multiplier • Horizontal multiplier • Load constant	Lifting	Yes	Single and multitask
Grieco et al. (1997) PDA template	• Vertical multiplier • Displacement modifier • Horizontal multiplier • Asymmetrical multiplier • Coupling multiplier • Frequency multiplier • Variable load constant	Lifting, pushing, pulling, carrying	Yes	Single task
Hidalgo et al. (1997) PDA template	• Horizontal multiplier • Vertical distance at origin multiplier • Vertical travel multiplier • Lifting frequency multiplier • Task duration multiplier • Twisting angle multiplier • Coupling multiplier • Heat stress multiplier • Age multiplier • Body weight multiplier • Variable load constant	Lifting	Yes	Single task
Shoaf et al. (1997) PDA template	• Carrying; frequency, traveled distance, vertical height • Lowering; frequency, horizontal distance, vertical distance • Pushing; frequency, traveled distance, vertical height • Pulling; frequency, traveled distance, vertical height • Age, body weight, and task duration multiplier • Variable load constant	Push, pull, lower, carry	Methodology presented only.	Single task
Marras et al. (2000) Low back disorder risk model	• Maximum load moment • Maximum lateral velocity • Average twisting velocity • Lifting frequency • Maximum sagittal trunk angle	Lift	Yes	Single task

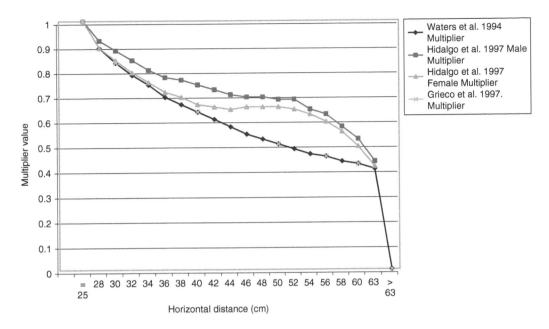

FIGURE 19.2
Comparison of horizontal multiplier values between lifting templates described by Waters et al. (1993), Hidalgo et al. (1997), and Grieco et al. (1997). Multiplier values of the templates compared were interpolated to enable comparison. Multiplier values presented for Hidalgo et al. (1997) were adapted from graphical form.

data set presented by Snook and Cirello (1991). Comparison of the discounting factors common to the templates proposed by Waters et al. (1993), Hidalgo et al. (1997), and Grieco et al. (1997) are presented in Figure 19.2 through Figure 19.5. Multiplier values of the templates compared were interpolated to enable comparison.

19.4.2.4 *Manual Handling Limits for Lowering, Pushing, Pulling, and Carrying Activities (Shoaf et al., 1997)*

Shoaf et al. (1997) describe a three-stage process used in developing a set of multiplicative mathematical models for manual lowering, pushing, pulling, and carrying tasks similar to the NIOSH equation. Initially, the psychophysical data set presented by Snook (1978) and Snook and Cirello (1991) was used to generate the multiplier values and recommended load capacities. The base weights generated via psychophysical data, for lowering and carrying, were revised based on Tichauer's (1973) biochemical lifting equivalent (BLE) equation to achieve a safe load standard based on the biomechanical integrity of the lower back. It was therefore determined by the authors that maximum acceptable weight of the lowering and carrying tasks (determined psychophysically) provided an empirical measure that integrates biomechanical and physiologic sources of stress (Karwowski and Ayoub, 1984). For pushing and pulling, it was determined that because of the short moment arm, the biomechanically derived forces were significantly higher than the psychophysically derived forces. Therefore, guidelines for pushing and pulling exertions determined psychophysically overestimated capacities of typical working populations. It is concluded by the authors that the hypothesis of Karwowski (1983) and Karwowski and Ayoub (1984) is valid only for tasks in which the compressive forces are critical but is not appropriate for tasks in which the shear forces are critical. Each model's frequency multiplier was tested for feasibility using the Garg (1976) energy expenditure equations and physiological fatigue limits developed by Asfour et al. (1991).

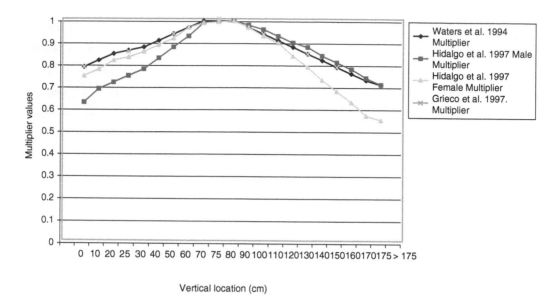

FIGURE 19.3

Comparison of vertical location multiplier values between lifting templates described by Waters et al. (1993), Hidalgo et al. (1997), and Grieco et al. (1997). Multiplier values of the templates compared were interpolated to enable comparison. Multiplier values presented for Hidalgo et al. (1997) were adapted from graphical form.

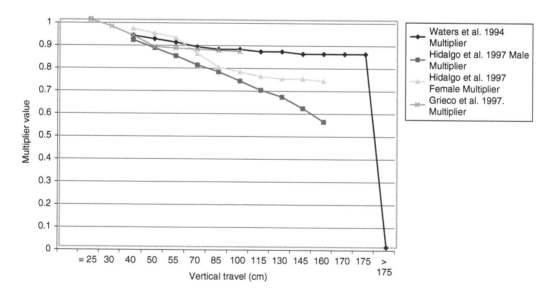

FIGURE 19.4

Comparison of vertical travel multiplier values between lifting templates described by Waters et al. (1993), Hidalgo et al. (1997), and Grieco et al. (1997). Multiplier values of the templates compared were interpolated to enable comparison. Multiplier values presented for Hidalgo et al. (1997) were adapted from graphical form.

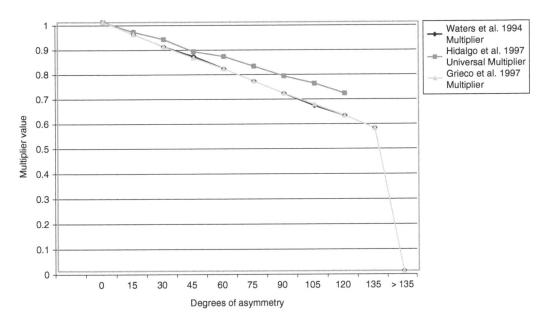

FIGURE 19.5

Comparison of asymmetry multiplier values between lifting templates described by Waters et al. (1993), Hidalgo et al. (1997), and Grieco et al. (1997). Multiplier values of the templates compared were interpolated to enable comparison. Multiplier values presented for Hidalgo et al. (1997) were adapted from graphical form.

19.4.2.5 *Low Back Disorder Model Using the Lumbar Motion Monitor (Marras et al., 1999a)*

Multiple authors have acknowledged the role of three-dimensional velocity and acceleration in the causation of low back injury. The model of low back disorder causation described by this author uses dynamic data recorded by a device utilizing electrogoniometers called the "lumbar motion monitor" (LMM). Using the LMM, high-risk group membership is predicted (those jobs associated with at least 12 injuries per 200,000 work hours of exposure) in repetitive manual materials-handling tasks. Acceleration, velocity, and range of motion are calculated in the sagittal, lateral, and twisting plane by the LMM. Maximum load moment, frequency of lift, sagittal flexion, twisting velocity, and lateral velocity are input into the low back disorder risk model to calculate the percentage likelihood that the job examined would be considered high risk. Likelihood of high-risk group membership is based upon data collected by Marras et al. (1993), which examined 403 industrial jobs from 48 manufacturing companies. Importantly, this model is limited to jobs involving repetitive tasks and no job rotation. When job rotation requires the worker to perform different tasks daily or weekly, the model loses the ability to correctly account for those variables, and thus predictive ability is affected. The job analyzed with this system must consist of a few repeatable consistently performed tasks (Marras, 1999a). Due to the special emphasis placed on trunk dynamics in this model, which resulted from repetitious jobs without rotation being examined, jobs involving lifting of heavy loads in awkward postures may escape identification (Mirka et al., 2000). Maximum duration of data collection may be limited to approximately 30 seconds, and relevant motion at the hip is not recorded (Li and Buckle, 1999). Lavender et al. (1999), in a comparison of five methods for quantifying work-related low back disorder risk in production jobs, found the lumbar motion monitor to be the second to the revised NIOSH equation as most likely to categorize jobs as high risk. As a result, the authors report that the lumbar motion monitor system is best utilized as a tool to predict injury resulting from cumulative load and not acute risk.

19.4.3 Upper Extremity Templates

Table 19.7 presents some of the upper extremity templates that have been presented in literature for the purpose of identifying risk of musculoskeletal injury based on the quantification of physical factors.

19.4.3.1 *The Strain Index (Moore and Garg, 1995)*

The "strain index" described by Moore and Garg (1995) considered multiple risk factors in determining the risk of development of distal upper extremity disorders. Risk factors are classified into five categories of increasing risk, and a multiplicative model is used to arrive at the final index of risk. Important limitations of the "strain index" are: it is not able to analyze multiple tasks, not designed to predict hand–arm vibration syndrome, hypothenar hammer syndrome (mechanical compression of distal upper extremity tissues by extrinsic sources), or disorders of the shoulder, shoulder girdle, neck or back. The physical factors used in the assessment of risk are briefly summarized below. Physiological, biomechanical, and epidemiologic models are used to justify values of multiplier variables used. Physiologic equations used in the relative weighting of multiplier values are presented below.

19.4.3.1.1 *Physiologic Model of Localized Muscle Fatigue*

% Maximum strength (MS) = $100 \times$ required strength/Workers maximal strength (task specific) (Moore and Garg, 1995)

Endurance time $_\text{DYNAMIC}$ (sec) = $324{,}487/(\%\text{MS})^{2.23}$ (Hagberg, 1981)

Endurance time $_\text{ISOMETRIC}$ (sec) = $341{,}123/(\%\text{MS})^{2.14}$ (Hagberg, 1981)

19.4.3.1.2 *Multipliers*

Intensity of exertion: Measured using verbal descriptor similar to the Borg scale estimated by the observer. The multiplier values reflect the rating values (1–5) raised to a power of 1.6. This relationship was selected because (a) the physiological, biomechanical, and epidemiological principles suggest a nonlinear relationship between intensity of exertion and manifestations of strain, and (b) psychophysical theory suggests that perceived effort is related to applied force by a similar relationship.

Duration of exertion: Calculated by: percentage duration of exertion = average duration of exertion per cycle divided by average exertion cycle time. The corresponding category is then selected and multiplier applied. Multiplier values are determined based on expert opinion.

Efforts per minute: Observed frequency of efforts is categorized and multiplier assigned by scale described. Categories of repetition and multiplier values used are based on expert opinion.

Hand–wrist posture: Categorized and multiplier assigned according to scale described. Multiplier values are reported to reflect decreased grip strength and increased intrinsic stresses to the contents of the flexor and extensor compartments with nonneutral postures. Discounting factors (multiplier values) are based on expert opinion.

TABLE 19.7

Upper Extremity Templates Examined

Template	Body Regions Examined	Physical Factors Examined	Occupation Group	Calculation of Risk	Static or Dynamic	Additional Factors Examined
Colombini (1998) Occhipinti (1998) Grieco (1998) OCRA (Concise Exposure Index)	• Shoulder • Elbow • Hand/wrist	• Force • Posture • Repetition • Lack of recovery • Additional factors	Manufacturing industries; ceramics, timber, automotive, meat and vegetable processing, tellers	Yes	Dynamic	Vibration, velocity and acceleration, precision, localized compression, exposure to cold, use of gloves, coupling, wrenching movements, return shock
Moore and Garg (1995) The strain index	• Hand/wrist	• Intensity of exertion • Duration of exertion • Efforts per minute • Hand/wrist posture • Speed of work • Duration per day	Pork processing, turkey processing, chair manufacturing	Yes	Dynamic	Category not applicable
Keyserling et al. (1993) JDA template	• Hand/wrist • Shoulder	• Repetitiveness • Forceful manual exertions • Awkward postures and hand tool usage	Metal plant, engine plant, parts distribution warehouse	Yes	Dynamic	Local mechanical contact stress, gloves, vibration, decreased temperature
Keyserling (1986) JDA template	• Trunk • Shoulder	• Posture	Automobile assembly	No	Dynamic	Category not applicable
Kilbom et al. (1986) VIRA. (Video technique for the analysis of postures and movements of the head, shoulder and upper arm)	• Head • Shoulder • Upper arm	• Posture • Subjective discomfort	Electronics manufacturing industry	No	Dynamic	Subjective rating of discomfort
James et al. (1997) PRRI (postural and repetitive risk factors index)	• Neck • Shoulder • Elbow • Wrist	• Static contraction • Repetition • Posture	VDT use (banking industry)	Yes	Dynamic	Category not applicable

(continued)

TABLE 19.7 (CONTINUED)

Upper Extremity Templates Examined

Template	Body Regions Examined	Physical Factors Examined	Occupation Group	Calculation of Risk	Static or Dynamic	Additional Factors Examined
Li and Buckle (1998) QEC (Quick exposure check)	• Back • Shoulder/arm • Wrist/hand	• Frequency • Posture • Force	Undefined "practical tasks," manual assembly (bolting), manual materials-handling (lifting), and VDU work, simulated nursing tasks	Yes	Dynamic	Vibration, visual demand, work pace, stress
Genaidy et al. (1993)	• Fingers • Wrist • Elbow/shoulder/neck	• Repetition • Force • Posture	Methodology presented only	Yes	Dynamic	Category not applicable
Latko et al. (1997)	Fingers Wrist	Repetition	Office furniture, spark plug and container, automotive components, manufacturing industries	No	Dynamic	Force, posture, and localized mechanical stress scales reported as present but not described

Speed of work: Categories are correlated to the methods time measurements system and perceived speed determined by the observer. Values are designed to reflect the reduction in maximum voluntary strength as speed increases and the theory that a worker's muscles do not fully relax between high speed, high frequency exertions. Multiplier values are based on expert opinion.

Duration of task per day: Intended to reflect the beneficial effect of job rotation and the detrimental effects of prolonged activity. Multiplier values are based on expert opinion.

19.4.3.2 Concise Exposure Index (Occhipinti, 1998; Colombini, 1998; Grieco, 1998)

The exposure assessment presented by Occhipinti (1998), Colombini (1998), and Grieco (1998) is based on the calculation of an exposure index similar to the NIOSH lifting equation. Observed values of the variables considered are classified into groups and multiplied with the appropriate discounting factor. The model proposed yields an index resulting from the calculation of the total number of technical actions actually performed during the shift divided by the total number of recommended technical actions. Risk of MSD precipitation and recommended action are based on this ratio. The number of recommended actions is based on a constant "action frequency factor" of 30 repetitions per minute and is applied to all regions examined. The action frequency factor is then discounted by the other variables considered (force posture, additional elements, and recovery periods). Recovery periods are assessed through organizational analysis describing task duration and recovery periods, both considering natural breaks (i.e., lunch) and in relation to control actions (considered recovery periods) and mechanical actions (considered repetitive periods).

19.4.3.2.1 Multipliers

Repetitiveness/frequency: Calculation of the total number of recommended technical actions per shift is a product of the interaction of all variables considered.

Force: The CR-10 rating of perceived exertion described by Borg (1982) is used to quantify effort or force. Collection of data and assignment of the appropriate force score is accomplished by observing the full cycle and then asking the worker to rate each relevant action within the cycle. The relative duration of each action within the cycle is then calculated and multiplied with the appropriate discounting factor. All actions requiring a significant level of force are then summed to yield the force score.

Posture and types of movements: Postures of the hand, wrist, elbow, and shoulder are described in relation to the static and dynamic movements exceeding or falling below a critical angle. Posture scores are further modified with respect to type, duration held, and type of movement (static or dynamic). Increased risk scores are therefore associated with posture in relation to articular range or grip type, duration of time spent in the posture, and lack of variation in the cycle.

Additional factors: Risk in relation to additional factors is assessed through dichotomous classification of the presence of the factor and the fraction of the cycle time present (e.g., 1/3, 2/3, 3/3).

442

Muscle Strength

TABLE 19.8

Calculation of Recovery Periods for Operations Requiring Isometric
Contractions (Equal to or Longer Than 20 s) for Applied Times and Forces

Force (Borg Scale)	Time held (s)	Recovery Period (s)	Percentage Recovery (%)
Up to 2 (20% MCV)	20	2	10
	30	3	10
	45	7	15
	120	60	50
	180	180	100
	240	480	200
	300	1200	400
	450	2700	600
About 3 (30%MCV)	20	10	50
	40	40	100
	60	120	200
	90	360	400
	120	720	600
	150	1200	800
About 4 (40% MCV)	20	20	100
	30	60	200
	50	200	400
	70	420	600
	90	720	800
Circa 5 (50% MCV)	20	40	50
	30	120	400
	40	240	600

Adapted from Colombini, D. 1998, *Ergonomics*, 41, 1261–1289.

19.4.3.2.2 Work Breaks and Duration of Recovery Periods

Dynamic activity: Calculation of risk is based on the Victorian Occupational HSC Draft code of practice (1988), in relation to occupational overuse syndromes. Within this Australian document the authors report a work–rest ratio of 5:1 is recommended. The analysis model used for calculating risk in dynamic activities associated with inadequate rest is based on this 5:1 work–rest interval criteria. In the procedure proposed, the daily job activities are examined, and the work–rest interval is calculated. Increasing risk is associated with higher proportions of work compared with rest and the number of hours daily with insufficient rest or in potential overload.

Static activity: The levels of contraction force, their RPE equivalent, required recovery period, and percentage recovery are presented in Table 19.8.

19.4.3.3 Exposure Scale (Genaidy et al., 1993)

Genaidy et al. (1993a) describe a method of determining risk of upper extremity and neck MSD based on the determination of daily action and maximum permissible limits for the neck and upper extremity. The maximum permissible limit is defined as 3 times the action limit for each region considered. Guidelines given are based on epidemiologic criteria for repetition and posture. For force limits, biomechanical data is used to describe the action limit, and epidemiologic data to describe the maximum permissible limit. Calculation of "the ergonomic stress index" considers the physical factors of repetition, force, and posture individually and interactively. The effect of physical factors individually and the interaction between factors are equally weighted in the calculation. Based on the value of the physical factor observed, a numerical value reflecting level of risk is assigned. Repetition categories are assigned by classifying the number of observed repetitions per day. Force is assigned through calculation of force as a percentage of maximum voluntary contraction. Posture is reported as a percentage of the total range of motion.

19.4.3.3.1 Additional Templates

Additional templates described by Keyserling et al. (1993), Li and Buckle (1998), and James et al. (1997) determine risk based on categorization of observed physical factors. The templates proposed by Keyserling et al. (1993), Keyserling (1986), Kilbom et al. (1986), and James et al. (1997) imply increased risk with increasing levels of the physical factors examined but do not supply a method of risk calculation. The "quick exposure check" described by Li and Buckle (1998) does describe a method of calculating risk based on the categorization of physical factors observed. The system described by Latko et al. (1997) is an observational scale, in which repetition or hand activity is characterized using a visual analog scale ranging form the lowest to the highest amount imaginable. No method of risk quantification is described by Latko et al. (1997). The presence of scales capable of characterizing force, posture, and mechanical stresses are reported in Latko et al. (1999). However, these scales have not been presented. Table 19.9 through Table 19.12 present the cut points used in determining risk by the upper extremity templates examined.

19.5 Summary and Conclusions

Further research is needed describing the interactive effects of the multiplier variables used in all templates proposed thus far. Epidemiologic studies examining the relative role of each risk factor category (e.g., force, repetition, posture, recovery) in the risk of musculoskeletal injury precipitation specific to each body region are also needed. Values used in the calculation of risk as multipliers or constants are commonly extrapolated from epidemiologic studies specific to worker population and body region and applied universally. This approach, while arguably necessary in facilitating proactive injury control and disability management efforts, is not valid. Studies examining the ability of templates to identify high-risk jobs based on previous claims experience are present only for those templates examining the low back. Comparison studies examining general and upper extremity templates are also needed.

Currently there is little consistency between either the cut points used or method of risk calculation in physical demands analysis templates described in scientific literature. Significant limitations exist in all templates described. These issues need to be conclusively resolved and validated. Determination of the most appropriate method for industrial application requires careful consideration of factors, including the industrial population the template has been developed for, body region(s) considered, and the mechanism(s) of injury accounted for.

TABLE 19.9

Posture Cut Points Used to Identify Risk of Musculoskeletal Injury by Template

Body Region	Colombini (1998); Occhipinti (1998); Grieco (1998)	Moore, and Garg (1995)	Keyserling (1986)	Kilbom et al. (1986)	James et al. (1997)	Li and Buckle (1998)	Genaidy et al. (1993)[a]
Hand/wrist	Extension: > 45° Flexion: > 45° Radial deviation: > 15° Ulnar deviation: > 20°	Extension: 0–10° 11–25° 26–40° 41–55° > 60° Flexion: 0–5° 6–15° 16–30° 31–50° > 50° Ulnar deviation: 0–10° 11–15° 16–20° 21–25° > 25°	Category not applicable	Category not applicable	Flexion/extension angles < 20° Flexion/extension angles > 20° Radial/ulnar deviation angles < 20° Radial/ulnar deviation angles > 20°	Almost a straight wrist Deviated or bent wrist position	0–5% 6–10% 11–20% 21–30% 31%+
Elbow	Supination: > 60° Pronation: > 60° Flexion/ extension range: > 60°	Category not applicable	Category not applicable	Category not applicable	Angle maintained between 60 and 90° of flexion Angle of flexion beyond ideal range	Category not applicable	0–5% 6–10% 11–20% 21–30% 31%+

Shoulder	Abduction: >45° Flexion: >80° Extension: >20°	Category not applicable	Standard shoulder postures: (flexion/abduction) Neutral (≤45°) Mild flexion/abduction(45 < to ≤90°) Severe Flexion/abduction (>90°)	Abduction 0–30° 30–60° 60–90° >90° Flexion: 0–30° 30–60° >60° Extension: >0°	Shoulder flexion <30° Shoulder flexion >30° Maintained shoulder flexion <45° Maintained shoulder flexion >45°	Shoulder/arm: At or below waist height? About chest height At or above shoulder height?	0–5% 6–10% 11–20% 21–30% 31%+
Neck	Category not applicable	Category not applicable	Category not applicable	Abduction 0–20° >20°	Category not applicable	Category not applicable	0–5% 6–10% 11–20% 21–30% 31%+
Trunk	Category not applicable	Category not applicable	Standard trunk postures: Stand extension (<20°) Stand neutral Stand–mild flexion (20 < to ≤45°) Stand severe flexion (>45°) Stand–twisted/bent (>20° in either direction) Lie on back or side Sit–neutral Sit–mild flexion Sit–twisted/bent	Category not applicable	Category not applicable	Almost neutral Moderately flexed, twisted or side bent Excessively twisted or side bent	Category not applicable

a Posture scores are given as a percentage of total range of motion. The described categories are consistent for motions across the back and shoulder as well as the hand and wrist.

TABLE 19.10

Force Cut Points Used to Identify Risk of Musculoskeletal Injury by Template

Author(s)	Cololmbini (1998); Occhipinti (1998); Grieco (1998)	Moore and Garg (1995)	Li and Buckle (1998)	Genaidy et al. (1993)
Description:	An upper extremity posture is considered static when it is held for more than 4 seconds Force factor: Mean force perceived/mean effort in percentage with respect to MVC $\geq 0.5/\geq 5$ 1/10 1.5/15 2/20 2.5/25 3/30 3.5/35 4/40 4.5/45 5/50	Rating criterion/% max. strength/perceived effort: Light/< 10%/Barely noticeable or relaxed effort Somewhat hard/10–29%/Noticeable or definite effort Hard/30–49%/Obvious effort; unchanged facial expression Very hard/50–79%/Substantial effort; changes facial expression Near Maximal/≥ 80%/Uses shoulder or trunk to generate force	Maximum weight handled: Light (5 kg or less) Moderate (6 to 10 kg) Heavy (11 to 20 kg) Very heavy (more then 20 kg) Maximum force exerted by one hand: Low (e.g., less than 1 kg) Medium (e.g., 1 to 4 kg) High (e.g., more than 4 kg)	% MVC static 0–1.6% 1.7–3.2% 3.3–6.4% 6.5–9.6% 9.7%+

TABLE 19.11

Repetition/Frequency Cut Points Used to Identify Risk of Musculoskeletal Injury by Template

Author(s)	Colombini (1998); Occhipinti (1998); Grieco (1998)	Moore and Garg (1995)	James et al. (1997)	Li and Buckle (1998)	Genaidy et al. (1993)	Latko et al. (1997)
Description	Calculation of the total number of recommended technical actions per shift is a product of the interaction of all variables considered	Efforts per minute: < 4 4–8 9–14 15–19 ≥ 20	Duration constituting static posture not specified	For manual materials-handling tasks only: is the movement of the back B1: In frequent (around 3 times per minute or less) B2: Frequent (around 8 times per minute) B3: Very frequent? (around 12 times per minute or more) Is the arm movement repeated? D1: Infrequently (some intermittent arm movement) D2: Frequently? (regular arm movement with some pauses) D3: Very frequently? (almost continuous arm movement) Is the task performed with similar repeated motion patterns? F1: 10 times per minute or less? F2: 11 to 20 times per minute? F3: More then 20 times per minute?	Repetitions per day: (0–0.5 Action limit) Fingers: (0–3656) Wrist: (0–1951) Elbow/Shoulder./Neck: (0–473) (0.6–1.0 Action limit) Fingers: (3657–7312) Wrist: (1952–3902) Elbow/Shoulder./Neck: (474–946) (1.1–2.0 Action limit) Fingers: (7,313–14,624) Wrist: (3903–7804) Elbow/Shoulder./Neck: (947–1893) (2.1–3.0 Action limit) Fingers: (14625–21936) Wrist: (7805-11706) Elbow/Shoulder./Neck: (1894–2838) (3.1 + Action limit) Fingers: (21937+) Wrist: (11707+) Elbow/Shoulder./Neck: (2839+)	Repetitions per cycle described in terms of duration and frequency of observed rest pauses and the speed of hand movements. Repetition or hand activity is characterized using a visual analog scale ranging form the lowest to the highest amount imaginable. The rating system consists of a 10 cm visual analog scale that ranges form 0 which corresponds to no hand activity to 10 the most possible hand activity. 0 — Hands idle most of the time; no regular exertions 2 — Consistent conspicuous, long pauses; or very slow motions 4 — Slow steady motion/exertion; frequent brief pauses 6 — Steady motion/exertion; infrequent pauses 8 — Rapid steady motion/exertion; no regular pauses 10 — Rapid steady motion/exertion; difficulty keeping up

TABLE 19.12

Additional Factors, Description and Cut Points Used to Identify Risk of Musculoskeletal Injury by Template

Author(s)	Cololmbini (1998); Occhipinti (1998); Grieco (1998)	Moore And Garg (1995)	Keyserling (1986)	Li And Buckle (1998)
Description	Grip scores: - Wide grip (4–5 cm) - Tight grip (1.5 cm) - Fine finger movements - Pinch - Palmer grip - Hook grip Risk due to additional factors (vibration, velocity and acceleration, precision, localized compression, exposure to cold, use of gloves, coupling, wrenching movements, return shock) quantified by dichotomous classification and percentage of cycle present (e.g., 1/3, 2/3, 3/3) Risk due to inadequate recovery calculated by applying the appropriate multiplier to the number of hours observed without adequate recovery	Speed of work: Rating criterion/MTM-1/ Percieved speed Very Slow/ ≤ 80%/Extremely relaxed pace Slow /81–90%/"taking one's own time" Fair/91–100%/"normal" speed of motion Fast/101–115%/Rushed, but able to keep up Very fast/> 115%/Rushed and barely able or unable to keep up Duration of exertion (percentage of cycle) < 10 10–29 30–49 50–79 ≥ 80 Duration per day ≤ 1 1–2 2–4 4–8 ≥ 8	Subjective discomfort: nonexistent or very slight slight moderate severe	Duration of time spent performing a task: Less than 2 hours 2 to 4 hours More than 4 hours Vibration exposure during work: Low (or no) Medium High Visual demand: Low (there is almost no need to view fine details) High (there is a need to view some fine details) Difficulty keeping up with this work? (work pace) Never Sometimes Often How stressful do you find this work? (work stress) Not at all Low Medium High

References

Asfour, S.S., Khalil, T.M., Genaidy, A.M., Akcin, M., Jomoah, I.M., Koshy, and J.G., Tritar, M. 1991. Ergonomics injury control in high frequency lifting tasks, Final Report, NIOSH Grant Nos. 5 R01 0H02591-01 and 5 R01 OH02591-02, Cincinnati, OH.

Borg, G. 1979. Psychophysical scaling with applications in physical work and the perception of exertion. *Scand. J. Work Environ. Health*, 16 (Suppl. 1), 55–58.

Borg, G. 1982. A category scale with ratio properties for intermodal and interindividual comparisons, in *Psychophysical Judgment and the Process of Perception*: *XXIInd International Congress of Psychology*, H. Geissler and P. Petzold, H.F.J.M. Buffart, and Y.M. Zabrodin, Eds. (New York: North-Holland Pub.), pp. 25–34.

Chen, J., Peacock, J.B., and Schlegel, R.E. 1989. An observational technique for physical work stress analysis, *Int. J. Ind. Ergon.*, 3, 167–176.

Colombini, D. 1998. An observational method for classifying exposure to repetitive movements of the upper limbs, *Ergonomics*, 41, 1261–1289.

Corlett, E.N., Madeley, S.J., and Manenica, I. 1979. Posture targeting: a technique for recording working postures, *Ergonomics*, 22, 357–366.

Cote-Gill, H.J. and Tunes, E. 1989. Posture recording: a model for sitting posture, *Ergonomics*, 20, 53–57.

Davis, P.R. 1999. The biological basis of physiological ergonomics requirements, *Int. J. Ind. Ergon.*, 23, 241–245.

Drury, C.G. 1987. A biomechanical evaluation of the repetitive motion injury potential of industrial jobs, *Semin. Occup. Med.*, 2, 41–49.

Drury, C.G. and Pfeil, R.E. 1975. A task based model of manual lifting performance, *Int. J. Prod. Res.*, 13, 137–148.

Foreman, T.K., Davies, J.C., and Troup, J.D.G. 1988. A posture and activity classification system using a micro-computer, *Int. J. Ind. Ergon.*, 2, 285–289.

Fransson-Hall, C., Gloria, R., Kilbom, A., Winkel, J., Karlqvist, L., and Wiktorin, C. 1995. A portable ergonomic observation method (PEO) for computerized on-line recording of postures and manual handling, *Appl. Ergon.*, 26, 93–110.

Garg, A. 1976. A Metabolic Tare Prediction Model for Manual Materials Handling Jobs, Doctoral dissertation, University of Michigan, Ann Arbor, MI.

Garg, A., Chaffin, D.B., and Herrin, G.D. 1978. Prediction of metabolic rates for manual materials handling jobs, *AIHA J.*, 39, 661–677.

Genaidy, A.M., Al-Shedi, A.A., and Shell, R.L. 1993a. Ergonomic risk assessment: Preliminary guidelines for analysis of repetition force and posture, *J. Hum. Ergol.*, 22, 45–55.

Genaidy, A.M., Waly, S.M., Khalil, T.M., and Hidalgo, J. 1993b. Spinal compression tolerance limits for the design of manual materials handling operations in the workplace, *Ergonomics*, 36, 415–434.

Grieco, A. 1998. Application of the concise exposure index (OCRA) to tasks involving repetitive movements of the upper limbs in a variety of manufacturing industries: preliminary validations, *Ergonomics*, 41, 1347–1356.

Grieco, A., Occhipinti, E., Colombini, D., and Molteni, G. 1997. Manual handling of loads: the point of view of experts involved in the application of EC directive 90/269, *Ergonomics*, 40, 1035–1056.

Hagberg, M. 1981. Electromyographic signs of shoulder muscular fatigue in two elevated arm positions, *Am. J. Phys. Med.*, 60, 111–121.

Havez, H.A. 1984. Manual Lifting under Hot Environmental Conditions, Ph.D. thesis, Texas Tech University, Lubbock, TX.

Hidalgo, J., Genaidy, A., Karwowski, W., Christensen, D., Huston, R., and Stambough, J. 1997. A comprehensive lifting model: beyond the NIOSH lifting equation, *Ergonomics*, 40, 916–927.

Hignett, S. and McAtamney, L. 2000. Rapid entire body assessment (REBA), *Appl. Ergon.*, 31, 201–205.

Holzmann, P. 1982. ARBAN: A new method for analysis of ergonomic effort. *Appl. Ergonomics*, 13, 82–86.

James, C.P.A., Haraburn, K.L., and Kramer, J.F. 1997. Cumulative trauma disorders in the upper extremities: reliability of the postural and repetitive risk-factors index, *Arch. Phys. Med. Rehabil.*, 78, 860–866.

Karhu, O., Kansi, P., and Kuorinka, I. 1977. Correcting working postures in industry: A practical method for analysis, *Appl. Ergon.*, 8, 199–201.

Karwowski, W. 1983. A pilot study of the interaction between physiological, biomechanical and psychophysical stresses involved in manual lifting tasks, in *Proceedings of the Ergonomics Society Conference*, (Cambridge: Taylor and Francis), pp. 95–100.

Karwowski, W. and Ayoub, M.M. 1984. Effect of frequency on the maximum acceptable weight of lift, in *Trends in Ergonomics/Human Factors 1*, Anil Mital, Ed. (New York: Elsevier Science), pp. 167–172.

Kemmlert, K. 1995. A method assigned for the identification of ergonomic hazards, *Appl. Ergon.*, 26, 199–211.

Keyserling, W.M. 1986. Postural analysis of the trunk and shoulders in simulated real-time, *Ergonomics*, 29, 569–583.

Keyserling, W.M., Stetson, D.S., Silverstein, B.A., and Brouer, M.L. 1993. A checklist for evaluating ergonomic risk factors associated with upper extremity cumulative trauma disorders, *Ergonomics*, 36, 807–831.

Kilbom, A., Perrson, J., and Jonsson, B.G. 1986. Disorders of the cervicobrachial region among female workers in the electronics industry, *Int. J. Ind. Ergon.*, 1, 37–47.

Kumar, S. 2001. Theories of musculoskeletal injury causation, *Ergonomics*, 44, 17–47.

Latko, W.A., Armstrong, T.J., Franzblau, A., Ulin, S.S., Werner, R.A., Albers, J.W. 1999. Cross-sectional study of the relationship between repetiive work and the prevalence of upper limb musculoskeletal disorders. *Am. J. Ind. Med.*, 36, 248–59.

Lavender, S.A., Oleske, D.M., Nicholson, L., Andersson, G.B.J., and Hahn, J. 1999. Comparison of five methods to determine low-back disorder risk in a manufacturing environment. *Spine*, 24, 1441–1448.

Leamon, T.B. 1994. Research to reality: a critical review of the validity of various criteria for the prevention of occupationally induced low back pain disability, *Ergonomics*, 37, 1959–1974.

Li, G. and Buckle, P. 1998. A practical method for the assessment of work-related musculoskeletal risks-quick exposure check, *Proceedings of the Human Factors and Ergonomics Society* 42nd Annual Meeting, October 5–9, Chicago, IL: Human Factors and Ergonomics Society, pp. 1351–1355.

Li, G. and Buckle, P. 1999. Current techniques for assessing physical exposure to work-related musculoskeletal risks, with emphasis on posture based methods, *Ergonomics*, 42, 674–695.

Marras, W.S., Allread, W.G., and Ried, R.G. 1999a. Occupational low back disorder risk assessment using the lumbar motion monitor, in *The Occupational Ergonomics Handbook*, W. Karwowski and W.S. Marras, Eds. (Boca Raton, FL: CRC Press), pp. 1075–1100.

Marras, W.S., Fine, L.J., Ferguson, S.A., and Waters, T.R. 1999b. The effectiveness of commonly used lifting assessment methods to identify industrial jobs associated with elevated risk of low-back disorders, *Ergonomics*, 42, 229–245.

Marras, W.S., Lavender, S.A., Leurgans, S.E., Rajulu, S.L., Allread, W.G., Fathallah, F.A., and Ferguson, S.A. 1993. The role of three dimensional motion in occupationally related low back disorders, *Spine*, 18, 617–628.

McAtamney, L. and Corlett, E.N. 1993. RULA: a survey method for investigation of work related upper limb disorders, *Appl. Ergon.*, 24, 91–99.

Mirka, G.A., Kelaher, D.P., Nay, T., and Lawrence, B.M. 2000. Continuous assessment of back stress (CABS): a new method to quantify low-back stress in jobs with variable biomechanical demands, *Hum. Factors*, 42, 209–225.

Moore, J.S. and Garg, A. 1994. Upper extremity disorders in a pork processing plant: relationships between job risk factors and morbidity. *AIHA J.*, 55, 703–715.

Moore, J.S. and Garg, A. 1995. The strain index: a proposed method to analyze jobs for risk of distal upper extremity disorders, *AIHA J.*, 56, 443–458.

National Research Council. 2001. *Musculoskeletal Disorders and the Workplace*, (Washington, DC: National Academy Press)

Occhipinti, E. 1998. OCRA: a concise index for the assessment of exposure to repetitive movements of the upper limbs, *Ergonomics*, 41, 1290–1311.

Priel, V.Z. 1974. A numerical definition of posture, *Hum. Factors*, 16, 576–584.

Ridd, J.E., Nicholson, A.S. and Motan, A.J. 1989. A portable microcomputer based system for on site activity and posture recording, in *Contemporary Ergonomics*, E.D. McGraw, Ed. (Taylor and Francis, London), pp. 366–371.

Shoaf, C., Genaidy, A., Karwowski, W., Waters, T., and Christensen, D. 1997. Comprehensive manual handling limits for lowering, pushing, pulling and carrying tasks, *Ergonomics*, 40, 1183–1200.

Snook, S.H. 1978. The design of manual materials handling tasks, *Ergonomics*, 21, 963–985.

Snook, S.H. and Ciriello, V.M. 1991. The design of manual materials handling tasks: revised tables of maximum acceptable weights and forces, *Ergonomics*, 34, 1197–1213.

Stetson, D.S., Keyserling, W.M., Silverstein, B.A., Leonard, J.A. 1991. Observational analysis of the hand and wrist: a pilot study. *Appl. Occup. Environ. Hyg.*, 6, 927–937.

Tichauer E.R. 1973. *The Industrial Environment: Its Evaluation and Control*, (Cincinnati, OH, DHHS, NIOSH).

Tichauer, E.R. 1978. *The Biomechanical Basis of Ergonomics* (New York: John Wiley & Sons).

U.S. Department of Health and Human Services, (NIOSH). 1997. Musculoskeletal Disorders and Work Place Factors, Pub. No. 97B141, U.S. Department of Health and Human Services, National Institute for Occupational Safety and Health, Cincinnati, OH.

Victorian Occupational HSC (Australia). 1988. Draft code of practice. Occupational overuse syndrome. Sydney.

Waters, T.R., Putz-Anderson, V., and Garg, A. 1994. Application Manual for the Revised NIOSH Lifting Equation, Pub. No. 94-110, U.S. Department of Health and Human Services, National Institute for Occupational Safety and Health, Cincinnati, OH.

Waters, T.R., Putz-Anderson, V., Garg, A., and Fine, L.J. 1993. Revised NIOSH equation for the evaluation and design of lifting tasks, *Ergonomics*, 36, 749–776.

Wells, R., Moore, A., Potvin, J., and Norman, R. 1994. Assessment of risk factors for development of work-related musculoskeletal disorders (RSI), *Ergonomics*, 25, 157–164.

Wiktorin, C., Mortimer, M., Ekenvall, L., Kilbom, A., and Hjelm, E.W. 1995. HARBO, a simple computer aided observation method for recording work postures, *Scand. J. Work Environ. Health*, 212, 440–449.

Appendix 1: Users Guide for the Rapid Entire Body Assessment (REBA)

Adapted from: Hignett, S. and McAtamney, L. 2000. Rapid entire body assessment (REBA). *Appl. Ergon.*, 31, 201–205.

Group A

Trunk

Movement	Score	Change Score
Upright	1	
0–20° flexion or extension	2	+1 if twisting or side flexed
20–60° flexion or > 20° extension	3	
> 60° flexion	4	

Neck

Movement	Score	Change Score
0–20° flexion	1	+1 if twisting or side flexed
> 20° flexion or in extension	2	

Legs

Position	Score	Change Score
Bilateral weight bearing walking or sitting	1	+1 if knee(s) are between 30–60° of flexion
Unilateral weight bearing, feather weight bearing or an unstable posture	2	+2 if knee(s) are > 60° flexion (not for sitting)

Group B

Upper Arms

Position	Score	Change Score
20 degrees extension to 20° flexion	1	+1 if arm is: abducted or rotated
> 20° extension 25–45° of flexion	2	+1 if shoulder is raised
45–90° of flexion	3	−1 if leaning supporting weight of arm or if posture is gravity assisted
> 90° flexion	4	

Lower Arms

Movement	Score
60–100° of flexion	1
< 60° flexion or > 100° flexion	2

Wrists

Movement	Score	Change Score
0–15° of flexion or extension	1	+1 if wrists are deviated or twisted
> 15° of flexion or extension	2	

Load/Force

0	1	2	+1
< 5 kg	5–10 kg	> 10 kg	Shock or rapid buildup of force

Coupling Score

Good (0)	Wel-fitted handle and mid-range power grip
Fair (1)	Hand hold acceptable but not ideal or coupling is acceptable via another part of the body
Poor (2)	Hand hold not acceptable although possible
Unacceptable (3)	Awkward, unsafe grip; no handles Coupling is unacceptable using other parts of the body

Activity Score

+1 1 or more body parts are static (held for longer than 1 minute)
+1 Repeated small range actions, e.g., repeated more than 4 times per minute (not including walking)
+1 Action causes rapid large range changes in posture or an unstable base

Table A

Trunk		Neck											
		1				2				3			
	Legs	1	2	3	4	1	2	3	4	1	2	3	4
	1	1	2	3	4	1	2	3	4	3	3	5	6
	2	2	3	4	5	3	4	5	6	4	5	6	7
	3	2	4	5	6	4	5	6	7	5	6	7	8
	4	3	5	6	7	5	6	7	8	6	7	8	9
	5	4	6	7	8	6	7	8	9	7	8	9	9

Table B

Upper Arm		Lower Arm					
		1			2		
	Wrist	1	2	3	1	2	3
1		1	2	2	1	2	3
2		1	2	3	2	3	4
3		3	4	5	4	5	5
4		4	5	5	5	6	7
5		6	7	8	7	8	8
6		7	8	8	8	9	9

Table C

Score A		Score B											
		1	2	3	4	5	6	7	8	9	10	11	12
	1	1	1	1	2	3	3	4	5	6	7	7	7
	2	1	2	2	3	4	4	5	6	6	7	7	8
	3	2	3	3	3	4	5	6	7	7	8	8	8
	4	3	4	4	4	5	6	7	8	8	9	9	9
	5	4	4	4	5	6	7	8	8	9	9	9	9
	6	6	6	6	7	8	8	9	9	10	10	10	10
	7	7	7	7	8	9	9	9	10	10	11	11	11
	8	8	8	8	9	10	10	10	10	10	11	11	11
	9	9	9	9	10	10	10	11	11	11	12	12	12
	10	10	10	10	11	11	11	11	12	12	12	12	12
	11	11	11	11	11	12	12	12	12	12	12	12	12
	12	12	12	12	12	12	12	12	12	12	12	12	12

REBA Action Levels

Action Level	REBA Score	Risk Level	Action (Including Further Assessment)
0	1	Negligible	None necessary
1	2–3	Low	May be necessary
2	4–7	Medium	Necessary
3	8–10	High	Necessary soon
4	11–15	Very High	Necessary now

Appendix 2: Users Guide for the Revised NIOSH Equation

Adapted from: Waters, T.R., Putz-Anderson, V., Garg, A., Fine, L.J. 1993. Revised NIOSH equation for the evaluation and design of lifting tasks, *Ergonomics*, 36, 749–776. Waters,

REBA SCORING METHOD

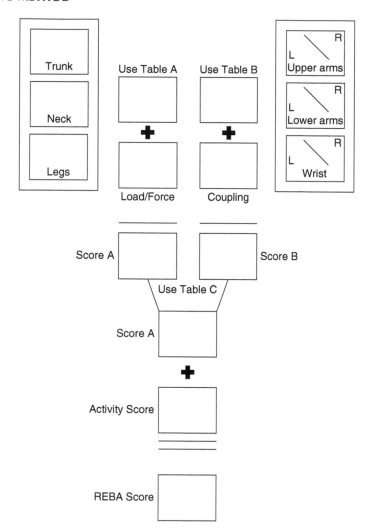

General limitations:

1. References not detailed for cut points used.

2. Expert opinion used to develop multiplier values. Two workshops involving 14 professional (PT, OT, nurses, ergonomists) to arrive at codes considering the revised NIOSH equation (Waters et al., 1993), Rated perceived exertion (Borg, 1985), OWAS (Karhu et al., 1977), body part discomfort survey (Corlett and Bishop, 1976), and RULA (McAtamney and Corlett, 1993).

T.R., Putz-Anderson, V., Garg, A. 1994. *Application manual for the revised NIOSH lifting equation,* Pub. No. 94-110, U.S. Department of Health and Human Services, National Institute for Occupational Safety and Health, Cincinnati, OH.

The revised NIOSH equation (equation and tables adapted from Waters et al., 1994).

Definition and Explanation of Terms

Recommended weight limit (RWL)	The RWL is defined for a specific set of task conditions as the weight of the load that nearly all healthy workers could perform over a substantial period of time (e.g., up to 8 hours) without an increased risk of developing lifting related LBP, where "healthy workers," is defined as workers who are free of adverse health conditions that would increase their risk of musculoskeletal injury (Waters et al., 1994).
Lifting index (LI)	The lifting index is a term that provides a relative estimate of the level of physical stress associated with a particular manual lifting task. The estimate of the level of physical stress is defined by the relationship of the weight of the load lifted and the recommended weight limit. The LI is defined by the following equation. $$\text{LI} = \text{Load weight/recommended weight limit (L/RWL)}$$ (Waters et al., 1994) From the NIOSH perspective, it is likely that lifting tasks with a LI > 1.0 pose an increased risk for lifting related low back pain for some fraction of the workforce (Waters et al., 1993). Hence the goal should be to design all lifting jobs to achieve a LI of 1.0 or less (Waters et al., 1994). Nearly all workers will be at an increased risk of work-related injury while performing highly stressful lifting tasks (i.e., lifting tasks what would exceed a LI of 3.0) (Waters et al., 1994).
Load weight (L)	Weight of the object to be lifted, in pounds or kilograms, including the container (Waters et al., 1994).
Horizontal location (H)	Distance of the hands away from the midpoint between the ankles, in inches or centimeters (measure at origin and destination of lift) (Waters et al., 1994). Horizontal multiplier values are described in the table provided. Should measurement of the H value not be possible, horizontal distance approximation equations are provided.
Vertical location (V)	Distance of the hands above the floor, in inches or centimeters (measure at origin and destination of lift). V is measured vertically from the floor to the midpoint between the hand grasps, as defined by the large middle knuckle (Waters et al., 1994). Vertical location multiplier values are described in the table provided.
Vertical travel distance (D)	Absolute value of the difference between the vertical heights at the destination and the origin of the lift, in inches or centimeters (Waters et al., 1994). Vertical travel distance multiplier values are described in the table provided.
Angle of asymmetry (A)	The angular measure of how far the object is displaced from the front (midsagittal plane) of the worker's body at the beginning or ending of the lift, in degrees (measure at the origin and destination of the lift). The asymmetry angle is defined by the location of the load relative to the workers midsagittal plane, as defined by the neutral body posture, rather than the position of the feet or the extent of body twist (Waters et al., 1994). Asymmetric angle multiplier values are described in the table provided.
Neutral body position	Describes the position of the body when the hands are directly in front of the body and there is minimal twisting at the legs, torso, or shoulders (Waters et al., 1994).
Frequency of lifting (F)	Average number of lifts per minute over a 15-min period. The frequency multiplier is defined by the number of lifts per minute (frequency) the amount of time engaged in the lifting activity (duration), and the vertical height of the lift from the floor. Lifting frequency for repetitive lifting may range from 0.2 lifts/min to a maximum frequency that is dependent on the vertical location of the object (V) and the duration of lifting (Waters et al., 1994). Frequency multiplier values are described in the tables provided.
Duration of lifting	Three-tiered classification of lifting duration specified by the distribution of work-time and recovery-time (work pattern). Duration is classified as either short (1 hour), moderate (1–2 hours), or long (2–8 hours), depending on work pattern (Waters et al., 1994). Selection of the correct duration (D) is assisted by the further explanation provided below and the table provided.
Coupling classification	Classification of the quality of the hand-to-object coupling (e.g., handles, cut-out, or grip). Coupling is classified as good, fair, or poor. Based on the coupling classification and vertical location of the lift (Waters et al., 1994). Selection of the Coupling Multiplier (CM) is assisted by the further explanation provided below and the table provided.
Significant control	Significant control is defined as a condition requiring "precision placement" of the load at the destination of the lift. This is usually the case when (1) the worker has to regrasp the load near the destination of the lift, or (2) the worker has to momentarily hold up the object at the destination, or (3) the worker has to carefully position or guide the load at the destination (Waters et al., 1994).

$$RWL = LC \times HM \times VM \times DM \times AM \times FM \times CM$$

where

Abbreviation	Definition	Metric	U.S. Customary
LC	Load constant	23 kg	51 lb.
HM	Horizontal multiplier	(25/H)	(10/H)
VM	Vertical multiplier	$1 - (.003\|V - 75\|)$	$1 - (.0075\|V - 30\|)$
DM	Distance multiplier	.82 + (4.5/D)	.82 + (1.8/D)
AM	Asymmetric multiplier	1 - (.0032A)	1 - (.0032A)
FM	Frequency multiplier	From Table	From Table
CM	Coupling multiplier	From Table	From Table

Horizontal distance approximation equations:

Metric (cm)	U.S. customary (inches)
H = 20 + W/2 for V ≥ 10 inches	H = 8 + W/2 for V ≥ 25 cmH
H = 25 + W/2 for V < 25 cm	H = 8 + W/2 for V < 10 inches

Where: W is the width of the container in the sagittal plane and V is the vertical location of the hands form the floor.

Horizontal multiplier:

H (in.)	HM	H (cm)	HM
≤ 10	1.0	≤ 25	1.0
11	.91	28	.89
12	.83	30	.83
13	.77	32	.78
14	.71	34	.74
15	.67	36	.69
16	.63	38	.66
17	.59	40	.63
18	.56	42	.60
19	.53	44	.57
20	.50	46	.54
21	.48	48	.52
22	.46	50	.50
23	.44	52	.48
24	.42	54	.46
25	.40	56	.45
> 25	.00	58	.43
		60	.42
		63	.40
		> 63	.00

Vertical multiplier:

V (in.)	VM	V (cm)	VM
0	.78	0	.78
5	.81	10	.81
10	.85	20	.84
15	.89	30	.87
20	.93	40	.90
25	.96	50	.93
30	1.0	60	.96
35	.96	70	.99
40	.93	80	.99
45	.89	90	.96
50	.85	100	.93
55	.81	110	.90
60	.78	120	.87
65	.74	130	.84
70	.70	140	.81
> 70	.00	150	.78
		160	.75
		170	.72
		175	.70
		> 175	.00

Distance multiplier

D (in.)	DM	D(cm.)	DM
≤ 10	1.00	≤ 25	1.00
15	.94	40	.93
20	.91	55	.90
25	.89	70	.88
30	.88	85	.87
35	.87	100	.87
40	.87	115	.86
45	.86	130	.86
50	.86	145	.85
55	.85	160	.85
60	.85	175	.85
70	.85	> 175	.00
> 70	.00		

Asymmetric multiplier

A (degree)	AM
0	1.0
15	9.5
30	.90
45	.86
60	.81
75	.76
90	.71
105	.66
120	.62
135	.57
> 135	.00

Frequency multiplier

Frequency: lifts/min	≤1 hr. V <75	≤1 hr. V ≥ 75	≤2 hr. V < 75	≤2 hr. V ≥ 5	≤8 hr. V < 75	≤8 hr. V ≥ 75
≤ 0.2	1.0	1.0	0.95	0.95	0.85	0.85
0.5	0.97	0.97	0.92	0.92	0.81	0.81
1	0.94	0.94	0.88	0.88	0.75	0.75
2	0.91	0.91	0.84	0.84	0.65	0.65
3	0.88	0.88	0.79	0.79	0.55	0.55
4	0.84	0.84	0.72	0.72	0.45	0.45
5	0.80	0.80	0.60	0.60	0.35	0.35
6	0.75	0.75	0.50	0.50	0.27	0.27
7	0.70	0.70	0.42	0.42	0.22	0.22
8	0.60	0.60	0.35	0.35	0.18	0.18
9	0.52	0.52	0.30	0.30	0.00	0.15
10	0.45	0.45	0.26	0.26	0.00	0.13
11	0.41	0.41	0.00	0.23	0.00	0.00
12	0.37	0.37	0.00	0.21	0.00	0.00
13	0.00	0.34	0.00	0.00	0.00	0.00
14	0.00	0.31	0.00	0.00	0.00	0.00
15	0.00	0.28	0.00	0.00	0.00	0.00
> 15	0.00	0.00	0.00	0.00	0.00	0.00

Determination of the Frequency Component (Waters et al., 1994)

Lifting frequency refers to the average number of lifts made per minute, as measured over a 15-minute period. If significant variation exists in the frequency of lifting over the course of the day, analysts should employ standard work sampling techniques to obtain a representative work sample for determining the number of lifts per minute. Selection of the appropriate frequency multiplier is based on determination of the lifting duration.

Determination of the Lifting Duration Component (Waters et al., 1994)

Lifting duration is classified into three categories: short duration, moderate duration, and long duration. These categories are based on the pattern of continuous work time and recovery time (i.e., light work) periods. A continuous work period is defined as a period of uninterrupted work. Recovery time is defined as the duration of light work activity following a period of continuous lifting. Examples of light work include activities such as sitting at a desk or table, monitoring operations, light assembly work, etc.

1. Short-duration lifting defines lifting tasks that have a work duration of 1 hour or less, followed by a recovery time equal to 1.2 times the work time [i.e., at least a 1.2 recovery-time to work-time ratio (RT/WT)]. If the required recovery time is not met for a job of one hour or less, and a subsequent lifting session is required, then the total lifting time must be combined to correctly determine the duration category and thus the frequency multiplier. If the recovery time was insufficient to meet the requirement, it is disregarded for the purposes of determining the appropriate category.

2. Moderate duration defines lifting tasks that have a duration of more than 1 hour, but not more than 2 hours, followed by a recovery period of at least 0.3 times the work time [i.e., at a least 0.3 recovery-time to work-time ratio (RT/WT)].

Muscle Strength

3. Long duration defines lifting tasks that have a duration of between 2 and 8 hours, with standard industrial rest allowances (e.g., morning, lunch, and afternoon rest breaks).

If the workers do not lift continuously during the 15-minute sampling period,

1. Compute the total number of lifts performed for the 15-minute period (i.e., lift rate times work time).
2. Divide the total number by 15.
3. Use the resulting value as the frequency (F) to determine the frequency multiplier form the table provided.

Using this procedure, rest periods that occur during the 15-minute sampling period are not considered as recovery periods for the purpose of determining the duration category.

If significant variation exists in the frequency of lifting over the course of the day, analysts should employ standard work sampling techniques to obtain a representative working sample for determining the number of lifts per day.

Hand-to-Container Coupling Classification

Good	Fair	Poor
For containers of optimal design, such as some boxes, crates, etc., a "good" hand-to-object coupling would be defined as handles or hand-hold cutouts of optimal design (see notes 1–3 below)	For containers of optimal design a "fair" hand-to-object coupling would be defined as handles or cutouts of less than optimal design (see notes 1–4 below)	Containers of less then optimal design or loose parts or irregular objects that are bulky, hard to handle, or have sharp edges (see note 5 below)
For loose parts or irregular objects which are not usually containerized such as castings, stock, and supply materials, a "good" hand hold would be defined as a comfortable grip in which the hand can be easily wrapped around the object (see note 6 below)	For containers of optimal design with no handles or hand-hold cutouts for loose parts or irregular objects, a "fair" coupling is defined as a grip in which the hand can be flexed about 90° (see note 4 below)	Lifting nonrigid bags (i.e., bags that sag in the middle)

Adapted from Waters et al., 1994.

1. An optimal handle design has 0.75–1.5 inches (1.9–3.8 cm) diameter, ≥ 4.5 inches (11.5 cm) length, 2 inches (5 cm) clearance, cylindrical shape, and a smooth nonslip surface.
2. An optimal hand-hold cutout has the following approximate characteristics: ≥1.5 inch (3.8 cm) height, 4.5 inch (11.5 cm) length, semi-oval shape, ≥2 inch (5 cm) clearance, smooth nonslip surface, and ≥.25 (.60 cm) container thickness (e.g., double thickness cardboard).
3. An optimal container design has ≤ 16 inches (40cm) frontal length, ≤ 12 inches (30cm) height, and a smooth, nonslip surface.
4. A worker should be capable of clamping the fingers at nearly 90° under the container, such as required when lifting a cardboard box from the floor.
5. A container is considered less then optimal if it has a frontal length > 16 inches (40 cm), height > 12 inches (30 cm), rough or slippery surfaces, sharp edges, asymmetric center of mass, unstable contents, or requires the use of gloves.

6. A worker should be able to comfortably wrap the hand around the object without causing excessive wrist deviation or awkward postures; the grip should not require excessive force.

COUPLING MODIFIER SELECTION (Adapted from Waters et al., 1994)

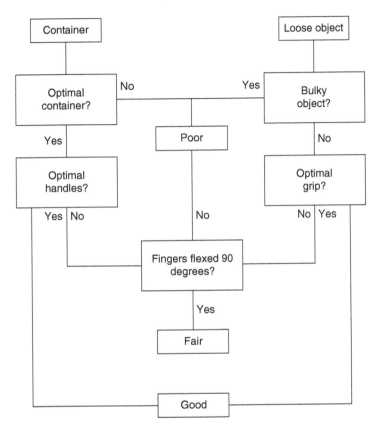

Selection of the Relevant Analysis Procedure (Waters et al., 1994)

Single-task manual lifting job:

- Defined as a lifting job in which the task variables do not significantly vary from task to task, or only one task is of interest (e.g., worst case task) because the other tasks do not have a significant effect on strength, localized muscle fatigue, or whole body fatigue. Analysis of single-task manual lifting jobs do not require any modification to the revised NIOSH equation described above.

Multitask manual lifting jobs:

- Defined as jobs in which there are significant differences in task variables between tasks. Analysis of multitask manual lifting jobs is based on the formation of a composite lifting index, which represents the collective demands of the job. The method for determining the combined demand is based on three assumptions:

1. Performing an additional task or multiple lifting tasks would result in increased physical/metabolic load, and this load should be reflected in a reduced recommended weight limit or a higher value for the lifting index

2. That an increase in the lifting index depends upon the characteristics of the additional lifting task

3. That an increase in the lifting index due to the addition of one or more tasks is independent of the lifting index of any of the preceding tasks (i.e., lifting indices from tasks already performed

The composite lifting index (CLI) is equal to the sum of the largest single task lifting index and the incremental increases as each subsequent task is added. The incremental increase in the CLI for a specific task is defined as the difference between the lifting index for that task at the cumulative frequency and the lifting index for that task at its actual frequency.

Example:

Using the CLI concept in a case where the two tasks are identical:

$$CLI = LI_{A,1} + (LI_{B,2} - LI_{B,1})$$

Where the numeric part of the subscript represents the frequency, such that $LI_{B,2}$ indicates the LI value for Task B at a frequency of 2 lifts/minute and $LI_{B,1}$ indicates the LI value for Task B at a frequency of 1 lift/minute.

Since Task A and B are identical, $LI_{A,1}$ and $LI_{B,1}$ cancel out and $CLI = LI_{B,2}$. As expected, the CLI for the job is equivalent to the LI value for the simple task being performed at a rate of 2 times/minute.

Using the CLI concept in a case where the two tasks are different:

$$CLI = LI_{A,1} + (LI_{B,2} - LI_{B,1})$$

In this case, $LI_{A,1}$ and $LI_{B,2}$ do not cancel each other out. The CLI is equal to the sum of $LI_{A,1}$, which refers to the demand of Task A, and the increment of demand for Task B, with the increment being equal to the increase in demand when the frequency for Task B is increased form 1 lift/minute (corresponding to the frequency of Task A) to a rate of 2 lifts/minute (corresponding to the sum of the frequencies if Task A and B). Thus, as each additional task is added, the CLI is increased appropriately.

Single- and Multitask Procedures (Waters et al., 1994)

A. Single-Task Procedure (Refer to Single-Task Worksheet)

Compute the Recommended Weight Limit (RWL) and Lifting Index (LI).

- Calculate the RWL at the origin for each lift. For lifting tasks that require significant control at the destination, calculate the RWL at both the origin and the destination of the lift. The latter procedure is required if (1) the worker has to regrasp the load near the destination of the lift, (2) the worker has to momentarily hold the object at the destination, or (3) the worker has to position or guide the load at the destination. The purpose of calculating the RWL at both the origin and destination of the lift is to identify the most stressful location of the lift. Therefore, the lower of the RWL values at the origin or destination should be used to compute the Lifting Index for the task, since this value would represent the limiting set of conditions.

- The assessment is completed on the single-task worksheet by determining the lifting index (LI) for the task of interest. This is accomplished by comparing the actual weight of the load (L) lifted with the RWL value obtained from the lifting equation.

B. Multitask Procedure (Refer to Multitask Worksheet)

1. Compute the frequency-independent recommended weight limit (FIRWL) and single-task recommended weight limit (STRWL) for each task.
2. Compute the frequency-independent lifting index (FILI) and single-task lifting index (STLI) for each task.
3. Compute the composite lifting index (CLI) for the overall job.

Compute the frequency-independent recommended weight limits (FIRWLs).

- Compute the FIRWL value for each task by using the respective task variables and setting the Frequency Multiplier to a value of 1.0. The FIRWL for each task reflects the compressive force and muscle strength demands for a single repetition of that task. If significant control is required at the destination for any individual task, the FIRWL must be computed at both the origin and the destination of the lift, as described above for a single-task analysis.

Compute the single-task recommended weight limit (STRWL).

- Compute the STRWL for each task by multiplying its FIRWL by its appropriate Frequency Multiplier (FM). The STRWL for a task reflects the overall demands of that task, assuming it was the only task being performed. Note, this value does not reflect the overall demands of the task when the other tasks are considered. Nevertheless, this value is helpful in determining the extent of excessive physical stress for an individual task.

Compute the frequency-independent lifting index (FILI).

- The FILI is computed for each task by dividing the maximum load weight (L) for that task by the respective FIRWL. The maximum weight is used to compute the FILI because the maximum weight determines the maximum biomechanical loads to which the body will be exposed, regardless of the frequency of occurrence. Thus, the FILI can identify individual tasks with potential strength problems for infrequent lifts. If any of the FILI values exceed a value of 1.0, then ergonomic changes may be needed to decrease the strength demands.

Compute the single-task lifting index (STLI).

- The STLI is computed for each task by dividing the average load weight (L) for that task by the respective STRWL. The average weight is used to compute the STLI, because the average weight provides a better representation of the metabolic demands, which are distributed across the tasks, rather than dependent on individual tasks. The STLI can be used to identify individual tasks with excessive physical demands (i.e., tasks that would result in fatigue). The STLI values do not indicate the relative stress of the individual tasks in the context of the whole job, but the STLI value can be used to prioritize the individual tasks according to the magnitude of their physical stress. Thus, if any of the STLI values exceed a value of 1.0, then ergonomic changes may be needed to decrease the overall physical demands of the task. Note, it may be possible to have a job in which all of the individual tasks have a STU less than 1.0 and still be physically demanding due to the combined demands of the tasks. In cases where the FILI exceeds the STLI for any task, the maximum weights may represent a significant problem and careful evaluation is necessary.

Compute the composite lifting index (CLI).

- The assessment is completed on the multi-task worksheet by determining the Composite Lifting Index (CLI) for the overall job. The CLI is computed as follows:

    a. The tasks are renumbered in order of decreasing physical stress, beginning with the task with the greatest STLI down to the task with the smallest STLI. The tasks are renumbered in this way so that the more difficult tasks are considered first.

    b. The CLI for the job is then computed according to the following formula:

$$CLI = STLI_1 + \cdot \ \Sigma\Delta \ LI$$

Where:

$$\Sigma\Delta LI = [FILI_2 \times (1/FM_{1,2} - 1/FM_1)]$$

$$+ [FILI_3 \times (1/FM_{1,2,3} - 1/FM_{1,2})]$$

$$+ [FILI_4 \times (1/FM_{1,2,3,4} - 1/FM_{1,2,3})]$$

$$\dots\dots$$

$$\dots\dots$$

$$+ [FILI_n \times (1/FM_{1,2,3,4,\dots n} - 1/FM_{1,2,3\dots(n-1)})]$$

Note, that (1) the numbers in the subscripts refer to the new task numbers; and (2) the FM values are determined from the frequency multiplier table, based on the sum of the frequencies for the tasks listed in the subscripts.

Example: The following example is provided to demonstrate the multitask procedure. Assume that an analysis of a typical three-task job provided the following results:

Task #	Load Weight (L)	Task Frequency (F)	FIRWL	FM	STRWL	FILI	STLI	New Task #
1	30	1	20	.94	18.8	1.5	1.6	1
2	20	2	20	.91	18.2	1.0	1.1	2
3	10	4	15	.84	12.6	.67	.8	3

To compute the composite lifting index (CLI) for this job, the tasks are renumbered in order of decreasing physical stress, beginning with the task with the greatest STLI down to the task with the smallest STLI. In this case, the task numbers do not change. Next, the CLI is computed according to the formula shown above. The task with the greatest CLI is Task 1 (STLI = 1.6). The sum of the frequencies for Tasks 1 and 2 is 1+2 or 3, and the sum of the frequencies for Tasks 1, 2, and 3 is 1 + 2 + 4 or 7. Then, from the table above, FM_1 is .94, $FM_{1,2}$ is .88, and $FM_{1,2,3}$ is .70. Finally, the CLI = 1.6 + 1.0 (1/.88–1/.94) + .67

$(1/.70–1/.88) = 1.6 + .07 + .20 = 1.9$. Note, that the FM values were based on the sum of the frequencies for the subscripts, the vertical height, and the duration of lifting.

General Limitations

The revised lifting equation does not apply if (Waters et al., 1994):

- Lifting/lowering with one hand
- Lifting/lowering for over 8 hours
- Lifting/lowering while seated or kneeling
- Lifting/lowering in a restricted work space
- Lifting/lowering unstable objects
- Lifting/lowering while carrying, pushing or pulling
- Lifting/lowering with wheel barrows or shovels
- Lifting/lowering with "high speed" motion (faster than about 30 inches/second)
- Lifting/lowering with unreasonable foot/floor coupling (< 0.4 coefficient of friction between the sole and the floor)
- Lifting lowering in an unfavorable environment (temperature significantly outside 66–79°F (19–26°C) range; relative humidity outside 35–50% range)
- Manual materials-handling activities other than lifting (pushing, pulling, carrying, holding, walking, climbing) account for more than about 10% of the total worker activity

Single-Task Manual Lifting Job Analysis Worksheet

Step 1. Measure and Record Task Variables

Object Weight (lbs)		Hand Location (In)				Vertical Distance (In)	Asymmetric Angle (Degrees)		Frequency Rate	Duration (Hrs)	Object Coupling
		Origin		Dest			Origin	Dest			
L(avg.)	L(max.)	H	V	H	V	D	A	A	F		C

Step 2. Determine the Multipliers and Compute the RWLs

| | RWL | = | L | × | H | × | V | × | D | × | A | × | F | × | C | | |
| | | | C | | M | | M | | M | | M | | M | | M | | |

Origin RWL = [51] × [ ] × [ ] × [ ] × [ ] × [ ] × [ ] = ____ lbs

Destination RWL = [51] × [ ] × [ ] × [ ] × [ ] × [ ] × [ ] = ____ lbs

Step 3. Compute the Lifting Index (LI)

Origin   LI   =   Object weight (L) ÷ RWL   =   _____ ÷ _____   =   [   ]

Destination   LI   =   Object weight (L) ÷ RWL   =   _____ ÷ _____   =   [   ]

Adapted from Waters et al. (1994).

Multitask Manual Lifting Job Analysis

Step 1. Measure and Record Task-Variable Data

	Object Weight (Lbs)		Hand Location (In)				Vertical Distance (In)	Asymmetric Angle (Degrees)		Frequency Rate	Duration (Hrs)	Object Coupling
			Origin		Dest			Origin	Dest			
Task#	L (avg.)	L (max.)	H	V	H	V	D	A	A	F		C
___												
___												
___												
___												
___												

Adapted from Waters et al. (1994).

Step 2. Compute Multipliers and FIRWL, STRWL, FILI, and STLI for Each Task

Task#	LC	×HM	×VM	×DM	×AM	×CM	FIRWL × FM	STRWL	FILI = L/FIRWL	STLI = L/STRWL	New task#	F
___	51											
___	51											
___	51											
___	51											
___	51											

Step 3. Compute the Composite Lifting Index for the Job (After Renumbering Tasks)

CLI=	STLI$_1$	+	ΔFILI$_2$	+	ΔFILI$_3$	+	ΔFILI$_4$	+	ΔFILI$_5$
		+	FILI$_2$(1/FM$_{1,2}$ $-$ 1/FM$_1$)	+	FILI$_3$(1/FM$_{1,2,3}$ $-$ 1/FM$_{1,2}$)	+	FILI$_4$(1/FM$_{1,2,3,4}$ $-$ 1/FM$_{1,2,3}$)	+	FILI$_5$(1/FM$_{1,2,3,4,5}$ $-$ 1/FM$_{1,2,3,4}$)
CLI=		+		+		+		+	

Adapted from Waters et al. (1994).

Appendix 3: Users Guide for the Strain Index

Adapted from Moore, S.J. and Garg, A. 1995. The strain index: a proposed method to analyze jobs for risk of distal upper extremity disorders, *AIHA J.*, 56, 443–458.

Step 1: Data Collection

1. Intensity of exertion is an estimate of the strength required to perform the task one time rated by the observer/rater. Guidelines for assigning a rating criterion are presented in the table below.

Force Rating Criterion

Rating Criterion	%Ms[a]	Borg Scale[b]	Perceived Effort
Light	< 10%	≤ 2	Barely noticeable or relaxed effort
Somewhat hard	10–29%	3	Noticeable or definite effort
Hard	30–49%	4–5	Obvious effort; unchanged facial expression
Very hard	50–79%	6–7	Substantial effort; changes facial expression
Near maximal	≥ 80%	> 7	Uses shoulder or trunk to generate force

[a] Percentage of maximal strength.
[b] Compared to the Borg CR-10 scale.

2. Duration of exertion is calculated by measuring the duration of all exertions during an observation period, then dividing the measured duration of exertion by the total observation time and multiplying by 100.

$$\% \ Duration \ of \ exertion = 100 \ \times \frac{duration \ of \ all \ exertions \ (sec)}{total \ observation \ time \ (sec)} = 100 \times \underline{\quad} = \underline{\quad}$$

3. Efforts per minute are measured by counting the number of exertions that occur during an observation period, then dividing the number of exertions by the duration of the observation period, measured in minutes.

$$Efforts \ per \ minute = \frac{number \ of \ exertions}{total \ observation \ time \ (min)} = \underline{\quad} = \underline{\quad}$$

4. Hand/wrist posture is an estimation of the position of the hand or wrist relative to neutral position. Guidelines for assessing a rating criterion are presented in the table below.

5. Speed of work is an estimate of how fast the worker is working. Guidelines for assigning a rating criterion are presented in the table below.

Step 2: Assign Ratings Values

Using the table provided below, find the rating values for each task variable. Select the appropriate entry for each variable, then find the corresponding rating value on the same row at the far left.

Posture Rating Criterion

Rating Criterion	Wrist Extension[a]	Wrist Flexion[a]	Ulnar Deviation[a]	Perceived Posture
Very good	0–10	0–5	0–10	Perfectly neutral
Good	11–25	6–15	11–15	Near neutral
Fair	26–40	16–30	16–20	Non neutral
Bad	41–55	31–50	21–25	Marked deviation
Very bad	> 60	> 50	> 25	Near extreme

[a] Derived from data presented in Stetson et al. 1991. Observational analysis of the hand and wrist: a pilot study. *Appl. Occ. Environ. Hyg.* 6(11), 927–937.

Speed of Work Criterion

Rating Criterion	Compared To MTM-1[a]	Perceived Speed
Very slow	≤ 80%	Extremely relaxed pace
Slow	81–90%	"Taking one's own time"
Fair	91–100%	"Normal" speed of motion
Fast	101–115%	Rushed, but able to keep up
Very fast	> 115%	Rushed and barely or unable to keep up

[a] The observed pace is divided by MTM-1's predicted pace and expressed as a percentage of predicted [see Barnes, R.M. 1980. *Motion and Time Study: Design and Measurement of Work* (New York: John Wiley & Sons)].

Table 1: Rating Criteria

Rating	Intensity of Exertion	Duration of Exertion (% of Cycle)	Efforts per Minute	Hand/Wrist Posture	Speed of Work	Duration per Day (h)
1	Light	< 10	< 4	Very good	Very slow	≤ 1
2	Somewhat hard	10–29	4–8	Good	Slow	1–2
3	Hard	30–49	9–14	Fair	Fair	2–4
4	Very hard	50–79	15–19	Bad	Fast	4–8
5	Near maximal	≥ 80	≥ 20	Very bad	Very fast	≥ 8

Step 3. Determine the Multipliers

Table 2: Multiplier Table

Rating	Intensity Of Exertion	Duration Of Exertion (% Of Cycle)	Efforts Per Minute	Hand/Wrist Posture	Speed of Work	Duration per Day (h)
1	1	0.5	0.5	1.0	1.0	0.25
2	3	1.0	1.0	1.0	1.0	0.50
3	6	1.5	1.5	1.5	1.0	0.75
4	9	2.0	2.0	2.0	1.5	1.0
5	13	3.0[a]	3.0	3.0	2.0	1.50

Note: If duration of exertion is 100%, then efforts per minute multiplier should be set to 3.0.

Data Entry Table

	Intensity of Exertion	Duration of Exertion	Efforts /Minute	Hand/Wrist Posture	Speed of Work	Duration per Day
Step 1: Rating criterion						
Step 2: Rating value						
Step 3: Multiplier						

Step 4: Calculate the SI Score

Insert the multiplier values for each of the six task variables into the spaces below, then multiply them all together.

Strain Index Calculation Table

Intensity of Exertion	×	Duration of Exertion	×	Efforts per Minute	×	Hand/Wrist Posture	×	Speed of Work	×	Duration of Task	=	SI Score
	×		×		×		×		×		=	

Step 5: Interpret the Result

Preliminary testing has revealed that jobs associated with distal upper extremity disorders had SI scores greater than 5. SI scores less than or equal to 3 are probably safe. SI scores greater or equal to 7 are probably hazardous. The strain index does not consider stresses related to localized mechanical compression. This risk factor should be considered separately.

General Limitations

- Many multiplier values are based on expert opinion
- Low subject numbers and lack of experimenter blinding in the original validation study (Moore and Garg, 1994)
- An examination of inter-rater and test-retest reliability has not been reported
- It is not clear whether or not the predictive validity of this assessment has been examined in multitask jobs

20

Job Accommodation

Donald S. Bloswick and Dee Hinckley

CONTENTS

20.1 Introduction

At the most fundamental level, the process of job accommodation deals with the relationship between the physical demands of the task and the capabilities of the worker. This can be thought of as the "fit" between the human and his or her work environment. A poor fit can cause unnecessary stress and can result in traumatic or cumulative injuries/illnesses. This fit may be pictured as the overlap between the capabilities of the individual and the requirements of the task as shown in Figure 20.1.

When the capabilities of the individual equal or exceed the physical demands of the task, the stresses tend to be low. When the capabilities of the individual are exceeded by the physical demands of the task, the stresses increase.

This chapter deals with (a) methods to quantify the physical demands of essential job functions, (b) methods to estimate the capability of the individual to accomplish these demands, (c) the regulatory framework within which these issues must be addressed, and

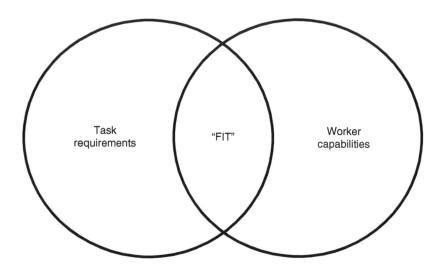

FIGURE 20.1
Fit between worker capabilities and task requirements must be optimized.

(d) use of contemporary ergonomic analysis tools to evaluate and optimize the "fit" between the worker and the job.

20.2 Quantification of Job Physical Requirements

Musculoskeletal strength plays an important role in assessing the physical status of workers or potential employees. It is also a key element in assessing an individual's overall functional capability level. Musculoskeletal strength can be assessed through Physical Ability Testing or Functional Capacity Evaluations. However, if comparisons are to be made between an individual's functional capabilities and an employment position, it is important to assess the job task requirements for the employment positions. Past legislation provides us with clear direction on making such assessments and comparisons.

20.2.1 The Americans with Disabilities Act

The Equal Employment Opportunity Commission (EEOC, Web site) and the Americans with Disability Act (ADA, Web site) both provide guidelines and regulations concerning job analyses. Specifically, an analysis must detail the essential functions of the job. The following issues must be considered when establishing the essential functions of a job.

1. Whether employees are actually required to perform the task
2. If removing the task would fundamentally change the job position
3. Whether the job exists to perform the task
4. Whether there are a limited number of other employees available to perform the task, or among whom the task can be distributed

5. Whether the task is highly specialized, and the person in the job has been hired for his or her special expertise and/or ability

The ADA suggests that any physical testing of a current or potential employee be partially based on the essential functions of the job. However, to complete accurate testing, more information about the job tasks is needed. Essential functions are often written in a broad sense, without specific details that provide information about the physical requirements of the function. For example, an essential function for a baker for ABC Grocers might read "Must be able to load the mixer with ingredients." While this indicates what the function is, it does not give us details such as the weight of the ingredients, the height of the mixer, the frequency that loading is required, and whether there is an "assist" available, either mechanized or human, when loading. Consequently, simply listing essential functions in a job analysis does not provide enough accurate information to develop adequate testing criteria.

Through breakdown and analysis of essential functions, information is obtained that gives an accurate picture of a job task's physical demands. This information provides specific, valid, reliable, and reproducible data needed to develop legal testing protocols for assessing employees and determining job accommodation. Also, the rate of occurrence for each task should be determined.

It should be noted that the ADA does not require that testing only apply to the tasks involved in the essential functions of the job. A test or selection criteria may evaluate all functions of a job, and employers can make hiring decisions based on an individual's performance of all job functions. However, if an individual with a disability completes the testing process and fails one or more test components, an employer must then make the essential function determination to evaluate the test components that were failed and make a hiring decision. Such a determination may lead to a need for reasonable accommodation on the employer's behalf.

The "Reasonable Accommodation" issue is seen as a key component of the ADA with respect to preventing discrimination. The following points should be remembered when discussing reasonable accommodation in relation to Physical Ability Testing.

1. An employer does not have to make an accommodation for an individual who is not otherwise qualified for a position.
2. An employer cannot deny an employment opportunity to a qualified applicant or employee because of the need to provide reasonable accommodation, unless it would cause undue hardship.
3. Generally it is the obligation of an individual with a disability to request an accommodation.

20.2.2 Essential Function vs. Physical Demands

Listing essential functions is the first step in developing a valid and reliable physical testing program. Essential functions, however, must then be broken down into specific job tasks. Using the example above, the essential functions of loading the mixer should be broken down and analyzed further and may look something like the following:

> Ingredients, weighing between 30 and 60 lb, are retrieved from a pallet in the cold storage room and placed on a 39-inch-high cart. The cart is then pushed 100 feet to the mixing area. The ingredients are opened with a box knife and then carried 10 feet, including two stairs, lifted to a 60-inch height, and poured into the mixer.

The physical demands of the essential job functions in the job noted above will then resemble the following:

Lifting	Vertical lift of 30–60 lb from 4 to 39 in.
	Vertical lift of 30–60 lb from 39 to 60 in.
Pushing	45 lb of force, 100 feet
Carrying	30–60 lb, 10 feet
Stair climbing	Two, while carrying (30–60 lb)
Grip	Box knife to open product
Reaching	To obtain boxes from back and sides of pallet

The frequencies of these physical demands must also be established.

Bloswick et al. (1998) present a detailed description of how job tasks were measured and quantified as part of the development of an ADA-compliant in-hire and promotion program in a lumber mill and Johns et al. (1994) present a process for analyzing fitness for duty under the ADA.

20.2.3 Obtaining Job Task Information

Information relating to the physical demands of the job tasks that relate to the essential functions of the job should be documented to the maximum possible extent. These items include (a) product and tool weights, (b) push/pull forces, and (c) walk, carry, and push/pull distances. Force gages, which allow the direct measurement of load weights and push/pull forces, are commercially available (Chatillon, Web site; Wagner, Web site). Video documentation should be accomplished whenever possible. The use of a digital video camera will facilitate the acquisition of static data (postures, etc.). Guidelines on effective video recording have been provided by Bloswick and Villnave (2000). A sample of current employees should be interviewed to assure that the observed physical task demands represent what is normally performed on the job. Supervisors and other appropriate management should complete the questionnaire.

In some cases, a more complex representation of the physical demands of the essential job functions may be needed. If a job consists of several different tasks which may very greatly from day to day, it may be difficult to define and replicate the exact job tasks in a clinical setting. One option is to estimate the effect of the external loads on the musculoskeletal system of the body and determine the applicant's ability to react to these musculoskeletal stresses. For example, in the case of a repair or maintenance activity in a plant, as part of the essential functions of the job, a worker may be required to exert a variety of different hand forces in different directions in different postures. It would be difficult to have the applicant demonstrate the ability to perform every different task. It would be possible, however, to estimate the shoulder moments required to perform the different tasks and select the maximum required moment, or at least the maximum for each posture. The maximum shoulder moment capability of the applicant (or maximum for each posture) can then be determined through a standard strength test and compared to the maximum required moment (or maximum required moment for each posture). The shoulder moments required to perform a task can be estimated using commercially available biomechanical models or the hand calculation method discussed later in this chapter.

20.3 Evaluation of Worker Capabilities

Prior to the ADA, employers used a variety of different methods for assessing the physical capabilities of applicants and employees. These methods were often used to screen out workers. Some employers used visual assessment, which involved making a judgment on the individual's ability to complete required tasks based on how the individual appeared (whether he or she was tall enough, muscular enough, etc.). Other employers required that a complete physician evaluation be performed (often including x-rays) and made decisions based on *appearances* of arthritis, limited range of motion, or decreased flexibility. Still other employers required a treadmill test, or evaluated individuals on a one-time lift of 100–150 pounds. With the inception of the ADA, hiring practices changed dramatically and the concept of job-specific testing was born. The ADA is intended not to screen out individuals with disabilities, but to encourage testing that facilitates the placement of disabled individuals in jobs that match their capabilities.

The ADA mandates that any test or selection criteria must be a legitimate measure or qualification of the specific job for which it is being used. For example, a qualification for a construction laborer job may be to push/pull wheelbarrows of cement, since this is a task often associated with construction. This is not job related, however, if the worker in the particular laborer job actually completes framing tasks. The ADA specifically states, "if a test ... excludes an individual with a disability because of the disability and does not relate to the essential functions of a job, it is not consistent with business necessity."

20.3.1 Physical Abilities Test

Using a well-defined job quantification of the physical demands of the essential job functions (discussed previously), employers can develop a testing standard and specific criteria against which they can measure potential employees. This measure is often referred to as a Physical Ability Test (PAT). It is important to understand the difference between PATs and medical examinations or inquiries that are prohibited by the ADA. The major differentiation is that while a PAT may be performed by a physician, PATs do not involve medical examinations or diagnoses. Physical Ability Testing consists of the evaluation of the applicant's ability to perform physical tasks that relate specifically to defined and measured job tasks. In addition, PATs must be given to all similarly situated applicants or employees, regardless of disability. PATs may, however, be administered after making a conditional job offer, and in this situation some medical information may be obtained. Physical Ability Testing can also be used to obtain baseline measurements of worker ability which, in case of worker injury, can be used in the planning of future medical treatment and/or rehabilitation programs. In summary, a PAT is an evaluation of an individual's physical capabilities that are directly related to physical demands of the essential job functions.

Physical Ability Testing is typically used as a post-offer, preemployment condition, not unlike drug testing. A job applicant is offered a position based on passing a PAT that evaluates his/her capabilities that are directly related to physical demands of the essential job functions. Occasionally industries such as fire and police departments will use testing as part of the application process. The ADA also allows employers to test existing employees if there appears to be a substantial risk or direct threat secondary to their ability to perform job tasks. Employers also use testing programs to determine "fit for duty" issues secondary to illness or injury.

Employers must develop policies before initiating a testing program. These must specify procedures for testing, including job categories to be tested, testing start dates, dealing with individuals who are unable to perform work tasks, working with individuals requesting accommodations, retesting applicants, etc. Once testing protocols have been developed, a random sampling of incumbent employees may be tested to document test validity and reliability. Protocols must be established detailing the frequency of reassessing physical demand loads. Initiating of a testing program can begin once these steps are completed. Some potential benefits are as follows:

1. Decreased worker's compensation injuries secondary to a "fit to work" employee population
2. Decreased time loss incidents, days away from work
3. Decreased worker's compensation costs secondary to decreased injuries
4. Increased productivity and quality
5. Increased employee morale
6. Increased rehabilitation efficiency
7. Reduced claims management costs
8. Decreased disability settlement and legal costs
9. Documented degrees of impairment
10. Established seriousness of company safety policies
11. Provision of realistic work scenario
12. Allowance for cumulative trauma screening

In addition, an aggressive testing program prevents overtreatment when injuries do occur by providing baseline levels and existing limitations.

20.3.2   Functional Capacity Evaluations

A Functional Capacity Evaluation (FCE) differs from a PAT in that it assesses all levels of an individual's physical capabilities. It is not necessarily limited to specific job tasks, although the individual's capability to perform specific job tasks can be addressed within the FCE. FCEs are also sometimes referred to as Physical Capacity Evaluations, Physical Capability Assessments, and Functional Capability Assessments. Components of an FCE should include a musculoskeletal evaluation, isometric consistency measurements, static lifting measurements, dynamic lift assessments (maximum and repetitive), and positional tolerance testing. Inconsistent measures should be documented. Again, job-specific testing can be added if a purpose of the FCE is to determine if the individual has the capability to return to a specific job. Other purposes of an FCE include (a) determining physical demand capacity levels for job placement, (b) determining whether medical stability has been reached, (c) assessing further medical treatment or rehabilitation needs, (d) determining short- or long-term disability status and, as previously mentioned, (e) determining if an individual is ready to return to work.

Many companies market a specific FCE protocol, and some use relatively sophisticated test machines to measure static forces, lifting, positional tolerances, etc. While these machines are not required for completing a successful FCE, they do allow more objective measurement of the individual's physical capabilities. The optimum FCE is one that gathers objective data, is based on peer-reviewed literature, evaluates as many of the above-listed components as possible, and documents inconsistencies in testing. Subjective information reported in the FCE should be substantiated by observable objective signs and behaviors. For example, the person performing the FCE may observe and note, "The patient appeared to be putting forth good effort, as evidenced by elevation in heart rate and sweating with tasks."

FCEs can be completed independently or in combination with an Independent Medical Evaluation (IME). An IME is usually completed by a physician through an examination and review of the individual's pertinent medical history. FCEs should be completed by an occupational therapist or physical therapist with assistance from a certified occupational therapy assistant, physical therapy assistant, or athletic trainer.

In summary, both PATs and FCEs are means of assessing an individual's capability and comparing this to the physical demands of the essential job functions.

20.3.3 Modification of Ergonomic Analysis Tools

As noted earlier, this chapter focuses on the relationship or "fit" between the physical demands of the task and the capabilities of the worker. This fit can be optimized by increasing the capabilities of the individual through, for example, physical conditioning or training or by decreasing the physical requirements of the task through analysis and redesign. Within the context of this chapter, the process of job accommodation deals primarily with the design of a job to fit within the capability of an individual who may be applying for a job or returning to the workplace with temporarily or permanently reduced physical abilities. For the purposes of this section, this type of applicant or worker will be referred to as a "disabled" worker.

In many cases, conventional ergonomic analysis methods can be modified to allow an evaluation of the "fit" between a disabled worker and a job by considering the specific disability of the worker when interpreting the output of the analytical tool. This concept will be presented for analytical tools relating to (a) low-back compressive force, (b) shoulder moment, (c) metabolic load, and (d) overall musculoskeletal stress during lifting/lowering.

20.3.4 Low-Back Compressive Force

Biomechanical analysis of the stress at the low back is possible with any of several different computer models. A simple hand-calculation method to estimate low-back compressive forces has been developed by one of the authors and has gained some acceptance. This method is shown in Figure 20.2. Back compressive forces estimated with this calculation tend to be within 10 to 15% of the results of computer models and may serve as an initial estimate when computer facilities are not available. This hand-calculation method also provides information about the components of the task that contribute most to the low-back hazard.

Biomechanical calculations are frequently used to determine the back compressive forces resulting from a material-handling task. For example, it has been established that a compressive force of 770 lb "defines an increased risk of low-back injury," and that even a limit as low as 770 lb may not be protective of the entire workforce (Waters et al., 1994). It would seem that this limit would certainly not be protective for a worker returning to work after a back surgery. It is proposed that the 770-lb limit be modified in accordance with guidance by a physician or other health care professional to represent the capability of the specific disabled worker. Johns et al. (1994) suggest that the 770-lb limit be reduced to 385 lb (50% of 770) for high-risk workers and to 578 lb (75% of 770 lb) for moderate-risk workers. In addition, the job should be redesigned to reduce the compressive forces in accordance with the relative task hazards indicated by the magnitudes of terms A and B (Figure 20.2). If term A is the highest (back muscle force reacting to upper body weight), the task should be redesigned and/or the worker should be trained to maintain the torso in a more upright posture. If term B is the highest (back muscle force reacting to load moment), the task should be redesigned to reduce the load and/or reduce the distance that the load is held away from the body, and/or the worker should be trained to keep the load as close to the body as possible

The worksheet can also give an indication of the shoulder stresses. If term A is low and term B is high, then the shoulder stresses are likely to be high. If the worker is returning to work after a shoulder surgery and does not have a history of back problems, then a minimization of the shoulder stress is most important. The job should be redesigned to reduce term B, even if it means that term A might have to increase slightly.

LOW-BACK FORCE COMPRESSIVE WORKSHEET

BW   =   Body weight   =   _____

L   =   Load in hands   =   _____

HB   =   Horizontal distance from hands to low back   =   _____

COS(h) = Cosine of torso angle with horizontal   =   _____

Fc   =   A + B + C

Where:

A   =   3(BW)cos(h)   =   .3 (___) × (___)   =   _____

B   =   .46(L × HB)   =   .46 (___) × (___)   =   _____

C   =   .8[(BW)/2 + L] [1−cos(h)] =   .8[(___)/2 + (___)] [1 − (___)] =   _____

Total Compressive Force Estimate (lbs.)   _____

Remember that:

1. A = 3(BW)cos = Back muscle force reacting to upper body weight. To lower this one must change the upper body angle with the horizontal.

2. B = .46(L × HB) = Back muscle force reacting to load moment. To lower this one must change the magnitude of the load or the distance that the load is held out from the body.

3. C = .8[(BW)/2 + L] [1−cos(h)] = Direct compressive component of upper body weight and load. To lower this one must change the magnitude of the load or upper body weight.

FIGURE 20.2
Estimation of low-back compressive force.

20.3.5 Shoulder Moment

Shoulder moment, which is one measure of stress at the shoulder, can also be directly estimated with computer. A simple hand-calculation method to estimate shoulder moment has been developed by one of the authors and has gained some acceptance. This method is shown in Figure 20.3. The moment at the shoulder depends on the weight of the load, arm weight (which is a function of body weight), and the distance that the load is located horizontally out from the point of rotation (shoulder).

SHOULDER MOMENT WORKSHEET

BW	=	Body weight (lbs)	_____
D	=	Horizontal distance from load to shoulder joint (in)	_____
L	=	Load weight (lbs)	_____
A	=	Included forearm elbow angle in degrees	_____
B	=	Included upper arm shoulder angle in degrees	_____

$M_t = M_b + M_f$

Where:
$$M_b = 0.0115 \times D \times BW = 0.0115 \times \text{\_\_\_\_} \times \text{\_\_\_\_} = \quad \text{\_\_\_\_}$$

$$M_f = 0.5 \times D \times L \quad = 0.5 \times \text{\_\_\_\_} \times \text{\_\_\_\_} = \quad \text{\_\_\_\_}$$

$$\mathbf{M_t = M_b + M_f} = \text{(in lbs)} \quad \text{\_\_\_\_}$$

Note that:

M_b = Moment at the shoulder due to the weight of the arm

M_f = Moment at the shoulder due to the weight of the load in the hands

M_t = Total moment at the shoulder = M_{task}

Substitute BW, D, L, into the above equation to estimate the total moment required at the shoulder (M_{task} expressed as in-lb). Table 20.1 indicates the maximum strength of an average male/female in different postures (upper arm angle, lower arm angle). Record the value from Table 20.1 based on gender and angles A and B (M_{cap}). The ratio of M_{task}/M_{cap} represents the required shoulder moment as percent of the maximum for the average male/female.

M_{task} = _____ (from above)

M_{cap} = _____ (from table based on angles A, B)

M_{task}/M_{cap} = _____ × 100.0 = percent maximum

Ratios (M_{task}/M_{cap}) below .5 will not present a hazard for most workers unless the frequency is quite high, and ratios above 1.0 will present a hazard for many members of the workforce.

FIGURE 20.3
Estimation of shoulder moment.

There are no generally accepted limits with which the estimated shoulder moment may be compared. Two of the variables that determine the shoulder moment capability on a specific task are gender and arm posture. Equations have been developed which provide an estimate of the maximum shoulder moment capability for an average male and female as a function of the included angle of the forearm and upper arm (Chaffin et al., 1999). Summaries of movement capabilities based on these equations for males and females with different arm postures are shown on Table 20.1. The metric proposed as a measure of the stress at the shoulder is the ratio of the shoulder moment required by the task (Mtask), as calculated by the worksheet, and the maximum strength of an average male/female in that posture (Mcap). While there are no empirically determined acceptable limits for this ratio, it is proposed that ratios below 0.5 (task-required shoulder moment is less than half of the maximum for the average male/female) will not present a hazard for most workers unless the frequency is quite high, and ratios above 1.0 (task required shoulder moment exceeds the maximum for the average male/female) will present a hazard for many members of the workforce. The relative contribution of the arm weight to the moment (Mb) and load weight to the moment (Mf) do not provide much meaningful information. In the case of a worker who is returning to work with reduced capability at the shoulder, the shoulder moment required by the task (Mtask) should be compared to the *actual* shoulder moment capability of the worker to determine if the fit is acceptable.

20.3.6 Metabolic Load

The fatigue resulting from a situation where the metabolic demands of the job exceed the worker's capacity may result in a direct cardiovascular threat or a change in work methods

TABLE 20.1

Shoulder Moment Capability (lb) of 50th Percentile Male and Female with Different Elbow and Shoulder Angles

50% Male

Included Upper Arm (Shoulder) Angle (B) (degrees)	*Included Forearm (Elbow) Angle (A) (degrees)*			
	45	90	135	180
0	632	691	751	810
45	598	658	717	777
90	565	624	684	743
135	531	591	650	710
180	498	557	617	676

50% Female

Included Upper Arm (Shoulder) Angle (B) (degrees)	*Included Forearm (Elbow) Angle (A) (degrees)*			
	45	90	135	180
0	332	363	395	426
45	314	346	377	408
90	297	328	359	391
135	279	310	342	373
180	262	293	324	355

that could increase musculoskeletal stress. In order to minimize this risk, it is important to estimate the energy requirements of the job and the work capacity of the worker.

One relatively simple method of estimating job energy requirements, developed by Thomas Bernard under contract with the American Automobile Manufacturers Association (Bernard, 1991), uses measures of arm use, walk distance, lift frequency and weight, and push/pull weight force and distance. This metabolic analysis method not only allows the determination of the overall metabolic rate associated with a job but also identifies the most stressful job components. A worksheet for the Bernard method of estimating energy expenditure is shown in Figure 20.4.

The worker's capacity is dependent on physical condition (sometimes called the physical fitness index) and on the work duration (Bink, 1962; Bonjer, 1962; Chaffin, 1966). This capacity to do work for a specific period of time is called the physical work capacity or PWC and is summarized in Table 20.2 for male and female workers of five different ages for four different work durations.

If the energy expenditure requirement of a job exceeds the physical work capacity of the worker, rest breaks need to be incorporated to avoid whole-body fatigue. In the case of a disabled/rehabilitated worker, the values in Table 20.2 will have to be reduced by (at least) a factor representing the worker's metabolic capability compared to a "normal" worker of the same age. For example, if a 50-year-old male worker were determined to have 70% of the metabolic capability of an average 50-year-old person, it is proposed that the values in Table 20.2 should be multiplied by 0.7 and reduced from the table values to those shown in Table 20.3.

The intermediate outputs of the Bernard method also indicate the energy associated with the arm work, walking, lifting, and push/pull involved in the task. This provides insight into task parameters that contribute the most to the energy requirements of the job, the modification of which have the greatest potential for significant reduction in metabolic load.

20.3.7 Overall Musculoskeletal Stress during Lifting/Lowering

The revised NIOSH RLE (Waters et al., 1994) recognizes the effect of metabolic energy expenditure, strength, and compressive forces on the low back. The Recommended Weight Limit or lifting limit established in the NIOSH Revised Lifting Equation is calculated by

$$51 \text{ lb} \times HM \times VM \times DM \times FM \times AM \times CM$$

where the six multipliers relate to physical parameters as follows:

> HM is based on the horizontal distance that the load is held out from the body (ankles)
>
> VM is based on the vertical distance of the load above the floor
>
> DM is based on the vertical distance that the load moves during the lift/lower
>
> FM is based on the frequency and duration of the lift
>
> AM is based on the torso rotational posture at the beginning and end of the lift
>
> CM is based on the type of grip between the hands and the load

In an optimum posture (load at waist level close to the front of the body), with good grip and low frequency and duration of lift, each of these factors has a value of 1.0, and

BERNARD PREDICTIVE METHOD
TO ASSESS METABOLIC LOAD (Kcal/hour).

Basal Metabolic Rate (in kcal/hour)

Arms <u>Description</u>   <u>Value</u>

Little hand/arm movement   0

Most hand movement within 20 in   1

Frequent hand movement more than 20 in   2

Bend, stoop, extend reach, etc.   3

_____ $\times$ **25** = _____
arm value

Walk <u>Description</u>

Distance (in feet) covered during walking or
carrying in one minute

_____ $\times$ **25** = _____
walk value

Lift <u>Description for weight</u>   <u>Value</u>

Most parts and tools less than 4 lb   1

Most parts and tools between 4 lb. and 11 lb   2

Most parts and tools greater than 11 lb   3

<u>Description for frequency</u>   <u>Value</u>

Less than 2 complete work cycles per min   1

Between 2 and 5 complete work cycles per min   2

More than 5 complete work cycles per min   3

_____$\times$_____$\times$ =_____$\times$ **4.4** = _____
arm value   weight value   frequency value lift value
(0,1,2,3 from above)

Push/Pull <u>Description for force</u>

Average force (lbs) exerted during push/pull

<u>Description for distance</u>

Distance (in feet) covered during walking or
carrying in one minute

(5.2 + 1.1 $\times$ _____ **)** $\times$ **(** _____ **)/3 =** _____
push/pull force   distance

Total Metabolism (kcal/hour)   = _____

Total Metabolism (kcal/min)   = _____

FIGURE 20.4
Estimation of metabolic requirements of a job.

TABLE 20.2

Physical Work Capacity (kcal/min) for Males and Females as a Function of Age and Work Duration

Male Age	PFI	Continual Work Duration			
		120 min	240 min	480 min	510 min
20	1.16	9.68	7.82	5.95	5.79
30	1.09	9.09	7.34	5.59	5.44
40	0.95	7.93	6.40	4.88	4.74
50	0.91	7.59	6.13	4.67	4.54
60	0.83	6.92	5.59	4.26	4.14
Female Age	PFI	120 min	240 min	480 min	510 min
20	1.16	7.26	5.86	4.46	4.34
30	1.09	6.82	5.51	4.20	4.08
40	0.95	5.94	4.80	3.66	3.56
50	0.91	5.69	4.60	3.50	3.41
60	0.83	5.19	4.19	3.19	3.11

TABLE 20.3

Physical Work Capacity (kcal/min) for a 50-Year-Old Male with 70% of the Metabolic Capacity of the Average 50-Year-Old Male

Age	PFI	120 min.	240 min.	480 min.	510 min.
50	0.91 × 0.7	5.31	4.29	3.27	3.18

the lifting limit is 51 lb. As the parameters involved in the lift/lower deviate from optimum, these factors decrease from 1.0 and the lifting limit decreases from the 51-lb value. The use of a 51-lb constant in the above NIOSH WPG formula is based on the assumption that the members of the workforce can lift 51 lb in the above-noted optimum posture. It is proposed that the 51-lb load constant used in the NIOSH WPG be reduced to recognize the actual lifting capacity of the disabled or rehabilitated worker for one lift/lower of 10 inches or less with a good grip in the "optimum" posture. As the actual task posture varies from the optimum, the factors will decrease from 1.0, and the lifting limit for the specific worker for specific tasks will also be reduced. While there is no empirical data to support this proposal, it certainly appears to be more appropriate than applying the 51-lb constant when analyzing tasks to be performed by a disabled/rehabilitated worker.

The multipliers in the NIOSH RLE also relate to different aspects of the hazard associated with the task and provide information relating to the redesign of the task for disabled/rehabilitated workers. The horizontal multiplier (HM) relates to the low-back hazard and, to a lesser extent, to the musculoskeletal hazard associated with the shoulder. If this factor is low, the task presents a higher risk for individuals with limitations in these areas, and the task should be redesigned to reduce this risk. If the vertical multiplier (VM) is low, and the actual vertical distance is high, there may be a musculoskeletal hazard associated with the shoulder, and the task presents higher risk for individuals with reduced capability in this area. The frequency multiplier (FM) relates to the metabolic load associated with the task. If this factor is low, the task presents a higher risk for individuals with cardiovascular limitations. In this way the NIOSH RLE can be used, not only to estimate job stresses, but to focus redesign efforts where they will most benefit workers with specific disabilities. The multipliers in the NIOSH RLE also provide information about where more focused analysis should be directed. For example if HM is low (horizontal distance high), a biomechanical analysis of the low back would likely be appropriate. If VM is low and the vertical distance is low, a biomechanical analysis of the shoulder is indicated. If FM is low (indicating a high frequency of lift/lower) a metabolic analysis is appropriate.

20.4 Future Directions

Future work to improve the process of job accommodation must deal with the relationship between the physical demands of the task and the capabilities of the worker, as illustrated earlier in Figure 20.1. Better methods of ergonomic analysis and task evaluation will assist in more precise quantification of the physical requirements of occupational tasks overall and also those which may be performed by the disabled or rehabilitated worker. In addition, as clinicians improve their ability to quantify the capabilities of humans, particularly those whose physical capabilities are not well defined by standard industrial norms, the fit between the worker or job applicant and the job can be more accurately evaluated.

20.5 Summary

The process of job accommodation deals with the relationship between the physical demands of the task and the capabilities of the worker, or the "fit" between the human and his/her work environment. A mismatch between task demands and worker capability generally results in musculoskeletal stress and can result in traumatic or cumulative injuries/illnesses. In this chapter, methods to quantify the physical demands of essential job functions and methods to estimate the capability of the individual to accomplish these demands were presented along with the regulatory framework within which these issues must be addressed. In addition, methods to adapt contemporary ergonomic analysis tools to allow the evaluation of the "fit" between the disabled/rehabilitated worker and the job were presented.

References

ADA Home Page. 2003. Information and technical assistance on the Americans with Disabilities Act, http://www.usdoj.gov/crt/ada/adahom1.htm.
Bernard, T.E. 1991. Metabolic Heat Assessment, *MVMA Agreement USF 9008-C0173*, University of South Florida, Tampa, FL.
Bink, B. 1962. The physical work capacity in relation to working time and age, *Ergonomics*, 5(1), 25–28.
Bloswick, D.S., Jeffries, D., Brakefield, S., and Dumas, M. 1998. Industrial setting case study: ergonomics and Title I of the Americans with Disabilities Act, in *Health Care and Rehabilitation Ergonomics*, V. Rice, Ed. (Newton, MA: Butterwork-Heinemann), pp. 307–315.
Bloswick, D.S. and Villnave, T. 2000. Ergonomics, in *Patty's Industrial Hygiene*, 5th ed., Vol. 4, R.L. Harris, Ed. (New York: John Wiley & Sons), pp. 2531–2638.
Bonjer, F. 1962. Actual energy expenditure in relation to the physical work capacity, *Ergonomics*, 5(1), 29–31.

Chaffin, D.B. 1966. The prediction of physical fatigue during manual labor. *J. Methods Time Meas.*, 11(5), 25–31.

Chaffin, D.B., Andersson, G.B.J. and Martin, B.J. 1999. *Occupational Biomechanics*, 3rd ed. (New York: John Wiley & Sons), p. 263.

Chatillon, AMETEK Test and Calibration Instruments Division, http://www.chatillon.com.

EEOC. 2003. U.S. Equal Employment Opportunity Commission (EEOC): Laws, regulations and policy guidance, http://www.eeoc.gov/policy/index.html.

Johns, R.E., Bloswick, D.S., Elegante, James M., and College, A.L. 1994. Chronic, recurrent low back pain: a methodology for analyzing fitness for duty and managing risk under the Americans with Disabilities Act, *J. Occup. Med.*, 36(5), 537–547.

Wagner Instruments, Wagner FDI Series Specs, http://www.wagnerforce.com/fdi.htm (accessed November 20, 2003).

Waters, T.R., Putz-Anderson, V., and Garg, A. 1994. *Applications Manual for the Revised NIOSH Lifting Equation*, DHHS (NIOSH) Publication No. 94-110.

Additional Reading

Bloswick, D.S., Villnave, T., and Joseph, B. 1998. Ergonomics, in *Sourcebook of Occupational Rehabilitation*, P.M. King, Ed. (New York: Plenum Press) pp. 145–165.

Demeter, S.L., Andersson, G.B.J., and Smith, G.M. 1996. *Disability Evaluation* (Chicago, IL: Mosley-Year Book).

Isernhagen, S.J. 1988. *Work Injury: Management and Prevention* (Gaithersburg, MD: Aspen Publishers).

Jacobs, K. and Bettencourt, C.M., Eds. 1995. *Ergonomics for Therapists* (Newton, MA: Butterworth-Heinemann Medical).

Kumar, S., Ed. 1997. *Perspectives in Rehabilitation Ergonomics* (London: Taylor and Francis).

Mital, A. and Karwowski, W., Eds. 1988. *Ergonomics in Rehabilitation* (Philadelphia, PA: Taylor and Francis).

Rice, V.J.B. 1998. *Ergonomics in Health Care and Rehabilitation* (Oxford, UK: Butterworth-Heinemann).

U.S. Department of Labor — Find It By Topic — Disability Resources — Job Accommodations, http://www.dol.gov/dol/topic/disability/jobaccommodations.htm (accessed November 20, 2003).

Wilson, A. 2002. *Effective Management of Musculoskeletal Injury: A Clinical Ergonomics Approach to Prevention, Treatment and Rehabilitation* (Philadelphia, PA: W. B. Saunders).

21

Strength and Disability

Alicia M. Koontz, Fabrisia Ambrosio, Aaron L. Souza, Mary Ellen Buning,
Julianna Arva, and Rory A. Cooper

CONTENTS

0-415-36953-3/04/$0.00+$1.50
© 2004 by CRC Press LLC

ABSTRACT There is great diversity among individuals with disabilities and as a result, muscle strength will be highly variable depending on the specific type of injury or illness, severity of the impairment, the age of disability symptoms, and length of time living with the disability. We begin this chapter with a brief overview of the prevalence of common disability types, physical activity as it relates to disability, adjusting to a disability, programmatic barriers to participating in an exercise program, and ergonomic relevance. Muscle strength will be discussed under several major headings: orthopedic disability, neurological disability, aging with a disability, age-related disabilities, mental impairments, metabolic disorders, respiratory disorders, coronary heart disease, and children with disabilities. We also present general considerations for developing and implementing a strength training program and address the environmental barriers that are faced by individuals with disabilities.

21.1 Prevalence of Disability

Under the Americans with Disabilities Act (ADA, 1990), disability has been defined as "a physical or mental impairment that substantially limits one or more of the major life activities." At the end of 1994, about 54 million people had some level of disability, and 26 million people had a severe disability (U.S. Census Bureau, 2002). Data from the 1992 National Health Interview Survey indicated heart disease as the condition most often named as the principal cause of disability, affecting 7.9 million Americans. Orthopedic impairments and disorders of the spine are second, limiting the activity of 7.7 million people. Next come arthritis (5.7 million), orthopedic impairments of the lower extremity (2.8 million), asthma (2.6 million), diabetes (2.6 million), mental illnesses (2.0 million), and learning disabilities and mental retardation (1.6 million) (LaPlante, 1996). Cancer ranks 10th affecting 1.3 million people, and cerebrovascular disease ranks 15th affecting 1.2 million. An estimated 1 million individuals have neurological disorders such as spinal cord injury, spina bifida, postpolio, cerebral palsy, and multiple sclerosis. Each year more than 80,000 Americans survive a hospitalization for traumatic brain injury (TBI); 5.3 million individuals today are living with a TBI (CDC, 2002).

21.2 Physical Activity and Disability

After acquiring a disability, the amount of physical activity is found to decrease rapidly, which leads to a loss of muscle mass and diminished level of strength (Janssen et al., 1994; Rimmer, 2001). Many people with disabilities fall into a sedentary lifestyle, which usually leads to additional medical problems, such as heart and lung disease, pressure sores, obesity, osteoporosis, joint pain, and arthritis (Janssen et al., 1994; Rimmer, 2001). Daily activities such as climbing stairs, getting in and out of an automobile, wheelchair, or bed, walking, and propelling a wheelchair, once thought to be easy, become much harder to perform as time and inactivity continue (Janssen et al., 1994; Rimmer, 2001).

For a person with a disability to perform functional activities requires adequate muscle strength and endurance. Studies on physical strain in the daily life of individuals with disability have shown that specific activities of daily living such as performing transfers, entering/exiting a car, ascending/descending a ramp, walking, and negotiating environmental obstacles provoke high levels of strain (Janssen et al., 1994). These tasks demand a relatively large effort from large muscle groups. Disturbances in muscle function due to disability will greatly impact the extent to which an individual can perform the task. Strength training regimens that focus on increasing strength could improve the ability of an individual with a disability to perform daily tasks. Many falls and osteoporotic fractures are partially related to muscle strength, and thus strengthening activities are particularly important for individuals with muscle weakness and diminishing bone mass due to disability. Flexibility also needs to be maintained along with strength to increase or maintain joint range of motion, prevent injury, and minimize muscle soreness.

21.3 Adjustment to Disability

On average, attitudes toward strength-building exercise will vary among people with disabilities, as they do in the general population. In addition to the time demands of modern life, consider that most daily activities like dressing, grooming, mobility, and transportation inherently take longer for individuals with disabilities. In order to include strengthening activities into weekly routines, an individual must have a high degree of commitment to the long- and short-term benefits. Intrinsically motivated individuals who are aware of the positive health consequences will work to balance time pressures. Others will need considerably greater assistance and extrinsic motivation in order to participate.

The location of the person in the process of adaptation to disability is an important variable. Adaptation is commonly thought to occur in stages (Livneh, 1991). Depending on the individual and his or her method of coping with or adapting to loss, interest in physical conditioning and strengthening may occur in an early or late phase of this process. "Working out" may be a way dealing with grief and loss. This implies that the later phases of accepting the permanence of disability or anger at fate and perceived dependence are still ahead. In contrast, when individuals have reached the last phase, where acceptance and adaptation occur, then they are ready to reintegrate socially and have regained a sense of positive self-worth. In either case, participation in individual or group fitness activities can contribute to the individual's progression. Personnel should be aware of this adjustment process and accept and support individuals wherever they are in adapting to disability.

Little research is available to document the contribution of strengthening programs to speeding the process of adaptation to disability. The area of strength programs for persons with disabilities is wide open for scientific inquiry (Rimmer et al., 1996). Given the positive correlation between strength and capacity to function despite disability or health conditions, it seems intuitive that participation would help.

Over the past 10 years, models of "able" disabled persons have become prevalent in our society. The media shows competitive athletes with disabilities, and it is common to see persons using wheelchairs and mobility devices in every aspect of community life, thanks to educational inclusion and public accommodation through the ADA. There may still be a shortage of role models in particular strength programs or settings, but it is clear that feeling shame about disability is a diminishing phenomenon. Individually, each person still needs to incorporate disability or chronic health problems into a revised body image. Until this occurs self-consciousness may be stronger than the desire to exercise. The social atmosphere of an exercise setting and its personnel will go far to reduce discomfort on this count.

21.4 Programmatic Barriers

People previously believed that many wheelchair users and people with disabilities could not benefit from exercise (Cooper, 1998, p. 99). Even the individual with a disability may assume that disability or a chronic health problem excuses them from seeking strength and fitness. Individuals and family members may feel overly concerned about a changed, more fragile self. They don't know how to adapt normal strength-building activities or are unsure about how hard to push in the presence of disability. Sensitive and knowledgeable trainers may need to help individuals modify progressions, find alternate gripping techniques, or modify body position. Trainers who have reasonable knowledge of disability, are familiar with safety procedures, and are able to see the person first and the disability second will do much to increase the comfort level and the ongoing participation of persons with disabilities in strength-building activities.

21.5 Ergonomic Relevance

Poor ergonomics can lead to many types of pain and injury including back pain, headaches/migraines, stiff neck/shoulders, and repetitive strain injuries and muscle fatigue. For individuals with an orthopedic or neurological disability resulting in physical limitations, working and living in an ergonomically healthy environment is important for enhancing productivity and preventing secondary problems. Ergonomics considers such factors as task frequency, length of time that a person can sustain a static posture, work surface heights, and reach requirements, which all depend on the individual's strength, range of motion, and endurance. Environments that have been ergonomically designed for individuals who have "normal" function may not work well for those who use wheelchairs or have limited upper extremity function such as poor grip strength, dexterity, and muscle control. Under the ADA (ADA, 1990), an employer is expected to make "reasonable accommodations" for employees with disabilities who can otherwise perform the essential functions of their jobs. This may involve modifying a workstation, making adaptations to tools and/or introducing assistive devices to enable the person to perform essential job functions. Specific recommendations for worksite accommodations (e.g., workstation height, clearances, sizes, and permissible reach distances) can be found in the Accessibility Guidelines of the ADA (ADA, 1990).

21.6 Orthopedic Disability

The following sections provide details on muscle strength and common orthopedic symptoms and disabilities.

21.6.1 Asymmetrical Weakness

Asymmetrical weakness is often found in people who have cerebral palsy or encounter a stroke (Rimmer, 2001). If there is weakness to one side of the body, the professional team (therapists, trainers, caregivers, and/or exercise physiologist) may want to make sure that active assistance is used when working with the affected side during resistance training (Rimmer, 2001). For instance, the professional should actively help the affected limb through the range of motion during the exercise. The weaker side should be focused on more readily, but not to the point of total fatigue.

21.6.2 Muscle Imbalance and Shoulder Musculoskeletal Injury

Muscle imbalance is often associated with people who have spinal cord injuries (SCI) and use manual wheelchairs for their primary means of transportation (Burnham et al., 1993). A person with a SCI utilizes arms much more than the unimpaired population, due to transfers, pressure relief, and repetitive motions involved with wheelchair propulsion and reaching overhead for items, which causes secondary conditions (Sie et al., 1992). With this increased use, people with SCI may acquire a muscle imbalance, which may further lead to loss of some range of motion, impingement syndrome, bursitis, and joint pain (Burnham et al., 1993; Sie et al., 1992).

Shoulder impingement has been projected to be the key musculoskeletal setback that manual wheelchair users endure. This is likely due to the shoulder's unstable structure (Bayley, 1987; Cohen and Williams, 1988; Nichols et al., 1979; Sie et al., 1992). Some researchers believe that the muscles become stronger in the drive phase of the propulsion stroke when utilizing a manual wheelchair (Burnham et al., 1993; Miyahara et al., 1998). The elevators (deltoids) are thought to become stronger, pulling the humeral head upward and, without a strong counterbalance from the shoulder depressors, causing possible damage to the immediate shoulder structures. Thus, muscle imbalance alters movement patterns, creating an environment that is conducive to injury.

Burnham and colleagues (1993) examined male wheelchair athletes (WA) and found that 26% of the WA shoulders were affected by rotator cuff impingement. This was attributed to weaker adduction, internal and external rotation strength. When investigating the abduction/adduction ratio, Burnham et al. (1993) discovered the subjects' exhibited weak shoulder adductor activity. Miyahara et al. (1998) found that their subjects (elite rugby players with quadriplegia) displayed weaker adduction strength when compared to the abduction muscles of the shoulder. Further, Nyland et al. (1997) and Bernard and Codine (1997) examined their subjects (wheelchair basketball players and people with paraplegia) and found that external rotation weakness was common.

People with spinal injuries are predisposed to shoulder weakness and pain. Establishing a resistance program at the onset of their rehabilitation program may provide some defense to support this volatile joint (Ballinger et al., 2000; Mayer et al., 1999; McCormack, 1977; Moreau and Moreau, 2001; Waring and Maynard, 1991). Although there may be weakness observed in upper back and posterior deltoid with people with SCI, as noted by Figoni (1993) and Groah and Lanig (2000), exercising the entire shoulder complex may generate a structure that is more resistive to injury. In addition, Curtis et al. (1999) noted that strengthening the posterior portion of the shoulder's musculature might be conducive in preventing further shoulder ailments.

21.6.3   Rheumatoid Arthritis

Whether to train or not to train people with rheumatoid arthritis is controversial (Lyngberg et al., 1988; van den Ende et al., 2000). In one study observing 64 patients with rheumatoid arthritis and an average age of 60 years, involvement in strength (isokinetic training 5 times per week) and cardiovascular (bicycle 3 times per week) training resulted in improvements in overall upper and lower body strength without adverse affects (van den Ende et al., 2000). Lyngberg et al. (1988) investigated the strength of 18 patients with a moderate case of rheumatoid arthritis. The authors implemented an 8-week strength program and aerobic series. After the conclusion of the training sessions, 35% of the patients recorded a decrease in pain sensation in swollen joints. Lyngberg and colleagues (1988) concluded that moderate levels of exercise may not aggravate people with moderate levels of rheumatoid arthritis. McCubbin (1990) agrees with van den Ende et al. (2000) and Lyngberg et al. (1988) that people who suffer from rheumatoid arthritis should be involved in a specialized exercise program designed specially for pain and swelling reduction with the benefits of an improved viewpoint on life. Häkkinen et al. (1999) implemented a 12-month strength training program with 32 subjects who suffered from the onset of rheumatoid arthritis. Each subject was encouraged to get involved in recreational activities and complete strength exercises 2 times per week at 50 to 70% of maximal load capability. At the end of the 12-month program, the subject's maximal strength was tested (knee extensors, trunk extensors and flexors, and hand grip). The researchers discovered a range of 22 to 35% strength gain in all muscles tested. Häkkinen and associates (1999) conclude that a minor program such as this demonstrates that people with the early onset of rheumatoid arthritis may benefit from resistance training.

21.6.4   Osteoarthritis

Osteoarthritis is a disease that affects the synovial joints and is one of the common forms of arthritis (Hurley, 1999). Many people who suffer from osteoarthritis find that hip problems are one of the major results of this disease. As people get older, they become more susceptible to decreased function and joint instability with local hip muscle weakness. Total hip arthroplasty is usually the prescribed means to correcting hip problems and relieving high levels of constant pain sensation (Hurley, 1999). Hurley (1999) suggests that a well-conditioned muscle will rebound much faster than an unconditioned one. Staying in shape and establishing and maintaining an exercise program throughout life, especially when old age is approaching, will aid in the recovery process of any muscular surgery.

Following hip arthroplasty surgery, a person feels a sense of pain relief after at least a 9-month postoperative period, but decreased mobility still exists (Sicard-Rosenbaum et al., 2002). The muscles in the legs were not fully strengthened back to their original levels following rehabilitation. The walking speeds of people who undergo arthroplasty surgery are much slower and household chores are cumbersome (Sicard-Rosenbaum et al., 2002). Because of this muscle weakness that still exists, Sicard-Rosenbaum et al. (2002) suggest that some type of intervention be implemented beyond the rehabilitation process. Reardon et al. (2001) examined the strength of the quadriceps in 19 patients who underwent total hip arthroplasty. The authors discovered that even after a full rehabilitation process to regain strength in the hip region after surgery, muscle weakness still persisted after 5 months postsurgery. Even though pain sensation was reduced, balance and coordination were still problems. Reardon and comrades (2001) recommended that different rehabilitation strategies need to be implemented and a post rehabilitation exercise program needs to be utilized to regain total hip strength.

Bertocci et al. (in review) evaluated differences in isokinetic hip flexion, extension, and abduction muscle performance between 20 older adults who underwent total hip arthroplasty and rehabilitation, and 22 similarly aged controls. Isokinetic performance was chosen because it has been shown to correlate with physical function and activities of daily living (Campbell et al., 1995; Wilk et al., 1994). Bertocci et al. (2002) found that isokinetic hip performance at 4 to 5 months postsurgery was significantly lower for hip patients than that of similarly aged healthy controls, suggesting that hip arthroplasty patients experience a deficit not only in strength, but in endurance as well. The researchers concluded that hip patients are not being restored to the same level of strength and muscular endurance as compared to similarly aged healthy adults. These findings may be useful in providing a preliminary rationale for revising current approaches in hip arthroplasty rehabilitation protocols.

21.6.5   Spinal Cord Injury

Preservation of the upper extremity function in SCI is essential in preventing or prolonging the onset of secondary disabilities. Participation in a resistance training program can help regain or maintain strength in the unaffected upper extremity. Davis and Shephard (1990) examined the elbow and shoulder flexion/extension and shoulder abduction/adduction muscular endurance in 11 individuals with paraplegia over an 8- to 16-week period. The investigators noted positive results (gains in power) from the high-intensity exercises, which were performed three days per week with bouts of 20 to 40 minutes. All shoulder exercises can be performed as long as they do not exacerbate current shoulder problems such as pain. Shoulder exercises above the head should be avoided, because this will aggravate existing shoulder problems. Fixation of the scapula or exercise involving scapular retraction should also be a priority, because this will help to stabilize the shoulder and prevent future or existing shoulder problems (Burnham et al., 1993; McCormack, 1977). Exercises including but not limited to front, side, and posterior arm raises, shrugs, and internal/external rotation can be performed using free weights, weight machines with cables, or resistive tubing or banding (Baechle, 1994). Duran et al. (2001) examined a weightlifting program consisting of various lifts (bench, military, and butterfly press, biceps, triceps, shoulder abductors, abdominal, and curl back neck) over a 3-week period consisting of 2-hour sessions. After completion, all of the subjects were noted to increase their strength in all areas. The investigators concluded that implementing a weightlifting program during and after the rehabilitation process is essential for people with spinal injuries to regain normal levels of strength needed to perform activities of daily living.

Implementing an exercise program has also been observed to increase not only strength but endurance in people who have SCI (Jacobs et al., 2001). Jacobs and Nash (2001) reviewed 149 articles dealing with the risks and benefits of exercise for persons with SCI. The authors found that upper-extremity resistive training increases strength and endurance levels. In addition, when Jacobs and Nash reviewed studies that examined the effects of arm and wheelchair ergometry training on people with SCI, improvements in arm endurance were discovered.

It is also important to consider maintaining muscle integrity of the paralyzed limbs to prevent muscle atrophy, lessen the risk of fractures, and enhance circulation. Recently, functional electrical stimulation (FES) has been a topic of interest for strengthening affected limbs in people with an SCI. A study published by Scremin et al. (1999) substantiated the hypothesis that FES is capable of increasing muscle cross-sectional area as well as the muscle-to-adipose tissue ration. Jacobs and Nash (2001) examined the affects of electrical stimulation of the lower body of individuals with SCI. Their review of the literature indicated that muscle mass and circulation were increased in common leg muscles with FES.

21.6.6   Fibromyalgia

Fibromyalgia is a complex, chronic condition which produces pain in the soft tissues located around joints and in the skin and organs throughout the body. Persons with fibromyalgia present with symptoms such as body stiffness, gastrointestinal complaints, myofascial trigger points (painful spots that form in taut bands in muscles or other connective tissue), leg sensations, swelling of the limbs, and depression and anxiety. Women are more likely to have fibromyalgia than men. A growing number of studies have investigated the effectiveness of aerobic, muscle strengthening, flexibility, and composite training in reducing the severity of symptoms in persons with fibromyalgia (Busch et al., 2002; Häkkinen et al., 2001; Jones et al., 2002). Jones et al. (2002) conducted a randomized controlled trial of muscle strengthening vs. flexibility training in 68 women with fibromyalgia. They found that patients were able to engage in the 12-week/2-days-per-week muscle strengthening program with improvements in disease activity and no significant exercise-induced pain. The magnitude of positive changes was more pronounced in those assigned to the muscle strengthening program as compared to the flexibility program. Häkkinen et al. (2001) examined a 21-week experimental strength training program, with 11 women with fibromyalgia randomly assigned to the program and 10 women with fibromyalgia assigned to a control group. The training was performed 2 times per week and included 6 to 8 exercises, such as squats, knee, and trunk extension/flexion exercises and bench press. The first 3 weeks involved high repetitions (15–20 reps) and low loads (40—60% of one repetition maximum) and gradually progressed to lower repetitions (5–10 reps) and higher loads (70–80%) in the remaining weeks. The cases with fibromyalgia showed significant improvement in pain, musculoskeletal performance, and psychological function but not in fatigue or sleep.

21.7   Neurologic Disability

We will now discuss the most common symptoms affecting strength as seen with a lesion or dysfunction of the nervous system. We will emphasize specific symptoms as they relate to strength in certain pathologies that have been most commonly investigated. A neurological limitation may have its origin at one or many different levels of the nervous system, making adequate diagnosis and consequent treatment planning complicated. For this reason, great debate often exists among clinicians as to finding an appropriate balance between strengthening in order to maintain everyday function and running the risk of overworking neurologically impaired muscles, leading to an increased level of disability. A good example is given in the case of spasticity.

21.7.1 Spasticity

Spasticity is formally defined as:

> A motor disorder characterized by a velocity-dependent increase in tonic stretch reflexes (muscle tone) with exaggerated tendon jerks, resulting from hyperexcitability of the stretch reflex, as a component of the upper motor neuron syndrome. (Lance, 1980.)

That is, there is an overactivity of the muscular stretch reflexes, and the resistance to passive movement is dependent on the speed at which the affected limb is moved. The stretch reflex is a monosynaptic reflex not dependent on higher centers. The purpose of this reflex is to maintain optimal skeletal muscle resting length. Spasticity is often a major determinant of disability, since the hypersensitivity of the stretch reflex often responds to normal movement with excessive muscle contraction. In addition, spasticity may be seen in resting muscle, in which case the stretch reflex is always "on." Due to the overexcitability of the reflex loop, spasticity is inevitably associated with a decreased ability to move. Therefore it is commonly seen in conjunction with muscle weakness and atrophy. However, the extent to which spasticity affects muscle strength is still unknown. Studies have revealed that there is no direct correlation between spasticity and muscle performance, and that the weakness and decreased dexterity is actually a symptom separate from the actual mechanism of spasticity itself (Sahrmann and Norton, 1977). Optimal treatment measures in cases of spasticity are still unclear.

21.7.1.1 Spasticity in Stroke

Traditionally, it was believed that individuals with spasticity should avoid resistive exercise and increased effort (Bobath, 1990). It was the belief of Bobath that a decreased function of the agonist muscle was a result of overactivity of the spastic antagonist muscle and not a function of weakness of the agonist muscle itself. With this in mind, he believed strength training would lead to an increased spastic tone and would actually reinforce abnormal activation patterns, essentially leading to an increased cocontraction. They argued that the effort involved in strengthening exercises would perpetuate not only spasticity but also widespread associated reactions. Emphasis was therefore placed on correcting abnormal postures and on normalizing tone.

Today, although this theory continues to be taught in many schools, an increased amount of literature is being published that supports the idea that spasticity may actually mask underlying muscle weakness and that the two are, in fact, separate mechanisms. More recent studies advocate strengthening protocols in individuals with spasticity. Sharp and Brouwer (1997) looked at changes in isokinetic torque in 15 individuals with a unilateral stroke before and after a 6-week training program. The objectives of the study were (a) to examine if isokinetic strengthening has an effect on the degree of spasticity and (b) to investigate whether strength improvements were possible in a spastic lower extremity. They found significant improvements in the torque production of the affected leg. Furthermore, these improvements were made in the absence of any detectable change in extensor spasticity. Recognizing the correlation between lower extremity muscle strength and gait performance (Bohannon and Walsh, 1992), Teixeira-Salmela and colleagues (1999) performed a similar study of physical conditioning in 13 individuals who sustained a stroke. Each subject participated in a 10-week combination program consisting of muscle strengthening and aerobic conditioning. Not only were significant gains in the total strength of the lower extremities found; they also found significant improvements in gait speed and in the rate of stair climbing. Finally, significant improvements were seen in quality-of-life scores, suggesting an overall increase in life satisfaction. These findings are important, considering that they demonstrate not only a physiological effect of strength training in individuals who have had a stroke, but also functional and psychosocial benefits as are seen in unimpaired populations. Another study investigating changes in strength of a spastic lower extremity in chronic stroke patients (Smith et al., 1999) has also provided valuable information regarding the functional benefit of an exercise program. This study focused on a "task-oriented" strengthening protocol in persons with residual strength impairments resulting from a stroke more than 6 months previously. Traditionally, it has been accepted that the window for maximal functional recovery is around 3 months from the time of the stroke. Not only did this study demonstrate significant strength improvements in the absence of any deleterious effects of increased spasticity, it also showed that these strength improvements were possible well beyond the aforementioned window (subjects were, on the average, 2 years post-stroke). This raises questions about the ability of the cortex to undergo sensorimotor plastic changes long after the time of the initial stroke. Studies regarding neural plasticity have cited it as being, at least in part, a consequence of motor experience (Greenough, 1988). Therefore, an argument may be made that the changes that are seen in stroke patients after repetitive training are in fact a result of relearning appropriate motor patterns for muscle activation.

21.7.1.2 *Spasticity in Cerebral Palsy*

The principle of relearning appropriate motor patterns is intuitive, since stroke may for these purposes be considered a traumatic event where motor ability is destroyed. This raises the question as to whether strengthening principles applied to spasticity in stroke patients are appropriate for individuals with spasticity resulting from a congenital malformation, such as spastic cerebral palsy (CP). Spastic CP results from a static lesion of the brain and often presents very differently from individual to individual, varying from spastic diplegia, to spastic hemiplegia, to spastic quadriplegia. Weakness is also commonly seen in this population, and for this reason a significant amount of research has investigated the effect of strengthening spastic muscles on overall functioning. A recent study has refuted the idea that performance of maximal effort resistance exercise increased quantifiable spasticity in children with cerebral palsy, suggesting that an increased force production in affected muscles does not necessarily contribute to a decreased overall function (Fowler et al., 2001). Damiano and Abel (1998) performed a study looking at the functional outcome of strength training in children with spastic CP. This study built on findings that individuals with CP have been shown to become stronger at a rate comparable to nondisabled controls (Damiano et al., 1995). The investigators measured functional changes seen in gait parameters after participating in a 6-week strength training program. Specifically, 11 children performed resistive exercises at 65% of their maximum isometric strength, four sets of five repetitions 3 times a week for 6 weeks. They found that all children had significant strength gains, concomitant with an increased gait velocity and cadence. Dodd et al. (2002), during a review of research pertaining to exercise and cerebral palsy, concluded that strength-training people with cerebral palsy would not exacerbate previous secondary conditions (spasticity or reduced range of motion) as long as the exercise regimen is not too vast in scale. Improvements in motor activity and strength are possible with no adverse effects (Dodd et al., 2002).

21.7.2 Mechanism behind Weakness in Spasticity

Considering the aforementioned studies looking at strength gains in spastic limbs, it is important to consider whether the weakness seen in these upper motor neuron lesions is a function of secondary effects on the contractile properties of the muscle or if it is due to a decreased activation from the higher centers (cerebral cortex and spinal cord). This was investigated by Landau and Sahrmann (2002) in a study regarding muscle stimulated directly through electrical stimulation in 17 patients with acute hemiplegia or hemiparesis and 14 patients with chronic hemiplegia/hemiparesis, all resulting from a cerebral infarction or hemorrhage. These two groups were compared to 13 healthy control subjects. Findings revealed that when compared to control subjects, the directly elicited force production resulting from the electrical stimulation was not significantly impaired in the subjects who had a stroke. Investigators concluded that the weakness seen is not a function of impaired muscle contractile competence, but rather is primarily an effect of impaired initiation and coordination. Therefore, it is hypothesized that strengthening protocols may be effective in this population because it results in a recruitment of an increased number of motor neurons and by an increased motor unit firing rate. This is consistent with the imaging findings of Karni and colleagues (1998) demonstrating that in an unimpaired population, motor practice results in an increased recruitment of motor units of a local network representing the task, ultimately lending to skill acquisition.

The next sections discuss muscle strength in progressive conditions such as multiple sclerosis and muscular dystrophy. We will also briefly summarize the literature on strength in individuals with traumatic brain injury.

21.7.3 Multiple Sclerosis

Multiple sclerosis (MS) is a chronic, unpredictable neurological disease of the central nervous system (brain, spinal cord, and optic nerves). In MS, myelin which surrounds the nerve fibers of the central nervous system is lost in multiple areas, leaving scar tissue called sclerosis. Persons with MS often experience muscle fatigue and are weaker than their healthy counterparts. In a study comparing 15 individuals with MS to 15 control subjects, Lambert et al. (2001) found that individuals with MS had a peak torque production 25.7 and 20.8% lower than controls for dominant and nondominant flexors of the lower extremity, respectively. Persons with MS also have difficulty performing activities of daily living. Ambrosio et al. (2002) found that individuals with MS were physically unable to propel their wheelchairs at a functional standard speed (1 m/sec). When asked to propel for a 5-minute period, they were unable to maintain this slow speed beyond the first two minutes.

Fay (2001) highlighted the need for more sensitive clinical measures of strength in individuals with MS. Their study compared the isokinetic strength of individuals with MS as determined with the use of a BioDex System 3 (Biodex Medical, Shirley, NY) against a manual muscle test (MMT) grade (American Spinal Injury Association and International Medical Society of Paraplegia, 1996) as determined by a physical or occupational therapist. They found that the MMT results were not correlated with isokinetic measures of strength, indicating that the MMT may not be the most appropriate test for quantifying strength in populations with MS. Additional research is therefore needed for the development of new, more effective clinical examinations that more accurately describe the strength of individuals in this population.

While the mechanism for muscle weakness in MS is still unclear, some findings have provided important information in this area. First, Kent-Braun and colleagues (1997) found a 26% reduced muscle fiber cross-sectional area relative to controls, a very comparable percentage to that seen in Lambert and colleagues' (2001) findings. Even so, Lambert et al. (2001) demonstrated that fat-free mass differences did not account for the differences in strength between individuals with MS and control subjects. In their study, there was no significant difference between the fat-free muscle masses of the two groups, despite significant differences in force production. Instead, they concluded that strength differences were attributed to a decreased muscle quality. Along these lines, Rice et al. (1992) conducted a study looking at the motor neuron firing rates during an isometric exercise protocol and found rates to be significantly lower than seen in control subjects. This indicates a decreased ability for muscle activation in this group of individuals. Finally, studies by Kent-Braun and colleagues (1997) have demonstrated a higher prevalence of fast twitch muscle fibers in the MS group when compared to control counterparts. Fast twitch muscle fibers have a greater reliance on anaerobic energy supply and demonstrate a reduced muscle oxidative capacity. Therefore, an increased percentage of this muscle fiber type would contribute to a decreased endurance and an increased fatigability.

Murray (1985) stated that up to 40% of individuals with MS claim that fatigue is the most serious symptom they experience. With this in mind, several studies have concentrated efforts to investigating the mechanism of fatigue in MS. Certain aspects of fatigue have been well established in this population. First, it is generally accepted to be strongly temperature dependent (Krupp et al., 1988). Also, people with MS commonly describe their fatigue as a general lack of physical energy rather than a weakness of a specific muscle group (Krupp et al., 1988). This often propels people with MS into a cycle of *fatigue → decreased activity →increased fatigue*. One study by Sheean et al. (1997) took an electrophysiological look at the mechanism of fatigue in MS. All subjects performed a fatiguing protocol while electromyographical data were collected. Results revealed evidence of an exercise-induced reduction in force generation. Interestingly, this reduction in strength was only evident with sustained contraction. Therefore it may be difficult to clinically detect fatigue in this population, since most standardized clinical evaluation tools do not incorporate prolonged measures of strength and ability. Finally, they found no correlation between the amount of inducible fatigue and the amount of fatigue experienced in everyday life, contributing further to the mystery of fatigue in this population.

Surprisingly similar to the nature of debate in regard to strengthening and spasticity, controversy has also existed regarding how much physical exercise is necessary and beneficial for people with MS. One study demonstrated that because sustained physical activity increases core body temperature, vigorous exercise may actually be detrimental (Rusk, 1977). Recently, however, it has become more widely accepted that people with MS should be encouraged to participate in regular physical activity for improved cardiovascular conditioning as well as for the psychosocial benefits that often accompany physical activity (Petajan et al., 1996; Stuifbergen, 1997) and that individuals with MS can, in fact, exercise with no detrimental effects (Ponichtera-Mulcare, 1993). Nonetheless, care must be taken to control external variables such as the level of exertion and environmental temperature when designing exercise protocols for individuals with this diagnosis.

21.7.4 Myopathies

This section addresses strength and neurologic deficits resulting from myopathies; specifically, muscular dystrophies. There are various forms of muscular dystrophy, including Duchenne's muscular dystrophy, involving a defect or absence of the muscle protein, dystrophin, limb-girdle muscular dystrophy, also involving a defect in membrane-associated proteins of the muscle, and myotonic dystrophy, a disorder resulting from a decreased synthesis of muscle proteins. The primary characteristic of these disorders is progressive muscle weakness associated with significant muscle atrophy. It is evident that studies examining the effect of strength training in this category of diseases are particularly important. As seems to be the theme for strength training in neurologic disability, much debate exists throughout the literature as to whether strengthening exercises are beneficial to individuals with these types of pathology, and if so, what are the general guidelines for training? Most importantly, what are the functional outcomes of strengthening exercises in individuals with muscular dystrophies?

Several reports have cited increasing weakness in this group of individuals as a result of overworking of the muscles (Johnson and Braddom, 1971; Wagner et al., 1986). Wagner et al. (1986) have published an interesting case reporting the reversal of a decline in muscle strength experienced by a patient with muscular dystrophy. The subject under study had noticed a decrease in strength over a relatively short period of time. The investigators hypothesized this decrease in strength to be a result of an excessive workload for severely diseased muscles. Through serial testing with an isokinetic dynamometer, changes in lower extremity peak torques were measured over time and with treatment. The treatment program consisted of a specified amount of daily activity, rest, and therapeutic exercise. Isokinetic testing revealed an increase in peak torque production concurrent with modifications limiting daily work and exercise levels. Investigators concluded that documentation of a patient's total daily activity and the quantification of muscle strength allows for the prescription of a treatment program that will maximize muscle function. More recent literature reviews conclude that exercise prescription should be concurrent with the individual characteristics and type of disease (Ansved, 2001). That is, in slowly progressing muscular dystrophies, strength training at submaximal to near-maximal levels may actually be beneficial at improving force production. This, however, is not necessarily the case for individuals with a rapidly progressive form of the disease. In these cases, high-resistance, eccentric exercises are generally not recommended (Ansved, 2001). This is supported by a study done by Smith et al. (1987) where high-resistance inspiratory exercise in patients with severely impaired lung function was shown to be potentially hazardous. On the other hand, these patients may benefit greatly from low-resistance types of training regimens. In four boys with DMD, isokinetic strength training at submaximal levels was shown to improve strength in the absence of any negative side effects (DeLateur and Giaconi, 1979). Ultimately, Ansved (2001) recommends beginning strength training in the early stages of the disease, before muscle tissue has been replaced with connective tissue.

21.7.5 Traumatic Brain Injury

Survivors of a traumatic brain injury (TBI) are often left with a multitude of physical, cognitive, and psychosocial-behavioral-emotional impairments. Depending on the level of severity, individuals with TBI may exhibit high tone, contractures, spasticity, and/or paralysis. A search of strength + TBI yielded only about 10 hits on a national medical journal search engine. Unfortunately, none of these investigate specifically the impact of strength training on this population of individuals. However, though studies considering the impact of aerobic training and conditioning are also far and few between, they do offer insight as to the overall effects of exercise on individuals with TBI. One such study measured disability and handicap as it relates to level of exercise participation considered "exercisers" vs. "nonexercisers" in 240 persons with a TBI and 139 individuals without a disability (Gordon et al., 1998). Using a series of outcome measures, investigators found a widespread range of benefits of exercise for individuals with TBI. Specifically, the TBI exerciser group reported fewer cognitive symptoms, less impairment, elevated mood, and perceptions of better health. They hypothesized these results to be due, at least in part, to a more efficient use of oxygen, a beneficial effect on the production of brain-derived neurotrophic factor (BDNF). BDNF is a growth factor involved in supporting the function and survival of neurons, and exercise has been shown to increase its availability (Neeper et al., 1995). Interestingly, the exerciser group was predominantly comprised of individuals who had sustained more severe brain injuries when compared to the nonexerciser group. Authors concluded that severity of brain injury might have little effect on a person's ability to participate in exercise. Another study performed by Jankowski and Sullivan (1990) looked at the effect of training on the oxidative capacity and gait abilities in individuals with traumatic brain injuries. It is known that an injury to the central nervous system generally presents with a subnormal aerobic capacity and an increased cost of locomotion (Findley and Agre, 1988; Olgiati et al., 1988). Therefore, the purpose of the study was to improve these deficits in individuals with a TBI. While aerobic capacity did improve significantly, there was no evidence of an increased efficiency of locomotion. Authors argued, however, that alterations in oxidative capacity are helpful for reducing fatigability in performing everyday tasks.

21.8 Aging with a Disability

Among the 9.9 million persons 75 to 84 years old living in the United States, 6.3 million (63.7%) have disabilities; of those, 4.1 million or 41.5% report having severe disabilities (Kraus et al., 1996). Figure 21.1 illustrates the tendency toward increasing disability with age. During the normal aging process, there is an associated loss of muscle mass and strength, particularly in muscles of the upper body and those that are not involved in weight-bearing (Gleberzon and Annis, 2000). Changes are thought to occur at earlier ages in persons with disabilities than for persons without long-term disabilities (Kailes, 1996).

Before starting an exercise program, one needs to consider chronological age and the length of time living with the disability. Individuals with major physical disabilities living at least 30 years with a disability note some tenderness and soreness in joints, muscles, and tendons from compensating for impaired physical function from other parts of the body (Trieschmann, 1987). There is little known about how people with disabilities maintain or lose function as they get older and to what degree they experience age-associated changes (Kailes, 1996).

In a study of 297 individuals who had sustained their initial SCI 20 to 47 years earlier, 22% reported the need for physical assistance with daily activities increased with age (Gerhart et al., 1993). Regular strength training appears to lessen the extent of age-related muscular changes (Gleberzon and Annis, 2000), enabling individuals to make significant gains in functional ability. However, having a disability makes it that much harder to exercise and remain active throughout the aging process. Excessive weight gain and inactivity cause the individuals with disabilities to have accelerated and significant problems with muscle integrity as they age. There isn't much information on aging and disability, because long survival after an injury or diagnosis of a progressive or neurological condition is a relatively recent development.

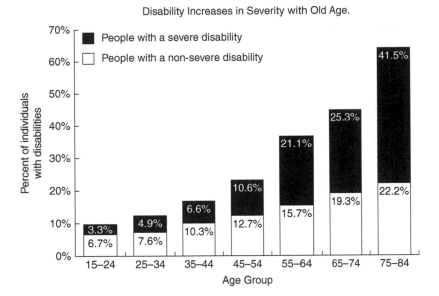

FIGURE 21.1

Percentage of individuals with disabilities according to age and level of severity. From Kraus, L.E., Stoddard, S., Gilmartin, D. 1996. Chartbook on Disability in the United States, prepared for U.S. Department of Education #H133D50017, National Institute on Disability and Rehabilitation Research, Washington, DC, http://www.infouse.com/disabilitydata/p16.textgfx.html.

Several reports have indicated that a majority of individuals with SCI have musculoskeletal problems in the upper extremities due to repetitive strain and overuse over the course of their lifetime. The prevalence of such problems as shoulder rotator cuff tears/tendonitis, shoulder pain, and carpal tunnel syndrome increase as the individual with SCI grows older (Aljure et al., 1985; Gellman et al., 1988; Nichols et al., 1979; Sie et al., 1992). Muscle atrophy is also a common occurrence among individuals with SCI that will continue to worsen with age and inactivity (Baldi et al., 1998). Changes in muscle fiber distribution occur early in paralyzed limbs. Muscle fiber distribution shifts towards Type IIB fibers (fast-twitch, fast fatigable) with the first 2 years of an SCI (Mohr et al., 1997). With electrically-induced muscle training of the gluteal, hamstring, and quadriceps muscles, a shift towards more fatigue-resistant muscle fibers (Type IIA) is possible. This change in muscle fiber distribution occurring in individuals with varying levels of SCI was independent of the number of SCI years. Thus, reversal of the effects of aging may be possible with electrically induced muscle training (Mohr et al., 1997).

21.9 Age-Related Disability

The population of elderly with disabilities is expected to increase at least 300% by 2049 (Boult et al., 1996). Older adults can minimize their risk of developing chronic disorders if they manage their symptoms, adhere to medication regimens, and maintain functional ability (Lorig et al., 1999). Light- to moderate-intensity physical activity can reduce risk factors of chronic diseases prevalent in the elderly population, improve psychological health, and optimize functional independence (O'Grady et al., 2000).

21.9.1 Parkinson's Disease

Parkinson's disease is a slowly progressive disease of the nervous system that has a peak onset of 60 years of age. Persons with Parkinson's disease usually exhibit rigidity of the limbs simulating arthritis, tremors, bradykinesia (difficulty initiating movement, slowness in movement, or incompleteness of movement), and postural instability due to impaired postural reflexes. The value of exercise training among individuals with Parkinson's disease is somewhat controversial. Some studies found either no improvement in motor disability or only short-term improvements (Comella et al., 1994; Gibberd et al., 1981). Other studies have shown that physical activity can influence mortality (Kuroda et al., 1992), improve motor disability (Formisano et al., 1992), and slightly lessen the risk of developing Parkinson's disease (Sasco et al., 1992). In Reuter et al.'s (1999) study, sixteen patients with mild to moderate idiopathic Parkinson's disease underwent a 16-week-long (2 days per week) intensive exercise program consisting of standardized sports activities. The program was designed specifically to increase muscle strength as a means to reduce trunk stiffness and rigidity as well as improve initiation of movements, gait, flexibility, coordination, and motor skills. Included in the program were roll exercises with a Pezzi ball and water exercises such as movement of arms and legs against water resistance and swimming exercises with boards/paddles. Clear improvements were noted in Parkinson's disease, specific motor disability, and sport-specific motor disability. At 6 weeks, upon termination of the program, only a minimal loss of regained motor skills was found.

21.9.2 Alzheimer's Disease

Dementia, a condition of declining mental abilities accompanied with symptoms such as memory loss, confusion, and inability to think through problems and follow instructions is common among the elderly. One of the leading causes for dementia is Alzheimer's disease, in which certain cells in the brain stop working and die. Being physically active in the younger years may protect individuals against Alzheimer's (Laurin et al., 2001), although little attention has been directed toward the effects of exercise on preventing or slowing down the progression of this disease. In addition to cognitive impairment, individuals with Alzheimer's disease often have poor physical health, have an increased risk for fall-related injuries/fractures, and are frail. Teri et al. (1998) conducted a study that documented the preliminary outcomes of a 12-week community-based program designed to increase balance, flexibility, strength, and endurance among 30 individuals with Alzheimer's disease. Caregivers were trained in the program and assisted the Alzheimer's subjects with the exercise program. The strength-training portion of the program mainly focused on lower-body strengthening. Specific exercises included knee extension and flexion, toe lifts and raises (ankle dorsi- and plantarflexion), marching (hip flexors), side lifts (abductors), and back leg lifts (extensors). Participants performed these strength exercises 3 days per week alternating with endurance-activity days (primarily walking). Flexibility training focused on the back, shoulder, hips, hamstrings, gastrocnemius/soleus/Achilles, neck, and hand. Adherence to the program was very good, indicating that Alzheimer's subjects and their caregivers can participate in a structured exercise program. While no specific data were presented on how effective the approach was for reducing physical disability in individuals with Alzheimer's, the caregivers reported that their Alzheimer's partners enjoyed doing the exercises and that mood and behavior was better on days when exercises were completed. In a different study, physical activity in the form of endurance exercises (walking and exercise bicycle) was found to improve cognitive function and nutritional status as well as decrease the risk of falls and behavioral problems (Rolland et al., 2000).

21.9.3 Arthritis and Sarcopenia

The aging process is associated with deterioration in several biologic systems. Arthritis and sarcopenia (reduced muscle mass) are common impairments in the older adult. Arthritis comes in many forms and presents with characteristics such as pain, swelling, limited range of motion, muscle weakness, instability, and poor cardiovascular fitness. A recent study indicated that quadriceps weakness may be a risk factor for knee osteoarthritis, further emphasizing the importance of regular exercise (Slemenda et al., 1997). Physical inactivity exacerbates disability in most patients with arthritis who adopt a sedentary lifestyle (Hettinger and Affable, 1994). Several studies provide evidence that involvement in regular physical activity can provide older adults with arthritis the same physical, psychologic, and functional benefits observed in the general population without exacerbating pain or disease activity (O'Grady et al., 2000). Specific exercise guidelines for range of motion (flexibility), strength training, and aerobic training for the older adult with joint disease are provided in O'Grady et al. (2000). Sarcopenia, a major cause of disability and frailty in the elderly, can be reversed with high-intensity progressive resistance exercise (Roubenoff, 2000).

21.10 Mental Impairments

Research investigating the benefits of strength training in individuals with mental impairments has mainly focused on mental illnesses and mental retardation.

21.10.1 Mental Illness

There is an association between physical fitness and psychological well-being. The level of physical fitness is negatively correlated with depression and anxiety and positively correlated with self-satisfaction and social adjustment (Heaps, 1978). A growing body of literature demonstrates that exercise promotes wellness, mental health, self-esteem, and improved body image. Psychiatric patients tend to be less physically fit and more obese than the general population (Hesso and Sorrenson, 1982; Pelham et al., 1993). Many psychiatric rehabilitation programs do not offer components that focus on the fitness needs of their patients (Pelham et al., 1993). Several fitness programs have been developed and the benefits researched for individuals carrying diagnoses such as schizophrenia, depression, anxiety, addiction (drug and alcohol), bipolar disorder, and chronic fatigue syndrome (Meyer and Brooks, 2000). Although many of the programs focus on aerobic endurance training, some programs have incorporated anaerobic training (muscle strengthening and flexibility exercises). Aerobic and nonaerobic forms of exercise appear to be effective in improving mood and depressive and anxiety symptoms in selected psychiatric patients (Chastin and Shapiro, 1987; Meyer and Brooks, 2000). Both aerobic and resistance exercise have also been shown to reduce disability and pain among individuals with knee osteoarthritis who also demonstrated high and low depressive symptomatology (Penninx et al., 2002). There appear to be no contraindications for individuals with psychiatric disorders to participate in exercise programs provided they are free from acute infectious and cardiovascular disease.

21.10.2   Mental Retardation

Body strength and competence in upper body muscular skills enables individuals with mental retardation (MR) to engage in recreational activities and perform activities of daily living and is a prerequisite for many available vocational opportunities (Pitetti et al., 1993). A positive correlation has been established between muscular strength and industrial work performance and level of independence in persons with MR (Nordgren, 1970, 1971; Nordgren and Backstrom, 1971). Obesity and cardiovascular disorders are prevalent in MR (Janicki and Jacobson, 1982; Janicki and MacEachron, 1984), and life expectancy is directly related to activity among individuals with profound MR (Eyman et al., 1990). Children with MR score lower on standardized physical fitness tests than children without MR (Rarick et al., 1970). More specific isometric muscle tests indicate that young adults with MR average 78% of expected upper body and 71% of expected lower body isometric strength when compared to adults without MR (Nordgren, 1970). Studies on adults with MR suggest that muscular strength and endurance decreases with age in adulthood (Fernhall, 1993). It is unknown if the underlying cause of poor strength levels among individuals with MR is retardation dependent, a result of inactivity, or caused by other factors.

Muscular strength and endurance training programs have been remarkably successful for individuals with MR (Fernhall, 1993). Specially structured public school physical education programs produced substantial chin-up and sit-up performance among children with MR (Giles, 1968; Solomon and Pangle, 1967). Similar results, although not as dramatic, have been discovered among adults with MR (Campbell, 1973; Corder, 1966;). Rimmer and Kelly (1991) found improved upper and lower body strength in adult men and women with MR who participated in a 9-week/2-days-per-week/1-hour-session traditional weight training program. Participants performed eight exercises (leg extension, leg curl, pectoral deck, shoulder abduction, pull over, pull-up, biceps curl, and triceps extension) using Nautilus Isokinetic Systems and a pull-up machine. The group performed 3 sets of exercises on each machine in a progressive manner (30, 60, and 70% of maximum weight that could be lifted in one repetition) for 8 to 10 repetitions per set. Improvements in strength at the end of the 9-week program ranged from 25 to 100% in each of the exercises. Resistance training of the upper body, concentrating on elbow extension, elbow flexion, shoulder transverse adduction, and shoulder abduction strength using surgical tubing resulted in increases in isometric strength among individuals with MR (Croce and Horvat, 1992). It is important to note that the strength training programs described above involved close supervision of each individual and often involved some kind of reward for adhering to the program.

21.11   Metabolic Disorders

The primary disorders under this category are diabetes and obesity.

21.11.1   Diabetes

Exercise training can lead to lower blood glucose levels and improve glucose intolerance and peripheral insulin sensitivity in diabetics (Young, 1995). There are two major types of diabetes. Type 1 (juvenile onset) is an insulin-dependent form of diabetes. Persons with this type require exogenous insulin by injections to sustain life (Young, 1995). Type 2 (maturity onset) is a non-insulin-dependent form of diabetes. Most patients with type 2 diabetes are obese, over the age of 35, and have high blood lipids and marked peripheral insulin resistance (Young, 1995). Young's (1995) summary of the literature on exercise and metabolic disorders yields an "exercise prescription" for individuals with both forms of diabetes. Exercises similar to those recommended for the general population can be pre-scribed to persons with type 1 or type 2 diabetes; however a thorough physical evaluation identifying any comorbid conditions (e.g., presence of proliferative retinopathy, sensory or autonomic neuropathy, and cardiovascular disease) and including a graded exercise stress test should first be performed. Type 1 diabetics who engage in anaerobic, power-type activities, such as weight lifting, show improvements in glucose disposal, but the main benefit is largely associated with an increase in muscle mass. Exercises to develop muscle endurance (e.g., high repetitions, low weight) can be included in the program, since they do not dramatically elevate blood pressure. Exercise intensity should be 50–70% of maximal aerobic capacity or 60–80% of maximal heart rate. Intensities below 30 to 50% of maximal capacity do not enhance glucose disposal (Maehlum and Pruett, 1973). Training sessions lasting more than 45 minutes increase the risk of hypoglycemia in type 1 diabetics. Persons prone to exercise-induced hypoglycemia should avoid exercise in the evening because delayed episodes of hypoglycemia can occur during sleep. Exercising in the morning before breakfast is recommended for optimal glycemic control. Individuals with type 1 and type 2 diabetes should perform exercises three times per week. Type 2 diabetics who are trying to lose weight should exercise more frequently (5 to 7 times per week). Exercises involving rhythmic contractions of large muscle groups are preferred for type 2 diabetics. There is substantial evidence that indicates that strength training can improve glucose homeostasis in men and that strength training as an intervention against insulin resistance may be just as effective as aerobic training (Hurley and Roth, 2000). A study focusing on gender differences in strength training on glucose and insulin responses found no significant improvements in older women, whereas improvements were signif-icant in older men (Napier et al., 1999).

21.11.2 Obesity

Obesity results when a person's caloric intake exceeds his or her energy expenditure, and the body stores the extra calories in fat cells present in adipose tissue. It is not clear whether obesity is caused strictly by fat accumulation or the metabolic disorders associated with obesity such as non-insulin-dependent diabetes, hypertension, hypercholesterolemia, hypertriglyceridemia, and hyperinsulinemia (NIHCD Panel, 1985). In any case, strength training exercises should be accompanied by low-impact, non-weight-bearing, high-inten-sity aerobic activity and a reduced caloric intake. Some studies however, have found that diet and aerobic exercise led to the same reduction in visceral fat in middle-aged obese men as diet and strength training (Hurley and Roth, 2000). It was estimated that the low-volume strength training program required less than one-third of the energy required for the aerobic training program. A possible explanation for the discrepancy in energy balance and the equivocal visceral fat loss could be an increase in resting metabolic rate with strength training (Hurley and Roth, 2000). To increase lean body mass, a broad regimen of 8 to 10 exercises involving all the major muscle groups of the body (arms, legs, hips, back, neck, and chest) should be performed at a moderate intensity in sets of 10 to 15 repetitions per exercise (Paluska, 2002). Strength training sessions should take place 2 to 3 times per week. Recent studies on controlling obesity in children included strength training as major component of a weight management program (Sothern et al., 2000; Sung et al., 2002). The exercises were performed "circuit style," working the upper body, back, legs, and abdominal muscles and could be performed in 20 to 30 minutes. At the completion of the programs, the children demonstrated greater lean body mass (Sung et al., 2002), a decrease in the low-density lipoprotein: high-density lipoprotein ratio (Sung et al., 2002), decrease in weight (Sothern et al., 2000), and a decrease in body mass index (Sothern et al., 2000). Many of the improvements noted after the completion of the weight management program described in Sothern et al. (2000) were still present at a 1-year followup examination.

21.12 Respiratory Disorders

Aerobic capacity and anaerobic threshold are abnormally low in individuals with chronic pulmonary diseases (Systrom et al., 1998; Tirdel et al., 1998). Most research related to exercise has dealt mostly with chronic obstructive pulmonary disease (COPD). COPD is an umbrella term used to describe airflow obstruction and is typically associated with chronic bronchitis and emphysema. Over 16 million people have been diagnosed with COPD, and an estimated 16 million more are undiagnosed (Petty, 1997). A common indication among persons with COPD is peripheral muscle weakness, particularly the muscles involved with ambulation (Bernard et al., 1998). A few reports have suggested that continued exercise intolerance observed with COPD may be attributed to peripheral muscle dysfunction (Storer, 2001). Hamilton et al. (1995) found that ergometer work rate in 785 pulmonary patients was 73% of that found in 919 healthy subjects, and that knee extensor strength was correlated with maximum work capacity in both groups of patients. Bernard et al. (1998) reported that maximum voluntary strength measures by the one-repetition maximum method for the quadriceps, pectoralis major, and latissimus dorsi muscles in 34 patients with moderate to severe COPD were 73, 84, and 84%, respectively, of 16 healthy subjects. Preferential loss of muscle strength in the lower extremity has also been documented (Bernard et al., 1998; Gosselink et al., 1996).

Resistance exercise training has recently been investigated as a means to reduce muscle dysfunction and improve responsiveness to exercise. Simpson et al. (1992) conducted a randomized controlled trial with a COPD group who performed resistance training consisting of 8-weeks/3 exercises/10 repetitions–3 sets/3 times per week with progressive increases in the percent loads of one-repetition maximum. Improvements in maximal voluntary strength ranged between 16 and 44% in the training group, with no significant change in the control group. In addition, endurance time, during cycling at 80% of pretraining peak work rates to fatigue, improved from 8.6 minutes to 15 minutes in the case group with no changes in the control group. Bernard et al. (1999) showed that individuals who performed aerobic activity (cycling 3 times per week) plus strength training demonstrated significant improvements in thigh muscle cross-sectional area and strength in the quadriceps and pectoralis major muscle groups, whereas individuals who performed only aerobic activity showed only slight improvement in quadriceps strength. The findings of these and other studies highlighted in Storer's (2001) review of the literature on COPD and strength suggest that resistance exercise training has the potential to improve functional performance and that COPD patients can maintain or further improve strength levels over a 3-year period. Storer (2001) also extrapolated from programs used to develop strength, power, and endurance in healthy individuals and the successful outcomes noted in studies on strength training and patients with COPD to develop a template for designing a program for COPD patients. Although further studies are needed to determine the components of a training program, Storer (2001) suggests starting with a single set of an exercise for each body part and adding additional sets as the patient progresses. Fewer rather than more repetitions per set (6 to 10) seem to be better tolerated by the COPD patient. Beginning the resistance training with 50 to 60% of the pretraining one-repetition maximum for major muscle groups is recommended to avoid excessively sore muscles. Loads as high as 80% of the pretraining one-repetition maximum have been successfully applied in COPD patients without ill effects. More specific recommendations and considerations regarding strength training programs among individuals with COPD can be found in Storer (2001).

21.13 Coronary Heart Disease

Heart disease is the leading cause of death of citizens in the United States and United Kingdom, accounting for 710,760 U.S. and 235,000 U.K. deaths in 2000 (National Vital Statistics Report, 2002; Peterson and Rayner, 2002). It is well established that aerobic exercise training can lead to substantial improvements in cardiovascular fitness, but the effects of strength training have not been typically viewed as a means of improving cardiovascular fitness. Traditional resistance exercise programs alone fail to demonstrate significant improvements in aerobic capacity among young and middle-aged individuals (Hurley and Roth, 2000). However, recent studies have shown that strength training may enhance cardiovascular fitness among elderly individuals (Hurley and Roth, 2000). In Hurley and Roth's (2000) review of the literature on strength training and elderly, individuals were noted to have decreased heart rates, blood pressure, rate pressure product (as an index of myocardial oxygen uptake), and improvements in treadmill walking endurance after participating in a strength training program. Vincent et al. (2002) examined the effect of 6 months of high- and low-intensity resistance exercise on aerobic capacity and treadmill time to exhaustion in adults between the ages of 60 and 83. The exercises included abdominal crunch, leg press, leg extension, leg curl, calf press, seated row, chest press, overhead press, biceps curl, seated dip, leg abduction, and lumbar extension. Participants in the "high-intensity" group performed eight repetitions at 80% of one-repetition maximum while the "low-intensity" group performed 13 repetitions at 50% of one-repetition maximum. A control group with comparative baseline strength was also included in the study. Loads were increased by 5% when their rating of perceived exertion dropped below 18 or "very hard" (Borg, 1982). Significant and similar improvements in aerobic capacity and treadmill time to exhaustion were found for both high- and low-intensity groups with no improvements in the control group. While the specific mechanisms for improvements in cardiovascular fitness from strength training are not well understood, they are believed to be related to changes to fiber type recruitment (e.g., greater rate of Type I and a reduced rate of Type II muscle fiber recruitment), less occlusion of blood flow, and increased lactate threshold (Hurley and Roth, 2000). Increased strength may allow aerobic exercise to be performed at a greater intensity or for a longer duration, resulting in improvements in aerobic capacity (Vincent et al., 2002).

21.14   Children with Disabilities

Children with disabilities need means of independent, safe, and efficient mobility as early as possible, just like their unimpaired peers (Tefft et al., 1999). Mobility is a basic variable contributing to normal child development. Children without disabilities will start standing at 9 months of age and will be walking by approximately 12 months of age. Mobility allows them to explore their environments, access objects to satisfy their curiosity, and participate in meaningful family and social activities (Butler, 1986; Douglas and Ryan, 1987). While it may be difficult for parents/guardians to accept wheelchair provision from a psychosocial perspective for a young child, this tendency quickly disappears when increased independence and participation are demonstrated by the child (Bottos et al., 2001).

However, in our current society ambulation is valued above mobility. When a child ambulates he is viewed as less disabled, even though he may not have the strength and endurance to truly achieve his participatory goals. For many children with various diagnoses (e.g., cerebral palsy, spina bifida), ambulation may require increased effort, as measured by energy expenditure (Rose et al., 1990). Speed of ambulation may also be seriously limited in pathological gait, for reasons of impaired equilibrium reactions and decreased strength and endurance (Duffy et al., 1996).

There is a tendency to assume that children are stronger and can tolerate more exercise than adults. However, energy expenditure of walking appears to be higher for smaller children than for adolescents (Waters et al., 1983). This means that smaller children will tire just as easily as their larger counterparts, if not more. This fact is often neglected, and children are pushed to their limits of endurance during walking exercises. Consequently, focusing children with mobility deficits on ambulation may take away their chances of participating in many other activities.

Manual wheelchair propulsion has been shown to require less effort than ambulation for children with pathological gait (Evans and Tew, 1981; Williams et al., 1983). However, even manual wheelchair propulsion may require 16% more effort than ambulatory locomotion of unimpaired children (Luna-Reyes et al., 1988). One contributing factor to this is the design of pediatric manual wheelchairs. Currently, even the lightest early intervention wheelchair weighs 15 lb. However, growth-adjustable frames, which weigh close to 40 lb, are more frequently prescribed for children. Due to the importance of accommodating growth, the seating systems used on pediatric wheelchairs are usually also modular and adjustable, further increasing weight. This means that children need to propel systems that closely approximate their body weight (equal to adults having to propel 150 to 180-lb manual wheelchairs).

Upper extremity repetitive strain injuries as a consequence of manual wheelchair propulsion have been widely studied in adults (Aljure et al., 1985; Gellman et al., 1988; Nichols et al., 1979; Sie et al., 1992). At this point, studies of this sort have not been conducted with children. However, it would be beneficial to start prevention as early as possible by improved manual wheelchair design for the pediatric population.

Due to the above problems, power wheelchairs are often the ideal solution for children with disabilities to achieve independence (Wiart and Darrah, 2002). By utilizing powered mobility, children can get to places and keep up with their peers, improving psychosocial well-being. Utilizing the most optimal movement strategies, children can increase participation as well as preserve energy for increased efficiency of therapeutic activities.

21.15 General Recommendations for Developing and Implementing a Strength Training Program

With over 61 million disabling conditions reported by Americans with disabilities (LaPlante, 1996), recommendations for improving strength must consider an individual's specific condition(s) and particular capacities. Determinations of which muscles are functioning properly and which are weak and/or overworked provide information required to forge a proper plan of action (Rimmer, 2001). The professional team must remember that each disability and lifestyle is different, so individualized physical fitness plans must be implemented. The professional team must also consider what activities are performed on a daily basis and the person's goals. When working with a person with a disability, considerations about the exercise selection and duration must be kept in mind. Overworking a person with a disability may cause muscle soreness to the point that daily activities cannot be performed, therefore defeating the purpose of exercise in the first place (Rimmer, 2001). Before starting a person with a disability on an exercise regimen, the professional team must consider the following:

1. Obtain a recommendation from the individual's physician. The recommendation should be specific and focus on the health/fitness goals of the individual and concomitant risk factors (Cooper et al., 1999).

2. Consider potential barriers such as transportation to a suitable exercise facility and the possibility of performing exercises at home because of the lesser cost, convenience, and potential to promote greater independence (Cooper et al., 1999).

3. Evaluate joint integrity, muscle weakness (paying particular attention to asymmetries), muscle imbalances, degree and areas of paralysis, spasticity or tone, presence of pain or arthritis, and cardiac status (Cooper et al., 1999; Dodd et al., 2002; Fowler et al., 2001; Rimmer, 2001; van der Ende et al., 2000).

4. Incorporate exercise strategies that protect any compromised joints and encourage muscle balance between the antagonistic and protagonistic muscle groups (Burnham et al., 1993) and highlight weight-bearing activities even if adaptive equipment is required (Cooper et al., 1999).

5. Establish short-term goals, promote positive reinforcement through periodic feedback of results, document individual achievements on progress charts, and compare an individual's achievements to past status rather than to established norms (Cooper et al., 1999).

6. Teach persons with disabilities about the importance of alternating rest and activity to minimize fatigue and maximize conditioning effects (Cooper et al., 1999).

7. Ensure that individual is receiving proper nutrition (Cooper et al., 1999).

8. Develop progressive programs that are sufficiently practical, accessible, and compatible with individual's life-style, so that exercise will be continued on a long-term basis (Cooper et al., 1999).

Unfortunately, there are no specific strength training guidelines for individuals with disabilities such as the American College of Sports Medicine's (ACSM) resistance exercise guidelines for healthy sedentary and physically active adults, elderly persons, people with cardiovascular disease, and children (Feigenbaum and Pollock, 1999). On the other hand, many health care providers advise their patients with disabilities to follow these guidelines, and many studies investigating strength training exercise protocols for individuals with disabilities have incorporated ACSM guidelines (Busch et al., 2002; O'Grady et al., 2000; Pitetti et al., 1993; Rimmer and Kelly, 1991).

The strength training regimen can be accomplished using various devices and techniques but must consider the capabilities and limitations of the individual. The traditional methods in the past have been the use of free weights and standard weight machines. More recently, computer-controlled (Lido, Biodex, etc.) machines that control the velocity (isokinetic and isometric) of desired movements have been implemented in the fitness world. Rehabilitation centers utilize even more unconventional objects such as variable resistive bands, tubing, and balls that also provide a means of enhancing strength. Some fitness centers offer adapted equipment designed to be used by those with physical disabilities (e.g., wheelchair users) (see Environmental Barriers and Adapted Fitness Equipment). If a fitness center is intimidating or cost and transportation to the center is a problem, then using resistive tubing or banding in the comfort of home may be the desired choice; just note that strength gains will be limited using this type of method. The common technique to improve endurance is to perform high repetitions (12 to 25) with low weight amounts utilizing circuit training (one set per exercise). Therefore, the lower the repetitions (6 to 8) with heavier weights using periodization training (multiple set exercise) produces increases in strength/power. However, many people with disabilities should not follow the traditional mode of training for strength, at least not initially. The person with a disability faces strength challenges already on a daily basis, and training just the customary way for strength would fatigue most (Davis et al., 1981; Mayer et al., 1999). Designing a short-term exercise program may be the key for success when working with people with disabilities. Feigenbaum and Pollock (1999) suggest that a single-set program consisting of 8 to 10 exercises with 15 repetitions for a minimum of 2 days per week may be beneficial for healthy individuals and most people with chronic diseases as long as doctor approval is obtained beforehand.

Stretching is an important part of overall health, and some believe that stretching increases flexibility and plays an integral part in the prevention of further injury (Baechle, 1994; Patrick et al., 2002). Performing a stretching schedule at least 2 to 3 times per week will increase flexibility (Baechle, 1994). When working with a person who needs assistance (for example, asymmetrical weakness), passive stretching may be needed. Passive stretching takes place when the professional or professional trainer supplies the needed force to stretch the muscle (Baechle, 1994). An active stretch occurs when the person can go through the range of motion on his/her own without assistance. The professional or person with a disability should move the limb through a range of motion to a point where a stretch is felt on the muscle. Execute a constant hold and stretch for 10 to 30 seconds with no bouncing (ballistic movement). Performing a static stretch such as this will reduce the chances of further injuring the muscle.

Below are a few techniques that people with disabilities may be able to perform on their own without assistance.

For the upper body:

Straight arms behind back and seated lean-back (deltoids and pectoralis major)

Behind-neck stretch "chicken wing" (triceps and latissimus dorsi)

Cross arm in front of chest (latissimus dorsi and teres major)

Arms straight up above head (latissimus dorsi and wrist flexors)

Side bend with straight or bent arms (external oblique, latissimus dorsi, serratus anterior and triceps), to alleviate some of the tightness throughout the upper body

Curtis et al. found that stretching the anterior deltoid, chest, and biceps may reduce some shoulder pain intensity for people who use manual wheelchairs (Curtis et al., 1999). Figoni supports this finding by suggesting that the chest and anterior part of the shoulder should be stretched at least twice a week with a strengthening factor (Curtis et al., 1999; Figoni, 1993).

For the lower body:

Spinal twist (internal and external oblique and spinal erectors)

Semi-leg straddle (spinal erectors)

Forward lunge (iliopsoas, rectus femoris)

Supine knee flex (hip extensors, gluteus maximus, and hamstring)

Side quadriceps stretch (quadriceps and iliopsoas), kneeling quadriceps stretch (quadriceps)

Sitting toe touch (hamstring, spinal erectors, and gastrocnemius)

Semistraddle (gastrocnemius, hamstring, and spinal erectors)

Straddle (gastrocnemius, hamstring, spinal erectors, adductors, and sartorius)

Butterfly (adductors and sartorius)

Bent-over toe raise (gastrocnemius and soleus)

Adhering to a stretching ritual will enhance the elasticity of the lower body. Finding an appropriate stretching routine is essential to achieving desired results for people with disabilities. For instance, people who suffer from cerebral palsy have a tight adductor muscle that pulls at the hip and in some cases causes dislocation. Choosing the correct stretch for a person with cerebral palsy may alleviate any future pain experiences due to chronic hip problems (Rimmer, 2001).

21.16 Environmental Barriers and Adapted Fitness Equipment

Given their special challenges to participation in strength programs, persons with disabilities have a few nonnegotiable requirements. Facilities and their programs must be architecturally and attitudinally accessible. The Accessibility Guidelines for buildings and facilities (U.S. Congress, 1991) offers standards for doorways, clearance, and hardware placement. While not mandated by law, attitudinal accessibility will have a direct relationship on participation as well. A programmatic atmosphere of welcome and inclusion and the presence of knowledgeable personnel will provide the "make or break" factor for participation for many individuals — at least when starting into strengthening programs or when disability is new.

In general, facilities must comply with the level of accessibility mandated by ADA (ADA, 1990), which covers parking, entrances, passageways, showers, and locker facilities. Gyms and weight rooms need to permit safe passage when individuals use crutches, canes, rollators, or wheelchairs or have wide-based gaits or low vision. Though the ADA mandates a level of general accessibility, facilities should have a process for addressing the feasibility of individual requests. Requests should be limited by safety for all patrons and due financial hardship.

Some individuals may be able to accomplish transfers from wheelchairs to the seats of weight equipment or may be able to handle some level of free weights while sitting in their wheelchair or on a floor mat. Even if the individual has a disability that allows use of typical weight equipment, some modifications to technique or position may be needed. When individuals have greater mobility impairment and consistently rely on wheelchairs, more comprehensive solutions must be identified. At present, it is the rare community athletic facility that has a wheelchair-accessible weight machine. Such equipment has mostly been seen in rehabilitation-oriented or disability-specific facilities rather than in a public site. Some individuals with financial resources may have this kind of equipment at home, and some progressive facilities are making this commitment.

Several manufacturers meet the demand for adaptive weight training equipment. Uppertone*, manufactured by GPK, offers a product designed for a person with tetraplegia, i.e., for individuals with poor grip. This means that resistance can be changed by sliding a lever rather than having to add or remove weights, and the frame can be modified to change dimensions or planes of movement for a variety of exercises. It is an ideal gym for home use. In contrast the Equalizer, 1000**, manufactured by Helm Distributing, can be set up for several users at one time — including persons without disabilities. This larger piece of equipment can be approached from several sides with stations like the seated overhead press, rowing, latissimus pull, and vertical butterfly, available simultaneously. In a variation, the VersaTrainer***, manufactured by Access to Recreation, offers resistance from "power rods" which allow for smooth progressive arcs of movement that range in resistance from 3 to 300 lb. Variable resistance arm ergometers, like the Saratoga Cycle****, offer the opportunity to gain moderate increases in strength by increasing resistance but contribute more to endurance and aerobic fitness.

21.17 Summary

Increasing or maintaining strength is important for everyone, but for individuals with a disability it becomes a necessity for accomplishing everyday activities and independence. There is no question that engaging in a strength training program can help individuals of varying disability types and degrees of severity to gain function and the ability to carry out desired activities. While the long-term benefits of engaging in a strength training program remain under investigation, there are definite short-term benefits that range from reducing the incidence of falls and fracture to gaining the physical capacity to engage in community and vocational opportunities. Besides the programmatic obstacles that have to be overcome, environmental barriers limit prospects to maintain strength. The lack of accessibility to various facilities (fitness centers, amusement parks, train stations, restaurants, shopping malls) limits the options available for venturing out and exercising muscles on a daily basis. Working with a team of trained professionals (medical doctor, physical therapist, athletic trainer, and certified exercise specialist) will aid people with disabilities

* Uppertone, GPK Inc., 535 Floyd Smith Drive, El Cajon, CA 92020; phone: (619) 593-7381 or (800) 468-8679; http://www.gpk.com/.
** The Equalizer, Helm Distributing, 911 Kings Point Road, Polson, MT 59860; phone: (406) 883-2147; http://www.accesstr.com/.
*** The VersaTrainer, Access to Recreation, 8 Sandra Court, Newbury Park, CA 91320; phone: (800) 634-4351; http://www.accesstr.com.
**** Saratoga Cycle, Rand-Scott Co., 401 Linden Center Drive, Fort Collins, CO 80524; Toll-Free: (800) 467-7967, Phone: (970) 484-7967; http://www.saratoga-intl.com/saratoga/.

to seek out more possibilities for preserving their strength and the most appropriate strength training regimen based on their disability type, abilities and needs (Edwards, 1996; Maurer et al., 1998; Noreau and Shepard, 1995; Rimmer, 2001). Formulation of an exercise plan should incorporate a brief learning period, so the person with the disability has time to get used to the stress placed upon the body, which will help to avoid overtaxing the body in the beginning of the exercise process (Davis et al., 1981). Knowing the problems that are associated with a certain disability type will make things easier for the individual and professional team, resulting in a successful workout plan.

References

Aljure, J., Eltorai, I., Bradley, W.E., Lin, J.E., and Johnson, B. 1985. Carpal tunnel syndrome in paraplegic patients, *Paraplegia*, 23, 182–186.

Ambrosio, F., Boninger, M.L., Fay, B.T., Souza, A.L., Fitzgerald, S.G., Koontz, A.M., and Cooper, R.A. 2002. A fatigue analysis during wheelchair propulsion in patients with multiple sclerosis, *Proceedings of the RESNA 2002 Annual Conference*, Minneapolis, MN, June 28–July 1, 2002, pp. 282–284.

American Spinal Injury Association and International Medical Society of Paraplegia 1996. *Reference Manual for the Standards for Neurologic and Functional Classification of Spinal Cord Injury* (Chicago: American Spinal Injury Association).

Americans with Disabilities Act (ADA). 1990. Public Law 336 of the 101st Congress, enacted July 26, 1990. U.S. Department of Justice.

Ansved, T. 2001. Muscle training in muscular dystrophies, *Acta Physiol. Scand.*, 171, 359–366.

Baechle, T.R. 1994. *Essentials of Strength Training and Conditioning*, National Strength and Conditioning Association (Champaign, IL: Human Kinetics).

Baldi, J.C., Jackson, R.D., Moraille, R., and Mysiw, W.J. 1998. Muscle atrophy is prevented in patients with acute spinal cord injury using electrical stimulation, *Spinal Cord*, 36(7), 463–469.

Ballinger, D.A., Rintala, D.H., and Hart, K.A. 2000. The relation of shoulder pain and range-of-motion problems to functional limitations, disability, and perceived health of men with spinal cord injury: a multifaceted longitudinal study, *Arch. Phys. Med. Rehabil.*, 81(12), 1575–1581.

Bayley, J.C. 1987. The weight-bearing shoulder. The impingement syndrome in paraplegics, *J. Bone Joint Surg. Am.*, 69(5), 676–678.

Bernard, P.L. and Codine, P. 1997. Isokinetic shoulder of paraplegics: observation of global and specific muscle ratio, *Int. J. Rehabil. Res.*, 20(1), 91–98.

Bernard, S., Leblanc, P., Whittom, F., Carrier, G., Jobin, J., Belleau, R., and Maltais, F. 1998. Peripheral muscle weakness in patients with chronic obstructive pulmonary disease, *Am. J. Respir. Crit. Care Med.*, 158, 629–634.

Bernard, S., Whittom, F., Leblanc, P., Jobin, J., Belleau, R., Berube, C., Carrier, G., and Maltais, F. 1999. Aerobic and strength training in patients with chronic obstructive pulmonary disease, *Am. J. Respir. Crit. Care Med.*, 159, 896–901.

Bertocci, G.E., Munin, M.C., Frost, K.L., Burdett, R., Fitzgerald, S.G., and Wassinger, C.A. 2004. Isokinetic performance following total hip replacement (THR): a comparison of THR patients and healthy controls, *Am. J. Phys. Med. Rehabil.*, 83, 1–9.

Bobath, B. 1990. *Adult Hemiplegia. Evaluation and Treatment*, 3rd ed. (Woburn, MA: Butterworth Heinemann).

Bohannon, R.W. and Walsh, S. 1992. Nature, reliability, and predictive value of muscle performance measures in patients with hemiparesis following stroke, *Arch. Phys. Med. Rehabil.*, 73(8), 721–725.

Borg, G. 1982. A category scale with ratio properties for intermodal and interindividual comparisons, in *Psychophysical Judgment and the Process of Perception*, H.-G. Geisler and P. Petzold, Eds. (Berlin: VEB Deutscher Verlag der Wissenschaft), pp. 25–34.

Boult, C., Altmann, M., Gilbertson, D., Yu, C., and Kane, R.L. 1996. Decreasing disability in the 21st century: the future effects of controlling six fatal and nonfatal conditions, *Am J. Public Health*, 86(10), 1388–1393.

Bottos, M., Bolcati, C., Sciuto, L., Ruggeri, C., and Feliciangel, A. 2001. Powered wheelchairs and independence in young children with tetraplegia, *Dev. Med. Child Neurol.*, 43(11), 769–777.

Burnham, R.S., May, L., Nelson, E., Steadward, R., and Reid, D.C. 1993. Shoulder pain in wheelchair athletes. The role of muscle imbalance, *Am. J. Sports Med.*, 21(2), 238–242.

Busch, A., Schachter, C.L., Peloso, P.M., and Bombardier, C. 2002. Exercise for treating fibromyalgia syndrome, *Cochrane Database Syst. Rev.*, 3, CD003786.

Butler, C. 1986. Effects of powered mobility on self-initiated behaviors of very young children with locomotor disability, *Dev. Med. Child Neurol.*, 28(3), 325–332.

Campbell, J. 1973. Physical fitness and the MR: a review of research, *Ment. Retard.*, 11, 26–29.

Campbell, M.H., Signorile, J.F., Suidmak, P., Miller, P., Puhl, J., and Perry, A. 1995. The correlation between isokinetic strength measures and functional performance in elderly population, *Med. Sci. Sports Exerc.*, 27, S232.

Centers for Disease Control and Prevention (CDC), National Center for Injury Prevention and Control, Last review, October 2003. Traumatic Brain Injury Fact Sheet http://www.cdc.gov/ncipc/factsheets/tbi.htm.

Chastin, P.B. and Shapiro, G.E. 1987. Physical fitness program for patients with psychiatric disorders: a clinical report, *Phys. Ther.*, 67(4), 545–548.

Cohen, R.B. and Williams, G.R. 1998. Impingement syndrome and rotator cuff disease as repetitive motion disorders, *Clin. Orthop.*, 351, 95–101.

Comella, C.L., Stebbins, G.T., Brown-Toms, N., and Goetz, C.G. 1994. Physical therapy and Parkinson's disease, *Neurology*, 44, 376–378.

Cooper, R. A. 1998. *Wheelchair Selection and Configuration* (New York: Demos Medical Publishing).

Cooper, R.A., Quatrano, L.A., Axelson, P.W., Harlan, W., Stineman, M., Franklin, B., Krause, J.S., Bach, J., Chambers, H., Chao, E.Y.S., Alexander, M., and Painter, P. 1999. Research on physical activity and health among people with disabilities: a consensus statement, *J. Rehabil. Res. Dev.*, 36(2), 142–154.

Corder, W.O. 1966. Effects of physical education on the intellectual, physical and social development of educable mentally retarded boys, *Exceptional Child.*, 32, 357–364.

Croce, R. and Horvat, M. 1992. Effects of reinforcement based exercise on fitness and work productivity in adults with mental retardation, *APAQ*, 9(2), 148–178.

Curtis, K.A., Tyner, T.M., Zachary, L., Lentell, G., Brink, D., Didyk, T., Gean, K., Hall, J., Hooper, M., Klos, J., Lesina, S., and Pacillas, B. 1999. Effect of a standard exercise protocol on shoulder pain in long-term wheelchair users, *Spinal Cord*, 37(6), 421–429.

Damiano, D.L. and Abel, M.F. 1998. Functional outcomes of strength training in spastic cerebral palsy, *Arch. Phys. Med. Rehabil.*, 79, 119–125.

Damiano, D., Vaughan, C.L., and Abel, M.F. 1995. Muscle response to heavy resistance exercise in spastic cerebral palsy, *Dev. Med. Child Neurol.*, 75, 658–671.

Davis, G.M., Kofsky, P.R., Kelsey, J.C., and Shephard, R.J. 1981. Cardiorespiratory fitness and muscular strength of wheelchair users, *CMAJ*, 125(12), 1317–1323.

Davis, G.M. and Shephard, R.J. 1990. Strength training for wheelchair users, *Br. J. Sports Med.*, 24(1), 25–30.

DeLateur, B.J. and Giaconi, R.M. 1979. Effect on maximal strength of submaximal exercise in Duchenne muscular dystrophy, *Am. J. Phys. Med.*, 58, 26–36.

Dodd, K.J., Taylor, N.F., and Damiano, D.L. 2002. A systematic review of the effectiveness of strength-training programs for people with cerebral palsy, *Arch. Phys. Med. Rehabil.*, 83, 1157–1164.

Douglas, J. and Ryan, M. 1987. A preschool severely disabled boy and his powered wheelchair: a case study, *Child Care Health Dev.*, 13(5), 303–309.

Duffy, C.M., Hill, A.E., Cosgrove, A.P. Corry, I.S., and Graham, H.K. 1996. Energy consumption in children with spina bifida and cerebral palsy: a comparative study, *Dev. Med. Child Neurol.*, 38(3), 238–243.

Duran, F.S, Lugo, L., Ramirez, L., and Eusse, E. 2001. Effects of an exercise program on the rehabilitation of patients with spinal cord injury, *Arch. Phys. Med. Rehabil.*, 82(10), 1349–1354.

Edwards, P.A. 1996. Health promotion through fitness for adolescents and young adults following spinal cord injury, *SCI Nurs.*, 13(3), 69–73.

Ettinger, W.H. and Afable, R.F. 1994. Physical disability from knee osteoarthritis: the role of exercise as an intervention, *Med. Sci. Sports Exerc.*, 26(12), 1435–1440.

Evans, E.P. and Tew, B. 1981. The energy expenditure of spina bifida children during walking and wheelchair ambulation, *Z. Kinderchir.*, 34(4), 425–427.

Eyman, R.K., Grossman, H.J., Chaney, R.H., and Call, T.L. 1990. The life expectancy of profoundly handicapped people with mental retardation, *N. Engl. J. Med.*, 323, 584–589.

Fay, B.T. 2001. Influence of Dynamical, Clinical, and Neuromotor Measures in Evaluating Individuals with Multiple Sclerosis for Manual Wheelchair Use, unpublished doctoral thesis, University of Pittsburgh, Pittsburgh, PA.

Feigenbaum, M.S. and Pollock, M.L. 1999. Prescription of resistance training for health and disease, *Med. Sci. Sports Exerc.*, 31(1), 38–45.

Fernhall, B. 1993. Physical fitness and exercise training of individuals with mental retardation, *Med. Sci. Sports Exerc.*, 24(4), 442–450.

Figoni, S. F. 1993. Exercise responses and quadriplegia, *Med. Sci. Sports Exerc.*, 25(4), 433–441.

Findley, T.W. and Agre, J.C. 1988. Ambulation in the adolescent with spina bifida. II. Oxygen cost of mobility, *Arch. Phys. Med. Rehabil.*, 69, 855–861.

Formisano, R., Pratesi, L., Modarelli, F.T., Bonifati, V., and Meco, G. 1992. Rehabilitation and Parkinson's disease, *Scand. J. Rehab. Med.*, 24, 157–160.

Fowler, E.G., Ho, T.W., Nwigwe, A.I., and Dorey, F.J. 2001. The effect of quadriceps femoris muscle strengthening exercises on spasticity in children with cerebral palsy, *Phys. Ther.*, 81, 1215–1223.

Gellman, H., Sie, I., and Waters, R.L. 1988. Late complications of the weight-bearing upper extremity in the paraplegic patient, *Clin. Orthop.*, 233, 132–135.

Gerhart, K.A., Bergstrom, E., Charlifue, S.W., Menter, R.R., and Whiteneck, G.G. 1993. Long-term spinal cord injury: functional changes over time, *Arch. Phys. Med. Rehabil.*, 74(10), 1030–1034.

Gibberd, F.B., Page, N.G.R., Spencer, K.M., Kinnear, E., and Hawksworth, J.B. 1981. Controlled trial of physiotherapy and occupational therapy for Parkinson's disease, *Br. Med. J.*, 282, 1196.

Giles, M.T. 1968. Classroom research leads to physical fitness for retarded youth, *Ed. Training Ment. Retard.*, 3, 67–74.

Gleberzon, D.C. and Annis, R.S. 2000. The necessity of strength training for the older patient, *J. Can. Chiropr. Assoc.*, 44(2), 98–102.

Gordon, W.A., Sliwinski, M., Echo, J., McLoughlin, M., Sheerer, M., and Meili, T.E. 1998. The benefits of exercise in individuals with traumatic brain injury: a retrospective study, *J. Head Trauma Rehabil.*, 13(4), 58–67.

Gosselink, R., Troosters, T., and Decramer, M. 1996. Peripheral muscle weakness contributes to exercise limitation in COPD, *Am. J. Respir. Crit. Care Med.*, 153, 976–980.

Greenough, W.T. 1988. *Message to Mind* (Massachusetts: Sinauer Associates Inc.).

Groah, S.L. and Lanig, I.S. 2000. Neuromusculoskeletal syndromes in wheelchair athletes, *Semin. Neurol.* 20(2), 201–208.

Häkkinen, A., Häkkinen, K., Hannonen, P., and Alen, M. 2001. Strength training induced adaptations in neuromuscular function of premenopausal women with fibromyalgia: comparison with healthy women, *Ann. Rheum. Dis.*, 60, 21–26.

Häkkinen, A., Sokka, T., Kotaniemi, A., Kautiainen, H., Jappinen, I., Laitinen, L., and Hannonen, P. 1999. Dynamic strength training in patients with early rheumatoid arthritis increases muscle strength but not bone mineral density, *J. Rheumatol.*, 26(6), 1257–1263.

Hamilton, A.L., Killian, K.J., Summers, E., and Jones, N.L. 1995. Muscle strength, symptom intensity, and exercise capacity in patients with cardiorespiratory disorders, *Am, J. Respir. Crit. Care Med.*, 152, 2021–2031.

Heaps, R.A. 1978. Relating physical and psychological fitness: a psychological point of view, *J. Sports Med. Phys. Fitness*, 18, 399–408.

Hesso, R. and Sorensen, M. 1982. Physical activity in the treatment of mental disorders, *Scand. J. Soc. Med., Suppl.*, 29, 259–264.

Hurley, M.V. 1999. The role of muscle weakness in the pathogenesis of osteoarthritis, *Rheum. Dis. Clin. North Am.*, 25(2), 283–297.

Hurley, B.F. and Roth, S.M. 2000. Strength training in the elderly: effects on risk factors for age-related diseases, *Sports Med.*, 30(4), 249–268.

Jacobs, P.L. and Nash, M.S. 2001. Modes, benefits, and risks of voluntary and electrically induced exercise in persons with a spinal cord injury, *J. Spinal Cord Med.*, 24(1), 10–18.

Jacobs, P.L., Nash, M.S., and Rusinowski, J.W. 2001. Circuit training provides cardiorespiratory and strength benefits in persons with paraplegia, *Med. Sci. Sports Exerc.*, 33(5), 711–717.

Janicki, M.P. and Jacobson, J.W. 1982. The character of developmental disabilities in New York State: preliminary observations, *Int. J. Rehabil. Res.*, 5, 191–202.

Janicki, M.P. and MacEachron, A.E. 1984. Residential, health, and social service needs of elderly developmentally disabled persons, *Geronologist*, 24, 128–137.

Jankowski, L.W. and Sullivan, S.J. 1990. Aerobic and neuromuscular training: effect on the capacity, efficiency, and fatigability of patients with traumatic brain injuries, *Arch. Phys. Med. Rehabil.*, 71, 500–504.

Janssen, T.W., van Oers, C.A., van der Woude, L.H., and Hollander, A.P. 1994. Physical strain in daily life of wheelchair users with spinal cord injuries, *Med. Sci. Sports Exerc.* 26(6), 661–670.

Johnson, E.W. and Braddom, R. 1971. Over-work weakness in facioscapulohumeral muscular dystrophy, *Arch. Phys. Med. Rehabil.*, 52, 333–336.

Jones, K.D., Burckhardt, C.S., Clark, S.R., Bennett, R.M., and Potempa, K.M. 2002. A randomized clinical trial of muscle strengthening versus flexibility training in fibromyalgia, *J. Rheumatol*, 29(5), 1041–1048.

Kailes, J.I. 1996. Resource List: Wellness, Self-Care, Exercise and Aging with Disability, prepared for Research and Training Center on Aging with a Disability, Downey, CA, http://www.usc.edu/dept/gero/RRTConAging/paper1.html.

Karni, A., Meyer, G., Rey-Hipolito, C., Jezzard, P., Adams, M.M., Turner, R., and Ungerleider, L.G. 1998. The acquisition of skilled motor performance: fast and slow experience-driven changes in primary motor cortex, *Proc. Natl. Acad. Sci. U.S.A.*, 95, 861–868.

Kent-Braun, J.A., Ng, A.V., Castro, M., Weiner, M.W., Gelinas, D., Dudley, G.A., and Miller, R.G. 1997. Strength, skeletal muscle composition, and enzyme activity in multiple sclerosis, *J. Appl. Physiol.*, 83(6), 1998–2004.

Kraus, L.E., Stoddard, S., and Gilmartin, D. 1996. Chartbook on Disability in the United States, prepared for U.S. Department of Education #H133D50017, National Institute on Disability and Rehabilitation Research, Washington, DC, http://www.infouse.com/disabilitydata/p16.text-gfx.html.

Krupp, L.B., Alvarez, L.A., LaRocca, N.G., and Scheinberg, L.C. 1988. Fatigue in multiple sclerosis, *Arch. Neurol.*, 45(4), 435–437.

Kuroda, K., Tatara, K., Takatorige, T., and Shinsho, F. 1992. Effect of physical exercise on mortality in patients with Parkinson's disease, *Acta Neurol. Scand.*, 86, 55–59.

Lambert, C.P., Archer, R.L., and Evans, W.J. 2001. Muscle strength and fatigue during isokinetic exercise in individuals with multiple sclerosis, *Med. Sci. Sports Exerc.*, 33(10), 1613–1619.

Lance, J.W. 1980. Symposium synopsis, in *Spasticity: Disordered Motor Control*, R.G. Feldman, R.R. Young, and W.P. Koella, Eds. (Chicago, IL: Year Book Medical), pp. 485–494.

LaPlante, M. P. 1996. Health conditions and impairments causing disability, *Disability Statistics Abstract*, No. 16, published by U.S. Department of Education, National Institute on Disability and Rehabilitation Research, http://dsc.ucsf.edu/UCSF/pdf/ABSTRACT16.pdf.

Landau, W.M. and Sahrmann, S.A. 2002. Preservation of directly stimulated muscle strength in hemiplegia due to stroke, *Arch. Neurol.*, 59, 1453–1457.

Laurin, D., Verreault, R., Lindsay, J., MacPherson, K., and Rockwood, K. 2001. Physical activity and risk of cognitive impairment and dimentia in elderly persons, *Arch. Neurol.*, 58(3), 498–504.

Livneh, H. 1991. A unified approach to existing models of adaptation to disability: a model of adaptation, in *The Psychological and Social Impact of Disability*, R.P. Marinelli and A.E. Dell Orto, Eds. (New York: Springer Publishing Company), pp. 111–138.

Lorig, K.R., Sobel, D.S., Stewart, A.L., Brown, B.W., Jr., Bandura, A., Ritter, P., Gonzalez, V.M., Laurent, D.D., and Holman, H.R. 1999. Evidence suggesting that a chronic disease self-management program can improve health status while reducing hospitalization. A randomized trial, *Med. Care*, 37(1), 5–14.

Luna-Reyes, O.B., Reyes, T.M., So, F.Y., Matty, B.M., Lardizabal, A.A. 1988. Energy cost of ambulation in healthy and disabled Filipino children, *Arch. Phys. Med. Rehabil.*, 69(11), 946–949.

Lyngberg, K., Danneskiold-Samsoe, B., and Halskov, O. 1988. The effect of physical training on patients with rheumatoid arthritis: changes in disease activity, muscle strength and aerobic capacity, a clinical controlled minimized cross-over study, *Clin. Exp. Rheumatol.*, 6(3), 253–260.

Maehlum, S. and Pruett, E.D.R. 1973. Muscular exercise and metabolism in male juvenile diabetes. Glucose tolerance after exercise, *Scand. J. Clin. Lab Invest.*, 32, 149–153.

Maurer, C., Floersheim, M., and Craig, P. 1998. Fitness training for a healthy lifestyle, *Rehab. Manag.*, 11(2), 36–39, 41, 43.

Mayer, F., Billow, H., Horstmann, T., Martini, F., Niess, A., Rocker, K., and Dickhuth, H.H. 1999. Muscular fatigue, maximum strength and stress reactions of the shoulder musculature in paraplegics, *Int. J. Sports Med.*, 20(7), 487–493.

McCormack, E. 1977. An exercise protocol for SCI, *Am. Correct. Ther. J.*, 31(4), 122–123.

McCubbin, J.A. 1990. Resistance exercise training for persons with arthritis, *Rheum. Dis. Clin. North Am.*, 16(4), 931–943.

Meyer, T. and Brooks, A. 2000. Therapeutic impact of exercise on psychiatric diseases: Guidelines for exercise testing and prescription, *Sports Med.*, 30(4), 269–279.

Miyahara, M., Sleivert, G.G., and Gerrard, D.F. 1998. The relationship of strength and muscle balance to shoulder pain and impingement syndrome in elite quadriplegic wheelchair rugby players, *Int. J. Sports Med.*, 19(3), 210–214.

Moreau, C.E., and Moreau, S.R. 2001. Chiropractic management of a professional hockey player with recurrent shoulder instability, *J. Manipulative Physiol. Ther.*, 24(6), 425–430.

Mohr, T., Anderson, J.L., Biering-Sorensen, F., Galbo, H., Bangsbo, J., Wagner, A., and Kjaer, M. 1997. Long-term adaptation to electrically induced cycle training in severe spinal cord injured individuals, *Spinal Cord*, 35(1), 1–16.

Murray, T.J. 1985. Amantadine therapy for fatigue in multiple sclerosis, *Can. J. Neurol. Sci.*, 12(3), 251–254.

Napier, J.R. Thomas, M.F., Sharma, M., Hodgkinson, S.C., and Bass, J.J. 1999. Insulin-like growth factor-1 protects myoblasts from apoptosis but requires other factors to stimulate proliferation, *J. Endocrinol.*, 163, 63–68.

National Institutes of Health Consensus Development (NIHCD) Panel 1985. Health implications of obesity, *Ann. Intern. Med.*, 103, 1073–1077.

National Vital Statistics Report 2002. 50(16), Sept. 16, 2002, p. 49, http://www.cdc.gov/nchs/fastats/pdf/nvsr50_16TB2.pdf.

Neeper, S.A., Gomez-Pinilla, F., Choi, J., and Cotman, C. 1995. Exercise and brain neurotrophins, *Nature*, 373, 109.

Nichols, P.J., Norman, P.A., and Ennis, J.R. 1979. Wheelchair user's shoulder? Shoulder pain in patients with spinal cord lesions, *Scand. J. Rehabil. Med*, 11, 29–32.

Nordgren, B. 1970. Physical capacities in a group of mentally retarded adults, *Scand. J. Rehabil. Med.*, 2, 125–132.

Nordgren, B. 1971. Physical capacity and training in a group of young adult mentally retarded persons, *Acta Paediatr. Scand. Suppl.*, 217, 119–121.

Nordgren, B. and Backstrom, L. 1971. Correlations between muscular strength and industrial work performance in mentally retarded persons, *Acta Paediatr. Scand. Suppl.*, 217, 122–126.

Noreau, L., and Shephard, R.J. 1995. Spinal cord injury, exercise and quality of life, *Sports Med*, 20(4), 226–250.

Nyland, J., Robinson, K., Caborn, D., Knapp, E., and Brosky, T. 1997. Shoulder rotator torque and wheelchair dependence differences of National Wheelchair Basketball Association players, *Arch. Phys. Med. Rehabil.*, 78(4), 358–363.

O'Grady, M., Fletcher, J., and Ortiz, S. 2000. Therapeutic and physical fitness exercise prescription for older adults with joint disease: an evidence-based approach, *Rheum. Dis. Clin. North Am.*, 26(3), 617–646.

Olgiati, R., Burgunder, J.-M., and Mumenthaler, M. 1988. Increased energy cost of walking in multiple sclerosis: effect of spasticity, ataxia, and weakness, *Arch. Phys. Med. Rehabil.*, 69, 846–849.

Paluska, S.A. 2002. The role of physical activity in obesity management, *Clin. Fam. Pract.*, 4(2), 369–389.

Patrick, J.H., Farmer, S.E., and Bromwich, W. 2002. Muscle stretching for treatment and prevention of contracture in people with spinal cord injury, *Spinal Cord*, 40(8), 421–422.

Pelham, T.W., Campagna, P.D., Ritvo, P.G., and Birnie, W.A. 1993. The effects of exercise therapy on clients in a psychiatric rehabilitation program, *Psychosoc. Rehabil. J.*, 16(4), 75–84.

Penninx, B.W., Rejeski, W.J., Pandya, J., Miller, M.E., Di Bari, M., Applegate, W.B., and Pahor, M. 2002. Exercise and depressive symptoms: a comparison of aerobic and resistance exercise effects on emotional and physical function in older persons with high and low depressive symptomatology, *J. Gerontol. B. Psychol. Sci. Soc. Sci.*, 57(2), P124–P132.

Petajan, J.H., Gappmaier, E., White, A.T., Spencer, M.K., Mino, L., and Hicks, R.W. 1996. Impact of aerobic training on fitness and quality of life in multiple sclerosis, *Ann. Neurol.*, 39(4), 432–441.

Petersen, S. and Rayner, M. 2002. Coronary heart disease statistics, British Heart Foundation Statistics Database, British Heart Foundation Health Promotion Research Group, Dept of Public Health, University of Oxford.

Petty, T.L. 1997. A new national strategy for COPD, *J. Resp. Dis.*, 18(4), 365–369.

Pitetti, K.H., Rimmer, J.H., Fernhall, B. 1993. Physical fitness and adults with mental retardation: an overview of current research and future directions, *Sports Med.*, 16(1), 23–56.

Ponichtera-Mulcare, J.A. 1993. Exercise and multiple sclerosis, *Med. Sci. Sports Exerc.*, 25(4), 451–465.

Rarick, G.L, Widdop, J.H., and Broadhead, G.D. 1970. The physical fitness and motor performance of educable mentally retarded children, *Exceptional Child.*, 36, 504–519.

Reardon, K., Galea, M., Dennett, X., Choong, P., and Byrne, E. 2001. Quadriceps muscle wasting persists 5 months after total hip arthroplasty for osteoarthritis of the hip: a pilot study, *Intern. Med. J.*, 31, 7–14.

Reuter, I., Engelhardt, M., Stecker, K., and Baas, H. 1999. Therapeutic value of exercise training in Parkinson's disease, *Med. Sci. Sports Exerc.*, 31(11), 1544–1549.

Rice, C.L., Vollmer, T.L., and Bigland-Ritchie, B. 1992. Neuromuscular responses of patients with multiple sclerosis, *Muscle Nerve*, 15(10), 1123–1132.

Rimmer, J.H. 2001. http://www.lifefitness.com/exercise_physical_therapy/physical_therapy.asp.

Rimmer, J.H., Braddock, D., and Pitetti, K. H. 1996. Research on physical activity and disability: an emerging national priority, *Med. Sci. Sports Exerc.*, 28(11), 1366–1372.

Rimmer, J.H. and Kelly, L.E. 1991. Effects of a resistance training program on adults with mental retardation, *APAQ*, 8(2), 146–153.

Rolland, Y., Rival, L., Pillard, F., Lafont, C., Rivere, D., Albarede, J., and Vellas, B. 2000. Feasibility of regular physical exercise for patients with moderate to severe Alzheimer disease, *J. Nutr., Health Aging*, 4(2), 109–113.

Rose, J., Gamble, J.G., Burgos, A., Medeiros, J., and Haskell, W.L. 1990. Energy expenditure index of walking for normal children and for children with cerebral palsy, *Dev. Med. Child Neurol.*, 32(4), 333–340.

Roubenoff, R. 2000. Sarcopenia: a major modifiable cause of frailty in the elderly, *J. Nutr., Health Aging*, 4(3), 140–142.

Rusk, H.A. 1977. Howard Rusk on rehabilitation: cardiac cases, cancer patients, chronic obstructive lung diseases, lower limb orthotics, *Med. Times*, 105(1), 64–75.

Sahrmann, S.A. and Norton, B.J. 1977. The relationship of voluntary movement to spasticity in the upper motor neuron syndrome, *Ann. Neurol.*, 2(6), 460–465.

Sasco, A., Paffenbarger, R.S., Gendre, I., and Wing, A.L. 1992. The role of physical exercise in the occurrence of Parkinson's disease, *Arch. Neurol.*, 49, 360–365.

Scremin, A.M.E., Kurta, L. Gentili, A., Wisemann, B., Perell, K., Kunkel, C., and Scremin, O.U. 1999. Increasing muscle mass in spinal cord injured persons with a functional electrical stimulation exercise program, *Arch. Phys. Med. Rehabil.*, 80, 1531–1536.

Sharp, S.A. and Brouwer, B.J. 1997. Isokinetic strength training of the hemiparetic knee: effects on function and spasticity, *Arch. Phys. Med. Rehabil.*, 78, 1231–1236.

Sheean, G.L., Murray, N.M., Rothwell, J.C., Miller, D.H., and Thompson, A.J. 1997. An electrophysiological study of the mechanism of fatigue in multiple sclerosis, *Brain*, 120(2), 299–315.

Sicard-Rosenbaum, L., Light, K.E., and Behrman, A.L. 2002. Gait, lower extremity strength, and self-assessed mobility after hip arthroplasty, *J. Gerontol. A Biol. Sci. Med. Sci.*, 57(1), M47–M51.

Sie, I.H., Waters, R.L., Adkins, R.H., and Gellman, H. 1992. Upper extremity pain in the postrehabilitation spinal cord injured patient, *Arch. Phys. Med. Rehabil.*, 73, 44–48.

Simpson, K., Killian, K., McCartney, N., Stubbing, D.G., and Jones, N.L. 1992. Randomized controlled trial of weightlifting exercise in patients with chronic airflow limitation, *Thorax*, 47, 70–75.

Slemenda, C., Brandt, K.D., Heilman, D.K., Mazzuca, S., Braunstein, E.M., Katz, B.P., Wolinsky, F.D. 1997. Quadriceps weakness and osteoarthritis of the knees, *Ann. Intern. Med.*, 127(2), 97–104.

Smith, G.V., Silver, K.H.C., Goldberg, A.P., Macko, R.F. 1999. "Task-oriented" exercise improves hamstring strength and spastic reflexes in chronic stroke patients, *Stroke*, 30, 2112–2118.

Smith, P.E.M, Calverley, P.M., Edwards, R.H.T., Evans, G.A., and Campbell, E.J.M. 1987. Practical considerations of respiratory care of patients with muscular dystrophy, *N. Engl. J. Med.*, 316, 1197–1205.

Solomon, A. and Pangle, R. 1967. Demonstrating physical fitness improvements in the EMR, *Exceptional Child.*, 34, 177–181.

Sothern, M.S., Loftin, J.M., Udall, J.N., Suskind, R.M., Ewing, T.L., Tang, S.C., and Blecker, U. 2000. Safety, feasibility, and efficacy of a resistance training program in preadolescent obese children, *Am. J. Med. Sci.*, 319(6), 370–375.

Storer, T.W. 2001. Exercise in chronic pulmonary disease: resistance exercise prescription, *Med. Sci. Sports Exerc.*, 33(7), S680–S686.

Stuifbergen, A.K. 1997. Physical activity and perceived health status in persons with multiple sclerosis, *J. Neurosci. Nurs.*, 29(4), 238–243.

Sung, R.Y.T., Yu, C.W., Chang, S.K.Y., Mo, S.W., Woo, K.S., and Lam, C.W.K. 2002. Effects of dietary intervention and strength training on blood lipid level in obese children, *Arch. Dis. Child.*, 86(6), 407–410.

Systrom, D.M., Pappagianopoulos, P., Fishman, R.S, Wain, J.C., and Ginns, L.C. 1998. Determinants of abnormal maximum oxygen uptake after lung transplantation for chronic obstructive pulmonary disease, *J. Heart Lung Transplant.*, 17, 1220–1230.

Tefft, D., Guerette, P., and Furumasu, J. 1999. Cognitive predictors of young children's readiness for powered mobility, *Dev. Med. Child Neurol.*, 41(10), 665–670.

Teixeira-Salmela, L.F., Olney, S.J., Nadeau, S., and Brouwer, B. 1999. Muscle strengthening and physical conditioning to reduce impairment and disability in chronic stroke survivors, *Arch. Phys. Med. Rehabil.*, 80, 1211–1218.

Teri, L., McCurry, S.M., Buchner, D.M., Logsdon, R.G., LaCroix, A.Z., Kukull, W.A., Barlow, W.E., and Larson, E.B. 1998. Exercise and activity level in Alzheimer's disease: a potential treatment focus, *J. Rehabil. Res. Dev.*, 35(4), 411–419.

Tirdel, G.B., Girgis, R., Fishman, R.S., and Theodore, J. 1998. Metabolic myopathy as a cause of the exercise limitation in lung transplant recipients, *J. Heart Lung Transplant.*, 17, 1231–1237.

Trieschmann, R. 1987. *Aging with a Disability* (New York: Demos Publication).

U.S. Census Bureau Report. 2002 . Americans with Disabilities: 1994–95, http://www.census.gov/hhes/www/disable/sipp/disab9495/asc9495.html.

U.S. Congress. 1991. Americans with Disabilities (ADA) accessibility guidelines for buildings and facilities (Federal Register No. 28 CFR Ch1), Washington, DC: U.S. Architectural & Transportation Barriers Compliance Board.

van den Ende, C.H., Breedveld, F.C., le Cessie, S., Dijkmans, B.A., de Mug, A.W., and Hazes, J.M. 2000. Effect of intensive exercise on patients with active rheumatoid arthritis: a randomized clinical trial, *Ann. Rheum. Dis.*, 59(8), 615–621.

Vincent, K.R., Braith, R.W., Feldman, R.A., Kallas, H.E., and Lowenthal, D.T. 2002. Improved cardiorespiratory endurance following 6 months of resistance exercise in elderly men and women, *Arch. Intern. Med.*, 162, 673–678.

Wagner, M., Vignos, P.J., and Fonow, D. 1986. Serial isokinetic evaluations used for a patient with scapuloperoneal muscular dystrophy: a case report, *Phys. Ther.*, 66, 1110–1113.

Waters, R.L., Hislop, H.J., Thomas, L., Campbell, J. 1983. Energy cost of walking in normal children and teenagers, *Dev. Med. Child Neurol.*, 25(2), 184–188.

Waring, W.P. and Maynard, F.M. 1991. Shoulder pain in acute traumatic quadriplegia, *Paraplegia*, 29(1), 37–42.

Wiart, L. and Darrah, J. 2002. Changing philosophical perspectives on the management of children with physical disabilities — their effect on the use of powered mobility, *Disabil. Rehabil.*, 24(9), 492–498.

Wilk, K.E., Romeniello, W.T., Soscia, S.M., Arrigo, C.A., and Andrews, J.R. 1994. The relationship between subjective knee scores, isokinetic testing and functional testing in ACL reconstructed knees, *J. Orthop. Sports Phys. Ther.*, 20, 60–73.

Williams, L.O., Anderson, A.D., Campbell, J., Thomas, L., Feiwell, E., and Walker, J.M. 1983. Energy cost of walking and of wheelchair propulsion by children with myelodysplasia: comparison with normal children, *Dev. Med. Child Neurol.*, 25(5), 617–624.

Young, J.C. 1995. Exercise prescription for individuals with metabolic disorders: practical considerations, *Sports Med.*, 19(1), 43–54.

22

Gaps in Knowledge and Future of the Field

Shrawan Kumar

CONTENTS

22.1   Gaps in Application Needing Work

22.1.1   Strength Capabilities in Various Activities

Even a cursory review of tasks carried out in industries on shop floors, assembly lines, in the field, and other places will reveal the enormity of different human activities performed. The scope is further enlarged due to differences in anthropometry and/or ethnicity. In the strength field, we stabilize the subjects and all other joints (other those than being studied) that could possibly modify or contribute to the effort and then measure the strength capability in an activity that is very narrowly defined. This is essential if we want to know the strength capability of that joint in that specific effort. In industry, however, rarely can one find an activity that requires such a narrow motion scope. Multiple joints are commonly involved in executing a certain kind of motion on the industrial shop floor. This is where our purist scientific approach falls short in telling us the varying strength capability of a joint as it undergoes complex motion — either alone or in concert with other joints. There is a significant need for such data in industry. Additionally, specific industrial tasks can benefit from measurement of task-specific strength. Such data may assist in design of tasks and/or injury control.

22.1.2   Strength Capabilities of Different Ethnic Groups

Large amounts of strength data on the North American population are currently available from many military and civilian documents (Ayoub et al., 1978; Chaffin et al., 1978; Clarke, 1966; Kamon et al., 1982; Kroemer, 1970; Kumar, 1991b, 1996, 2001; Kumar and Garand, 1992; Kumar et al., 1988, 1995a,b, 1997, 1998; Laubach and McConville, 1966; Mital, 1978, 1984; Mital and Faard, 1990; Mital et al., 1991, 1995; Snook, 1978; Snook and Irvine, 1967; Webb Associates, 1978; Yates and Kamon, 1980; and numerous others). There are many European reports as well (Davids and Stubbs, 1980; Kroemer and Grandjean, 1997). Consequently, most product and process designs are based on the available data. Needless to say, there is significant variation between the races and nationalities. It is extremely important that the design data should come from the population that is the end user. Development of national data registries has been slow. However, many nations are developing their own data sets. A very comprehensive and advanced data system for Japanese people exists in Japan, which was developed through the Japanese government. Unfortunately, it is expensive and not readily available. Similar efforts have been undertaken in Korea (Kim and Wang, 1997; Park et al., 1999) for the Korean population and in India for the people of that subcontinent (Chakrabarti, 1997). Additional reports from other countries continue to be published (e.g., Luk et al., 2003; Thailand — Mamansari and Salokhe, 1997). However, it appears that there are a large number of ethnic groups for whom data are still unavailable.

There has been a significant move by many countries to adopt standards that have been developed in industrialized countries and to use them in their own context. Two prime examples are: (a) the weight limit for lifting by male and female workers proposed by WHO and (b) the NIOSH lifting guidelines. However, the suitability of either of these standards remains to be validated, even to the extent to which these apply to North American and European populations.

22.1.3   Contour Development with Prediction for Intervening Points

This concept is based on the fact that most industrial activities require motion, and with motion the orientation of the joint and mechanical relationship of tendon insertion may change. Hence, it may be advantageous to develop strength capability of an individual in the full range of motion around the joint in several axes. Such contours for lifting activities have been developed for arm lifting (Kumar, 1991b) and for stoop and squat lifts (Kumar and Garand, 1992). These contours give a comprehensive and overall picture of capability of the person around the point of interest. However, such an effort may not cover every point on the three-dimensional contour. This can be overcome by development of regression equations for predicting capability in the intervening spaces. Such a set of equations has been developed by Kumar (1995) for lifting activities. From such a database, the lowest common value can be taken as a design limitation.

22.1.4   Margin of Safety and Strength-Based Design

It is important to understand the relationship between strength capability and the limits of safety. Since direct dose–response experiments must not be carried out on humans to the point of injury precipitation, prospective epidemiological study will be extremely valuable in understanding the margin of safety. Based on published literature, Kumar and Mital (1992) proposed this concept and examined the same, using various approaches for manual materials-handling activity. Such understanding, along with the appropriate databases, can go a long distance in achieving the injury control. Furthermore, this information can be used proactively to design the tasks.

22.2 Gaps in Knowledge for Understanding

22.2.1 Mechanism of Muscle Contraction

Knowledge of the mechanism of muscle contraction is fundamental to an understanding of strength generation. Ever since the series of papers by Huxley (1957, 1969) and Huxley and Simmons (1971), our thinking of muscle contraction has been guided by the sliding-filament and cross-bridge-formation theories. Lately, however, questions have been raised as to whether biased Brownian motion may be responsible for skeletal muscle contraction (Geeves, 2002; Tanaka et al., 2002). Hatze (2002) states that the myosin motion is still a secret. The contractile elements are situated within the multitude of globular heads (or subfragments —S1) of the heavy meromyosin of the myosin molecules. These S-1 sub-fragments provide the connection between the myosin and the actin filaments. There is no contraction in the muscle without cross-linking action and myosin (Hatze, 2002). The contraction was thought to be achieved by rotation at the attachment point of the cross-bridge. This has been challenged by a couple of studies published by Rayment and colleagues (1993a,b) that suggest instead that the rotation occurs through a conformational change in light meromyosin about a hinge within the myosin head. Thus, understanding of the phenomenon of muscle contraction is in a flux and until this is entirely sorted out, further progress in many aspects of muscle physiology may remain tentative.

22.2.2 Skeletal Muscle Organization and Architecture

Until recently, motor units were generally thought to be homogenous. That is, a one-motor neuron, its axon, and its branches supply muscle fibers of only one kind (i.e., Type II). Muscle fibers belonging to one motor unit occupy a certain territory within one muscle and are intermingled with the muscle fibers of other muscles units. However, Unguez et al. (1995) demonstrated that the muscle fibers belonging to a motor unit need not be of the same muscle type or of uniform diameter. This challenges our current understanding, and the new knowledge is still to emerge.

22.2.3 Transmission and Distribution of Forces

Until recently we believed that upon contraction the muscle fiber shortens and transmits its force longitudinally through its collagen harness to the tendon. However, the phenomenon of lateral transmission of force in muscle structure was experimentally demonstrated by Street (1983). He showed that in a muscle preparation where at one end there were many fibers, and from among them one muscle fiber extended a much longer distance. In one preparation, the single long fiber was fixed at both ends and stimulated maximally. In the other preparation, the single fiber was fixed only at the insertion end with all other fibers, and the other end was left free. Instead, he fixed the lateral fibers at the other end. A maximal stimulation of the single long fiber, away from all other fibers, produced a tension that varied between 76 and 100% of the first preparation. This convincing proof of lateral transmission has not only changed our thinking but also made it a sounder scientific concept to explain why we do not see more internal injuries within muscles themselves. Thus in each muscle, the lateral myofascial and serial sarcomere-to-sarcomere contractions occur simultaneously. These facts have profound implications for muscle modeling. For these reasons, it has been proclaimed by Hatze (2002) that the contemporary models of skeletal muscles *in vivo* are inadequate. These models are required in both forward and inverse simulations of musculoskeletal system models in orthopaedics, rehabilitation, ergonomics, and sports (Hatze, 2002). It has been emphasized that it is essential that the muscle models must mimic the myodynamic phenomenon as realistically as possible (Hatze, 2000). Since that is not the case, all models of the system are at best inadequate, and in the worst case, inaccurate. In either case, our trust in biomechanical models is misplaced.

22.2.4 Load Deformation Characteristics of Tissues *In Vivo*

Our understanding of the mechanical properties of biological tissues has significantly increased over the last half century. The pioneering works of a number of researchers (Evans, 1969; Fung, 1981; Nachemson, 1960; Noyes, 1977; Viidik, 1979; Woo and Buckwalter, 1988) have significantly contributed to the field. However, in almost all works, of necessity, we have obtained these properties *in vitro*. Given the nature of the connective tissues, perhaps it may not be very different *in vivo*. But what will be significantly different is the geometrical configuration of these tissues and, hence, the pattern of their load distribution. To understand the mechanical behavior of tissues may not be enough until we can put it to the use in injury control efforts. Almost none of the biomechanical models applied in the field of ergonomics even differentiate or take into account different tissues. Instead they consider the biological organism as a black box. As long as an application of inverse dynamics results in balancing input and output, the models are considered valid. However, it must be pointed out that the complexity of the compositional, anatomical, geometrical, and distributional details, if unaccounted for, may render these biomechanical models meaningless apart from an indication of the force going through the total system. Because injuries are localized and sustained by one of the several tissues, the strategy of treating the entire biological system as a homogenous mass with uniform mechanical property is erroneous. Unfortunately, our biomechanical models have not passed this stage.

In light of the foregoing discussion, future efforts should be directed to develop strength-based models of subsystems, with progressive integration till we reach the level of the entire organism. It will be only at that stage that we will have a meaningful predictive capability and a control arsenal. The efforts carried out so far provide a launching ground for such activities. It is time that we indulge in a serious rethinking of criteria, variables, and strategies of strength-based biomechanical models to begin to approach our stated goal. The future is interesting and enticing and, at the same time, challenging. With the industrial and technological revolutions we have created an extremely challenging environment for our bodies to work in. Because the evolutionary process was not driven by modern and artificial constructs based on economic prosperity, the organizational, biomechanical, and physiological traits that constitute our genetic endowment could not be more ill-adapted (Kumar, 2001). Since we have embarked on the path of industrialization, it is an imperative that we work toward optimizing the match between what human beings have and what the society demands of them.

References

Ayoub, M., Bethea, N.J., Deivanayangam, S., Asfour, S.S., Baker, G.M., Liles, D., Mital, A., and Sherif, M. 1978. Determination and modeling of lifting capacity, HEW (NIOSH) Report. Grant 5 R01 0H 00545-02.

Chaffin, D.B., Herrin, G.D., and Keyserling, W.M. 1978. Pre-employment strength testing: an updated position, *J. Occup. Med.* 20, 403–408.

Chakrabarti, D. 1997. *Indian Anthropometric Dimensions* (Ahemdabad, India: National Institute of Design).

Clarke, H.H. 1966. *Muscular Strength and Endurance in Man* (New Jersey: Prentice Hall).

Davis, P.R. and Stubbs, D.A. 1980. *Force Limits in Manual Work* (Guildford, UK: IPC Science and Technology Press).

Evans, J.H. 1969. Biomechanical study of human lumbar ligamentum flavum, *J. Anat.*, 105, 188–189.

Fung, Y.C. 1981. *Biomechanics: Mechanical Properties of Living Tissues* (New York: Springer-Verlag).

Geeves, M.A. 2002. Stretching the lever-arm theory, *Nature*, 415, 129–131.

Hatze, H. 2000. The inverse dynamics problem of neuromuscular control, *Biol. Cybern.*, 82, 133–141.

Hatze, H. 2002. The fundamental problem of myoskeletal inverse dynamics and its implications, *J. Biomech.*, 35, 109–115.

Huxley, A.F. 1957. Muscle structure and theories of contraction, *Prog. Biophys. Chem.*, 7, 255–318.

Huxley, A.F. 1969. The mechanism of muscular contraction, *Science*, 261, 58–65.

Huxley, A.F. and Simmons, R.M. 1971. Proposed mechanism of force generation in striated muscle, *Nature*, 233, 533–538.

Kamon, E., Kiser, D., and Pytel, J. 1982. Dynamic and static lifting capacity and muscular strength of steelmill workers, *AIHA J.*, 43, 853–857.

Kim, J.H. and Whang, M.C. 1997. Development of a set of Korean manikins, *Appl. Ergon.*, 28, 407–410.

Kroemer, K.H.E. 1970. Human strength: terminology, measurements and interpretation of data, *Hum. Factors*, 12, 297–313.

Kroemer, K.H.E. and Grandjean, E. 1997. *Fitting the Task to Man* (London: Taylor and Francis).

Kumar, S. 1991a, *A Research Report on Functional Evaluation of Human Back* (Edmonton, AB: University of Alberta).

Kumar, S. 1991b. Arm lift strength in work space, *Appl. Ergon.*, 22, 317–328.

Kumar, S. 1995. Development of predictive equations for lifting strengths, *Appl. Ergon.*, 26, 327–341.

Kumar, S. 1996. Isolated planar trunk strengths measurement in normals: Part III — results and database, *Int. J. Ind. Ergon.*, 17, 103–111.

Kumar, S. 2001. Theories of musculoskeletal injury causation, *Ergonomics*, 44, 17–47.

Kumar, S. and Garand, D. 1992. Static and dynamic strength at different reach distances on symmetrical and asymmetrical planes, *Ergonomics*, 35, 861–880.

Kumar, S. and Mital, A. 1992. Margin of safety for the human back: a probable consensus based on published studies, *Ergonomics*, 35, 769–781.

Kumar, S., Chaffin, D.B., and Redfern, M. 1988. Isometric and isokinetic back and arm lifting strengths: device and measurement, *J. Biomech.*, 21, 35–44.

Kumar, S., Dufresne, R.M., and Schoort, T.V. 1995a. Human trunk strength profile in flexion and extension, *Spine*, 20, 160–168.

Kumar, S., Dufresne, R.M., and Schoort, T.V. 1995b. Human trunk strength profile in lateral flexion and axial rotation, *Spine*, 20, 169–177.

Kumar, S., Narayan, Y., and Zedka, M. 1998. Trunk strength in combined motions of rotation and flexion extension in normal young adults, *Ergonomics*, 41, 835–852.

Laubach, L.L. and McConville, J.T. 1966. Muscle strength, flexibility, and body size of adult males, *Res. Q.*, 37, 384–392.

Luk, K.D.K., Lu, W.W., Kwan, W.W., Hu, Y., Wong, Y.W., Law, K.K.P., and Leong, J.C.Y. 2003. Isometric and isokinetic lifting capacity of Chinese in relation to the physical demand of the job, *Appl. Ergon.*, 34, 201–204.

Mamansari, D.V. and Salokhe, V.M. 1996. Static strength and physical work capacity of agricultural labourers in the central plain of Thailand, *Appl. Ergon.*, 27, 53–60.

Mital, A. 1978. Strength and lifting capacity: data norms and prediction models, Technical Report, Texas Technical University, Lubbock, TX.

Mital, A. 1984. Prediction of maximum weights of lift acceptable to male and female industrial workers, *J. Occup. Accid.*, 5, 223–231.

Mital, A. and Faard, H.F. 1990. Effects of sitting and standing, reach distance, and arm orientation on isokinetic pull strengths in the horizontal plane, *Int. J. Ind. Ergon.*, 6, 241–248.

Mital, A., Bishu, R.R., and Manjunath. 1991. Review and evaluation of techniques for determining fatigue allowances, *Int. J. Ind. Ergon.*, 8, 165–178.

Mital, A., Kopavdekar, P., and Motorwala, A. 1995. Isokinetic pull strength in the vertical plane: effects of speed and arm angle, *Clin. Biomech.*, 10, 110–112.

Nachemson, A. 1960. Lumbar intradiscal pressure: Experimental studies on post-mortem material, *Acta Orthop. Scand.*, 43. Supplement.

Noyes, F.R. 1977. Functional properties of knee ligaments and alterations induced by immobilization: a correlative biomechanical and histological study in primates, *Clin. Orthop.*, 123, 210–242.

Park, S.J., Park, S.C., Kim, J.M., and Kim, C.B. 1999. Biomechanical parameters on body segments of Korean adults, *Int. J. Ind. Ergon.*, 23, 23–31.

Rayment, I., Holden, H.M., Whittaker, M., Yohn, C.B., Lorenz, M., Holmes, K.C., and Milligan, R.A. 1993a. Structure of the actin-myosin complex and its implications for muscle contraction, *Science*, 261, 58–65.

Rayment, I., Rypniewski, W.R., Schmidt-Base, K., Smith, R., Tomchich, D.R., Benning, M.M., Winkelmann, D.A., Wesenberg, G., and Holden, H.M. 1993b. Three-dimensional structure of myosin subfragment-1: a molecular motor, *Science*, 261, 50–58.

Snook, S.H. 1978. The design of manual handling tasks, *Ergonomics*, 21, 963–983.

Snook, S.H. and Irvine, C.H. 1967. Maximum acceptable weight of lift, *AIHA J.*, 28, 322–339.

Street, S.F. 1983. Lateral transmission of tension in frog myofibers: a myofibrillar network and transverse cytoskeletal connections are possible transmitters, *J. Cell. Physiol.*, 114, 346–364.

Tanaka, H., Homma, K., Iwane, A.H., Katayama, E., Ikebe, R., Slato, J., Yanagida, T., and Ikebe, M. 2002. The motor domain determines the large step of myosin-V, *Nature*, 192–195.

Unguez, G.A., Roy, R.R., Pierotti, D.J., Bodine-Fowler, S., and Edgerton, V.R. 1995. Further evidence of incomplete neural control of muscle properties in cat tibialis anterior motor units, *Am. J. Physiol.*, 268, 527–534.

Viidik, A. 1979. Connective tissues: possible implications of the temporal changes for the aging process, *Mech. Ageing Dev.*, 9, 267–285.

Webb Associates. 1978. *Anthropometric Source Book*, NASA RP-1024, Washington, D.C.

Woo, S.L.Y. and Buckwalter, J.A. 1988. *Injury and Repair of the Musculoskeletal Soft Tissues* (Park Ridge, IL: American Academy of Orthopaedic Surgeons).

Yates, J.W., Kamon, E., Rodgers, S.H., and Champney, P.C. 1980. Static lifting strength and maximal isometric voluntary contractions of back, arm and shoulder muscles, *Ergonomics*, 23, 37–47.

Index

M